Crystal Structures

CRYSTAL STRUCTURES

Second Edition

Ralph W. G. Wyckoff, *University of Arizona, Tucson, Arizona*

VOLUME 6

The Structure of Benzene Derivatives

Part 2.
Molecules Containing More than One Benzene Ring

INTERSCIENCE PUBLISHERS
a division of John Wiley & Sons, Inc.
New York · London · Sydney · Toronto

QD951
W8
1963

Physics

sd 12/13/71

Preface

In recent years structures containing cyclic organic molecules have been determined at a steadily mounting rate. Instead of the single volume given over to them in the loose-leaf edition, it will now require more than three for even the brief descriptions attempted here. The first of these volumes, covering Part 1 of Chapter XV, deals with compounds having a single benzene ring per molecule.

An effort has been made to arrange these compounds in a way that will associate together those that are chemically related. The classification chosen, based on the number of substitutions in the ring, departs from that used in the loose-leaf edition. It is straightforward except for chelates, where the molecule can be taken either as the entire complex or as one of the chelated groups. For the present purposes it has seemed better to take the second alternative which permits grouping together, for instance, all the derivatives of salicylic acid. If this, perhaps arbitrary, step is borne in mind, it should not be too difficult to locate a compound in the text even without the use of the master table or the index.

The complexity of many of the organic compounds now being analyzed has made it necessary to omit packing drawings for most of the more recently determined structures. From many standpoints this seems regrettable but there are arguments favoring the omission. Such drawings lose value as structures become more complex, they are very costly to prepare and they occupy space which can more usefully be given to molecular drawings that state the principal bond dimensions.

As more and more complicated compounds are dealt with the critical evaluation that can be given in the kind of condensed description being attempted here becomes more restricted and the description itself becomes stereotyped. This has happened in the makeup of the present volume. It inevitably gives rise to a monotonous presentation but since a compendium of this sort would only be read in fragments this dullness in style is perhaps unimportant.

Most organic structures have their molecules distributed according to one or another of a very few space groups, notably C_{2h}^5 and C_{2h}^6. This means that, except for the very differently shaped cells needed, the molecular distributions fall into a limited number of groups. In the loose-leaf edition this fact was taken into account in grouping the compounds for description. It did not, however, lead to an easy comparison of chemically related structures.

v

Contents of Volume 6, Part 2, Second Edition

Molecules Containing More than One Benzene Ring

Contents

Chapter XV

THE STRUCTURE OF BENZENE DERIVATIVES

B. COMPOUNDS CONTAINING TWO BENZENE RINGS

So many structures have now been determined for compounds containing two benzene rings that it has proved convenient to describe them under several headings. This grouping has been based on the number and character of the atoms separating the two rings. In this section, therefore, structures are arranged as follows:

I. Derivatives of biphenyl
II. Compounds in which the rings are separated by one bonding atom
III. Compounds in which the rings are separated by two bonding atoms
IV. Compounds in which the rings are separated by a chain of three atoms
V. Compounds in which the rings are separated by a chain of four or more atoms
VI. Miscellaneous compounds containing two benzene rings

In each of these subgroupings the first crystals to be considered have fourth-column (C, Si, etc.) atoms uniting the rings. These are followed by crystals with fifth-column (N, As, etc.) and sixth-column (O, S, etc.) bonding atoms.

I. Derivatives of Biphenyl

XV,bI,1. *Biphenyl,* (diphenyl) $(C_6H_5)_2$, is monoclinic with a naphthalenelike (**XV,f1**) bimolecular unit of the dimensions:

$$a_0 = 8.12(2) \text{ A.}; \quad b_0 = 5.63(1) \text{ A.}; \quad c_0 = 9.51(2) \text{ A.}; \quad \beta = 95.1(3)°$$

The space group is $C_{2h}^5 (P2_1/a)$ with atoms in the positions:

$$(4e) \quad \pm (xyz; x+\tfrac{1}{2}, \tfrac{1}{2}-y, z)$$

1

Recent parameters are those of Table XVB,1, the values from 1961, T being in parentheses.

TABLE XVB,1

Parameters of the Atoms in Biphenyl

Atom	x	y	z
C(1)	0.0355 (0.0357)	−0.0042 (−0.0007)	0.0759 (0.0764)
C(2)	−0.0093 (−0.0042)	0.1694 (0.1738)	0.1707 (0.1690)
C(3)	0.0592 (0.0648)	0.1636 (0.1723)	0.3130 (0.3130)
C(4)	0.1665 (0.1719)	−0.0161 (−0.0040)	0.3569 (0.3607)
C(5)	0.2119 (0.2103)	−0.1825 (−0.1795)	0.2666 (0.2672)
C(6)	0.1420 (0.1416)	−0.1792 (−0.1770)	0.1241 (0.1253)
H(2)	(−0.091)	(0.308)	(0.150)
H(3)	(0.033)	(0.314)	(0.385)
H(4)	(0.222)	(−0.005)	(0.468)
H(5)	(0.284)	(−0.308)	(0.297)
H(6)	(0.197)	(−0.317)	(0.057)

The resulting structure, as seen in Figure XVB,1, closely resembles that of naphthalene and the several hydrocarbons of similar packing. The molecules themselves are planar and have centers of symmetry; their bond dimensions from the most recent study are those of Figure XVB,2. The shortest separations between molecules are C–C distances of 3.72 A.

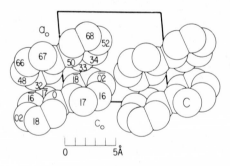

Fig. XVB,1.   The monoclinic structure of biphenyl projected along its b_0 axis.

XV,bI,2.   A structure has been determined for the triclinic modification   of   *bis(tricarbonylchromium)biphenyl*, $[Cr(CO)_3]_2(C_6H_5)_2$.   The

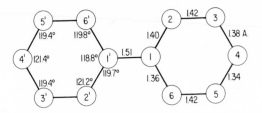

Fig. XVB,2. Bond dimensions in the molecule of biphenyl.

dimensions of its unimolecular cell are:

$$a_0 = 7.17(4) \text{ A.}; \quad b_0 = 6.91(4) \text{ A.}; \quad c_0 = 8.43(5) \text{ A.}$$
$$\alpha = 80°18(60)'; \quad \beta = 77°12(60)'; \quad \gamma = 80°48(60)'$$

The space group is $C_i^1 (P\bar{1})$ with all atoms in the positions $(2i) \pm (xyz)$. Determined parameters are listed in Table XVB,2.

TABLE XVB,2

Parameters of the Atoms in $[Cr(CO)_3]_2(C_6H_5)_2$

Atom	x	y	z
Cr	0.670	0.731	0.235
O(1)	0.259	0.953	0.298
O(2)	0.625	0.646	−0.095[a]
O(3)	0.791	0.113	0.060
C(1)	0.417	0.867	0.275
C(2)	0.643	0.680	0.031
C(3)	0.744	0.963	0.128
C(4)	0.901	0.452	0.246
C(5)	0.712	0.412	0.328
C(6)	0.599	0.517	0.458
C(7)	0.672	0.663	0.506
C(8)	0.860	0.710	0.420
C(9)	0.972	0.600	0.294

[a] It is assumed that z for O(2) should be -0.095 and not the 0.095 of the original; otherwise an unreasonable structure is obtained.

The structure is shown in Figure XVB,3. In it a $Cr(CO)_3$ group is approximately centered over each phenyl group, with the (CO) radicals pointed away from the rings and with the chromium atoms approximately equidistant from the six carbon atoms of the adjacent ring

(Cr–C = 2.18 A.). In the biphenyl molecule the ring C–C = 1.40 A. and the bond between rings is 1.51 A. In the $Cr(CO)_3$ component, Cr–C = 3.01(3) A., C–O = 1.15 A. and O–C–Cr–C'–O' = about 89°.

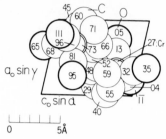

Fig. XVB,3. The triclinic structure of bis(tricarbonylchromium)biphenyl projected along its b_0 axis.

XV,bI,3. Crystals of *4-acetyl-2'-fluorobiphenyl*, $FC_6H_4 \cdot C_6H_4C(O)$-CH_3, are monoclinic with a tetramolecular unit of the dimensions:

$$a_0 = 13.687(5) \text{ A.}; \quad b_0 = 5.971(3) \text{ A.}; \quad c_0 = 14.766(5) \text{ A.}$$
$$\beta = 116°10(5)'$$

TABLE XVB,3

Parameters of the Atoms in $FC_6H_4 \cdot C_6H_4C(O)CH_3$

Atom	x	y	z
F	0.3757	0.8924	0.4910
O	−0.1636	0.7933	0.3252
C(1)	0.2287	0.5682	0.4937
C(2)	0.1686	0.3891	0.4340
C(3)	0.0568	0.4046	0.3771
C(4)	0.0018	0.5981	0.3809
C(5)	0.0609	0.7719	0.4398
C(6)	0.1704	0.7592	0.4959
C(7)	0.3478	0.5501	0.5550
C(8)	0.3953	0.3651	0.6171
C(9)	0.5069	0.3463	0.6701
C(10)	0.5745	0.5131	0.6648
C(11)	0.5288	0.6970	0.6042
C(12)	0.4190	0.7108	0.5525
C(13)	−0.1197	0.6197	0.3216
C(14)	−0.1857	0.4274	0.2603

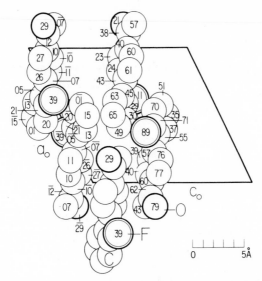

Fig. XVB,4.   The monoclinic structure of 4-acetyl-2′-fluorobiphenyl projected along its b_0 axis.

The space group is $C_{2h}^5(P2_1/c)$, with atoms in the positions:

$$(4e)\quad \pm(xyz;\ x,{}^1\!/_2-y,z+{}^1\!/_2)$$

Parameters are listed in Table XVB,3; values for hydrogen are given in the original article.

The structure is shown in Figure XVB,4. Its molecules have the bond dimensions of Figure XVB,5. Their rings, planar to less than 0.01 A., make an angle of 50.5° with one another; the acetyl plane is turned through 2.5° with respect to its attached phenyl group. Between

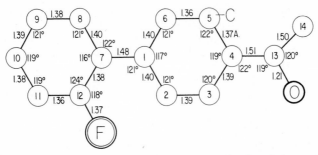

Fig. XVB,5.   Bond dimensions found for the molecule of 4-acetyl-2′-fluoro-biphenyl.

molecules the shortest C–F = 3.33 A.; the shortest C–O and C–C are both 3.51 A.

XV,bI,4. Crystals of *4-acetyl-2′-chlorobiphenyl*, $ClC_6H_4 \cdot C_6H_4C(O)$-$CH_3$, have a structure different from that of the fluorine derivative (**XV,bI,3**). The symmetry is monoclinic with a tetramolecular unit of the dimensions:

$$a_0 = 4.00(1) \text{ A.}; \quad b_0 = 38.51(1) \text{ A.}; \quad c_0 = 7.52(1) \text{ A.}$$
$$\beta = 100.07(8)°$$

The space group is $C_{2h}^5(P2_1/c)$ with all atoms in the positions:

$$(4e) \quad \pm(xyz; x, 1/2-y, z+1/2)$$

The determined parameters are listed in Table XVB,4, those proposed for the atoms of hydrogen being given in the original article.

TABLE XVB,4

Parameters of the Atoms in $ClC_6H_4 \cdot C_6H_4C(O)CH_3$

Atom	x	y	z
Cl	−0.1891	0.2028	0.0168
C(1)	0.1693	0.1542	0.2474
C(2)	0.0723	0.1900	0.2154
C(3)	0.1768	0.2143	0.3484
C(4)	0.3807	0.2060	0.5101
C(5)	0.4703	0.1715	0.5443
C(6)	0.3653	0.1463	0.4113
C(7)	0.0641	0.1264	0.1115
C(8)	0.1124	0.1298	−0.0663
C(9)	0.0200	0.1026	−0.1900
C(10)	−0.1247	0.0721	−0.1347
C(11)	−0.1594	0.0692	0.0421
C(12)	−0.0707	0.0955	0.1653
C(13)	−0.2333	0.0437	−0.2714
C(14)	−0.3532	0.0125	−0.2139
O	−0.2399	0.0506	−0.4410

The resulting structure, shown in Figure XVB,6, is composed of molecules having the bond dimensions of Figure XVB,7. In them the angle between the two planar benzene rings is 49° and between the acetyl radical and its attached phenyl group, 8°.

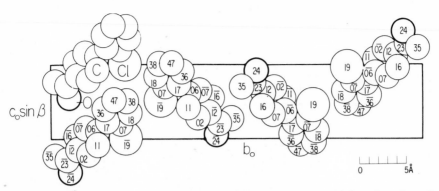

Fig. XVB,6. The monoclinic structure of 4-acetyl-2'-chlorobiphenyl projected along its short a_0 axis.

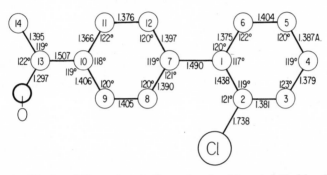

Fig. XVB,7. The bond dimensions in the molecule of 4-acetyl-2'-chlorobiphenyl.

XV,bI,5. Crystals of *4,4'-dinitrobiphenyl*, $(C_6H_4NO_2)_2$, are monoclinic, pseudo-orthorhombic, with a bimolecular unit of the dimensions:

$$a_0 = 3.753(3) \text{ A.}; \quad b_0 = 9.5840(5) \text{ A.}; \quad c_0 = 15.5080(15) \text{ A.}$$
$$\beta = 90°0(30)'$$

The space group is the low symmetry $C_s^2(Pc)$. The parameters given in Table XVB,5 do not apply to coordinates as customarily chosen but to those in which the origin has been displaced by $1/4b_0$. Values assigned to the hydrogen atoms are stated in the original article.

The resulting molecules have the bond dimensions of Figure XVB,8. In this compound their two benzene rings are turned by 34° with respect to one another. The plane of the N(1')–O(1',2') group is coincident with that of its phenyl ring, but at the other end of the molecule the N(1)–O(1,2) group is twisted by 11° from the plane of its benzene nucleus.

TABLE XVB,5

Parameters of the Atoms in $(C_6H_4NO_2)_2$

Atom	x	y	z
C(1)	0.056	0.068	0.024
C(2)	0.196	0.052	0.110
C(3)	0.288	0.178	0.155
C(4)	0.253	0.302	0.119
C(5)	0.120	0.324	0.035
C(6)	0.006	0.209	−0.013
C(1′)	−0.073	−0.057	−0.025
C(2′)	−0.028	−0.059	−0.114
C(3′)	−0.097	−0.177	−0.161
C(4′)	−0.220	−0.282	−0.116
C(5′)	−0.280	−0.293	−0.027
C(6′)	−0.196	−0.176	0.018
N(1)	0.341	0.438	0.165
N(1′)	−0.352	−0.415	−0.162
O(1)	0.510	0.406	0.231
O(2)	0.261	0.538	0.131
O(1′)	−0.305	−0.414	−0.242
O(2′)	−0.475	−0.513	−0.128

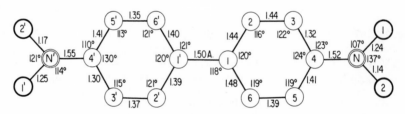

Fig. XVB,8. Bond dimensions in the molecule of 4,4′-dinitrobiphenyl.

XV,bI,6. Crystals of *m-tolidine*, $[NH_2C_6H_3(CH_3)]_2$, are orthorhombic. Their unit cell, containing eight molecules, has the edge lengths:

$$a_0 = 7.60 \text{ A.;} \quad b_0 = 14.81 \text{ A.;} \quad c_0 = 21.05 \text{ A.}$$

Atoms are in general positions of $V_h^{15}(Pcab)$

(8c) $\pm (xyz;\ x,y+1/2,1/2-z;\ x+1/2,1/2-y,z;\ 1/2-x,y,z+1/2)$

with the parameters of Table XVB,6.

TABLE XVB,6

Parameters of the Atoms in m-Tolidine

Atom	x	y	z
C(1)	0.055	0.127	0.140
C(2)	−0.050	0.118	0.091
C(3)	−0.005	0.118	0.028
C(4)	0.164	0.142	0.009
C(5)	0.270	0.159	0.060
C(6)	0.218	0.152	0.119
CH$_3$	−0.230	0.083	0.106
N	0.214	0.142	−0.056
CH$_3'$	0.089	−0.035	0.221
N′	−0.110	0.142	0.405
C(1′)	−0.014	0.130	0.209
C(2′)	0.030	0.055	0.249
C(3′)	−0.009	0.060	0.313
C(4′)	−0.080	0.136	0.337
C(5′)	−0.100	0.213	0.308
C(6′)	−0.075	0.207	0.242

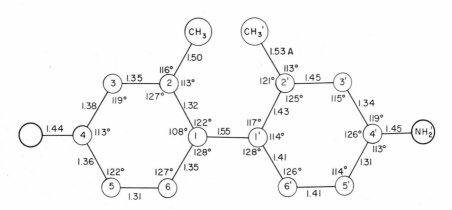

Fig. XVB,9.   Bond dimensions in the molecule of m-tolidine.

The molecule that results (Fig. XVB,9) is, as would be expected, similar to that found in the hydrochloride (**XV,bI,7**). The angle between the planes of the two benzene rings is, however, appreciably greater (86° rather than 71°). In the crystal these molecules lie in sheets that are parallel to the b-face (Fig. XVB,10).

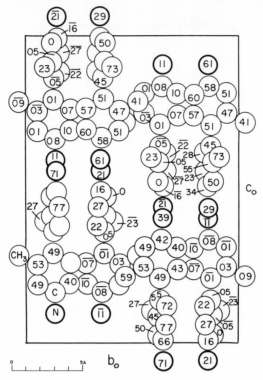

Fig. XVB,10a.   The orthorhombic structure of m-tolidine projected along its a_0 axis. Left-hand axes.

TABLE XVB,7

Parameters of the Atoms in m-Tolidine Dihydrochloride

Atom	x	y	z
C(1)	0.0598	−0.024	0.030
C(2)	0.9925	−0.180	0.072
C(3)	0.0889	−0.162	0.125
C(4)	0.2785	−0.003	0.140
C(5)	0.3412	0.147	0.101
C(6)	0.2369	0.133	0.047
C(7)	0.7752	−0.343	0.060
NH₂	0.3934	0.005	0.198
Cl	0.5761	0.501	0.216

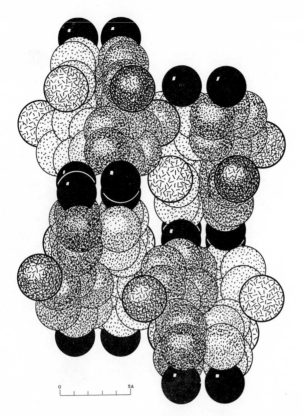

Fig. XVB,10b.   A packing drawing of the orthorhombic *m*-tolidine structure viewed along its a_0 axis. The nitrogen atoms are black, the methyl groups are heavily outlined and line shaded. Phenyl carbon atoms are dotted. Left-hand axes.

XV,bI,7.   Crystals of *m-tolidine dihydrochloride*, $[NH_2C_6H_3(CH_3)]_2 \cdot$ 2HCl, are monoclinic with a bimolecular unit of the dimensions:

$$a_0 = 4.9475 \text{ A.}; \quad b_0 = 6.1796 \text{ A.}; \quad c_0 = 23.246 \text{ A.}; \quad \beta = 91°20'$$

The chosen space group is C_2^3, so that with the body-centering that prevails here (axial orientation $I2$) all atoms have the coordinates:

$$(4c) \quad xyz; \bar{x}y\bar{z}; \quad \text{B.C.}$$

The determined parameters are those of Table XVB,7.

As Figure XVB,11 indicates, this is a layer structure in which sheets of the elongated organic cations are separated in the *c*-direction by

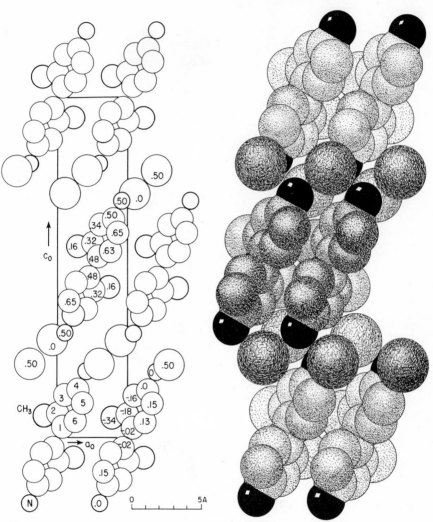

Fig. XVB,11a. The monoclinic structure of *m*-tolidine dihydrochloride
projected along its b_0 axis. The largest
circles are chlorine; the phenyl carbons
have the same size as the methyl
groups but are lightly outlined. Left-
hand axes.

Fig. XVB,11b. A packing drawing of
the monoclinic *m*-tolidine dihydrochloride arrangement seen along its b_0
axis. The nitrogen atoms are black,
the chloride ions large and line shaded.
All carbon atoms are dotted, those of
the methyl groups being somewhat
larger than the others. Left-hand axes.

layers of chloride ions. Its molecule, having a twofold axis of symmetry, is not planar, though all atoms of a half-molecule lie in a plane that is turned through 70°36' with respect to the plane of the other half [about the line through C(4)–C(1)–N]. The established bond angles and lengths, accurate to ca. 0.02 A., are shown in Figure XVB,12.

Fig. XVB,12. Bond dimensions in the cation of *m*-tolidine dihydrochloride.

XV,bI,8. The compound *2,2'-dichlorobenzidine*, $[ClC_6H_3(NH_2)]_2$, crystallizes with orthorhombic symmetry. Its tetramolecular unit has the dimensions:

$$a_0 = 7.51 \text{ A.;} \quad b_0 = 15.20 \text{ A.;} \quad c_0 = 10.40 \text{ A.}$$

All atoms are in general positions of the space group $V_h{}^{14}$ in the axial orientation *Pnca*:

$$(8d) \quad \pm(xyz; \, x+{}^1/_2, {}^1/_2-y, {}^1/_2-z; \, x, {}^1/_2-y, z+{}^1/_2; \, x+{}^1/_2, y, \bar{z})$$

The parameters are those of Table XVB,8.

TABLE XVB,8

Parameters of the Atoms in 2,2'-Dichlorobenzidine

Atom	x	y	z
C(1)	0.236	0.050	0.365
C(2)	0.120	0.091	0.272
C(3)	0.084	0.174	0.269
C(4)	0.153	0.225	0.373
C(5)	0.269	0.192	0.452
C(6)	0.305	0.097	0.460
NH_2	0.125	0.316	0.360
Cl	0.033	0.028	0.150

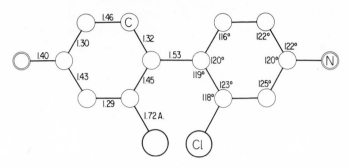

Fig. XVB,13.   Bond dimensions in the molecule of 2,2′-dichlorobenzidine.

They result in a molecule which has the bond lengths indicated in Figure XVB,13. It is not planar, though each of its benzene rings is; they are turned through 36° to place the two chlorine atoms 3.36 A. apart. The molecular packing in the crystal is shown in Figure XVB,14. Between molecules there is an unusually short Cl–Cl = 3.27 A. and one pair of carbon atoms is separated by only 3.45 A.

XV,bI,9.   Crystals of *4,4′-diamino-3,3′-dimethyl biphenyl* (*o*-tolidine), [CH$_3$C$_6$H$_3$(NH$_2$)]$_2$, are orthorhombic with a tetramolecular unit of the edge lengths:

$$a_0 = 7.44(1) \text{ A.};   b_0 = 23.70(5) \text{ A.};   c_0 = 6.47(2) \text{ A.}$$

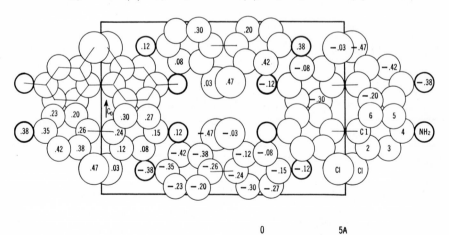

0   5A

Fig. XVB,14a.   The orthorhombic structure of 2,2′-dichlorobenzidine projected along its a_0 axis. Left-hand axes.

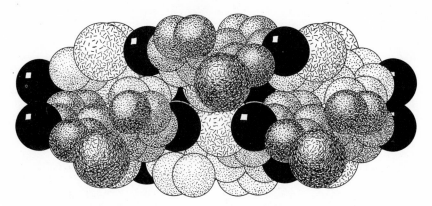

Fig. XVB,14b. A packing drawing of the orthorhombic 2,2′-dichlorobenzidine structure viewed along its a_0 axis. The nitrogen atoms are black; those of chlorine are larger and line shaded. The atoms of carbon are dotted. Left-hand axes.

The space group is $V^4(P2_12_12_1)$ with atoms in the positions:

$$(4a) \quad xyz;\; 1/2-x,\bar{y},z+1/2;\; x+1/2,1/2-y,\bar{z};\; \bar{x},y+1/2,1/2-z$$

The determined parameters, including those assigned to the hydrogen atoms, are listed in Table XVB,9.

The atomic arrangement that results is shown in Figure XVB,15. Its molecules have the bond dimensions of Figure XVB,16. Each benzene ring with its attached CH_3 and NH_2 radicals is planar, and these two halves of the molecule are turned by 41° about the bond joining them. Between molecules, the shortest NH_2–NH_2 distance is 3.50 A., the shortest NH_2–CH_3 = 3.69 A., and the shortest distance between ring carbons is 3.44 A.

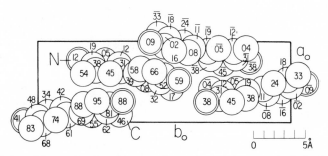

Fig. XVB,15. The orthorhombic structure of o-tolidine projected along its c_0 axis.

TABLE XVB,9

Parameters of the Atoms in *o*-Tolidine

Atom	x	y	z
C(1)	0.4119	0.1115	0.1097
C(2)	0.4880	0.0705	0.2452
C(3)	0.5158	0.0151	0.1772
C(4)	0.4633	−0.0010	−0.0162
C(5)	0.3900	0.0390	−0.1605
C(6)	0.3647	0.0941	−0.0827
C(7)	0.5968	−0.0275	0.3271
N(1)	0.4847	−0.0566	−0.0944
C(1′)	0.3775	0.1701	0.1897
C(2′)	0.4060	0.2152	0.0515
C(3′)	0.3646	0.2710	0.1197
C(4′)	0.2962	0.2791	0.3111
C(5′)	0.2688	0.2346	0.4536
C(6′)	0.3095	0.1796	0.3809
C(7′)	0.3989	0.3181	−0.0374
N(1′)	0.2535	0.3353	0.3820
H(1)	0.331	0.120	−0.179
H(2)	0.360	0.023	−0.290
H(3)	0.421	−0.068	−0.232
H(4)	0.546	−0.076	0.027
H(5)	0.569	−0.031	0.464
H(6)	0.568	−0.063	0.277
H(7)	0.650	−0.014	0.465
H(8)	0.520	0.077	0.404
H(1′)	0.298	0.151	0.490
H(2′)	0.202	0.234	0.595
H(3′)	0.177	0.335	0.509
H(4′)	0.313	0.351	0.259
H(5′)	0.303	0.321	−0.121
H(6′)	0.384	0.356	−0.058
H(7′)	0.459	0.345	0.079
H(8′)	0.433	0.210	−0.108

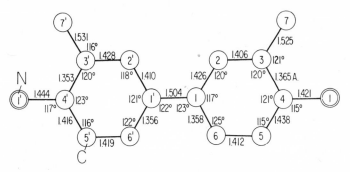

Fig. XVB,16. Bond dimensions in the molecule of *o*-tolidine.

XV,bI,10. Many years ago a structure was proposed for the addition compound *dinitrobiphenyl-hydroxybiphenyl (3 : 1)*, $3(NO_2C_6H_4 \cdot C_6H_4NO_2) \cdot (C_6H_5 \cdot C_6H_4OH)$. Its symmetry is monoclinic with a bimolecular unit of the dimensions:

$$a_0 = 20.06 \text{ A.}; \quad b_0 = 9.46 \text{ A.}; \quad c_0 = 11.13 \text{ A.}; \quad \beta = 99°39'$$

The space group was given as $C_{2h}^3(C2/m)$ with atoms in the positions:

$$(8j) \quad \pm (xyz; x\bar{y}z; x+{}^1/_2,y+{}^1/_2,z; x+{}^1/_2,{}^1/_2-y,z)$$

The approximate parameters that were assigned are those of Table XVB,10.

In this arrangement the dinitrobiphenyl molecules are distributed in a face-centered array to leave voids in which are found the molecules of hydroxybiphenyl. This, then, is a cage-like structure somewhat like that of the quinol clathrates (**XV,aIII,73**).

Probably the analogous dinitrobiphenyl-biphenyl (3 : 1) complex, $3(NO_2C_6H_4 \cdot C_6H_4NO_2) \cdot (C_6H_5 \cdot C_6H_5)$ is isostructural. Its bimolecular monoclinic unit has been given the dimensions:

$$a_0 = 19.9 \text{ A.}; \quad b_0 = 9.50 \text{ A.}; \quad c_0 = 11.0 \text{ A.}; \quad \beta = 99°30'$$

Clearly more work is needed to establish the structure described here.

II. Compounds in Which the Rings Are Separated by One Bonding Atom

XV,bII,1. Crystals of *diphenyl mercury*, $(C_6H_5)_2Hg$, are monoclinic with a bimolecular unit of the dimensions:

$$a_0 = 11.56(1) \text{ A.}; \quad b_0 = 8.30(1) \text{ A.}; \quad c_0 = 5.59(1) \text{ A.}; \quad \beta = 112°20'$$

TABLE XVB,10

Parameters Assigned the Atoms in Dinitrobiphenyl–hydroxybiphenyl (3 : 1)

Atom	x	y	z
Hydroxybiphenyl molecules			
C(1)	0.007	0.0	0.068
C(2)	0.073	0.0	0.131
C(3)	0.088	0.0	0.258
C(4)	0.034	0.0	0.323
C(5)	−0.032	0.0	0.260
C(6)	−0.047	0.0	0.132
O(1)	0.048	0.0	0.445
Dinitrobiphenyl molecules			
C(7)	0.038	0.500	0.010
C(8)	0.073	0.628	0.018
C(9)	0.073	0.372	0.018
C(10)	0.144	0.628	0.036
C(11)	0.144	0.372	0.036
C(12)	0.179	0.500	0.045
C(13)	0.069	0.500	0.363
C(14)	0.105	0.628	0.376
C(15)	0.105	0.372	0.376
C(16)	0.176	0.628	0.402
C(17)	0.176	0.372	0.402
C(18)	0.211	0.500	0.414
C(19)	−0.006	0.500	0.337
C(20)	−0.041	0.628	0.324
C(21)	−0.041	0.372	0.324
C(22)	−0.112	0.628	0.298
C(23)	−0.112	0.372	0.298
C(24)	−0.147	0.500	0.286
O(2)	0.285	0.600	0.103
O(3)	0.285	0.400	0.103
O(4)	0.317	0.600	0.430
O(5)	0.317	0.400	0.430
O(6)	−0.253	0.600	0.235
O(7)	−0.253	0.400	0.235
N(1)	0.256	0.500	0.088
N(2)	0.288	0.500	0.427
N(3)	−0.224	0.500	0.247

The space group is $C_{2h}^5(P2_1/a)$. The atoms of mercury are in the special positions (2a) 000; $^1/_2$ $^1/_2$ 0. All the other atoms are in the general positions:

$$(4e) \quad \pm (xyz; x + ^1/_2, ^1/_2 - y, z)$$

with the assigned parameters of Table XVB,11.

The structure is shown in Figure XVB,17.

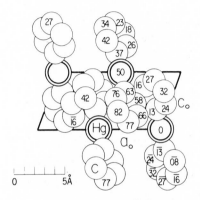

Fig. XVB,17.   The monoclinic structure of diphenyl mercury projected along its b_0 axis.

TABLE XVB,11

Parameters of Atoms in $(C_6H_5)_2Hg$

Atom	x	y	z
C(1)	0.084	0.126	0.355
C(2)	0.027	0.238	0.444
C(3)	0.081	0.322	0.659
C(4)	0.198	0.267	0.845
C(5)	0.254	0.159	0.765
C(6)	0.197	0.084	0.535

XV,bII,2. The compound *2-bromo-1,1-diphenyl-1-propene*, $(C_6H_5)_2C =$ $C(Br)CH_3$, forms monoclinic crystals with a tetramolecular unit of the dimensions:

$a_0 = 5.97(1)$ A.; $b_0 = 16.97(2)$ A.; $c_0 = 12.63(1)$ A.; $\beta = 103.7(1)°$

The space group is $C_{2h}^5(P2_1/c)$ with atoms in the positions:

$$(4e) \quad \pm (xyz; x, ^1/_2 - y, z + ^1/_2)$$

The stated parameters are those of Table XVB,12, values for the atoms of hydrogen being given in the original article.

TABLE XVB,12

Parameters of the Atoms in $(C_6H_5)_2C{=}CBrCH_3$

Atom	x	y	z
Br	0.1470(1)	0.1531(< 1)	0.0329(1)
C(1)	0.0552(11)	0.2158(3)	0.2297(4)
C(2)	0.1709(11)	0.2385(3)	0.1348(4)
C(3)	0.2706(10)	0.3050(3)	0.1211(4)
C(4)	0.3881(11)	0.3255(3)	0.0335(4)
C(5)	0.5502(12)	0.2758(4)	0.0038(5)
C(6)	0.6658(13)	0.2991(5)	−0.0735(6)
C(7)	0.6186(15)	0.3715(5)	−0.1240(6)
C(8)	0.4609(14)	0.4196(4)	−0.0975(5)
C(9)	0.3431(11)	0.3971(4)	−0.0192(5)
C(10)	0.2729(11)	0.3699(3)	0.2011(4)
C(11)	0.0715(13)	0.4122(4)	0.2039(6)
C(12)	0.0750(14)	0.4731(4)	0.2762(6)
C(13)	0.2822(16)	0.4931(4)	0.3477(5)
C(14)	0.4802(14)	0.4518(5)	0.3459(5)
C(15)	0.4753(11)	0.3921(4)	0.2746(5)

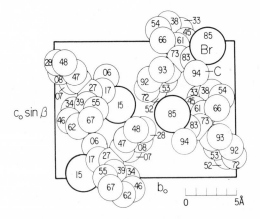

Fig. XVB,18. The monoclinic structure of 2-bromo-1,1-diphenyl-1-propene projected along its a_0 axis.

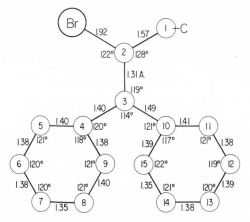

Fig. XVB,19.   Bond dimensions in the molecule of 2-bromo-1,1-diphenyl-1-propene.

The structure is illustrated in Figure XVB,18. Its molecules have the bond dimensions of Figure XVB,19. The benzene ring that is *trans* to the bromine atom is rotated through 70.7° and that which is *cis*, through 47.2° with respect to the propene plane.

XV,bII,3.   Crystals of *2-bromo-1,1-di-p-tolylethylene*, $(CH_3C_6H_4)_2C$= CHBr, are orthorhombic with a tetramolecular cell of the edge lengths:

$$a_0 = 16.89(1) \text{ A.}; \quad b_0 = 13.07(1) \text{ A.}; \quad c_0 = 6.26(1) \text{ A.}$$

The space group is $V^4(P2_12_12_1)$ with atoms in the positions:

$$(4a) \quad xyz; \; 1/2-x,\bar{y},z+1/2; \; x+1/2,1/2-y,\bar{z}; \; \bar{x},y+1/2,1/2-z$$

The determined parameters are those of Table XVB,13. Calculated positions for the hydrogen atoms, except for those in the methyl radical are stated in the original article.

The structure, shown in Figure XVB,20, has molecules with the bond dimensions of Figure XVB,21. Their shape can be described in terms of three planes, one defined by the ethylenic group, the other two by the tolyl components. The latter are turned, one by 67.9°, the other by 24.4°, with respect to the first.

XV,bII,4.   *Octacarbonyl diphenylvinylidene di-iron*, $(CO)_8Fe_2C$= $C(C_6H_5)_2$, forms monoclinic crystals with a unit that contains eight molecules. Its dimensions are:

$$a_0 = 37.08(5) \text{ A.}; \quad b_0 = 8.67(2) \text{ A.}; \quad c_0 = 13.35(3) \text{ A.}; \quad \beta = 90°14(10)'$$

TABLE XVB,13

Parameters of the Atoms in $(CH_3C_6H_4)_2C{=}CHBr$

Atom	x	y	z
Br	0.1172(1)	0.2463(1)	0.3738(2)
C(1)	0.0464(7)	0.2834(8)	0.1574(22)
C(2)	0.0407(6)	0.3790(7)	0.0799(16)
C(3)	−0.0177(6)	0.4015(7)	−0.0941(15)
C(4)	−0.0445(7)	0.3266(7)	−0.2350(19)
C(5)	−0.0996(7)	0.3518(8)	−0.3939(18)
C(6)	−0.1261(6)	0.4521(10)	−0.4162(22)
C(7)	−0.0987(6)	0.5272(9)	−0.2767(20)
C(8)	−0.0449(5)	0.5019(7)	−0.1151(17)
C(9)	0.0913(6)	0.4674(7)	0.1602(19)
C(10)	0.0808(7)	0.5042(9)	0.3672(18)
C(11)	0.1274(8)	0.5855(8)	0.4393(22)
C(12)	0.1840(7)	0.6292(7)	0.3049(21)
C(13)	0.1942(6)	0.5919(7)	0.0977(20)
C(14)	0.1478(6)	0.5104(9)	0.0254(19)
C(15)	−0.1847(8)	0.4779(12)	−0.5898(24)
C(16)	0.2341(9)	0.7162(9)	0.3853(24)

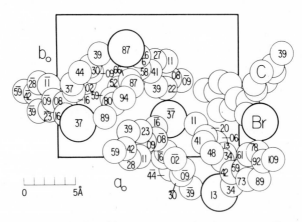

Fig. XVB.20.   The orthorhombic structure of 2-bromo-1,1-di-p-tolylethylene projected along its short c_0 axis.

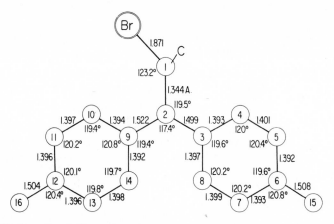

Fig. XVB,21.   Bond dimensions in the molecule of 2-bromo-1,1-di-*p*-tolylethy-
lene.

The space group is $C_{2h}^6(C2/c)$ with atoms in the positions:

(8*f*) $\pm\,(xyz;\ x,\bar{y},z+\tfrac{1}{2};\ x+\tfrac{1}{2},y+\tfrac{1}{2},z;\ x+\tfrac{1}{2},\tfrac{1}{2}-y,z+\tfrac{1}{2})$

The parameters are listed in Table XVB,14. Values for the hydrogen
atoms are given in the original article.

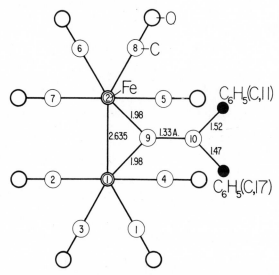

Fig. XVB,22.   Some bond lengths in the molecule of octacarbonyl diphenylvinylidene di-iron.

TABLE XVB,14

Parameters of the Atoms in $(CO)_8Fe_2C{=}C(C_6H_5)_2$

Atom	x	y	z
Fe(1)	0.08533(4)	−0.05347(25)	0.12173(12)
Fe(2)	0.06999(4)	−0.34489(27)	0.08511(12)
C(1)	0.1103(4)	0.0899(25)	0.1853(11)
O(1)	0.1246(3)	0.1939(18)	0.2251(9)
C(2)	0.0669(4)	−0.1100(26)	0.2393(12)
O(2)	0.0538(4)	−0.1351(20)	0.3155(9)
C(3)	0.0444(4)	0.0521(21)	0.0875(13)
O(3)	0.0193(3)	0.1240(20)	0.0678(10)
C(4)	0.1075(4)	−0.0133(24)	0.0044(11)
O(4)	0.1243(3)	0.0154(17)	−0.0645(7)
C(5)	0.0832(4)	−0.4170(24)	0.2091(10)
O(5)	0.0917(3)	−0.4665(14)	0.2810(8)
C(6)	0.0212(4)	−0.3711(22)	0.1074(9)
O(6)	−0.0075(2)	−0.3881(19)	0.1206(9)
C(7)	0.0635(3)	−0.2667(23)	−0.0401(13)
O(7)	0.0589(3)	−0.2246(17)	−0.1177(7)
C(8)	0.0799(4)	−0.5259(24)	0.0313(12)
O(8)	0.0866(3)	−0.6518(15)	0.0038(9)
C(9)	0.1165(3)	−0.2372(18)	0.1057(8)
C(10)	0.1516(3)	−0.2692(17)	0.1005(8)
C(11)	0.1627(3)	−0.4253(21)	0.0581(9)
C(12)	0.1581(3)	−0.4524(21)	−0.0435(10)
C(13)	0.1691(4)	−0.5985(24)	−0.0835(11)
C(14)	0.1860(4)	−0.6997(23)	−0.0279(11)
C(15)	0.1924(4)	−0.6660(25)	0.0734(13)
C(16)	0.1809(3)	−0.5318(20)	0.1151(10)
C(17)	0.1810(3)	−0.1705(20)	0.1363(9)
C(18)	0.2109(3)	−0.1461(21)	0.0743(10)
C(19)	0.2395(4)	−0.0542(24)	0.1068(12)
C(20)	0.2378(4)	0.0140(26)	0.1993(12)
C(21)	0.2084(4)	−0.0025(27)	0.2636(11)
C(22)	0.1802(3)	−0.0999(22)	0.2284(10)

The molecules in this structure have the bond lengths shown in Figure XVB,22. The iron-to-carbonyl carbon distances lie between 1.77 and 1.85 A. and C–O = 1.09–1.18 A.

XV,bII,5.   Crystals of *3,3'-dichloro-4,4'-dihydroxy diphenylmethane*, $Cl(OH)C_6H_3CH_2C_6H_3(OH)Cl$, are monoclinic with a tetramolecular unit of the dimensions:

$$a_0 = 22.3 \text{ A.}; \quad b_0 = 4.93 \text{ A.}; \quad c_0 = 11.76 \text{ A.}; \quad \beta = 112°36'$$

Atoms are in the following positions of C_{2h}^6 $(C2/c)$:

(4e) $\pm (0 \; u \; {}^1\!/_4; \; {}^1\!/_2, u + {}^1\!/_2, {}^1\!/_4)$

(8f) $\pm (xyz; \; x, \bar{y}, z + {}^1\!/_2; \; x + {}^1\!/_2, y + {}^1\!/_2, z; \; x + {}^1\!/_2, {}^1\!/_2 - y, z + {}^1\!/_2)$

with the parameters of Table XVB,15.

TABLE XVB,15

Positions and Parameters of the Atoms in 3,3'-Dichloro-4,4'-Dihydroxy Diphenyl Methane

Atom	x	y	z
C(1)	0.061	0.024	0.264
C(2)	0.059	0.217	0.184
C(3)	0.115	0.349	0.195
C(4)	0.173	0.290	0.288
C(5)	0.173	0.097	0.370
C(6)	0.119	−0.032	0.358
C(7)	0	−0.134	${}^1\!/_4$
OH	0.230	0.420	0.303
Cl	0.113	0.600	0.084
Calculated hydrogen parameters			
H(2)[a]	0.014	0.262	0.108
H(5)	0.221	0.053	0.445
H(6)	0.119	0.180	0.422
H(7)	0.012	0.268	0.332

[a] Parenthesized numbers identify the carbon atoms to which the hydrogens are attached. The assumed C–H separation is 1.10 A.

These lead to a molecule having the bond lengths shown in Figure XVB,23. It is not planar; instead the benzene rings are turned through 52° in opposite senses about the C(1)–C(7) bonds. In the crystal (Fig. XVB,24) the shortest atomic separations, O–O = 2.97 A., can be interpreted as being due to weak hydrogen bonds.

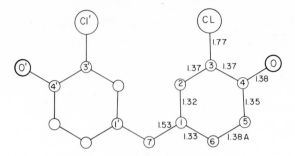

Fig. XVB,23. Bond lengths in the molecule of 3,3′-dichloro-4,4′-dihydroxy diphenyl methane.

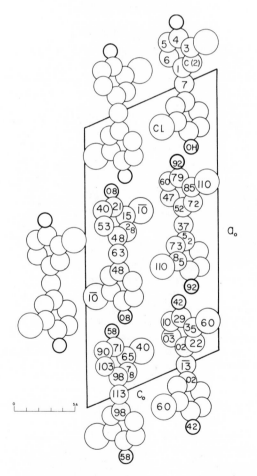

Fig. XVB,24a. The monoclinic structure of 3,3′-dichloro-4,4′-dihydroxy diphenyl methane projected along its b_0 axis. Left-hand axes.

Fig. XVB,24b.   A packing drawing of the monoclinic 3,3'-dichloro-4,4'-dihydroxy diphenyl methane arrangement viewed along its b_0 axis. The chlorine atoms are black, the nitrogens are heavily outlined and line shaded. Atoms of carbon are dotted. Left-hand axes.

XV,bII,6. Crystals of *bis hydroxydurylmethane*, $[C_6(CH_3)_4OH]_2CH_2$, have a bimolecular monoclinic unit of the dimensions:

$a_0 = 22.87(9)$ A.; $b_0 = 4.94(2)$ A.; $c_0 = 7.42(3)$ A.; $\beta = 91°0(25)'$

The space group is $C_2^3(C2)$ with the central carbon, C(0), in:

$\quad$ (2a) $0u0$; $^1/_2, u + {}^1/_2, 0$ $\qquad$ with $u = 0$ (arbitrary)

The other atoms are in:

$\quad$ (4c) xyz; $\bar{x}y\bar{z}$; $x + {}^1/_2, y + {}^1/_2, z$; ${}^1/_2 - x, y + {}^1/_2, \bar{z}$

Determined parameters are listed in Table XVB,16; probable values for the hydrogen atoms are stated in the original article.

TABLE XVB,16

Parameters of Atoms in $[C_6(CH_3)_4OH]_2CH_2$

Atom	x	y	z
C(1)	0.058	0.174	0.020
C(2)	0.096	0.121	0.173
C(3)	0.148	0.247	0.190
C(4)	0.165	0.437	0.061
C(5)	0.130	0.485	−0.088
C(6)	0.077	0.360	−0.104
C(7)	0.079	−0.085	0.305
C(8)	0.191	0.200	0.352
C(9)	0.149	0.696	−0.218
C(10)[a]	0.042	0.402	0.280
O	0.219	0.549	0.080

[a] It would appear that z for this atom should be −0.280 and not the 0.280 of the original article.

The structure is shown in Figure XVB,25. Its molecules, which have a twofold axis of symmetry passing through C(0), have the bond dimensions of Figure XVB,26. The planes of their two benzene rings are turned through 86° with respect to one another. The unusually long C(0)–C(1) = 1.60 A. is considered as index of strain.

XV,bII,7. There is a monoclinic modification of *d-methadone hydrobromide*, $(C_6H_5)_2C[C(O)C_2H_5][CH_2CH(CH_3)N(CH_3)_2]\cdot HBr$, which

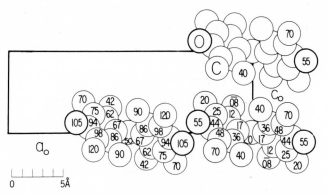

Fig. XVB,25.   The monoclinic structure of bis hydroxydurylmethane projected along its b_0 axis.

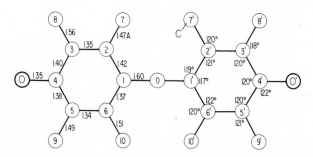

Fig. XVB,26.   Bond dimensions in the molecule of bis hydroxydurylmethane.

has a bimolecular unit of the dimensions:

$$a_0 = 10.69(3) \text{ A.}; \quad b_0 = 8.74(2) \text{ A.}; \quad c_0 = 10.74(3) \text{ A.}; \quad \beta = 94°36(12)'$$

The space group $C_2{}^2(P2_1)$ has its atoms in the positions:

$$(2a) \quad xyz; \; \bar{x}, y+{}^1/_2, \bar{z}$$

The parameters, given in Table XVB,17, express a determination of absolute configuration.

The structure, as shown in Figure XVB,27, is composed of molecules having the bond dimensions of Figure XVB,28. The bromide ion is closest to nitrogen, with Br–N = 3.22 A.; it is 3.62 and 3.70 A. from the adjacent carbon atoms. Between molecules the shortest atomic separation is 3.50 A.

TABLE XVB,17

Parameters of the Atoms in Methadone Hydrobromide

Atom	x	y	z
Br	0.9197	0.2500	0.1082
O	0.5840	0.4169	0.0838
N	0.8091	0.7296	−0.0040
C(1)	0.5900	0.1342	0.2179
C(2)	0.5577	0.2911	0.2809
C(3)	0.5893	0.4330	0.1961
C(4)	0.6131	0.5892	0.2579
C(5)	0.6556	0.7074	0.1628
C(6)	0.7956	0.6862	0.1338
C(7)	0.8879	0.7963	0.2086
C(8)	0.7583	0.6111	−0.0936
C(9)	0.7622	0.8807	−0.0496
C(10)	0.4774	0.6313	0.2919
C(11)	0.3913	0.6962	0.2019
C(12)	0.2645	0.7180	0.2272
C(13)	0.2265	0.6748	0.3456
C(14)	0.3111	0.6107	0.4326
C(15)	0.4369	0.5821	0.4089
C(16)	0.7104	0.5891	0.3732
C(17)	0.7083	0.6997	0.4620
C(18)	0.8059	0.7104	0.5584
C(19)	0.9010	0.6023	0.5665
C(20)	0.9033	0.4910	0.4781
C(21)	0.8055	0.4816	0.3836

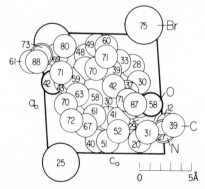

Fig. XVB,27.   The structure of monoclinic d-methadone hydrobromide projected along its b_0 axis.

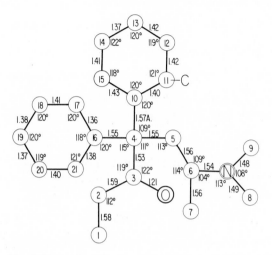

Fig. XVB,28.   Bond dimensions in the cation of d-methadone hydrobromide.

TABLE XVB,18

Parameters of the Atoms in Benzophenone

Atom	x	y	z
C(0)	0.206	0.227	0.284
C(1)	0.306	0.127	0.291
C(2)	0.282	0.032	0.202
C(3)	0.387	−0.061	0.218
C(4)	0.482	−0.035	0.322
C(5)	0.500	0.065	0.421
C(6)	0.408	0.146	0.391
C(1′)	0.063	0.194	0.252
C(2′)	−0.002	0.098	0.314
C(3′)	−0.135	0.082	0.282
C(4′)	−0.199	0.163	0.189
C(5′)	−0.137	0.244	0.119
C(6′)	−0.008	0.277	0.147
O	0.242	0.312	0.288

XV,bII,8. A preliminary structure has been described for *benzophenone*, $(C_6H_5)_2CO$. Its crystals are orthorhombic with a tetramolecular unit of the edge lengths:

$$a_0 = 10.26 \text{ A.}; \quad b_0 = 12.09 \text{ A.}; \quad c_0 = 7.88 \text{ A.}$$

Atoms are in the following positions of V^4 ($P2_12_12_1$):

(4a) $xyz;\ 1/2-x,\bar{y},z+1/2;\ x+1/2,1/2-y,\bar{z};\ \bar{x},y+1/2,1/2-z$

The parameters are those of Table XVB,18.

A further study and refinement of the structure is stated to be in progress.

XV,bII,9. According to a preliminary note, crystals of *3,3'-dibromobenzophenone*, $(BrC_6H_4)_2CO$, are orthorhombic with a tetramolecular unit of the edge lengths:

$$a_0 = 3.99 \text{ A}; \quad b_0 = 11.70 \text{ A.}; \quad c_0 = 24.69 \text{ A.}$$

The space group is V_h^{14} (*Pbcn*) with atoms in the positions:

(4c) $\pm (0\ u\ 1/4;\ 1/2,u+1/2,1/4)$
(8d) $\pm (xyz;\ x+1/2,y+1/2,1/2-z;\ 1/2-x,y+1/2,z;\ x,\bar{y},z+1/2)$

The positions and parameters are those of Table XVB,19.

TABLE XVB,19

Positions and Parameters of the Atoms in 3,3'-Dibromobenzophenone

Atom	Position	x	y	z
C(1)	(8d)	0.025	0.075	0.196
C(2)	(8d)	−0.100	−0.038	0.194
C(3)	(8d)	−0.081	−0.092	0.142
C(4)	(8d)	0.055	−0.029	0.098
C(5)	(8d)	0.174	0.079	0.102
C(6)	(8d)	0.158	0.138	0.151
C(7)	(4c)	0	0.133	$1/4$
O	(4c)	0	0.240	$1/4$
Br	(8d)	0.3400	0.1573	0.0448

The molecules thus defined have planar benzene rings which make an angle of 40° with one another. Each is twisted through 22.4° in relation to the C(1)–C(1')–C(7) – O plane.

XV,bII,10.   Crystals of *p,p'-dimethoxybenzophenone*, $(CH_3OC_6H_4)_2$-CO, are monoclinic. Their unit, containing eight molecules, has the dimensions:

$$a_0 = 16.43 \text{ A.}; \quad b_0 = 16.03 \text{ A.}; \quad c_0 = 9.62 \text{ A.}; \quad \beta = 100°15'$$

The space group is C_{2h}^5 $(P2_1/a)$ with atoms in the positions:

$$(4e) \quad \pm(xyz; x + 1/2, 1/2 - y, z)$$

The parameters, recently refined, are listed in Table XVB,20.

TABLE XVB,20

Parameters of the Atoms in $(CH_3OC_6H_4)_2CO$

Atom	x		y		z	
	Mol I	Mol II	Mol I	Mol II	Mol I	Mol II
O(1)	0.4327	0.4340	0.6809	0.6871	0.0815	0.5833
O(2)	0.2258	0.2290	0.3699	0.3764	0.2202	0.7347
O(3)	0.4344	0.4242	0.0629	0.0701	0.0680	0.5466
C(1)	0.4310	0.3860	0.7496	0.7612	0.1752	0.5648
C(2)	0.3878	0.4207	0.0153	0.0023	0.0522	0.6318
C(3)	0.2890	0.2907	0.3702	0.3776	0.1719	0.6818
C(4)	0.3980	0.3955	0.6080	0.6148	0.1100	0.6075
C(5)	0.3596	0.3110	0.5937	0.6076	0.2257	0.6106
C(6)	0.3250	0.2793	0.5158	0.5297	0.2426	0.6334
C(7)	0.3283	0.3281	0.4509	0.4584	0.1484	0.6507
C(8)	0.3687	0.4122	0.4663	0.4671	0.0341	0.6480
C(9)	0.4020	0.4468	0.5440	0.5447	0.0144	0.6281
C(10)	0.3268	0.3287	0.2895	0.2974	0.1365	0.6484
C(11)	0.2768	0.3258	0.2177	0.2287	0.1161	0.7353
C(12)	0.3105	0.3571	0.1404	0.1527	0.0926	0.7071
C(13)	0.3950	0.3922	0.1352	0.1432	0.0924	0.5882
C(14)	0.4464	0.3955	0.2048	0.2099	0.1135	0.4989
C(15)	0.4112	0.3653	0.2817	0.2867	0.1344	0.5296

The resulting structure is shown in Figure XVB,29. There are small, probably not significant, differences between the two crystallographically dissimilar molecules in the unit; their averaged bond lengths are given in Figure XVB,30. The benzene rings and attached OCH_3 groups

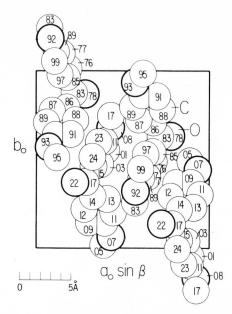

Fig. XVB,29. The contents of the bottom half of the monoclinic unit of *p,p'*-dimethoxy benzophenone projected along its c_0 axis. Molecules in the upper half would be almost superposed on those shown here.

are nearly planar, the maximum departure of the CH_3 radicals from their planes being 0.05 A. Between molecules the shortest distances are O–O = 3.38 A. and C–O = 3.41 A.

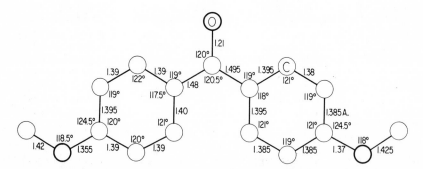

Fig. XVB,30. Averages of the bond dimensions in the two types of molecule in crystals of *p,p'*-dimethoxy benzophenone.

XV,bII,11. Crystals of *4,4'-dihydroxythiobenzophenone monohydrate*, $(C_6H_4OH)_2CS \cdot H_2O$, have monoclinic symmetry. Their tetramolecular cells have the dimensions:

$$a_0 = 5.62(1) \text{ A.;} \quad b_0 = 10.95(3) \text{ A.;} \quad c_0 = 20.24(6) \text{ A.}$$
$$\beta = 103°30(15)'$$

The space group is C_{2h}^5 $(P2_1/c)$ with atoms in the positions:

$$(4e) \quad \pm(xyz; x, 1/2 - y, z + 1/2)$$

Parameters are given in Table XVB,21.

TABLE XVB,21

Parameters of the Atoms in $(C_6H_4OH)_2CS \cdot H_2O$

Atom	x	y	z
S	0.1098	−0.1913	0.1941
O(1)	0.6395	0.2330	0.0443
O(2)	0.2640	0.1088	0.4874
O(H₂O)	−0.0352	0.3459	0.4746
C(1)	0.2408	−0.0610	0.2226
C(2)	0.3596	0.0166	0.1765
C(3)	0.5160	−0.0352	0.1393
C(4)	0.6110	0.0341	0.0954
C(5)	0.5489	0.1596	0.0881
C(6)	0.4050	0.2117	0.1257
C(7)	0.3060	0.1441	0.1700
C(8)	0.2446	−0.0123	0.2908
C(9)	0.0576	−0.0498	0.3234
C(10)	0.0693	−0.0078	0.3906
C(11)	0.2648	0.0678	0.4233
C(12)	0.4500	0.1039	0.3907
C(13)	0.4325	0.0640	0.3240

The structure, illustrated in Figure XVB,31, has molecules with the bond dimensions of Figure XVB,32. The plane through C(2)–S–C(8) contains the axes of the benzene rings through C(5)–O(1) and C(11)–O(2) to within 0.06 A., and the planes through the rings C(2)-to-C(7) and C(8)-to-C(13) are turned 47 and 30° with respect to this plane. Molecules are considered to be held together by a system of hydrogen bonds that spiral upward parallel to a_0; between O(1) and O(2) of neighboring

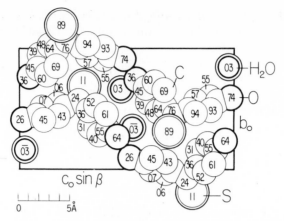

Fig. XVB,31.   The monoclinic structure of 4,4'-dihydroxythiobenzophenone
monohydrate projected along its a_0 axis.

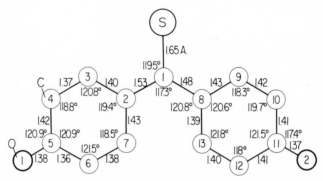

Fig. XVB,32.   Bond dimensions in the molecule of 4,4'-dihydroxythiobenzo-
phenone.

molecules they have a length of 2.76 A. There are also bonds between
these atoms and the water molecules and between the latter and sulfur.
Their distances are O(1)–H–H_2O = 2.70 A., O(2)–H–H_2O = 3.07 A.
and S–H–H_2O = 3.37 A. Parameters for the hydrogen atoms are given
in the original.

XV,bII,12.   *Auramine   perchlorate* $[C_6H_4N(CH_3)_2]_2C(NH_2)ClO_4$,
forms triclinic crystals that have a bimolecular unit of the dimensions:

$$a_0 = 8.86(2) \text{ A.}; \quad b_0 = 9.69(2) \text{ A.}; \quad c_0 = 11.27(2) \text{ A.}$$
$$\alpha = 97°40(10)'; \quad \beta = 99°34(10)'; \quad \gamma = 106°16(10)'$$

The space group is C_i^1 ($P\bar{1}$) with all atoms in the positions $(2i)$ $\pm(xyz)$. The determined parameters are those of Table XVB,22; values for hydrogen atoms are stated in the original article.

TABLE XVB,22

Parameters of the Atoms in Auramine Perchlorate

Atom	x	y	z
C(1)	0.7249(10)	0.9232(8)	0.8585(7)
C(2)	0.7503(10)	0.0314(8)	0.6685(7)
C(3)	0.5602(10)	0.7863(8)	0.6579(7)
C(4)	0.4724(10)	0.6777(8)	0.7125(7)
C(5)	0.5187(10)	0.7700(8)	0.5282(7)
C(6)	0.3956(10)	0.6506(8)	0.4597(7)
C(7)	0.3519(10)	0.5589(8)	0.6433(7)
C(8)	0.3092(10)	0.5427(8)	0.5150(7)
C(9)	0.1731(10)	0.4185(8)	0.4434(7)
C(10)	0.0785(10)	0.4290(8)	0.3284(7)
C(11)	−0.0050(10)	0.2991(8)	0.2417(7)
C(12)	0.0559(10)	0.5595(8)	0.3027(7)
C(13)	−0.0477(10)	0.5646(8)	0.1981(7)
C(14)	−0.1090(10)	0.3031(8)	0.1356(7)
C(15)	−0.1336(10)	0.4342(8)	0.1137(7)
C(16)	−0.2674(10)	0.5731(8)	−0.0162(7)
C(17)	−0.3428(10)	0.2996(8)	−0.0690(7)
N(1)	0.6838(7)	0.9046(7)	0.7245(6)
N(2)	0.1368(7)	0.2959(7)	0.4859(6)
N(3)	−0.2476(7)	0.4377(7)	0.0142(6)
O(1)	0.3001(10)	0.2287(7)	0.7186(6)
O(2)	0.1594(10)	0.0692(7)	0.8297(6)
O(3)	0.3774(10)	0.0248(7)	0.7553(6)
O(4)	0.1335(10)	−0.0076(7)	0.6189(6)
Cl	0.2436(2)	0.0773(2)	0.7314(2)

The structure, shown in Figure XVB,33, is ionic in nature with hydrogen bonds between the amino N(2) atom and oxygens of the perchlorate anion. The hydrogen atoms of these bonds, as located through difference maps, have the parameters:

$$H(N,1) : x = 0.200, y = 0.280, z = 0.525$$
$$H(N,2) : x = 0.065, y = 0.217, z = 0.460$$

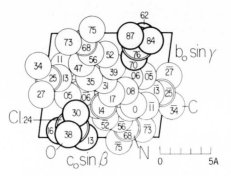

Fig. XVB,33.   The triclinic structure of auramine perchlorate projected along
its a_0 axis.

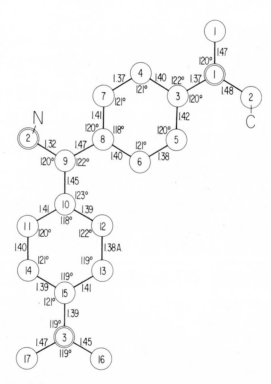

Fig. XVB,34.   Bond dimensions in the complex cation of auramine perchlorate.

The N(2) and oxygen atoms they unite are distant from one another by N–O = 3.02 and 3.06 A. In the tetrahedral ClO_4 ion Cl–O = 1.416–1.448 A. In the auramine cation the bond lengths are those of Figure XVB,34. It can be described in terms of five atomic planes, the NO_2 planes being turned through 6.8 and 5.3° with respect to their benzene planes. Each phenyl plane is rotated through 27.5° with respect to that through C(9) and the bonds it forms.

XV,bII,13. The large orthorhombic unit of *tetracarbonylcobalt diphenyltin manganesepentacarbonyl*, $Co(CO)_4 \cdot Sn(C_6H_5)_2 \cdot Mn(CO)_5$, contains eight molecules and has the edges:

$$a_0 = 8.38(2) \text{ A}; \quad b_0 = 17.25(6) \text{ A}; \quad c_0 = 33.47(10) \text{ A}.$$

The space group is V_h^{15} (*Pbca*) with all atoms in the positions:

$$(8c) \quad \pm (xyz; \, 1/2 - x, y + 1/2, z; \, x, 1/2 - y, z + 1/2; \, x + 1/2, y, 1/2 - z)$$

The parameters are listed in Table XVB,23; proposed values for hydrogen are to be found in the original article.

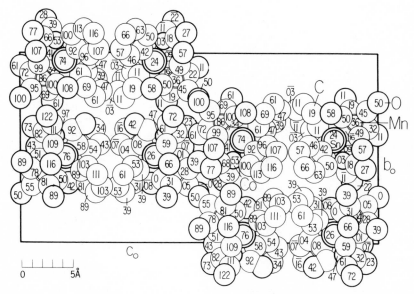

Fig. XVB,35.   The orthorhombic structure of tetracarbonylcobalt diphenyltin manganesepentacarbonyl projected along its a_0 axis.

CRYSTAL STRUCTURES

TABLE XVB,23

Parameters of the Atoms in $Co(CO)_4 \cdot Sn(C_6H_5)_2 \cdot Mn(CO)_5$

Atom	x	y	z
Sn	0.2608	0.0453	0.1246
Mn	0.1393	0.1369	0.0653
Co	0.4699	−0.0605	0.0994
O(1)	−0.070	−0.002	0.045
O(2)	0.388	0.072	0.007
O(3)	0.390	0.253	0.089
O(4)	−0.085	0.188	0.132
O(5)	0.000	0.244	0.002
O(6)	0.720	−0.167	0.074
O(7)	0.233	−0.107	0.038
O(8)	0.661	0.082	0.083
O(9)	0.425	−0.126	0.180
C(1)	0.007	0.056	0.054
C(2)	0.283	0.100	0.029
C(3)	0.309	0.198	0.078
C(4)	−0.002	0.172	0.102
C(5)	0.053	0.206	0.031
C(6)	0.606	−0.127	0.086
C(7)	0.317	−0.086	0.064
C(8)	0.589	0.025	0.091
C(9)	0.420	−0.098	0.148
C(10)	0.386	0.121	0.167
C(11)	0.307	0.186	0.183
C(12)	0.388	0.237	0.209
C(13)	0.532	0.213	0.226
C(14)	0.610	0.147	0.211
C(15)	0.530	0.098	0.185
C(16)	0.076	−0.021	0.157
C(17)	0.039	0.011	0.194
C(18)	−0.072	−0.025	0.219
C(19)	−0.159	−0.089	0.205
C(20)	−0.126	−0.120	0.167
C(21)	0.002	−0.090	0.145

The structure is shown in Figure XVB,35. Its molecules have the bond dimensions of Figure XVB,36. Coordination about the central tin atom is that of a distorted tetrahedron with two apices occupied by the other metallic atoms. About manganese the coordination is octahedral, whereas about cobalt it is described as that of a distorted trigonal bipyramid.

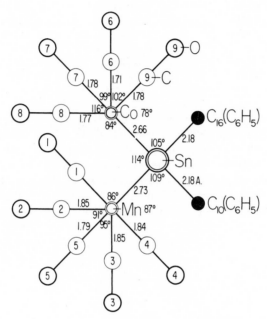

Fig. XVB,36. Bond dimensions in the molecule of tetracarbonylcobalt diphenyltin manganesepentacarbonyl.

XV,bII,14. *Diphenyllead dichloride*, $(C_6H_5)_2PbCl_2$, forms ortho-rhombic crystals whose bimolecular unit has the dimensions:

$$a_0 = 17.965(5) \text{ A.;} \quad b_0 = 8.565(3) \text{ A.;} \quad c_0 = 4.005(3) \text{ A.}$$

The space group has been determined as V_h^{12} (*Pnnm*) with atoms in the positions:

$$\text{Pb:} \quad (2a) \quad 000; \quad {}^1/_2\,{}^1/_2\,{}^1/_2$$

All the other atoms have been placed in

$$(4g) \quad \pm (uv0; \, u+{}^1/_2, {}^1/_2-v, {}^1/_2)$$

with the parameters of Table XVB,24.

TABLE XVB,24

Parameters of Atoms in $(C_6H_5)_2PbCl_2$

Atom	u	v
Cl	-0.4235	0.3385
C(1)	0.0876	-0.1662
C(2)	0.0709	-0.3259
C(3)	0.1293	-0.4367
C(4)	0.2044	-0.3879
C(5)	0.2220	-0.2271
C(6)	0.1636	-0.1163

The structure (Fig. XVB,37) is unusual for an organic compound in placing all atoms in special positions of its space group. $Pb(C_6H_5)_2$ groups, lying in the a_0b_0 planes, are separated by pairs of chlorine atoms which together form chains running along the c_0 axis. In the structure each lead atom is surrounded by an octahedron of four chlorine and two carbon atoms. In it Pb–Cl = 2.80 A., Pb–C = 2.12 A., Cl–Pb–Cl = 88.5° and C–Pb–Cl = 88 or 92°. In the chains Pb–Cl–Pb = 91.5°.

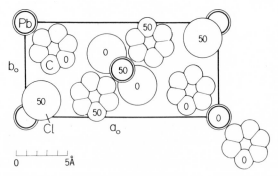

Fig. XVB,37.   The orthorhombic structure of diphenyllead dichloride projected along its c_0 axis.

XV,bII,15. Crystals of *p,p'-dichlorodiphenylamine*, $(ClC_6H_4)_2NH$, are monoclinic with a tetramolecular unit of the dimensions:

$$a_0 = 14.36(5) \text{ A.}; \quad b_0 = 12.63(8) \text{ A.}; \quad c_0 = 6.05(5) \text{ A.}$$
$$\beta = 96.0(5)°$$

The space group is C_{2h}^5 ($P2_1/a$) with atoms in the positions:

$$(4e) \quad \pm (xyz; x+{}^1/_2, {}^1/_2-y, z)$$

The parameters are those of Table XVB,25.

TABLE XVB,25

Parameters of the Atoms in $(ClC_6H_4)_2NH$

Atom	x	y	z
Cl(1)	0.1386	0.1312	0.8601
Cl(2)	0.1344	0.9461	0.3011
N	0.1060	0.5947	0.9315
C(1)	0.114	0.481	0.914
C(2)	0.075	0.431	0.720
C(3)	0.083	0.322	0.704
C(4)	0.129	0.263	0.880
C(5)	0.168	0.312	0.073
C(6)	0.160	0.421	0.089
C(1′)	0.112	0.681	0.777
C(2′)	0.160	0.666	0.595
C(3′)	0.166	0.748	0.447
C(4′)	0.126	0.846	0.480
C(5′)	0.079	0.861	0.666
C(6′)	0.072	0.778	0.814

The structure is shown in Figure XVB,38. The planar benzene rings have normal bond lengths with the average values: Cl–C = 1.67 A., C–N = 1.45 A., and C–C = 1.39 A. Between molecules the shortest Cl–Cl = 3.55 A. and Cl–C = 3.69 A.

XV,bII,16. The monoclinic crystals of *4,4′-bis(dimethylamino)-diphenylamine perchlorate*, $[(CH_3)_2NC_6H_4]_2NH \cdot ClO_4$, have a tetramolecular unit of the dimensions:

$$a_0 = 15.39 \text{ A.}; \quad b_0 = 13.11 \text{ A.}; \quad c_0 = 8.82 \text{ A.}; \quad \beta = 107°36'$$

Their space group is C_{2h}^6 ($C2/c$) with atoms in the positions:

N(1): (4e) $\pm (0\ u\ {}^1/_4; {}^1/_2, \dot{u}+{}^1/_2, {}^1/_4)$ with $u = 0.4767$
Cl: (4e) with $u = 0.1762$
H(1): (4e) with $u = 0.38$

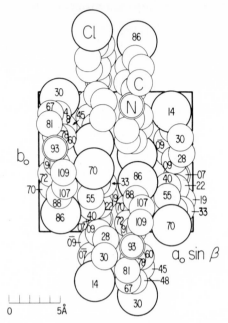

Fig. XVB,38.   The monoclinic structure of p,p'-dichlorodiphenylamine projected along its c_0 axis.

All other atoms are in

$$(8f) \quad \pm(xyz;\, x,\bar{y},z+{}^1\!/_2;\, x+{}^1\!/_2,y+{}^1\!/_2, z;\, x+{}^1\!/_2,{}^1\!/_2-y,z+{}^1\!/_2)$$

with the parameters of Table XVB,26.

The cations of this structure have the bond dimensions of Figure XVB,39. Each benzene ring and attached nitrogen atoms is planar, with the plane of $N(2)$–$(CH_3)_2$ rotated through 8°50′ with respect to the ring plane. The planes of the two halves of the cation are turned by 45°

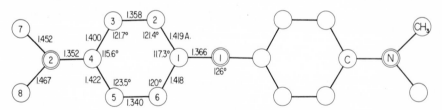

Fig. XVB,39.   Bond dimensions in the cation of 4,4′-bis(dimethylamino)diphenyl-amine perchlorate.

TABLE XVB,26

Parameters of the Atoms in $[(CH_3)_2NC_6H_4]_2NH \cdot ClO_4$

Atom	x	y	z
C(1)	0.4309	0.0240	0.1371
C(2)	0.3482	0.0299	0.5699
C(3)	0.2769	0.0135	0.9582
C(4)	0.2815	0.1131	0.9038
C(5)	0.3666	0.1637	0.9663
C(6)	0.4373	0.1238	0.0794
C(7)	0.1239	0.1058	0.7330
C(8)	0.2211	0.2565	0.7227
N(2)	0.2118	0.1562	0.7899
O(1)	0.4532	0.3786	0.8306
O(2)	0.4419	0.2559	0.6418
H(2)	0.16	0.39	0.39
H(3)	0.22	0.03	0.42
H(4)	0.13	0.26	0.08
H(5)	0.50	0.17	0.13
H(6)	0.36	0.48	0.27
H(7)	0.42	0.35	0.33
H(8)	0.42	0.39	0.18
H(9)	0.23	0.25	0.34
H(10)	0.35	0.22	0.37
H(11)	0.26	0.19	0.18

from one another. There is a bond C(7)–O(1) = 3.19 A. between the two ions. In the tetrahedral anion, Cl–O = 1.42 and 1.47 A.

This structure is closely related to the arrangement in the corresponding iodide (**XV,bII,17**).

XV,bII,17. Crystals of *4,4′-bis(dimethylamino)diphenylamine iodide*, $[(CH_3)_2NC_6H_4]_2NH \cdot I$, are monoclinic with a tetramolecular unit of the dimensions:

$$a_0 = 12.97 \text{ A.}; \quad b_0 = 14.22 \text{ A.}; \quad c_0 = 9.60 \text{ A.}; \quad \beta = 109°54'$$

Atoms are in the following positions of the space group C_{2h}^6 $(C2/c)$:

I: (4e) $\pm (0\ u\ ^1/_4; \ ^1/_2, u + ^1/_2, ^1/_4)$ with $u = 0.2041$

N(1):   (4e)   with $u = 0.4622$

The other atoms are in

$$(8f) \quad \pm(xyz; x,\bar{y},z+{}^1/_2; x+{}^1/_2,y+{}^1/_2,z; x+{}^1/_2,{}^1/_2-y,z+{}^1/_2)$$

with the parameters of Table XVB,27. Calculated values for the hydrogen atoms are to be found in the original.

TABLE XVB,27

Parameters of Atoms in $[(CH_3)_2NC_6H_4]_2NH \cdot I$

Atom	x	y	z
C(1)	0.4286	0.0018	0.1245
C(2)	0.3502	0.0599	0.5270
C(3)	0.2729	0.0265	0.4018
C(4)	0.2709	0.0711	0.8637
C(5)	0.3521	0.1305	0.9590
C(6)	0.4286	0.0959	0.0818
C(7)	0.1121	0.0423	0.6390
C(8)	0.1872	0.2065	0.7063
N(2)	0.1940	0.1045	0.7389

The structure is shown in Figure XVB,40. Its complex cations have the bond dimensions of Figure XVB,41. The benzene ring and attached N(2) and C(7) atoms are coplanar. The atom C(8) departs by 0.13 A.

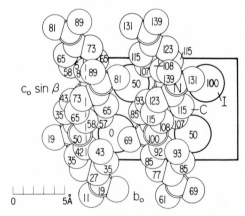

Fig. XVB,40. The monoclinic structure of 4,4′-bis(dimethylamino)diphenyl-amine iodide projected along its a_0 axis.

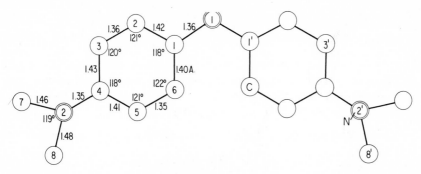

Fig. XVB,41. Bond dimensions in the cation of 4,4′-bis(dimethylamino)-
diphenylamine iodide.

TABLE XVB,28

Parameters of the Atoms in the Yellow Form of N-Picryl-p-iodoaniline

Atom	x	y	z
I	0.135	0.045	0.051
N(1)	0.665	0.26	0.304
N(2)	0.630	0.58	0.517
N(3)	0.420	0.96	0.328
NH(4)	0.518	0.63	0.242
O(1)	0.667	0.31	0.246
O(2)	0.703	0.07	0.331
O(3)	0.699	0.44	0.539
O(4)	0.587	0.71	0.551
O(5)	0.340	0.92	0.349
O(6)	0.440	0.12	0.289
C(1)	0.262	0.89	0.113
C(2)	0.359	0.96	0.105
C(3)	0.444	0.85	0.148
C(4)	0.429	0.67	0.195
C(5)	0.333	0.60	0.203
C(6)	0.249	0.71	0.162
C(7)	0.545	0.62	0.312
C(8)	0.623	0.46	0.340
C(9)	0.650	0.44	0.407
C(10)	0.600	0.59	0.448
C(11)	0.524	0.76	0.420
C(12)	0.494	0.77	0.352

from this plane which is turned through 12.5° with reference to the central plane defined by C(1)–N(1)–C(1′). The distance between I′ and N(1) is 3.67 A.

XV,bII,18. Many years ago a structure was described for the yellow form of *N-picryl-p-iodoaniline*, $(NO_2)_3C_6H_2 \cdot NHC_6H_4I$. It is monoclinic with a tetramolecular unit of the dimensions:

$$a_0 = 13.45 \text{ A.}; \quad b_0 = 5.15 \text{ A.}; \quad c_0 = 20.65 \text{ A.}; \quad \beta = 98°30'$$

The space group is C_{2h}^5 ($P2_1/a$) with atoms in the positions:

$$(4e) \quad \pm(xyz; x+{}^1/_2, {}^1/_2 - y, z)$$

The parameters then determined are stated in Table XVB,28, those for y being only approximate.

Molecular dimensions resulting from this structure (Fig. XVB,42) are those of Figure XVB,43.

Of the two other crystalline forms of this substance that have been studied, the first (red) variety also has a tetramolecular monoclinic cell and a structure based on C_{2h}^5. Its unit has

$$a_0 = 14.15 \text{ A.}; \quad b_0 = 5.85 \text{ A.}; \quad c_0 = 18.30 \text{ A.}; \quad \beta = 113°$$

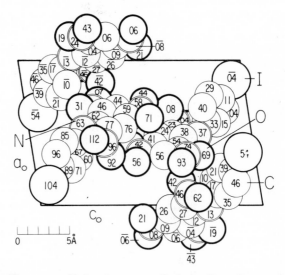

Fig. XVB,42.   The monoclinic structure of the yellow form of *N*-picryl-*p*-iodoaniline projected along its b_0 axis.

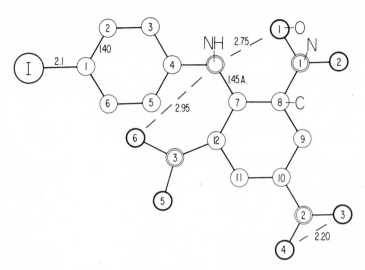

Fig. XVB,43.   Some bond lengths in the molecule of N-picryl-p-iodoaniline.

The axial orientation is $P2_1/c$ and, though details of a structure were discussed, no atomic parameters were published.

XV,bII,19.   Crystals of p-*methoxyindophenol-N-oxide*, $CH_3OC_6H_4$-$N(O)C_6H_4O$, are orthorhombic with a unit containing eight molecules and having the edge lengths:

$a_0 = 24.0$ A.; $b_0 = 12.53$ A.; $c_0 = 7.49$ A.   (room temperature)
$a_0 = 23.91(1)$ A.; $b_0 = 12.15(1)$ A.; $c_0 = 7.403(5)$ A.   ($-180°C.$)

The space group is V_h^{15} (*Pbca*) with all atoms in the positions:

(8c) $\pm(xyz; x+{}^1/_2,{}^1/_2-y,\bar{z}; \bar{x},y+{}^1/_2,{}^1/_2-z; {}^1/_2-x,\bar{y},z+{}^1/_2)$

The determined parameters are given in Table XVB,29; those calculated for hydrogen are stated in the original article.

The structure, shown in Figure XVB,44, is composed of molecules with the bond dimensions of Figure XVB,45. These molecules can be described in terms of three planes, one containing the quinone ring (II), another the benzene ring (I), and the third (III) the atoms C(1), C(7), N and O(1). The angle between I and II is 64° and between I and III, 60°. The methyl C(13) atom is 0.11 A. and the O(3) atom, 0.02 A. from the I plane.

TABLE XVB,29

Parameters of the Atoms in $CH_3OC_6H_4N(O)C_6H_4O$

Atom	x	y	z
N	0.1470	0.1200	0.1874
O(1)	0.1563	0.1179	0.0190
O(2)	0.3175	0.1222	0.6726
O(3)	−0.0776	0.0915	0.3817
C(1)	0.1887	0.1236	0.3117
C(2)	0.2454	0.1321	0.2490
C(3)	0.2884	0.1307	0.3689
C(4)	0.2781	0.1228	0.5621
C(5)	0.2203	0.1150	0.6223
C(6)	0.1775	0.1143	0.5019
C(7)	0.0888	0.1167	0.2414
C(8)	0.0573	0.0271	0.1865
C(9)	0.0011	0.0217	0.2365
C(10)	−0.0227	0.1064	0.3394
C(11)	0.0090	0.1963	0.3906
C(12)	0.0653	0.2016	0.3410
C(13)	−0.1038	0.1752	0.4992

XV,bII,20. Crystals of the relatively stable free radical *di-p-anisyl-N-oxide*, $(CH_3OC_6H_4)_2NO$, are orthorhombic. Their tetramolecular cell has the dimensions:

$$a_0 = 7.33 \text{ A.;} \quad b_0 = 26.8 \text{ A.;} \quad c_0 = 6.25 \text{ A.}$$

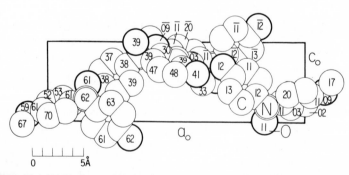

Fig. XVB,44. Half the molecules in the orthorhombic unit of *p*-methoxyindophenol-*N*-oxide projected along its b_0 axis. The other four are obtained by inverting these through a center of symmetry.

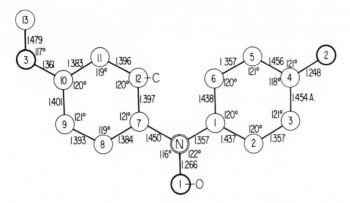

Fig. XVB,45. Bond dimensions in the molecule of p-methoxyindophenol-N-oxide.

Atoms have been placed in the following positions of C_{2v}^{17} (Aba):

(4a) $00u$; $\frac{1}{2}\,\frac{1}{2}\,u$; $0,\frac{1}{2},u+\frac{1}{2}$; $\frac{1}{2},0,u+\frac{1}{2}$

(8b) xyz; $\bar{x}\bar{y}z$; $\frac{1}{2}-x,y+\frac{1}{2},z$; $x+\frac{1}{2},\frac{1}{2}-y,z$;
        $x,y+\frac{1}{2},z+\frac{1}{2}$; $\bar{x},\frac{1}{2}-y,z+\frac{1}{2}$; $\frac{1}{2}-x,y,z+\frac{1}{2}$; $x+\frac{1}{2},\bar{y},z+\frac{1}{2}$

Positions and determined parameters are those of Table XVB,30.

TABLE XVB,30

Positions and Parameters of the Atoms in di-p-Anisyl-N-oxide

Atom	Position	x	y	z
O(1)	(4a)	0	0	0.384
O(2)	(8b)	−0.008	0.190	−0.186
N	(4a)	0	0	0.187
C(1)	(8b)	0.010	0.047	0.078
C(2)	(8b)	0.080	0.088	0.190
C(3)	(8b)	0.075	0.135	0.099
C(4)	(8b)	−0.008	0.142	−0.093
C(5)	(8b)	−0.083	0.102	−0.200
C(6)	(8b)	−0.074	0.054	−0.112
C(7)	(8b)	−0.104	0.203	−0.381

Note: According to a private communication from the author the z parameters for O(1) and C(7) are those of this table rather than of the original paper.

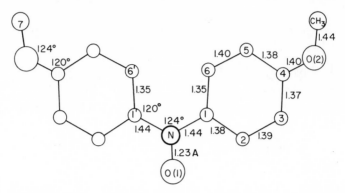

Fig. XVB,46.   Bond dimensions in the molecule of di-*p*-anisyl-*N*-oxide.

Bond dimensions in the molecule thus defined are shown in Figure XVB,46. It has a twofold axis through N–O(1) and the plane of the benzene ring is turned 33° about C(1)–C(4). The four atoms O(1), N, C(1), C(4) lie within 0.07 A. of being coplanar. The structure arising through the packing of these molecules is illustrated in Figure XVB,47.

XV,bII,21.   The orthorhombic crystals of *N,N-diphenylacetamide*, $(C_6H_5)_2NC(O)CH_3$, have a unit that contains eight molecules and has the dimensions:

$$a_0 = 11.0103(4) \text{ A.;} \quad b_0 = 12.0754(4) \text{ A.;} \quad c_0 = 17.2944(2) \text{ A.}$$

The space group is V_h^{15} (*Pbca*) with all atoms in the positions:

$$(8c) \quad \pm(xyz;\, x+\tfrac{1}{2},\tfrac{1}{2}-y,\bar{z};\, \bar{x},y+\tfrac{1}{2},\tfrac{1}{2}-z;\, \tfrac{1}{2}-x,\bar{y},z+\tfrac{1}{2})$$

The parameters listed in Table XVB,31 include those assigned the atoms of hydrogen.

The structure, shown in Figure XVB,48, has molecules with the bond dimensions of Figure XVB,49. In them the acetyl group and the two phenyl rings are planar. The primed ring makes 77.2° and the other ring 61.7° with respect to the acetyl plane. It is thought that the departure from 120° of the bond angle about C(2) is significant.

XV,bII,22.   The monoclinic crystals of *μ-hydrido-μ-diphenylphosphido-bis(tetracarbonyl manganese)*, $(CO)_4Mn(H)\cdot P(C_6H_5)_2\cdot Mn(CO)_4$, have a tetramolecular unit of the dimensions:

$$a_0 = 16.76(2) \text{ A.;} \quad b_0 = 8.15(1) \text{ A.;} \quad c_0 = 17.03(2) \text{ A.}$$
$$\beta = 110°46(2)'$$

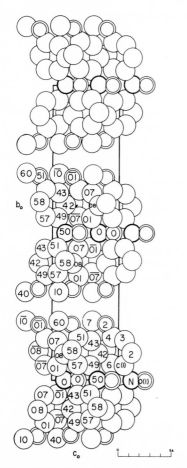

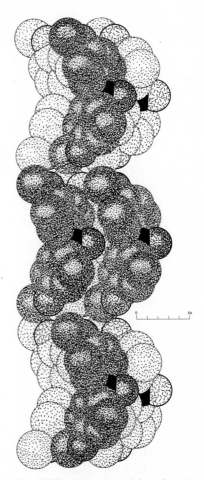

Fig. XVB,47a. The orthorhombic structure of di-p-anisyl-N-oxide projected along its a_0 axis. Left-hand axes.

Fig. XVB,47b. A packing drawing of the orthorhombic structure of di-p-anisyl-N-oxide viewed along its a_0 axis. The nitrogen atoms are black, those of oxygen heavily outlined and line shaded. The carbons are dotted. Left-hand axes.

The space group is C_{2h}^6 in the orientation $I2/c$. The phosphorus atoms thus are in the special positions

$$(2e) \quad \pm (0 \; u \; 1/4; \quad 1/2, u + 1/2, 3/4) \qquad \text{with } u = 0.2305(8)$$

TABLE XVB,31

Parameters of the Atoms in N,N-Diphenylacetamide

Atom	x	y	z
N	0.0947	0.2625	0.4071
O	−0.0134	0.2215	0.3005
C	0.0067	0.2048	0.3687
C(CH$_3$)	−0.0633	0.1200	0.4141
C(1)	0.1277	0.2390	0.4861
C(2)	0.2053	0.1526	0.5032
C(3)	0.2344	0.1319	0.5804
C(4)	0.1854	0.1968	0.6387
C(5)	0.1078	0.2826	0.6206
C(6)	0.0800	0.3044	0.5446
C(1′)	0.1605	0.3483	0.3682
C(2′)	0.1000	0.4394	0.3389
C(3′)	0.1639	0.5221	0.3011
C(4′)	0.2885	0.5140	0.2929
C(5′)	0.3490	0.4224	0.3215
C(6′)	0.2850	0.3397	0.3591
H(1,CH$_3$)	−0.098	0.155	0.456
H(2,CH$_3$)	−0.105	0.090	0.380
H(3,CH$_3$)	−0.016	0.059	0.426
H(2)	0.243	0.102	0.458
H(3)	0.294	0.065	0.594
H(4)	0.208	0.180	0.698
H(5)	0.070	0.333	0.666
H(6)	0.020	0.371	0.531
H(2′)	0.003	0.446	0.346
H(3′)	0.117	0.593	0.279
H(4′)	0.338	0.578	0.264
H(5′)	0.446	0.416	0.315
H(6′)	0.332	0.269	0.382

and it is considered that the hydrogen atoms attached to manganese are in (2e) also, with $u = -0.125$. All other atoms are in

$$(8f) \quad \pm (xyz;\, x,\bar{y},z + \tfrac{1}{2}); \quad \text{B.C.}$$

with the parameters of Table XVB,32.

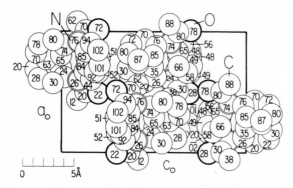

Fig. XVB,48.   The orthorhombic structure of N,N-diphenylacetamide projected along its b_0 axis.

In the molecules of this structure the manganese atoms, as Figure
XVB,50 indicates, are bridged by both the phosphorus and the
hydrogen atoms; their bond lengths and certain others are given in
the figure.

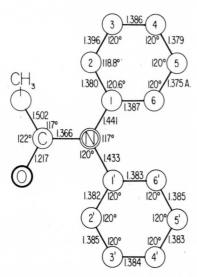

Fig. XVB,49.   Bond dimensions in the molecule of N,N-diphenylacetamide.

TABLE XVB,32

Parameters of Atoms in $(CO)_4Mn(H)P(C_6H_5)_2 \cdot Mn(CO)_4$

Atom	x	y	z
Mn	0.05748(15)	0.01594(38)	0.20194(14)
C(1)	0.1517(10)	0.0080(26)	0.3001(9)
O(1)	0.2094(6)	0.0057(20)	0.3587(7)
C(2)	−0.0325(9)	0.0197(27)	0.1022(10)
O(2)	−0.0858(7)	0.0166(20)	0.0378(6)
C(3)	0.1179(11)	0.1363(20)	0.1554(11)
O(3)	0.1553(7)	0.2238(16)	0.1270(8)
C(4)	0.0872(12)	−0.1800(25)	0.1733(11)
O(4)	0.1106(8)	−0.3012(18)	0.1521(8)
	Phenyl groups		
C(5)	0.0714	0.3723	0.3258
C(6)	0.0548	0.4142	0.3980
C(7)	0.1060	0.5298	0.4541
C(8)	0.1738	0.6034	0.4378
C(9)	0.1904	0.5614	0.3656
C(10)	0.1392	0.4459	0.3096
H(6)	0.0021	0.3571	0.4107
H(7)	0.0930	0.5623	0.5102
H(8)	0.2135	0.6930	0.4813
H(9)	0.2431	0.6186	0.3529
H(10)	0.1522	0.4134	0.2534

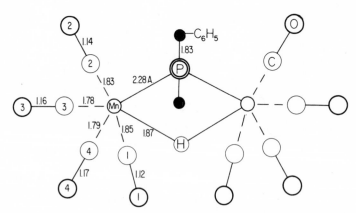

Fig. XVB,50.　Some bond lengths in the molecule of μ-hydrido-μ-diphenyl-phosphido-bis(tetracarbonyl manganese).

XV,bII,23. Crystals of *methyl diphenyl thiophosphinite*, $(C_6H_5)_2$-$P(S)OCH_3$, are monoclinic with a tetramolecular unit of the dimensions:

$$a_0 = 11.87(5) \text{ A.;} \quad b_0 = 12.70(5) \text{ A.;} \quad c_0 = 9.49(5) \text{ A.;} \quad \beta = 112°$$

Atoms are in the following positions of the space group C_{2h}^5 $(P2_1/a)$:

$$(4e) \quad \pm(xyz; x+1/2, 1/2-y, z)$$

According to a preliminary announcement, the atoms have the approximate parameters of Table XVB,33.

TABLE XVB,33

Parameters of the Atoms in $(C_6H_5)_2P(S)OCH_3$

Atom	x	y	z
S	0.190	0.427	0.217
P	0.237	0.307	0.121
O	0.366	0.318	0.101
CH_3	0.384	0.404	0.010
C(1)	0.267	0.185	0.230
C(2)	0.367	0.120	0.242
C(3)	0.384	0.026	0.329
C(4)	0.297	−0.006	0.392
C(5)	0.197	0.060	0.380
C(6)	0.175	0.158	0.296
C(1′)	0.119	0.274	−0.069
C(2′)	0.012	0.329	−0.124
C(3′)	−0.077	0.301	−0.272
C(4′)	−0.045	0.223	−0.354
C(5′)	0.066	0.164	−0.294
C(6′)	0.153	0.193	−0.147

XV,bII,24. *Bromodiphenylarsine*, $(C_6H_5)_2AsBr$, forms monoclinic crystals whose tetramolecular units have the dimensions:

$$a_0 = 11.00(4) \text{ A.;} \quad b_0 = 8.56(3) \text{ A.;} \quad c_0 = 12.03(4) \text{ A.}$$
$$\beta = 93°0(30)'$$

The space group is C_{2h}^5 $(P2_1/a)$ with atoms in the positions:

$$(4e) \quad \pm(xyz; x+1/2, 1/2-y, z)$$

The parameters are those of Table XVB,34.

TABLE XVB,34

Parameters of the Atoms in $(C_6H_5)_2AsBr$ and $(C_6H_5)_2AsCl$ (in parentheses)

Atom	x		y		z	
As	−0.0140	(−0.0130)	0.1636	(0.1673)	0.2652	(0.2659)
Br[Cl]	0.0793	(0.0715)	0.4000	(0.3857)	0.2033	(0.2000)
C(1)	0.082	(0.083)	0.024	(0.026)	0.175	(0.178)
C(2)	0.023	(0.025)	−0.098	(−0.102)	0.114	(0.125)
C(3)	0.090	(0.089)	−0.200	(−0.200)	0.051	(0.063)
C(4)	0.213	(0.213)	−0.176	(−0.163)	0.046	(0.053)
C(5)	0.267	(0.271)	−0.049	(−0.033)	0.105	(0.106)
C(6)	0.204	(0.206)	0.051	(0.062)	0.168	(0.165)
C(7)	0.073	(0.071)	0.167	(0.167)	0.418	(0.418)
C(8)	0.031	(0.031)	0.286	(0.275)	0.488	(0.489)
C(9)	0.082	(0.083)	0.286	(0.275)	0.593	(0.599)
C(10)	0.167	(0.176)	0.167	(0.167)	0.619	(0.638)
C(11)	0.210	(0.213)	0.047	(0.059)	0.550	(0.559)
C(12)	0.161	(0.159)	0.047	(0.059)	0.444	(0.450)

The structure is shown in Figure XVB,51. Its molecules have the bond dimensions of Figure XVB,52. Between the two planar phenyl groups of these molecules the angle is 61°.

The analogous *chlorodiphenylarsine*, $(C_6H_5)_2AsCl$, is isostructural with a unit of the dimensions:

$$a_0 = 11.09(4) \text{ A.}; \quad b_0 = 8.55(3) \text{ A.}; \quad c_0 = 11.93(4) \text{ A.}$$
$$\beta = 95°0(30)'$$

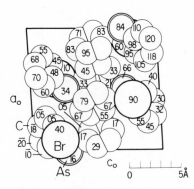

Fig. XVB,51.   The monoclinic structure of bromodiphenylarsine projected along its b_0 axis.

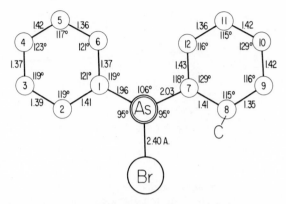

Fig. XVB,52.   Bond dimensions in the molecule of bromodiphenylarsine.

The determined parameters, very close to those for the bromide, are given in parentheses in the table. The molecular dimensions also are like those for the bromide, but with As–Cl = 2.26 A.

XV,bII,25. Crystals of *diphenyl trichlorostibine monohydrate,* $(C_6H_5)_2SbCl_3 \cdot H_2O$, are orthorhombic with a tetramolecular unit of the edge lengths:

$$a_0 = 17.78(4) \text{ A.}; \quad b_0 = 9.71(3) \text{ A.}; \quad c_0 = 8.50(2) \text{ A.}$$

The chosen space group is V_h^{14} (*Pbcn*). Three sets of atoms are in the

TABLE XVB,35

Positions and Parameters of the Atoms in $(C_6H_5)_2SbCl_3 \cdot H_2O$

Atom	Position	x	y	z
Sb	(4c)	0	−0.220	$1/4$
Cl(1)	(4c)	0	−0.460	$1/4$
Cl(2)	(8d)	0.577	0.707	0.506
C(1)	(8d)	0.399	0.693	0.615
C(2)	(8d)	0.346	0.592	0.652
C(3)	(8d)	0.281	0.573	0.574
C(4)	(8d)	0.277	0.650	0.440
C(5)	(8d)	0.324	0.746	0.379
C(6)	(8d)	0.385	0.780	0.471
O	(4c)	0	0.006	$1/4$

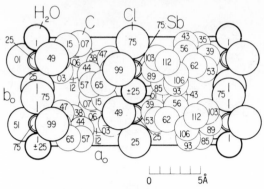

Fig. XVB,53. The orthorhombic structure of diphenyl trichlorostibine mono-hydrate projected along its c_0 axis.

positions:

$$(4c) \quad \pm (0\ u\ {}^1\!/_4;\ {}^1\!/_2,u+{}^1\!/_2,{}^1\!/_4)$$

All others are in:

$$(8d) \quad \pm (xyz;\ x+{}^1\!/_2,y+{}^1\!/_2,{}^1\!/_2-z;\ {}^1\!/_2-x,y+{}^1\!/_2,z;\ x,\bar{y},z+{}^1\!/_2)$$

Their parameters are stated in Table XVB,35.

The structure, as shown in Figure XV.B,53, is composed of molecules with the bond dimensions of Figure XVB,54. Coordination about antimony is octahedral, with water one of the six ligands. In the molecule a twofold axis passes through antimony, water, and one of the chlorine atoms.

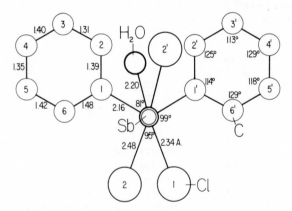

Fig. XVB,54. Bond dimensions in the molecule of diphenyl trichlorostibine monohydrate.

TABLE XVB,36

Parameters of the Atoms in $(CH_3 \cdot C_6H_4)_2X$, where X = S, Se and Te

Atom	x			y			z		
	S	Se	Te	S	Se	Te	S	Se	Te
S,Se,Te	0.1391	0.1360	0.1304	0.2233	0.2328	0.2529	0.1701	0.1381	0.1106
C(1)	0.0822	0.0754	0.0690	0.1881	0.1977	0.2195	0.3392	0.3299	0.3267
C(2)	0.0860	0.0809	0.0759	0.0977	0.1089	0.1303	0.5431	0.5353	0.5239
C(3)	0.0409	0.0366	0.0343	0.0693	0.0826	0.1103	0.6776	0.6748	0.6708
C(4)	0.9920	0.9876	0.9846	0.1314	0.1452	0.1798	0.6071	0.6081	0.6250
C(5)	0.9876	0.9821	0.9776	0.2220	0.2365	0.2681	0.4029	0.4030	0.4266
C(6)	0.0323	0.9263	0.0196	0.2502	0.2621	0.2882	0.2698	0.2643	0.2797
C(7)	0.9438	0.9405	0.9391	0.1020	0.1061	0.1575	9.7498	0.7567	0.7839
C(1')	0.1950	0.1962	0.1941	0.2404	0.2475	0.2529	0.3454	0.3401	0.3234
C(2')	0.1906	0.1918	0.1941	0.3263	0.3311	0.3576	0.5510	0.5479	0.5066
C(3')	0.2360	0.2362	0.2378	0.3402	0.3409	0.3576	0.6918	0.6921	0.6534
C(4')	0.2850	0.2841	0.2807	0.2698	0.2678	0.2529	0.6267	0.6275	0.6122
C(5')	0.2893	0.2876	0.2807	0.1845	0.1852	0.1508	0.4225	0.4214	0.4307
C(6')	0.2441	0.2437	0.2378	0.1685	0.1735	0.1508	0.2789	0.2772	0.2838
C(7')	0.3330	0.3317	0.3279	0.2811	0.2778	0.2529	0.7796	0.7840	0.7689

XV,bII,26. The *di-p-tolyl sulfide, selenide,* and *telluride* are ortho-rhombic with tetramolecular cells of the dimensions:

For $(p\text{-}CH_3\cdot C_6H_4)_2S$:

$$a_0 = 25.07 \text{ A.}; \quad b_0 = 7.92 \text{ A.}; \quad c_0 = 5.81 \text{ A.}$$

For $(p\text{-}CH_3\cdot C_6H_4)_2Se$:

$$a_0 = 25.12 \text{ A.}; \quad b_0 = 7.99 \text{ A.}; \quad c_0 = 5.88 \text{ A.}$$

For $(p\text{-}CH_3\cdot C_6H_4)_2Te$:

$$a_0 = 25.33 \text{ A.}; \quad b_0 = 8.05 \text{ A.}; \quad c_0 = 6.01 \text{ A.}$$

They are isostructural, with all atoms in the general positions of V^4 ($P2_12_12_1$):

$$(4a) \quad xyz; \ ^1\!/_2-x,\bar{y},z+^1\!/_2; \ x+^1\!/_2,^1\!/_2-y,\bar{z}; \ \bar{x},y+^1\!/_2,^1\!/_2-z$$

The chosen parameters are those of Table XVB,36.

The structure thus defined is illustrated in Figure XVB,55. The phenyl groups are planar by assumption, with C–C = 1.40 A. For the sulfide, S–C = 1.76 and 1.74 A.; the angle C–S–C = 109°. For the selenide, Se–C = 1.92 and 1.93 A.; C–Se–C = 106°. For the telluride, Te–C = 2.05 A. and C–Te–C = 101°. The angles between normals to the phenyl groups of a molecule (measuring the twist of these groups about the X–C bonds) are 56, 55, and 62° for the sulfide, selenide, and telluride, respectively.

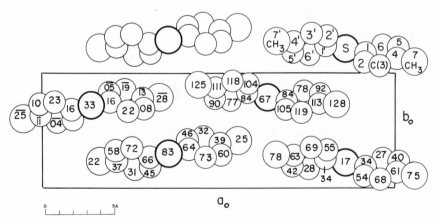

Fig. XVB,55a. The orthorhombic structure of di-*p*-tolyl sulfide projected along its c_0 axis. The large heavily ringed circle is the sulfur atom. Left-hand axes.

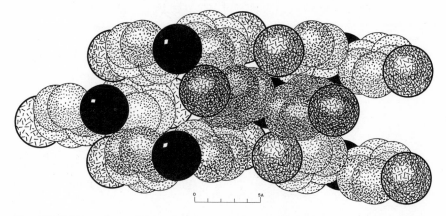

Fig. XVB,55b. A packing drawing of the orthorhombic di-p-tolyl sulfide arrangement viewed along its c_0 axis. The black circles are sulfur. Methyl carbons are heavily outlined and line shaded; the ring carbons are dotted. Left-hand axes.

TABLE XVB,37

Parameters of the Atoms in $(C_6H_5)_2SO$

Atom	x	y	z
S	0.1813	0.2047	0.2240
O	0.0481	0.2010	0.0883
C(1)	0.2854	0.0999	0.2106
C(2)	0.4349	0.0923	0.2996
C(3)	0.5187	0.0085	0.2893
C(4)[a]	0.4571	0.9378	0.1754
C(5)[a]	0.3055	0.9458	0.0763
C(6)	0.2255	0.0302	0.0984
C(7)	0.1127	0.1750	0.4026
C(8)	0.9652	0.1376	0.3904
C(9)	0.9101	0.1224	0.5351
C(10)	0.9999	0.1397	0.6931
C(11)	0.1424	0.1762	0.6910
C(12)	0.2031	0.1949	0.5519

[a] The parameters for these two atoms as stated in the original article do not lead to a reasonable structure. The values given here, reversed in signs for x and z, do so.

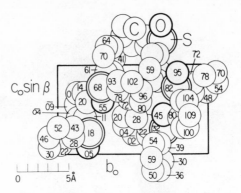

Fig. XVB,56.   The monoclinic structure of diphenyl sulfoxide projected along
its a_0 axis.

XV,bII,27.   *Diphenyl sulfoxide*, $(C_6H_5)_2SO$, forms monoclinic crystals
whose tetramolecular unit has the dimensions:

$$a_0 = 8.90(2) \text{ A.}; \quad b_0 = 14.08(3) \text{ A.}; \quad c_0 = 8.32(2) \text{ A.}$$
$$\beta = 101°7(10)'$$

The space group is C_{2h}^5 in the orientation $P2_1/n$ with atoms in the positions:

$$(4e) \quad \pm(xyz; x + 1/2, 1/2 - y, z + 1/2)$$

The parameters are those of Table XVB,37.

The structure is that of Figure XVB,56. Its molecules have the bond
dimensions of Figure XVB,57. In them the dihedral angle between the
two planar benzene rings is 75°50'; between the C–S–C plane and its
ring it is 80°36' and 83°18'. Between molecules the atomic separations
are normal.

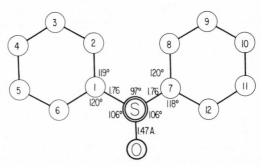

Fig. XVB,57.   Some bond dimensions in the molecule of diphenyl sulfoxide.

The analogous selenium compound $(C_6H_5)_2SeO$ is isostructural with a unit of the dimensions:

$$a_0 = 8.95 \text{ A.}; \quad b_0 = 14.1 \text{ A.}; \quad c_0 = 8.35 \text{ A.}; \quad \beta = 102°20'$$

XV,bII,28. Crystals of *4,4'-dichloro diphenyl sulfone*, $(ClC_6H_4)_2SO_2$, are monoclinic with a tetramolecular unit of the dimensions:

$$a_0 = 20.204(10) \text{ A.}; \quad b_0 = 5.009(10) \text{ A.}; \quad c_0 = 12.259(10) \text{ A.}$$
$$\beta = 90.57(25)°$$

The space group is C_{2h}^6 in the orientation $I2/a$ with atoms in the positions:

$$\text{S: } (4e) \quad \pm(1/4 \; u \; 0; \; 3/4, \; u + 1/2, 1/2) \qquad \text{with } u = 0.1546$$

The other atoms are in the general positions:

$$(8f) \quad \pm(xyz; \; x + 1/2, \bar{y}, z); \text{ B.C.}$$

TABLE XVB,38

Parameters of Atoms in $(ClC_6H_4)_2SO_2$[a]

Atom	x	y	z
Cl	0.0327 (0.0343)	0.9381	0.1621 (0.1636)
O	0.2244 (0.2234)	0.0126	0.9073 (0.9064)
C(1)	0.1868 (0.1883)	0.3696	0.0456 (0.0456)
C(2)	0.1352 (0.1345)	0.4367	0.9760 (0.9738)
C(3)	0.0877 (0.0867)	0.6117	0.0112 (0.0114)
C(4)	0.0926 (0.0923)	0.7173	0.1164 (0.1164)
C(5)	0.1431 (0.1420)	0.6448	0.1867 (0.1874)
C(6)	0.1910 (0.1919)	0.4743	0.1503 (0.1509)

[a] The values in parentheses were determined by neutron diffraction.

TABLE XVB,39

Parameters of Hydrogen in $(ClC_6H_4)_2SO_2$[a]

Atom	x	y	z
H(2)	0.1356 (0.132)	0.3537 (0.345)	0.8979 (0.894)
H(3)	0.0462 (0.047)	0.6663 (0.670)	0.9576 (0.956)
H(5)	0.1465 (0.146)	0.7257 (0.718)	0.2685 (0.271)
H(6)	0.2308 (0.231)	0.4198 (0.415)	0.2020 (0.204)

[a] Values in parentheses were determined in the x-ray study.

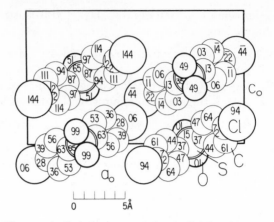

Fig. XVB,58.   The monoclinic structure of 4,4'-dichlorodiphenyl sulfone projected along its b_0 axis.

Parameters have been determined by using both x-rays and neutrons. The values for the heavier atoms are given in Table XVB,38, those from neutron diffraction being in parentheses. For the hydrogen atoms the neutron positions are more precise; they are stated in Table XVB,39 with the x-ray parameters in parentheses.

The structure is shown in Figure XVB,58. The bond dimensions in its molecules are those of Figure XVB,59. From the neutron data, C–H = 1.02–1.08 A. The benzene rings and their attached chlorine atoms are coplanar to less than 0.02 A., with sulfur 0.06 A. outside the plane. The dihedral angle between the plane of the benzene ring and the plane C(1)–S–C(1') is 84.7°. The shortest intermolecular distances are C–O = 3.21, 3.24, and 3.27 A.

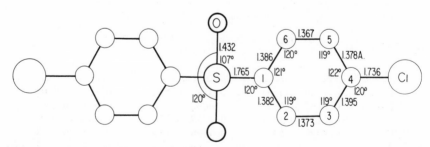

Fig. XVB,59.   Bond dimensions in the molecule of 4,4'-dichlorodiphenyl sulfone.

XV,bII,29. The monoclinic crystals of *4,4'-diiodo diphenyl sulfone,* $(IC_6H_4)_2SO_2$, possess a tetramolecular unit of the dimensions:

$$a_0 = 21.37 \text{ A.}; \quad b_0 = 4.92 \text{ A.}; \quad c_0 = 14.37 \text{ A.}; \quad \beta = 116°42'$$

The space group is C_{2h}^6 $(C2/c)$ with sulfur atoms in the positions:

$$(4e) \quad \pm (0 \; u \; {}^1/_4; \; {}^1/_2, u + {}^1/_2, {}^1/_4) \quad \text{with } u = 0.611$$

The other atoms are in:

$$(8f) \quad \pm (xyz; \; x, \bar{y}, z + {}^1/_2; \; x + {}^1/_2, y + {}^1/_2, z; \; x + {}^1/_2, {}^1/_2 - y, z + {}^1/_2)$$

Parameters are listed in Table XVB,40; except for the value for iodine, the y parameters were calculated on the assumption of a regular hexagonal benzene ring, together with the separations I–C = 2.10 A., S–C = 1.70 A., and S–O = 1.43 A.

TABLE XVB,40

Parameters of Atoms in $(IC_6H_4)_2SO_2$

Atom	x	y	z
I	0.200	−0.195	0.636
O	0.042	0.789	0.220
C(1)	0.051	0.405	0.348
C(2)	0.120	0.320	0.367
C(3)	0.153	0.148	0.430
C(4)	0.135	0.063	0.513
C(5)	0.066	0.148	0.494
C(6)	0.034	0.320	0.431

XV,bII,30. According to a preliminary note, crystals of *diamino diphenyl sulfone,* $(C_6H_4NH_2)_2SO_2$, are orthorhombic with a tetramolecular unit of the edge lengths:

$$a_0 = 25.57(1) \text{ A.}; \quad b_0 = 8.07(1) \text{ A.}; \quad c_0 = 5.77(1) \text{ A.}$$

The space group is V^4 $(P2_12_12_1)$ with atoms in the positions:

$$(4a) \quad xyz; \; {}^1/_2 - x, \bar{y}, z + {}^1/_2; \; x + {}^1/_2, {}^1/_2 - y, \bar{z}; \; \bar{x}, y + {}^{.1}/_2, {}^1/_2 - z$$

The parameters are listed in Table XVB,41.

TABLE XVB,41

Parameters of the Atoms in $(C_6H_4NH_2)_2SO_2$

Atom	x	y	z
S	0.374	0.331	0.126
O(1)	0.369	0.313	0.376
O(2)	0.357	0.195	−0.016
N(1)	0.600	0.443	−0.076
N(2)	0.277	0.961	−0.183
C(1)	0.442	0.365	0.059
C(2)	0.462	0.302	−0.150
C(3)	0.513	0.332	−0.198
C(4)	0.546	0.418	−0.027
C(5)	0.526	0.477	0.177
C(6)	0.471	0.450	0.231
C(7)	0.342	0.516	0.038
C(8)	0.348	0.654	0.176
C(9)	0.326	0.799	0.098
C(10)	0.297	0.807	−0.119
C(11)	0.291	0.660	−0.244
C(12)	0.314	0.513	−0.174

The resulting structure is shown in Figure XVB,60.

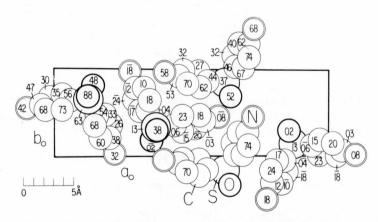

Fig. XVB,60.   The orthorhombic structure of diamino diphenyl sulfone projected along its c_0 axis.

XV,bII,31. Crystals of *diphenyl selenium dichloride*, $(C_6H_5)_2SeCl_2$, have a structure unlike that of the bromide (**XV,bII,32**). Their large orthorhombic unit containing eight molecules has the edge lengths:

$$a_0 = 7.59 \text{ A.}; \quad b_0 = 17.97 \text{ A.}; \quad c_0 = 17.77 \text{ A.}$$

The space group is V_h^{15} (*Pbca*) with all atoms in the general positions:

$$(8c) \quad \pm (xyz; x+1/2, 1/2-y, \bar{z}; \bar{x}, y+1/2, 1/2-z; 1/2-x, \bar{y}, z+1/2)$$

The parameters are those of Table XVB,42.

TABLE XVB,42

Parameters of the Atoms in Diphenyl Selenium Dichloride

Atom	x	y	z
Se	0.055	0.100	0.130
Cl(1)	0.28	0.152	0.062
Cl(2)	−0.17	0.049	0.201
C(1)	0.055	0.19	0.19
C(2)	0.055	0.26	0.16
C(3)	0.055	0.32	0.21
C(4)	0.055	0.31	0.28
C(5)	0.055	0.24	0.31
C(6)	0.055	0.18	0.27
C(7)	−0.08	0.10	0.04
C(8)	−0.01	0.07	0.98
C(9)	−0.12	0.07	0.92
C(10)	−0.29	0.10	0.92
C(11)	−0.35	0.13	0.98
C(12)	−0.25	0.13	0.04

In the resulting structure (Fig. XVB,61) the nonplanar molecules resemble those of the bromide. The Se–Cl separation is 2.30 A. Between molecules the Cl–Cl distance is 4.12 A.; positions of the carbon atoms C(7)-to-C(12) composing one benzene ring were not established with great accuracy.

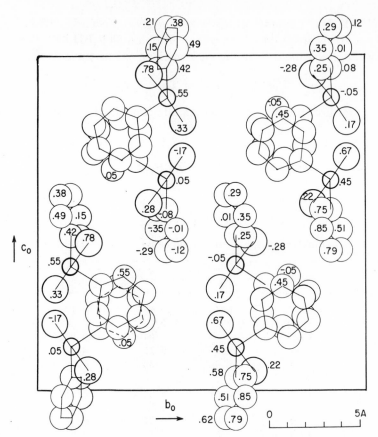

Fig. XVB,61a. The orthorhombic structure of diphenyl selenium dichloride projected along its a_0 axis. The small heavily ringed circles are selenium, the largest circles are chlorine. Atoms of carbon are intermediate in size.

XV,bII,32. *Diphenyl selenium dibromide,* $(C_6H_5)_2SeBr_2$, is orthorhombic with a tetramolecular unit having the edges:

$$a_0 = 13.95 \text{ A.}; \quad b_0 = 5.78 \text{ A.}; \quad c_0 = 15.40 \text{ A.}$$

The space group is V_h^{14} (*Pbcn*). Selenium atoms are in special positions:

$$(4c) \quad \pm(0\,u\,^1/_4; \,^1/_2, u+^1/_2, ^1/_4) \qquad \text{with } u = 0.080$$

All other atoms are in general positions:

$$(8d) \quad \pm(xyz; \,^1/_2-x, ^1/_2-y, z+^1/_2; \, x+^1/_2, ^1/_2-y, \bar{z}; \, \bar{x}, y, ^1/_2-z)$$

Their parameters are listed in Table XVB,43.

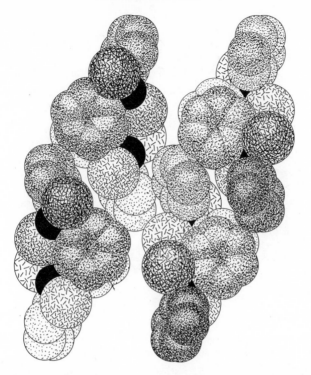

Fig. XVB,61b. A packing drawing of the orthorhombic structure of diphenyl selenium dichloride viewed along its a_0 axis. The selenium atoms are black, those of chlorine are large and line shaded. The carbons are dotted.

TABLE XVB,43

Parameters of Atoms in Diphenyl Selenium Dibromide

Atom	x	y	z
Br	0.149	0.080	0.157
C(1)	0.062	0.274	0.332
C(2)	0.126	0.461	0.304
C(3)	0.172	0.601	0.368
C(4)	0.155	0.545	0.454
C(5)	0.090	0.354	0.482
C(6)	0.044	0.219	0.422

In the resulting structure (Fig. XVB,62) adjacent molecules are separated by Br–Br = 4.04 A., by an unusually close CH–CH = 3.10 A.,

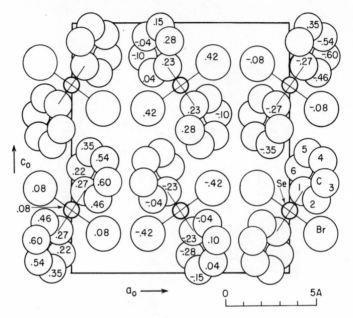

Fig. XVB,62a. The orthorhombic structure of diphenyl selenium dibromide projected along its b_0 axis. Left-hand axes.

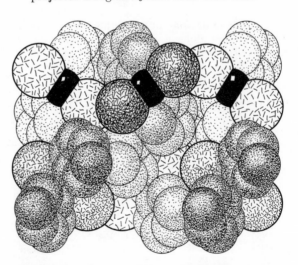

Fig. XVB,62b. A packing drawing of the orthorhombic structure of diphenyl selenium dibromide seen along its b_0 axis. The selenium atoms are black, the bromines are large, heavily outlined, and line shaded. Atoms of carbon are dotted. Left-hand axes.

and by other CH–CH distances from 3.44 A. upward. Within the molecule, which has a twofold axis of symmetry, Se–Br = 2.52 A. and Se–C = 1.91 A. It may be noted that this Se–Br distance is substantially the same as that found in K_2SeBr_6 (2.54 A.).

XV,bII,33. Crystals of *di-p-tolyl selenium dichloride*, $(CH_3C_6H_4)_2$-$SeCl_2$, and of the corresponding *dibromide* are isostructural. Their bimolecular orthorhombic units have the dimensions:

For $(CH_3C_6H_4)_2SeCl_2$: $a_0 = 14.20$ A.; $b_0 = 8.43$ A.; $c_0 = 5.81$ A.
For $(CH_3C_6H_4)_2SeBr_2$: $a_0 = 14.60$ A.; $b_0 = 8.64$ A.; $c_0 = 5.82$ A.

Selenium atoms are in special positions of V^3 ($P2_12_12$)

$$(2a) \quad 00u; \; 1/2 \; 1/2 \; \bar{u}$$

and all other atoms in general positions:

$$(4c) \quad xyz; \; \bar{x}\bar{y}z; \; 1/2-x,y+1/2,\bar{z}; \; x+1/2,1/2-y,\bar{z})$$

The chosen parameters are those of Table XVB,44.

TABLE XVB,44

Parameters of the Atoms in Di-p-tolyl Selenium Dichloride and Dibromide[a]

Atom	Position	x	y	z
Se	(2a)	0	0	0.069 (0.058)
Cl[Br]	(4c)	0.012 (0.008)	0.281 (0.294)	0.078 (0.070)
C(1)	(4c)	0.108 (0.107)	−0.021 (−0.023)	0.870 (0.863)
C(2)	(4c)	0.102 (0.101)	−0.112 (−0.112)	0.672 (0.665)
C(3)	(4c)	0.180 (0.178)	−0.128 (−0.129)	0.529 (0.525)
C(4)	(4c)	0.264 (0.259)	−0.052 (−0.056)	0.584 (0.583)
C(5)	(4c)	0.270 (0.265)	0.039 (0.032)	0.782 (0.781)
C(6)	(4c)	0.192 (0.188)	0.054 (0.049)	0.925 (0.921)
C(7)	(4c)	0.350 (0.344)	−0.069 (−0.075)	0.426 (0.428)

[a] Parameters for the dibromide are enclosed in parentheses.

Though the arrangement (Fig. XVB,63) is different, the molecules in this structure are of the same general shape as those of the corresponding diphenyl compounds (**XV,bII,31,32**). The halogen atoms are nearly opposite one another across a selenium atom (X–Se–X = 177°). In the chloride, Se–Cl = 2.38 A.; in the bromide, Se–Br = 2.55 A. In both crystals, Se–C = ca. 1.94 A. The angle C–Se–C is ca. 107° and the angle between this plane and that of the benzene rings is ca. 86°.

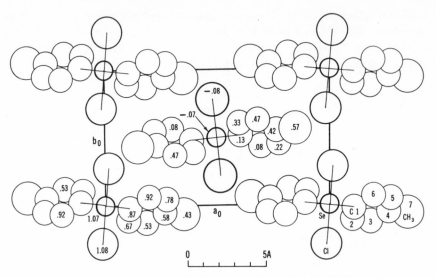

Fig. XVB,63a. The orthorhombic structure of di-p-tolyl selenium dichloride projected along its c_0 axis.

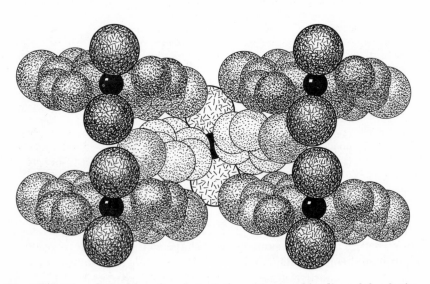

Fig. XVB,63b. A packing drawing of the orthorhombic di-p-tolyl selenium dichloride arrangement viewed along its c_0 axis. The small black circles are selenium. The chlorine atoms are large and line shaded. Carbon atoms are dotted, those of the methyl groups being larger than the others.

XV,bII,34. *Diphenyl tellurium dibromide*, $(C_6H_5)_2TeBr_2$, is tetragonal with a tetramolecular unit of the edge lengths:

$$a_0 = 11.421(10) \text{ A.}, \quad c_0 = 9.817(10) \text{ A.}$$

The space group is C_4^6 ($I4_1$). The tellurium atoms are in:

(4a) $00u; 0,^1/_2,u+^1/_4;$ B.C.   with $u = 0$ (arbitrary)

All the other atoms are in

(8b) $xyz; \bar{x}\bar{y}z; \bar{y},x+^1/_2,z+^1/_4; y,^1/_2-x,z+^1/_4;$ B.C.

with the parameters of Table XVB,45.

TABLE XVB,45

Parameters of Atoms in $(C_6H_5)_2TeBr_2$[a]

Atom	x	y	z
Br	0.2346	0.9906	0.0048
C(1)	0.0053	0.1394	0.8488
C(2)	0.0918	0.2256	0.8600
C(3)	0.0914	0.3166	0.7653
C(4)	0.0019	0.3208	0.6734
C(5)	0.9129	0.2327	0.6603
C(6)	0.9183	0.1453	0.7534

[a] Values in this table are simple averages of the final Fourier and least squares refinements.

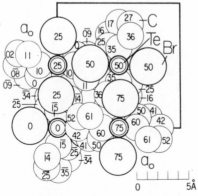

Fig. XVB,64.   The tetragonal structure of diphenyl tellurium dibromide projected along its c_0 axis.

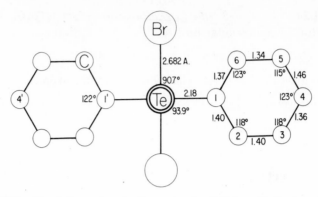

Fig. XVB,65.   Bond dimensions in the molecule of diphenyl tellurium dibromide.

The structure is shown in Figure XVB,64. Its molecules have the bond dimensions of Figure XVB,65. Their shapes are to be compared with those found for the related compounds described in **XV,bII,31–33.** In this compound the tellurium and halogens are almost in a straight line and the benzene groups are about normal to one another in a plane through tellurium at right angles to this line; there are atomic departures from this plane of up to 0.05 A. The plane of the benzene rings is rotated through 56° with respect to the plane defined by C(1)–Te–C(1′).

XV,bII,35. Crystals of *bis(p-chlorophenyl) tellurium diiodide*, $(p\text{-}ClC_6H_4)_2TeI_2$, are triclinic with a bimolecular unit of the dimensions:

$$a_0 = 9.751(6) \text{ A.}; \quad b_0 = 10.681(6) \text{ A.}; \quad c_0 = 9.531(6) \text{ A.}$$
$$\alpha = 115.7(1)°; \qquad \beta = 87.2(1)°; \qquad \gamma = 116.4(1)°$$

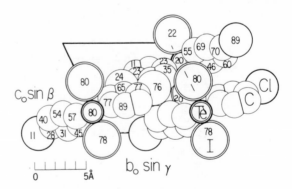

Fig. XVB,66.   The triclinic structure of bis(p-chlorophenyl) tellurium diiodide
projected along its a_0 axis.

All atoms are in the positions $(2i) \pm (xyz)$ of C_i^1 $(P\bar{1})$, with the parameters listed in Table XVB,46.

TABLE XVB,46

Parameters of the Atoms in $(ClC_6H_4)_2TeI_2$

Atom	x	y	z
Te	0.8042(1)	0.9834(1)	0.2412(1)
I(1)	0.7757(1)	0.9351(1)	0.9130(1)
I(2)	0.8022(1)	0.0378(1)	0.5692(1)
Cl(1)	0.1066(6)	0.3705(7)	0.9802(8)
Cl(2)	0.7637(9)	0.6410(7)	0.4802(9)
C(1)	0.5693(21)	0.8027(22)	0.1802(23)
C(2)	0.5437(22)	0.6762(23)	0.2068(25)
C(3)	0.3993(23)	0.5458(24)	0.1519(25)
C(4)	0.2831(23)	0.5442(24)	0.0685(25)
C(5)	0.3063(22)	0.6715(23)	0.0469(25)
C(6)	0.4543(22)	0.8038(23)	0.0989(25)
C(7)	0.7724(17)	0.1826(17)	0.3099(19)
C(8)	0.6468(19)	0.1961(20)	0.3796(22)
C(9)	0.6500(29)	0.3423(30)	0.4240(32)
C(10)	0.7658(24)	0.4619(25)	0.4060(27)
C(11)	0.8876(22)	0.4485(23)	0.3338(25)
C(12)	0.8913(20)	0.3047(20)	0.2865(22)

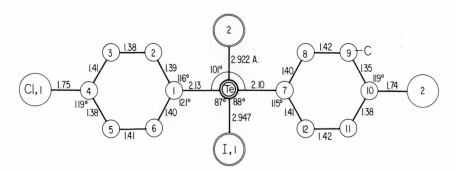

Fig. XVB,67.   Bond dimensions in the molecule of bis(p-chlorophenyl) tellurium diiodide.

The structure is shown in Figure XVB,66. Its molecules have the bond dimensions of Figure XVB,67. Their two phenyl groups are planar but the attached chlorine and tellurium atoms lie outside this plane by amounts which, for tellurium, are as great as 0.28 A. and for chlorine 0.20 A.

XV,bII,36. *Ditoluene chromium iodide,* $(CH_3C_6H_5)_2CrI$, forms monoclinic crystals. Their bimolecular unit has the dimensions:

$$a_0 = 7.94 \text{ A.}; \quad b_0 = 6.77 \text{ A.}; \quad c_0 = 12.55 \text{ A.}; \quad \beta = 104°22'$$

The space group is C_{2h}^3 in the orientation $I2/m$. Atoms are in the positions:

$$
\begin{aligned}
&\text{Cr} : (2a) \quad &&000; {}^1/_2 \, {}^1/_2 \, {}^1/_2 \\
&\text{I} : (2c) \quad &&0 \, 0 \, {}^1/_2; {}^1/_2 \, {}^1/_2 \, 0 \\
&\text{C(1)} : (4i) \quad &&\pm(u0v; u+{}^1/_2, {}^1/_2, v+{}^1/_2) \\
& &&\text{with } u = -0.077, v = 0.148 \\
&\text{C(4)} : (4i) \quad &&\text{with } u = 0.253, v = 0.098 \\
&\text{C(5)} : (4i) \quad &&\text{with } u = -0.244, v = 0.182 \\
&\text{C(2)} : (8j) \quad &&\pm(xyz; x\bar{y}z); \text{B.C.} \\
& &&\text{with } x = 0.006, y = 0.181, z = 0.135 \\
&\text{C(3)} : (8j) \quad &&\text{with } x = 0.171, y = 0.181, z = 0.110
\end{aligned}
$$

The resulting structure is shown in Figure XVB,68. As it indicates, the cation of this saltlike compound has its chromium atom at the midpoint of the line through the centers of the two rings, with all the C–Cr distances 2.08 A. In each toluene group, C–C = 1.42(3) A. and C–CH$_3$ = 1.49 A.

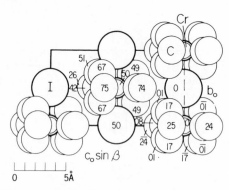

Fig. XVB,68.   The monoclinic structure of ditoluene chromium iodide projected along its a_0 axis.

III. Compounds in Which the Rings Are Separated by Two Bonding Atoms

XV,bIII,1. *Dibenzyl*, $(C_6H_5CH_2)_2$, forms monoclinic crystals with bimolecular units of the dimensions:

$$a_0 = 12.77 \text{ A.}; \quad b_0 = 6.12 \text{ A.}; \quad c_0 = 7.70 \text{ A.}; \quad \beta = 116°$$

The atoms are in the following positions of C_{2h}^5 $(P2_1/a)$:

$$(4e) \quad \pm (xyz; x+\text{}^1/_2, \text{}^1/_2 - y, z)$$

Their parameters are those of Table XVB,47.

TABLE XVB,47

Parameters of the Atoms in Dibenzyl and 1,2-Diphenyltetrafluoroethane
(in parentheses)

Atom	x	y	z
C(1)	0.0249 (0.0329)	0.0970 (0.0510)	−0.0298 (−0.0406)
C(2)	0.1472 (0.1474)	0.1542 (0.0700)	0.1213 (0.1122)
C(3)	0.1661 (0.1829)	0.3378 (0.2629)	0.2337 (0.2375)
C(4)	0.2782 (0.2871)	0.3933 (0.2727)	0.3656 (0.3782)
C(5)	0.3706 (0.3523)	0.2631 (0.0937)	0.3859 (0.3931)
C(6)	0.3531 (0.3157)	0.0790 (−0.0944)	0.2763 (0.2665)
C(7)	0.2399 (0.2120)	0.0229 (−0.1117)	0.1458 (0.1263)
F(1)	— (0.0165)	— (−0.0904)	— (−0.2042)
F(2)	— (−0.0048)	— (0.2692)	— (−0.1091)

The structure is illustrated by Figure XVB,69. Its molecules, possessing a center of symmetry, have the bond lengths of Figure XVB,70a; bond angles of the rings are $120 \pm 1°$. Each ring and its attached CH_2 carbon is planar but the two halves of the molecule are not coplanar. The molecular distribution is the same as in naphthalene (**XV,f1**) but in this crystal the long axis of the molecule is diagonal to the a_0 and c_0 axes.

Crystals of *1,2-diphenyl tetrafluoroethane*, $(C_6H_5CF_2)_2$, are isostructural with a cell of the dimensions:

$$a_0 = 15.08 \text{ A.}; \quad b_0 = 5.776 \text{ A.}; \quad c_0 = 7.705 \text{ A.}; \quad \beta = 120°16'$$

The determined parameters are listed in parentheses in Table XVB,47; values for x and z are close to those for dibenzyl but for y there are

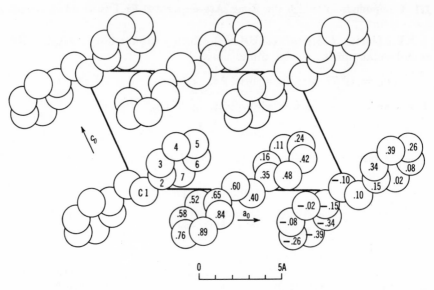

Fig. XVB,69a.   The monoclinic structure of dibenzyl projected along its b_0 axis. Left-hand axes.

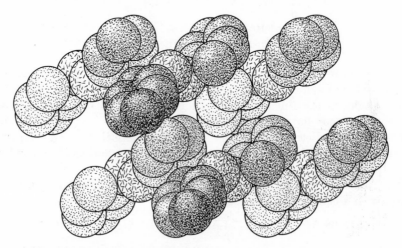

Fig. XVB,69b.   A packing drawing of the monoclinic dibenzyl arrangement seen along its b_0 axis. Left-hand axes.

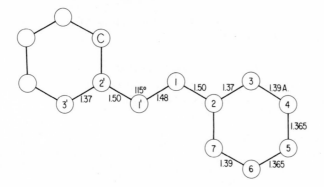

Fig. XVB,70a. Bond lengths in the molecule of dibenzyl.

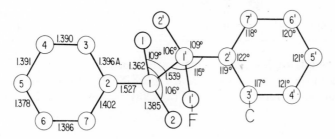

Fig. XVB,70b. Bond dimensions in the molecule of 1,2-diphenyl tetrafluoro-ethane.

important differences. Calculated parameters for the hydrogen atoms are stated in the original article.

The molecules that are present have the bond dimensions of Figure XVB,70b. The angle between the planes of the two benzene rings and their attached carbon atoms is 85°19′, more nearly a right angle than the 71°36′ found in dibenzyl.

XV,bIII,2. Crystals of *4,4′-dimethyl dibenzyl*, $(CH_3C_6H_4CH_2)_2$, are monoclinic. Their tetramolecular unit has the dimensions:

$$a_0 = 23.36 \text{ A.}; \quad b_0 = 6.10 \text{ A.}; \quad c_0 = 9.16 \text{ A.}; \quad \beta = 100°33′$$

The space group is C_{2h}^6 in the orientation $I2/a$. All atoms thus are in the positions:

$$(8f) \quad \pm (xyz; \bar{x},y,{}^1\!/_2-z); \quad \text{B.C.}$$

The approximate parameters found years ago (Table XVB,48) lead to a molecule like that of dibenzyl. It has a center of symmetry in the middle of the C(1)–C(1′) bond and the planes of its two benzene rings are parallel at different levels.

TABLE XVB,48

Parameters of the Atoms in 4,4′-Dimethyl Dibenzyl

Atom	x	y	z
C(1)	0.028	0.04	0.976
C(2)	0.072	0.15	0.095
C(3)	0.060	0.36	0.144
C(4)	0.100	0.46	0.252
C(5)	0.152	0.36	0.310
C(6)	0.163	0.15	0.263
C(7)	0.124	0.05	0.156
C(8)	0.195	0.48	0.428

The bond angles and lengths are those of Figure XVB,71. The packing in the crystal, as shown in Figure XVB,72, gives rise to shortest intermolecular separations of C(3)–C(3) = 3.69 A. and C(7)–C(4) = 3.77 A.

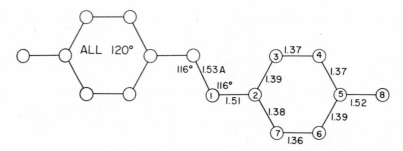

Fig. XVB,71.   Bond dimensions in the molecule of 4,4′-dimethyl dibenzyl.

XV,bIII,3. *Stilbene*, $C_6H_5CH{=}CHC_6H_5$, forms monoclinic crystals which have a tetramolecular unit of the dimensions:

$$a_0 = 12.35 \text{ A.}; \quad b_0 = 5.70 \text{ A.}; \quad c_0 = 15.92 \text{ A.}; \quad \beta = 114°0′$$

The space group is C_{2h}^5 ($P2_1/a$) with atoms in the positions:

$$(4e) \quad \pm(xyz; \ x+1/2, 1/2-y, z)$$

The determined parameters are those of Table XVB,49.

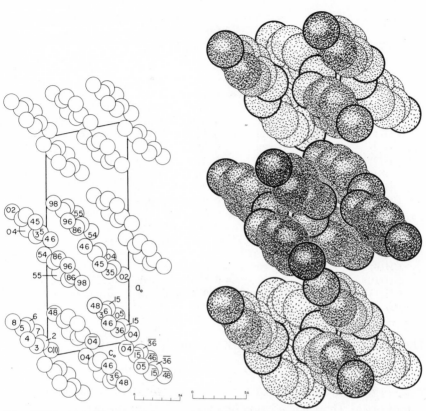

Fig. XVB,72a. The monoclinic structure of 4,4'-dimethyl dibenzyl projected along its b_0 axis. Left-hand axes.

Fig. XVB,72b. A packing drawing of the monoclinic 4,4'-dimethyl dibenzyl structure viewed along its b_0 axis. The non-ring carbons are heavily outlined. Left-hand axes.

They are so similar to the values found for *trans*-azobenzene (**XV,bIII,14**) that the structure for this substance is well reproduced by the figures that illustrate the other compound (Figs. XVB,90 and 91).

XV,bIII,4. Crystals of *trans-α,β-dicyanostilbene*, $[C_6H_5C(CN)=]_2$, are monoclinic with a tetramolecular unit of the dimensions:

$$a_0 = 17.98(5) \text{ A.}; \quad b_0 = 3.96(2) \text{ A.}; \quad c_0 = 17.23(5) \text{ A.}$$
$$\beta = 91°0(30)'$$

TABLE XVB,49

Parameters of the Atoms in Crystals of $C_6H_5CH{=}CHC_6H_5$ (I) and
$C_6H_5C{\equiv}CC_6H_5$ (II)

Atom	x		y		z	
	I	II	I	II	I	II
C(1)	0.025	0.049	0.096	0.031	0.024	0.024
C(2)	0.004	0.027	−0.075	−0.012	0.469	0.476
CH(1)	0.150	0.166	0.140	0.100	0.076	0.075
CH(2)	0.186	0.192	0.334	0.308	0.133	0.125
CH(3)	0.307	0.308	0.376	0.369	0.183	0.177
CH(4)	0.391	0.397	0.225	0.223	0.175	0.181
CH(5)	0.354	0.369	0.031	0.013	0.118	0.131
CH(6)	0.234	0.252	−0.011	−0.048	0.068	0.076
CH(7)	0.076	0.089	−0.082	−0.034	0.417	0.421
CH(8)	0.061	0.079	−0.261	−0.232	0.354	0.367
CH(9)	0.131	0.145	−0.267	−0.262	0.305	0.316
CH(10)	0.216	0.222	−0.095	−0.080	0.318	0.316
CH(11)	0.231	0.234	0.084	0.116	0.381	0.372
CH(12)	0.161	0.167	0.090	0.139	0.430	0.423

The space group is C_{2h}^6 ($C2/c$) with atoms in the general positions:

$$(8e) \quad \pm(xyz;\ x,\bar{y},z+\tfrac{1}{2};\ x+\tfrac{1}{2},y+\tfrac{1}{2},z;\ x+\tfrac{1}{2},\tfrac{1}{2}-y,z+\tfrac{1}{2})$$

The parameters of Table XVB,50 are considered to be no more than approximate, those for y being less accurate than the others.

TABLE XVB,50

Parameters of the Atoms in $[C_6H_5C(CN){=}]_2$

Atom	x	y	z
C(1)	0.005	0.000	0.042
C(2)	−0.050	−0.200	0.085
C(3)	0.079	0.055	0.083
C(4)	0.075	0.184	0.158
C(5)	0.142	0.241	0.197
C(6)	0.208	0.170	0.164
C(7)	0.214	0.043	0.090
C(8)	0.148	−0.015	0.049
N	−0.095	−0.370	0.123

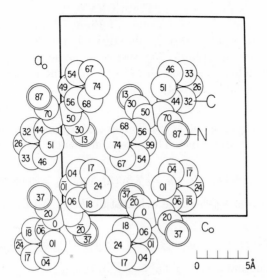

Fig. XVB,73.   The monoclinic structure of *trans-α,β*-dicyanostilbene projected along its b_0 axis.

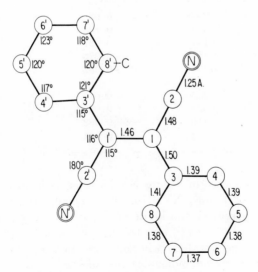

Fig. XVB,74.   Bond dimensions in the molecule of *trans-α,β*-dicyanostilbene.

The structure is illustrated in Figure XVB,73. Its centrosymmetric molecules have the bond dimensions of Figure XVB,74. In them the group C(1), C(3), C(1'), C(3') is planar with the linear C(1)–C(2)–N turned by 24° to this plane. The benzene planes are twisted about C(3)–C(6) by roughly the same amount (25°) to increase to 2.96 A. the distance between C(2) and C(4).

XV,bIII,5. Crystals of *p,p'-dibromo-α,α'-difluorostilbene*, [BrC$_6$H$_4$-C(F)=]$_2$, are orthorhombic. Their tetramolecular unit has the edge lengths:

$$a_0 = 28.32(14) \text{ A.}; \quad b_0 = 7.36(3) \text{ A.}; \quad c_0 = 6.08(2) \text{ A.}$$

The space group is V_h^{10} in the orientation *Pccn*. All atoms are in the positions

$$(8e) \quad \pm(xyz; \; x+1/2, y+1/2, \bar{z}; \; 1/2-x, y, z+1/2; \; x, 1/2-y, z+1/2)$$

with the parameters listed in Table XVB,51.

TABLE XVB,51

Parameters of the Atoms in [BrC$_6$H$_4$C(F)=]$_2$

Atom	x	y	z
Br	0.2078	0.0381	0.4819
F	0.027	−0.125	−0.206
C(1)	0.151	0.021	0.328
C(2)	0.149	−0.060	0.122
C(3)	0.106	−0.082	0.015
C(4)	0.065	−0.015	0.107
C(5)	0.068	0.065	0.316
C(6)	0.110	0.088	0.419
C(7)	0.022	−0.034	−0.017

The structure is shown in Figure XVB,75. Its molecules, having the bond dimensions of Figure XVB,76, are only approximately planar. Departures from planarity can be described by saying that the planes through C(4), C(7), C(7'), C(4') and through F, C(7), C(7'), F' make an angle of 7° with one another, whereas the benzene plane makes 8° with the former. Between molecules the shortest C–C = 3.68 A. and Br–Br = 3.93 A.

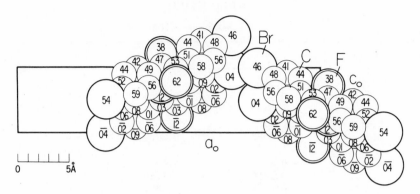

Fig. XVB,75. The elongated orthorhombic structure of p,p'-dibromo-α,α'-difluorostilbene projected along its b_0 axis.

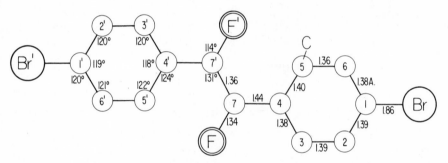

Fig. XVB,76. Bond dimensions in the molecule of p,p'-dibromo-α,α'-difluorostilbene.

XV,bIII,6. The symmetry of p,p'-$dimethyl$-α,α'-$difluorostilbene$, $[CH_3C_6H_4C(F)=]_2$, is monoclinic. Its tetramolecular unit has the dimensions:

$$a_0 = 11.56(2) \text{ A.}; \quad b_0 = 7.38(2) \text{ A.}; \quad c_0 = 16.14(3) \text{ A.}; \quad \beta = 111°21'$$

All atoms are in the following positions of C_{2h}^5 ($P2_1/c$):

$$(4e) \quad \pm (xyz; \, x, 1/2 - y, z + 1/2)$$

The parameters are those of Table XVB,52; values assigned to hydrogen atoms are to be found in the original.

CRYSTAL STRUCTURES

TABLE XVB,52

Parameters of the Atoms in $[CH_3C_6H_4C(F){=}]_2$

Atom	x	y	z
C(1)	-0.2483	0.2795	-0.3832
C(2)	-0.1676	0.2764	-0.2861
C(3)	-0.0492	0.3565	-0.2585
C(4)	0.0292	0.3555	-0.1702
C(5)	-0.0082	0.2757	-0.1046
C(6)	-0.1282	0.1964	-0.1325
C(7)	-0.2046	0.1980	-0.2200
C(8)	0.0703	0.2674	-0.0108
C(9)	0.5164	0.1704	0.3981
C(10)	0.4297	0.1968	0.3012
C(11)	0.3147	0.2758	0.2791
C(12)	0.2358	0.2995	0.1922
C(13)	0.2739	0.2445	0.1224
C(14)	0.3917	0.1661	0.1449
C(15)	0.4695	0.1437	0.2327
C(16)	0.1929	0.2651	0.0288
F(1)	0.0081	0.2438	0.0444
F(2)	0.2588	0.2716	-0.0267

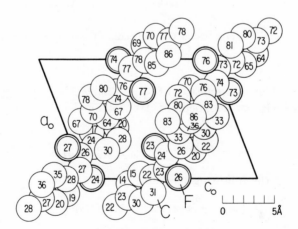

Fig. XVB,77.   The monoclinic structure of p,p'-dimethyl-α,α'-difluorostilbene projected along its b_0 axis.

The structure is that of Figure XVB,77. The principal bond dimensions in its molecules are given in Figure XVB,78. In these molecules the two benzene rings are rotated in opposite directions about the lines through C(2)–C(5) and through C(10)–C(13) and there is a small slope to the line C(9)–C(16).

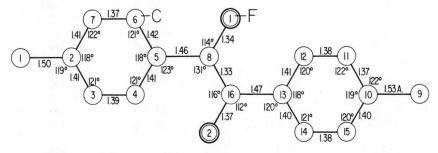

Fig. XVB,78.   Bond dimensions in the molecule of *p,p′*-dimethyl-*α,α′*-difluoro-stilbene.

XV,bIII,7.   Crystals   of   *2,6-dibromo-α-cyanostilbene*, $Br_2C_6H_3$-$C(CN){=}CHC_6H_5$, are pseudo-monoclinic, triclinic, with a bimolecular unit of the dimensions:

$$a_0 = 9.63(3) \text{ A.}; \quad b_0 = 9.69(2) \text{ A.}; \quad c_0 = 7.21(2) \text{ A.}$$
$$\alpha = 90.0(5)°; \quad \beta = 99.5(5)°; \quad \gamma = 90.0(5)°$$

The space group is C_1^1 $(P\bar{1})$ with atoms in the positions $(2i)$ $\pm(xyz)$. The parameters are listed in Table XVB,53.

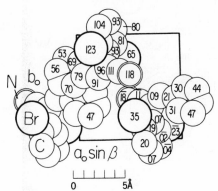

Fig. XVB,79.   The triclinic, pseudomonoclinic structure of 2,6-dibromo-α-cyanostilbene projected along its c_0 axis.

TABLE XVB,53

Parameters of the Atoms in $Br_2C_6H_3C(CN){=}CHC_6H_5$

Atom	x	y	z
Br(1)	0.4550	0.7704	0.6527
Br(2)	0.1333	0.8686	0.2340
C(1)	0.297	0.838	0.933
C(2)	0.231	0.932	0.016
C(3)	0.239	0.074	0.043
C(4)	0.311	0.116	0.929
C(5)	0.373	0.032	0.801
C(6)	0.367	0.891	0.805
Cα	0.292	0.693	0.965
Cβ	0.191	0.598	0.906
C(1′)	0.064	0.617	0.785
C(2′)	0.028	0.741	0.688
C(3′)	0.895	0.762	0.529
C(4′)	0.804	0.658	0.565
C(5′)	0.843	0.538	0.562
C(6′)	0.968	0.518	0.705
C	0.409	0.642	0.112
N	0.504	0.606	0.184

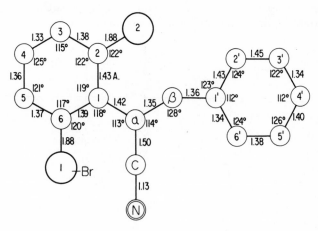

Fig. XVB,80. The bond dimensions in the molecule of 2,6-dibromo-α-cyano-stilbene.

The structure shown in Figure XVB,79 has molecules with the bond dimensions of Figure XVB,80. Atoms are in or near two planes which make 84° with one another. One contains the $C_6H_3Br_2C(\alpha)$ group; the other includes all the remaining atoms [plus $C(\alpha)$] except the carbon and nitrogen of the cyanide radical, which are 0.16 and 0.18 A. outside it. Between molecules there is a close C–C = 3.14 A. and N–N = 3.37 A.

XV,bIII,8. Crystals of *2-bromo-4′-dimethylamino-α-cyanostilbene*, $BrC_6H_4C(CN){=}CHC_6H_4N(CH_3)_2$, have orthorhombic symmetry. Their tetramolecular unit has the edge lengths:

$$a_0 = 14.61(2) \text{ A.}; \quad b_0 = 13.61(2) \text{ A.}; \quad c_0 = 7.75(2) \text{ A.}$$

The space group V^4 ($P2_12_12_1$) places all atoms in the positions:

$$(4a) \quad xyz; \ 1/2-x,\bar{y},z+1/2; \ x+1/2,1/2-y,\bar{z}; \ \bar{x},y+1/2,1/2-z$$

The determined parameters are listed in Table XVB,54.

TABLE XVB,54

Parameters of the Atoms in $BrC_6H_4C(CN){=}CHC_6H_4N(CH_3)_2$

Atom	x	y	z
Br	0.188	0.369	0.143
N(1)	0.311	0.261	0.595
N(2)	0.110	0.496	0.820
C(1)	0.268	0.175	0.610
C(2)	0.178	0.172	0.682
C(3)	0.111	0.094	0.682
C(4)	0.151	0.015	0.595
C(5)	0.235	0.020	0.507
C(6)	0.290	0.097	0.507
C(7)	0.267	0.337	0.682
C(8)	0.396	0.280	0.507
C(9)	0.485	0.101	0.000
C(10)	0.407	0.071[a]	0.085
C(11)	0.017	0.300	0.037
C(12)	0.099	0.272	0.105
C(13)	0.132	0.180	0.140
C(14)	0.064	0.114	0.114
C(15)	0.479	0.363	0.968
C(16)	0.454	0.270	0.000
C(17)	0.059	0.453	0.900

[a]The value 0.171 in the original is an error. It is to be presumed that the 0.071 given here was intended.

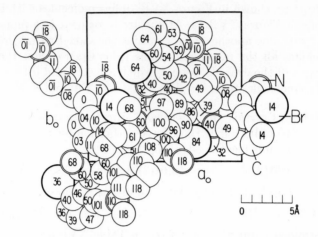

Fig. XVB,81.   The orthorhombic   structure   of   2-bromo-4′-dimethylamino-α-
cyanostilbene projected along its c_0 axis.

The structure that results (Fig. XVB,81) is composed of molecules
with the bond dimensions of Figure XVB,82. The benzene rings and
their attached atoms are coplanar. The angle between these two planes
is 58°, a value presumably influenced by repulsion between bromine
and cyanide (Br–N = 3.25 A.). The shortest intermolecular distance
is CH_3–N = 3.40 A.

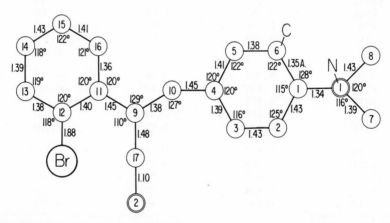

Fig. XVB,82.   Bond dimensions in the molecule of 2-bromo-4′-dimethylamino-α-
cyanostilbene.

XV,bIII,9.   Crystals of *4,4′-dichloro-α,β-diethyl stilbene*, [ClC$_6$H$_4$-C(C$_2$H$_5$)=]$_2$, are monoclinic with a bimolecular unit of the dimensions:

$$a_0 = 19.53(8) \text{ A.}; \quad b_0 = 5.49(1) \text{ A.}; \quad c_0 = 8.16(2) \text{ A.}; \quad \beta = 111.3(2)°$$

The space group is C_{2h}^5 ($P2_1/a$) with atoms in the positions:

$$(4e) \quad \pm (xyz; x+\tfrac{1}{2}, \tfrac{1}{2}-y, z)$$

The chosen parameters are those of Table XVB,55.

TABLE XVB,55

Parameters of the Atoms in 4,4′-Dichloro-α,β-diethyl Stilbene

Atom	x	y	z
Cl(1)	0.1581	0.2727	0.1055
C(2)	0.476	0.494	0.540
C(3)	0.491	0.459	0.731
C(4)	0.470	0.223	0.792
C(5)	0.396	0.434	0.430
C(6)	0.378	0.223	0.326
C(7)	0.305	0.181	0.226
C(8)	0.249	0.338	0.230
C(9)	0.267	0.555	0.334
C(10)	0.348	0.591	0.434
H(1)	0.46	0.62	0.74
H(2)	0.54	0.53	0.82
H(3)	0.51	0.10	0.89
H(4)	0.47	0.08	0.68
H(5)	0.42	0.14	0.80
H(6)	0.42	0.11	0.32
H(7)	0.29	0.02	0.14
H(8)	0.22	0.66	0.34
H(9)	0.35	0.75	0.51

The resulting structure (Fig. XVB,83) is built up of molecules having the configuration of Figure XVB,84. In them there is a center of symmetry midway along the bond C(2)–C(2′) and the planes of the two benzene rings are displaced by 1.02 A. with respect to one another.

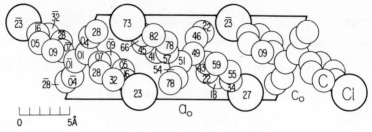

Fig. XVB,83. The monoclinic structure of 4,4'-dichloro-α,β-diethylstilbene projected along its b_0 axis.

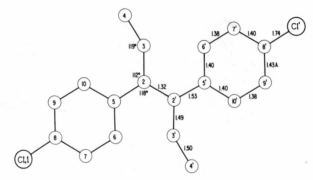

Fig. XVB,84. Bond dimensions in the molecule of 4,4'-dichloro-α,β-diethyl-stilbene.

XV,bIII,10. *Tolane*, $C_6H_5C{\equiv}CC_6H_5$, has a structure almost identical with that of stilbene **(XV,bIII,3)**. Its monoclinic unit has the dimensions:

$$a_0 = 12.75 \text{ A.}; \quad b_0 = 5.73 \text{ A.}; \quad c_0 = 15.67 \text{ A.}; \quad \beta = 115°12'$$

Atoms in the general positions of the space group C_{2h}^5 ($P2_1/a$) have the coordinates:

$$(4e) \quad \pm (xyz; x + {}^1\!/_2, {}^1\!/_2 - y, z)$$

The parameters given in Table XVB,49 are substantially the same as those of stilbene. They result from a recent redetermination and differ very little from those found many years ago.

The molecular arrangement is well represented by the figures for *trans*-azobenzene **(XV,bIII,14)**, with much the same, though less accurately determined, bond dimensions (Figs. XVB,90 and 91).

XV,bIII,11. The compound $C_6H_5C_2C_6H_5 \cdot Fe_2(CO)_6$ forms triclinic crystals. Their bimolecular unit has the dimensions:

$$a_0 = 7.211(7) \text{ A.}; \quad b_0 = 8.702(6) \text{ A.}; \quad c_0 = 15.860(15) \text{ A.}$$
$$\alpha = 73.9(1)°; \quad \beta = 103.8(1)°; \quad \gamma = 101.4(3)°$$

Atoms are in the positions $(2i)$ $\pm(xyz)$ of C_i^1 $(P\bar{1})$ with the parameters of Table XVB,56.

TABLE XVB,56

Parameters of the Atoms in $C_6H_5C_2C_6H_5 \cdot Fe_2(CO)_6$

Atom	x	y	z
Fe(1)	0.2641(3)	0.3400(2)	0.7006(1)
Fe(2)	0.2258(3)	0.6002(2)	0.7310(1)
C(1)	0.1982(18)	0.7966(14)	0.6649(8)
C(2)	0.1355(18)	0.6333(14)	0.8177(8)
C(3)	0.0085(18)	0.5044(14)	0.6838(7)
C(4)	0.1420(15)	0.3184(11)	0.5942(6)
C(5)	0.1203(17)	0.1641(13)	0.7518(7)
C(6)	0.4614(15)	0.2389(11)	0.7120(6)
C(7)	0.5007(17)	0.5470(12)	0.8078(7)
C(8)	0.5298(16)	0.6271(12)	0.7185(7)
C(9)	0.4061(15)	0.5521(11)	0.6507(6)
C(10)	0.4136(17)	0.6345(13)	0.5599(7)
C(11)	0.5366(17)	0.7701(13)	0.5376(7)
C(12)	0.6593(19)	0.8458(15)	0.6058(8)
C(13)	0.6549(17)	0.7709(13)	0.6965(7)
C(14)	0.3724(15)	0.4048(11)	0.8140(6)
C(15)	0.3465(16)	0.3095(12)	0.9048(7)
C(16)	0.5101(19)	0.2562(15)	0.9646(8)
C(17)	0.4821(21)	0.1573(17)	0.0513(9)
C(18)	0.3056(21)	0.1154(17)	0.0731(9)
C(19)	0.1468(22)	0.1703(17)	0.0162(9)
C(20)	0.1697(19)	0.2630(15)	0.9286(8)
O(1)	0.1723(15)	0.9252(12)	0.6200(7)
O(2)	0.0691(16)	0.6505(13)	0.8730(7)
O(3)	−0.1579(14)	0.4682(11)	0.6528(6)
O(4)	0.0603(13)	0.2953(10)	0.5235(6)
O(5)	0.0323(14)	0.0486(11)	0.7818(6)
O(6)	0.5950(14)	0.1707(12)	0.7194(7)

The general scheme and most important bond lengths in the molecule thus defined are given in Figure XVB,85. Dimensions within the benzene rings are normal.

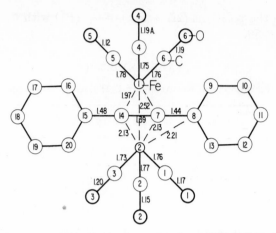

Fig. XVB,85.   Bond lengths in the molecule of $C_6H_5C_2C_6H_5 \cdot Fe_2(CO)_6$.

XV,bIII,12. *Benzil*, $(C_6H_5 \cdot CO)_2$, is hexagonal with a trimolecular unit of the edge lengths:

$$a_0 = 8.376(9) \text{ A.}, \quad c_0 = 13.700(8) \text{ A.} \qquad \text{(at } 20° \text{ C.)}$$
$$a_0 = 8.267(6) \text{ A.}, \quad c_0 = 13.407(26) \text{ A.} \qquad \text{(at } -92° \text{ C.)}$$

The space group is one or the other of the enantiomorphic pair D_3^4 ($P3_121$) or D_3^6 ($P3_221$). Choosing for description the former, atoms are in the positions:

$$(6c) \quad xyz; \; \bar{y}, x-y, z+\tfrac{1}{3}; \; y-x, \bar{x}, z+\tfrac{2}{3};$$
$$yx\bar{z}; \; \bar{x}, y-x, \tfrac{1}{3}-z; \; x-y, \bar{y}, \tfrac{2}{3}-z$$

The parameters are those of Table XVB,57.

TABLE XVB,57

Parameters of the Atoms in Benzil

Atom	x	y	z
C(1)	0.2315	0.0222	0.0769
C(2)	0.2849	0.0029	0.1701
C(3)	0.2870	−0.1568	0.1955
C(4)	0.2341	−0.2954	0.1250
C(5)	0.1793	−0.2766	0.0306
C(6)	0.1779	−0.1168	0.0056
C(7)	0.2272	0.1930	0.0529
O	0.2594	0.3121	0.1134

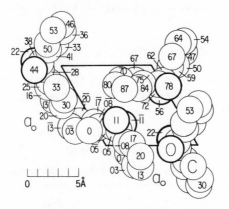

Fig. XVB,86. The hexagonal structure of benzil projected along its c_0 axis. In this drawing molecules stacked one above another are shown at different corners of the cell.

The structure is illustrated in Figure XVB,86. Its molecules, with a twofold axis of symmetry normal to the midpoint of the C(7)–C(7′) bond, have the bond lengths (corrected for angular oscillations) shown in Figure XVB,87.

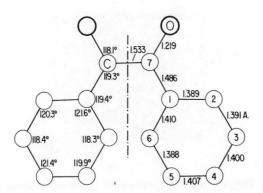

Fig. XVB,87. Bond dimensions of the molecule of benzil. The position of its twofold axis of symmetry is shown as a dot-and-dash line.

XV,bIII,13. Crystals of *cis-azobenzene*, $(C_6H_5N{=})_2$, have a tetramolecular orthorhombic unit of the edge lengths:

$$a_0 = 7.57 \text{ A.}; \quad b_0 = 12.71 \text{ A.}; \quad c_0 = 10.30 \text{ A.}$$

All atoms are in general positions of V_h^{14} in the axial orientation $Pbcn$:

$$(8d) \quad \pm(xyz; \; x+\tfrac{1}{2}, y+\tfrac{1}{2}, \tfrac{1}{2}-z; \; \tfrac{1}{2}-x, y+\tfrac{1}{2}, z; \; \bar{x}, y, \tfrac{1}{2}-z)$$

The parameters are those of Table XVB,58. The more recent determination recorded in this table has not greatly altered the values originally established.

TABLE XVB,58

Parameters of the Atoms in *cis*-Azobenzene

Atom	x	y	z
N	0.034	0.185	0.301
C(1)	0.116	0.281	0.359
CH(2)	0.213	0.356	0.292
CH(3)	0.273	0.446	0.356
CH(4)	0.236	0.460	0.487
CH(5)	0.139	0.385	0.554
CH(6)	0.079	0.295	0.490

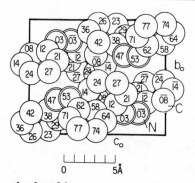

Fig. XVB,88. The orthorhombic structure of *cis*-azobenzene projected along its a_0 axis.

The resulting structure is shown in Figure XVB,88. Its molecules, with bond lengths shown in Figure XVB,89, are not planar, though of course each benzene ring is. The twist of these planes can be defined by saying that each is turned by ca. 34° about the line N–C with respect to the plane N–N′–C.

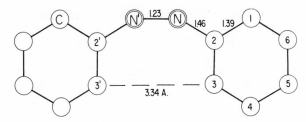

Fig. XVB,89. Some bond lengths in the molecule of *cis*-azobenzene.

XV,bIII,14. Crystals of *trans-azobenzene*, $(C_6H_5N{=})_2$, have a monoclinic tetramolecular unit of the dimensions:

$$a_0 = 12.144 \text{ A.}; \quad b_0 = 5.756 \text{ A.}; \quad c_0 = 15.396 \text{ A.}; \quad \beta = 114°8'$$

TABLE XVB,59

Parameters of the Atoms in *trans*-Azobenzene

Atom	x	y	z
C(1)	0.1409	0.1338	0.0678
C(2)	0.1732	0.3451	0.1142
C(3)	0.2940	0.3898	0.1709
C(4)	0.3808	0.2304	0.1780
C(5)	0.3482	0.0214	0.1290
C(6)	0.2285	−0.0275	0.0741
N(1)	0.0145	0.1027	0.0114
C(7)	0.0705	−0.0469	0.4211
C(8)	0.0621	−0.2299	0.3615
C(9)	0.1310	−0.2320	0.3096
C(10)	0.2072	−0.0521	0.3163
C(11)	0.2151	0.1338	0.3756
C(12)	0.1471	0.1369	0.4284
N(2)	−0.0082	−0.0744	0.4720
H(2)	0.1052	0.4741	0.1064
H(3)	0.3192	0.5513	0.2099
H(4)	0.4743	0.2669	0.2218
H(5)	0.4167	−0.1034	0.1340
H(6)	0.2032	−0.1904	0.0363
H(8)	0.0014	−0.3721	0.3557
H(9)	0.1247	−0.3767	0.2635
H(10)	0.2611	−0.0547	0.2756
H(11)	0.2747	0.2769	0.3807
H(12)	0.1534	0.2813	0.4753

The space group is C_{2h}^5 ($P2_1/a$) with atoms in the positions:

$$(4e) \quad \pm (xyz; \; x+\tfrac{1}{2}, \tfrac{1}{2}-y, z)$$

A recent redetermination has led to the parameters of Table XVB,59; they are more precise than but not greatly different from the original values.

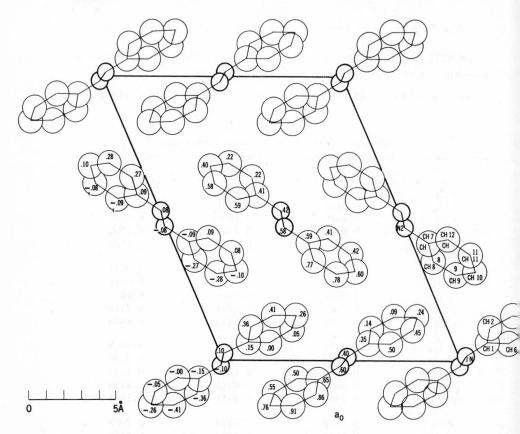

Fig. XVB,90a. The monoclinic structure of *trans*-azobenzene projected along its b_0 axis. Left-hand axes.

The structure shown in Figure XVB,90 results in two crystallographically different molecules with almost the same bond dimensions (Fig. XVB,91). This arrangement is similar to those of stilbene (**XV,bIII,3**) and tolane (**XV,bIII,10**).

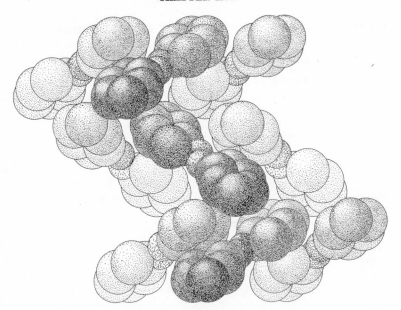

Fig. XVB,90b. A packing drawing of the monoclinic *trans*-azobenzene structure seen along its b_0 axis. The nitrogen atoms are smaller than the dotted carbon atoms and are line shaded. Left-hand axes.

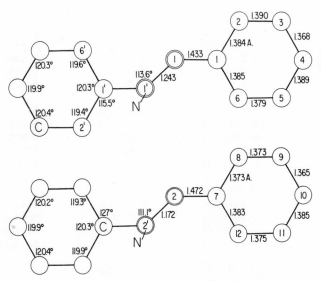

Fig. XVB,91a and b. Bond dimensions in the two crystallographically different molecules of *trans*-azobenzene.

XV,bIII,15. The compound μ-N,N'-dehydrosemidinatobis(tricarbonyl iron), $C_6H_5NNC_6H_5 \cdot Fe_2(CO)_6$, forms triclinic crystals whose bimolecular unit has the dimensions:

$$a_0 = 8.59(2) \text{ A.}; \quad b_0 = 13.92(3) \text{ A.}; \quad c_0 = 7.85(2) \text{ A.}$$
$$\alpha = 92°14(15)'; \quad \beta = 93°24(15)'; \quad \gamma = 90°24(15)'$$

The space group is C_i^1 ($P\bar{1}$) with atoms in the positions $(2i)$ $\mp(xyz)$. The determined parameters are listed in Table XVB,60.

TABLE XVB,60

Parameters of the Atoms in $C_6H_5NNC_6H_5 \cdot Fe_2(CO)_6$

Atom	x	y	z
Fe(1)	0.2250(2)	0.3728(1)	0.3168(2)
Fe(2)	0.4064(2)	0.2905(1)	0.1493(2)
C(1)	0.5448(14)	0.3784(8)	0.0959(17)
O(1)	0.6367(13)	0.4344(7)	0.0693(15)
C(2)	0.4445(18)	0.2055(10)	−0.0255(22)
O(2)	0.4688(15)	0.1498(8)	−0.1289(18)
C(3)	0.5422(15)	0.2507(8)	0.3104(18)
O(3)	0.6321(13)	0.2296(7)	0.4156(15)
C(4)	0.0230(17)	0.3988(9)	0.3472(24)
O(4)	−0.1035(15)	0.4146(8)	0.3656(20)
C(5)	0.2986(15)	0.4913(8)	0.3325(18)
O(5)	0.3472(13)	0.5698(7)	0.3425(15)
C(6)	0.2866(16)	0.3603(8)	0.5334(19)
O(6)	0.3343(14)	0.3527(7)	0.6738(16)
C(7)	0.0974(13)	0.2985(7)	0.0058(16)
C(8)	−0.0116(15)	0.3085(8)	−0.1288(19)
C(9)	−0.1263(18)	0.2366(9)	−0.1534(21)
C(10)	−0.1266(17)	0.1589(9)	−0.0536(20)
C(11)	−0.0146(15)	0.1495(8)	0.0857(19)
C(12)	0.0977(12)	0.2214(6)	0.1076(15)
C(13)	0.2243(14)	0.1517(7)	0.3581(17)
C(14)	0.3113(15)	0.0701(8)	0.3223(18)
C(15)	0.3091(16)	−0.0059(8)	0.4275(19)
C(16)	0.2314(16)	−0.0026(8)	0.5764(19)
C(17)	0.1468(17)	0.0796(9)	0.6144(20)
C(18)	0.1407(16)	0.1553(8)	0.5062(19)
N(1)	0.2207(12)	0.3643(6)	0.0644(14)
N(2)	0.2175(10)	0.2314(5)	0.2463(12)

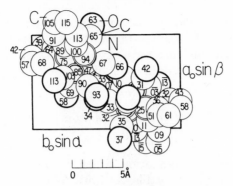

Fig. XVB,92.   The triclinic structure of $C_6H_5NNC_6H_5 \cdot Fe_2(CO)_6$ projected along its c_0 axis. The smaller doubly ringed circles are the iron atoms.

A general idea of the way the atoms are tied together in the molecule of this structure (Fig. XVB,92) can be gained from Figure XVB,93. The benzene ring C(7)-to-C(12) and its two attached nitrogen atoms are coplanar to within ca. 0.04 A. and so are the atoms in the other benzene ring C(13)-to-C(18). These two planes are almost normal to one another. In both the benzene rings and (CO) radicals the atomic separations have their usual values.

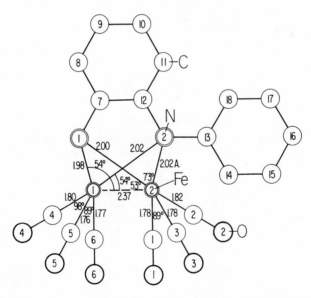

Fig. XVB,93.   Some bond dimensions in the molecule of $C_6H_5NNC_6H_5 \cdot Fe_2(CO)_6$.

XV,bIII,16. Crystals of *p-azotoluene*, $(CH_3C_6H_4N=)_2$, show an important disorder which suggests that the molecules have two orientations that they assume indeterminately. The symmetry is monoclinic and there are two molecules in a cell of the dimensions:

$$a_0 = 11.914 \text{ A.}; \quad b_0 = 4.850 \text{ A.}; \quad c_0 = 9.713 \text{ A.}; \quad \beta = 91°0'$$

The space group is C_{2h}^5 $(P2_1/a)$ with all atoms in the general positions:

$$(4e) \quad \pm (xyz; \ x+{}^1/_2,{}^1/_2-y,z)$$

An approximate determination which ignores the disorder gave atoms the parameters of Table XVB,61. When the molecules thus defined were turned in opposite directions by 5°, better agreement with the experimental data was found; the parameters of the molecules thus separated are those of Table XVB,62. It is presumed that chance determines which of these two orientations is taken by a particular molecule. The average bond dimensions are given in Figure XVB,94.

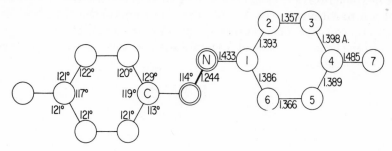

Fig. XVB,94. Averaged bond dimensions in the differently oriented molecules considered to be present in crystals of *p*-azotoluene.

TABLE XVB,61

Parameters of the Unseparated Molecules in *p*-Azotoluene

Atom	x	y	z
C(1)	0.0463	0.2558	0.1160
C(2)	−0.0280	0.3647	0.2108
C(3)	0.0059	0.5573	0.3040
C(4)	0.1163	0.6554	0.3062
C(5)	0.1891	0.5489	0.2094
C(6)	0.1554	0.3546	0.1157
C(7)	0.1539	0.8655	0.4082
N	0.0415	0.0064	−0.0184

TABLE XVB,62
Parameters Chosen for the Separated Molecules in p-Azotoluene
(one set in parentheses)

Atom	x	y	z
C(1)	0.0628 (0.0305)	0.2546 (0.2571)	0.1082 (0.1256)
C(2)	−0.0181 (−0.0367)	0.3636 (0.3657)	0.1946 (0.2281)
C(3)	0.0092 (0.0037)	0.5568 (0.5578)	0.2908 (0.3174)
C(4)	0.1191 (0.1136)	0.6550 (0.6558)	0.3048 (0.3078)
C(5)	0.1986 (0.1791)	0.5484 (0.5494)	0.2164 (0.2035)
C(6)	0.1717 (0.1390)	0.3536 (0.3556)	0.1195 (0.1136)
C(7)	0.1494 (0.1583)	0.8649 (0.8661)	0.4104 (0.4056)
N	0.0469 (−0.0294)	0.0562 (0.0580)	0.0003 (0.0437)
H(2)	−0.1040 (−0.1224)	0.2936 (0.2958)	0.1849 (0.2366)
H(3)	−0.0548 (−0.0498)	0.6355 (0.6362)	0.3575 (0.3971)
H(5)	0.2841 (0.2649)	0.6215 (0.6223)	0.2243 (0.1933)
H(6)	0.2354 (0.1919)	0.2767 (0.2789)	0.0518 (0.0329)
H(7a)	0.0786 (0.0875)	0.9904 (0.9916)	0.4390 (0.4342)
H(7b)	0.2112 (0.2200)	0.9932 (0.9944)	0.3615 (0.3567)
H(7c)	0.1888 (0.1977)	0.7615 (0.7627)	0.4970 (0.4922)

XV,bIII,17. The monoclinic crystals of *trans-p,p′-dibromoazobenzene*, $(BrC_6H_4N=)_2$, have a bimolecular unit of the dimensions:

$$a_0 = 4.01 \text{ A.}; \quad b_0 = 5.88 \text{ A.}; \quad c_0 = 24.69 \text{ A.}; \quad \beta = 92°36′$$

The space group is C_{2h}^5 $(P2_1/c)$ with atoms in the positions:

$$(4e) \quad \pm(xyz; \ x, ^1/_2 - y, z + ^1/_2)$$

Parameters are stated in Table XVB,63.

TABLE XVB,63
Parameters of the Atoms in $(BrC_6H_4N=)_2$

Atom	x	y	z
Br	0.9172(4)	0.5945(3)	0.7977(5)
N	0.470(3)	−0.0338(18)	0.9757(4)
C(1)	0.782(3)	0.4019(23)	0.8540(5)
C(2)	0.626(3)	0.1957(24)	0.8408(5)
C(3)	0.523(3)	0.0611(22)	0.8831(5)
C(4)	0.587(3)	0.1236(21)	0.9368(5)
C(5)	0.745(3)	0.3353(22)	0.9489(5)
C(6)	0.842(3)	0.4681(22)	0.9074(5)

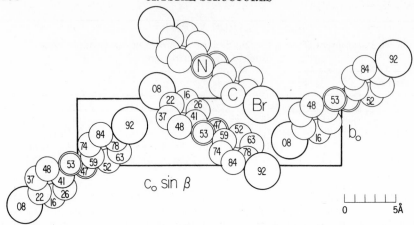

Fig. XVB,95.   The   monoclinic   structure   of   *trans-p,p'*-dibromoazobenzene
projected along its a_0 axis.

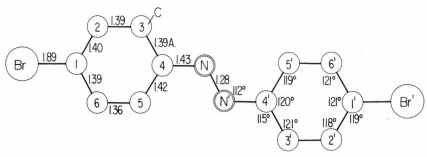

Fig. XVB,96.   Bond   dimensions   in   the   molecule   of   *trans-p,p'*-dibromoazo-
benzene.

The structure shown in Figure XVB,95 is built of planar centrosymmetric molecules that possess the bond dimensions of Figure XVB,96.

XV,bIII,18.   The   approximately   determined   structure   for   the
dimeric *tribromonitrosobenzene*, $(Br_3C_6H_2NO)_2$, is based on the orthorhombic space group V_h^{14}. Its tetramolecular unit has:

$$a_0 = 9.32 \text{ A.}; \quad b_0 = 23.5 \text{ A.}; \quad c_0 = 8.59 \text{ A.}$$

All atoms have been placed in general positions according to the axial
orientation *Pbna*:

(8d) $\pm (xyz; \ x+1/2, 1/2-y, z+1/2; \ x+1/2, y, 1/2-z; \ x, 1/2-y, \bar{z})$

Parameters (Table XVB,64) were accurately determined for the
bromine atoms; the light atoms were given positions that place the
benzene ring, with C–C = 1.38 A., in a plane with the bromine atoms.

TABLE XVB,64

Parameters of the Atoms in Tribromonitrosobenzene (Dimer)

Atom	x	y	z
Br(1)	0.225	−0.023	0.123
Br(2)	−0.053	0.183	0.133
Br(3)	0.550	0.178	−0.125
O(1)	0.250	0.240	−0.200
N(1)	0.252	0.230	−0.046
C(1)	0.258	0.061	0.064
C(2)	0.133	0.096	0.099
C(3)	0.135	0.153	0.068
C(4)	0.255	0.172	0.006
C(5)	0.375	0.139	−0.028
C(6)	0.376	0.084	0.001

The molecule that results, shown in Figure XVB,97, possesses a twofold axis.

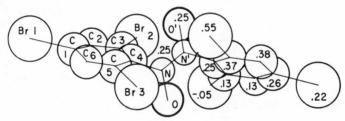

Fig. XVB,97a. The molecule of tribromonitrosobenzene seen in projection. Numbers at the right are its x parameters.

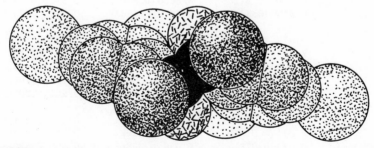

Fig. XVB,97b. A packing of the molecule of tribromonitrosobenzene as shown in Fig. XVB,97a. The nitrogen atoms are black, the oxygen atoms are line shaded. The dotted bromine atoms are larger than those of carbon.

XV,bIII,19. Crystals of *N-5-chlorosalicylideneaniline*, ClC_6H_3-$(OH)CH \cdot NC_6H_5$, are orthorhombic with a tetramolecular unit of the edge lengths:

$a_0 = 12.319$ A.; $b_0 = 4.527$ A.; $c_0 = 19.48$ A.   (at 20° C.)

$a_0 = 12.175$ A.; $b_0 = 4.480$ A.; $c_0 = 19.25$ A.   (at ca. 90° K.)

Atoms are in the positions

$$(4a)\quad xyz;\ \bar{x},\bar{y},z+{}^1/_2;\ {}^1/_2-x,y,z+{}^1/_2;\ x+{}^1/_2,\bar{y},z$$

of the space group C_{2v}^5 $(Pca2_1)$. The parameters are those of Table XVB,65, values in parentheses applying to the lower temperature.

TABLE XVB,65

Parameters of the Atoms in $ClC_6H_3(OH)CH \cdot NC_6H_5$

Atom	x	y	z
C(1)	0.1925 (0.1931)	0.4652 (0.4600)	0.1575 (0.1600)
C(2)	0.3038 (0.3051)	0.3863 (0.3737)	0.1575 (0.1600)
C(3)	0.3425 (0.3455)	0.1813 (0.1699)	0.1097 (0.1119)
C(4)	0.2749 (0.2764)	0.0636 (0.0457)	0.0615 (0.0631)
C(5)	0.1644 (0.1652)	0.1396 (0.1259)	0.0617 (0.0627)
C(6)	0.1239 (0.1232)	0.3322 (0.3277)	0.1087 (0.1103)
C(7)	0.1492 (0.1487)	0.6737 (0.6712)	0.2065 (0.2088)
C(8)	0.1687 (0.1696)	0.0152 (0.0194)	0.2970 (0.3001)
C(9)	0.0598 (0.0591)	0.1102 (0.1167)	0.2992 (0.3024)
C(10)	0.0282 (0.0275)	0.3090 (0.3223)	0.3481 (0.3512)
C(11)	0.1022 (0.1019)	0.4274 (0.4455)	0.3942 (0.3987)
C(12)	0.2088 (0.2122)	0.3408 (0.3511)	0.3915 (0.3960)
C(13)	0.2435 (0.2449)	0.1343 (0.1433)	0.3424 (0.3469)
N	0.2097 (0.2115)	0.8060 (0.8085)	0.2494 (0.2524)
O	0.3730 (0.3759)	0.5072 (0.4964)	0.2033 (0.2070)
Cl	0.0784 (0.0785)	−0.0188 (−0.0315)	0.0000 (0.0000)
H(3)	0.425 (0.425)	0.143 (0.113)	0.111 (0.115)
H(4)	0.301 (0.300)	−0.109 (−0.117)	0.029 (0.025)
H(6)	0.049 (0.047)	0.395 (0.394)	0.109 (0.109)
H(7)	0.075 (0.071)	0.714 (0.731)	0.204 (0.205)
H(9)	0.002 (−0.004)	−0.002 (0.013)	0.267 (0.271)
H(10)	−0.044 (−0.050)	0.368 (0.389)	0.353 (0.356)
H(11)	0.073 (0.075)	0.556 (0.598)	0.432 (0.439)
H(12)	0.256 (0.264)	0.395 (0.440)	0.425 (0.433)
H(13)	0.327 (0.326)	0.130 (0.118)	0.337 (0.344)

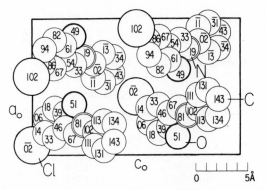

Fig. XVB,98.   The   orthorhombic   structure   of   *N*-5-chlorosalicylideneaniline   projected along its b_0 axis.

The structure is shown in Figure XVB,98. Its molecule has, at room temperature, the bond dimensions of Figure XVB,99; the slightly different values prevailing at low temperature are stated in the original article. The substituted ring with its attached Cl, O, and C(7) atoms is planar. The nitrogen atom, lying in the plane of the other benzene ring, is 0.10 A. outside this plane. Intermolecular distances range upward from 3.26 A.

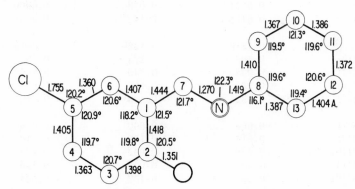

Fig. XVB,99.   Bond dimensions in the molecule of *N*-5-chlorosalicylindeneaniline.

XV,bIII,20.   Crystals of *2-chloro-N-salicylideneaniline*, $C_6H_4(OH)\cdot CH\cdot NC_6H_4Cl$, are orthorhombic with a tetramolecular unit of the edge lengths:

$$a_0 = 13.528(1) \text{ A.}; \quad b_0 = 12.185(1) \text{ A.}; \quad c_0 = 6.871(1) \text{ A.}$$

The space group is V^4 ($P2_12_12_1$) with atoms in the positions:

(4a) xyz; $1/2-x,\bar{y},z+1/2$; $x+1/2,1/2-y,\bar{z}$; $\bar{x},y+1/2,1/2-z$

Parameters are stated in Table XVB,66.

TABLE XVB,66

Parameters of the Atoms in $C_6H_4(OH)CH \cdot NC_6H_4Cl$

Atom	x	y	z
C(1)	0.7229	−0.1204	0.6147
C(2)	0.7070	−0.1531	0.4234
C(3)	0.7832	−0.1998	0.3148
C(4)	0.8754	−0.2128	0.4001
C(5)	0.8938	−0.1788	0.5865
C(6)	0.8177	−0.1328	0.6939
C(7)	0.6447	−0.0756	0.7357
C(8)	0.4812	−0.0254	0.7973
C(9)	0.4675	−0.0763	0.9754
C(10)	0.3920	−0.0464	0.0957
C(11)	0.3294	0.0397	0.0439
C(12)	0.3411	0.0908	0.8668
C(13)	0.4169	0.0572	0.7437
N	0.5571	−0.0596	0.6680
O	0.6164	−0.1427	0.3381
Cl	0.4310	0.1229	0.5212
H(3)	0.769	−0.223	0.166
H(4)	0.938	−0.245	0.324
H(5)	0.955	−0.198	0.667
H(6)	0.832	−0.103	0.846
H(7)	0.665	−0.060	0.886
H(9)	0.513	−0.138	0.016
H(10)	0.378	−0.082	0.223
H(11)	0.281	0.070	0.128
H(12)	0.296	0.152	0.824
H(0)	0.570	−0.102	0.458

The structure is that of Figure XVB,100. The molecules, with dimensions similar to those of the N-5-chloro compound (**XV,bIII,19**), have the bonds of Figure XVB,101. In them the ring C(8)-to-C(13) and its attached nitrogen and chlorine atoms are planar. The other ring

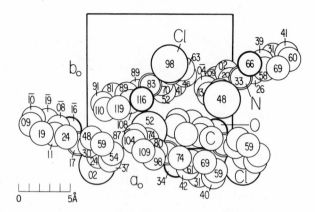

Fig. XVB,100. The orthorhombic structure of 2-chloro-N-salicylideneaniline projected along its c_0 axis.

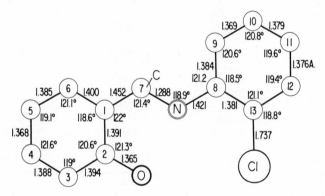

Fig. XVB,101. Bond dimensions in the molecule of 2-chloro-N-salicylideneani-line.

C(1)-to-C(6) and its oxygen atoms are also planar; nitrogen is 0.04 A. and the C(7) atom 0.06 A. outside this second plane.

XV,bIII,21. There has been a brief description of the approximate structure for *2,2'-diaminodiphenyl disulfide*, $(NH_2C_6H_4S)_2$. Its crystals are orthorhombic, pseudohexagonal, with a unit containing eight molecules and having the cell edges:

$$a_0 = 8.2111(4) \text{ A.}; \quad b_0 = 13.144(5) \text{ A.}; \quad c_0 = 22.766 \text{ A.}$$

The space group has been selected as V_h^{15} (*Pbca*) with atoms in the positions:

$$(8c) \quad \pm (xyz; \; 1/2-x,y+1/2,z; \; x,1/2-y,z+1/2; \; x+1/2,y,1/2-z)$$

The assigned parameters are listed in Table XVB,67.

TABLE XVB,67

Parameters of the Atoms in $(NH_2C_6H_4S)_2$

Atom	x	y	z
S(1)	0.321(1)	0.5132(6)	0.1644(3)
S(2)	0.528(1)	0.5611(6)	0.2073(4)
N(1)	0.145(2)	0.715(1)	0.179(1)
N(2)	0.653(2)	0.347(1)	0.230(1)
C(1)	0.308(3)	0.602(2)	0.100(1)
C(2)	0.198(3)	0.693(2)	0.120(2)
C(3)	0.174(3)	0.761(2)	0.067(1)
C(4)	0.232(3)	0.738(3)	0.018(2)
C(5)	0.321(3)	0.655(2)	0.001(1)
C(6)	0.359(4)	0.576(2)	0.044(2)
C(7)	0.698(4)	0.496(2)	0.166(2)
C(8)	0.740(6)	0.392(2)	0.186(3)
C(9)	0.876(26)	0.353(46)	0.150(14)
C(10)	0.942(8)	0.417(9)	0.108(2)
C(11)	0.911(21)	0.513(17)	0.089(10)
C(12)	0.763(23)	0.558(14)	0.123(6)

XV,bIII,22. *Diphenyl diselenide*, $(C_6H_5Se)_2$, forms orthorhombic crystals that have a tetramolecular unit of the dimensions:

$$a_0 = 24.07 \text{ A.}; \quad b_0 = 8.27 \text{ A.}; \quad c_0 = 5.64 \text{ A.}$$

Atoms are in general positions of V^4 $(P2_12_12_1)$ which for the purpose of this description were chosen as:

$$(4a) \quad xyz; \; \bar{x},\bar{y},z+1/2; \; x+1/2,1/2-y,1/2-z; \; 1/2-x,y+1/2,\bar{z}$$

The assigned parameters are those of Table XVB,68. These parameters for selenium were accurately established; the carbon positions chosen have involved the assumptions of a planar phenyl hexagon, with $C-C = 1.39$ A. and an axis passing symmetrically through each ring and the selenium atom nearer it. The Se–Se distance is 2.29 A. and the accepted Se–C = 1.93 A.

TABLE XVB,68

Parameters of the Atoms in Diphenyl Diselenide

Atom	x	y	z
Se	0.0962	0.041	0.075
C(1)	0.1730	0.079	0.157
C(2)	0.2057	0.178	0.016
C(3)	0.2610	0.206	0.075
C(4)	0.2835	0.134	0.276
C(5)	0.2508	0.035	0.417
C(6)	0.1955	0.007	0.358

The general shape of these molecules and the way they pack in the crystal can be judged from Figure XVB,102.

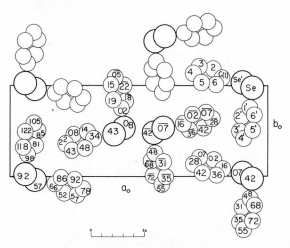

Fig. XVB,102a.   The orthorhombic structure of diphenyl diselenide projected along its c_0 axis. Left-hand axes.

XV,bIII,23.   Crystals of *p,p'-dichlorodiphenyl diselenide*, $(ClC_6H_4Se)_2$, have a monoclinic tetramolecular unit of the dimensions:

$$a_0 = 14.09(5) \text{ A.}; \quad b_0 = 6.48(3) \text{ A.}; \quad c_0 = 14.55(5) \text{ A.}$$
$$\beta = 102°48(12)'$$

The space group is C_{2h}^5 $(P2_1/n)$ with atoms in the positions:

$$(4e) \quad \pm(xyz; x+1/2, 1/2-y, z+1/2)$$

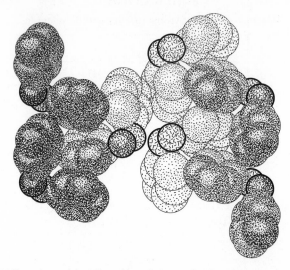

Fig. XVB,102b.   A packing drawing of the orthorhombic structure of diphenyl diselenide seen along its c_0 axis. The selenium atoms are heavily outlined and lined shaded. Atoms of carbon are dotted. Left-hand axes.

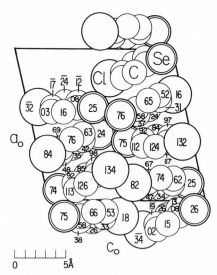

Fig. XVB,103.   The monoclinic structure of p,p'-dichlorodiphenyl diselenide projected along its b_0 axis.

The determined parameters are given in Table XVB,69, those for carbon involving the assumption that the benzene rings are regular hexagons.

TABLE XVB,69

Parameters of the Atoms in $(ClC_6H_4Se)_2$ and in Parentheses $(ClC_6H_4Te)_2$

Atom	x	y	z
Se(1)[Te(1)]	0.4458 (0.4321)	0.2444 (0.239)	0.4137 (0.4161)
Se(1′)	0.6123 (0.6209)	0.2534 (0.260)	0.4185 (0.4203)
Cl(1)	0.3032 (0.3153)	0.8445 (0.863)	0.0616 (0.067)
Cl(1′)	0.6182 (0.6101)	0.6807 (0.674)	0.0625 (0.0658)
C(1)	0.402 (0.393)	0.424 (0.446)	0.307 (0.297)
C(2)	0.339 (0.326)	0.352 (0.383)	0.226 (0.226)
C(3)	0.308 (0.302)	0.482 (0.512)	0.149 (0.152)
C(4)	0.340 (0.345)	0.686 (0.705)	0.154 (0.148)
C(5)	0.403 (0.412)	0.759 (0.770)	0.235 (0.218)
C(6)	0.434 (0.436)	0.629 (0.640)	0.311 (0.292)
C(1′)	0.614 (0.617)	0.082 (0.081)	0.312 (0.307)
C(2′)	0.579 (0.575)	0.157 (0.150)	0.221 (0.225)
C(3′)	0.581 (0.572)	0.032 (0.023)	0.144 (0.150)
C(4′)	0.617 (0.613)	0.832 (0.827)	0.157 (0.157)
C(5′)	0.651 (0.656)	0.757 (0.757)	0.247 (0.239)
C(6′)	0.650 (0.658)	0.881 (0.883)	0.325 (0.314)

The structure is shown in Figure XVB,103. As it suggests, the molecule is U-shaped, with C(1)–Se–Se′ = 102° and Se–Se′–C(1′) = 100°. The dihedral angle between the planes defined by these two sets of atoms is 74.5°. The Se–Se′ separation is 2.333 A., the Se–C distances are 1.92 and 1.94 A., and, with the assigned carbon parameters, C–Cl = 1.69 A.

The corresponding *tellurium* compound $(ClC_6H_4Te)_2$ is isostructural, with:

$$a_0 = 14.34(2) \text{ A.}; \quad b_0 = 6.47(1) \text{ A.}; \quad c_0 = 15.09(2) \text{ A.}$$
$$\beta = 101°30(30)'$$

Parameters are given in parentheses in the table. In the molecules of this crystal Te–Te = 2.702 A., Te–C = 2.10 and 2.16 A. and C–Cl = 1.92 and 1.67 A.

XV,bIII,24. According to a preliminary announcement, crystals of *N(1-bromo-3,5-dimethylphenyl)benzenesulfonamide*, $BrC_6H_2(CH_3)_2NHSO_2C_6H_5$, are monoclinic with a tetramolecular unit of the dimensions:

$$a_0 = 14.51(5) \text{ A.}; \quad b_0 = 19.70(5) \text{ A.}; \quad c_0 = 5.10(2) \text{ A.}; \quad \beta = 92°$$

The space group is C_{2h}^5 in the axial orientation $P2_1/n$. Atoms therefore are in the positions:

$$(4e) \quad \pm (xyz; x + \tfrac{1}{2}, \tfrac{1}{2} - y, z + \tfrac{1}{2})$$

Chosen parameters are listed in Table XVB,70.

TABLE XVB,70

Parameters of the Atoms in $BrC_6H_2(CH_3)_2NHSO_2C_6H_5$

Atom	x	y	z
Br	0.0824	0.1637	0.1905
S	0.3448	−0.1402	0.4524
C(1)	0.1585	0.0904	0.3389
C(2)	0.2491	0.0848	0.2254
C(3)	0.3052	0.0328	0.3245
C(4)	0.2708	−0.0130	0.5105
C(5)	0.1831	−0.0065	0.6093
C(6)	0.1214	0.0471	0.5101
C(7)	0.1411	−0.0513	0.8254
C(8)	0.4062	0.0297	0.2256
N	0.3297	−0.0693	0.6144
O(1)	0.3416	−0.1222	0.1833
O(2)	0.4281	−0.1721	0.5672
C(9)	0.2508	−0.1951	0.5177
C(10)	0.2617	−0.2369	0.7242
C(11)	0.1813	−0.2800	0.7694
C(12)	0.1012	−0.2806	0.5927
C(13)	0.0928	−0.2343	0.3952
C(14)	0.1684	−0.1908	0.3309

XV,bIII,25. A preliminary note indicates that crystals of *N(4-bromo-2-methylphenyl)benzenesulfonamide*, $BrC_6H_3(CH_3)NHSO_2C_6H_5$, are triclinic with a bimolecular unit of the dimensions:

$$a_0 = 8.29(3) \text{ A.}; \quad b_0 = 10.40(2) \text{ A.}; \quad c_0 = 8.36(4) \text{ A.}$$
$$\alpha = 97°; \quad \beta = 95°; \quad \gamma = 79°$$

All atoms are in the positions $(2i) \pm (xyz)$ of C_i^1 $(P\bar{1})$. Assigned parameters have the values of Table XVB,71.

TABLE XVB,71

Parameters of the Atoms in $BrC_6H_3(CH_3)NHSO_2C_6H_5$

Atom	x	y	z
Br	0.314	0.247	0.324
S	0.855	0.336	0.014
C(1)	0.447	0.324	0.491
C(2)	0.609	0.325	0.465
C(3)	0.703	0.364	0.583
C(4)	0.661	0.406	0.747
C(5)	0.490	0.410	0.800
C(6)	0.377	0.362	0.651
C(7)	0.849	0.376	0.529
N	0.766	0.457	0.893
O(1)	0.943	0.413	0.124
O(2)	0.719	0.285	0.080
C(8)	0.970	0.206	0.908
C(9)	0.906	0.102	0.842
C(10)	0.040	−0.010	0.755
C(11)	0.182	−0.001	0.744
C(12)	0.246	0.093	0.836
C(13)	0.125	0.211	0.921

The structure is shown in Figure XVB,104.

XV,bIII,26. According to a preliminary note, *N(p-bromophenyl)-benzenesulfonamide*, $BrC_6H_4NHSO_2C_6H_5$, forms orthorhombic crystals whose unit contains eight molecules and has the edge lengths:

$$a_0 = 10.98(3) \text{ A.}; \quad b_0 = 23.87(3) \text{ A.}; \quad c_0 = 9.80(3) \text{ A.}$$

Atoms are in the positions

$(8c) \quad \pm (xyz; \; x,y+\frac{1}{2}, \frac{1}{2}-z; \; x+\frac{1}{2},\frac{1}{2}-y,z; \; \frac{1}{2}-x,y,z+\frac{1}{2})$

of the space group V_h^{15} (*Pcab*). The parameters are listed in Table XVB,72.

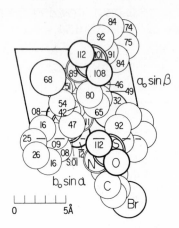

Fig. XVB,104.   The triclinic structure of N(4-bromo-2-methylphenyl)benzene-
sulfonamide projected along its c_0 axis.

TABLE XVB,72

Parameters of the Atoms in N(p-Bromophenyl)benzenesulfonamide

Atom	x	y	z
Br	0.115	0.061	0.541
S	0.729	0.112	0.705
C(1)	0.286	0.064	0.556
C(2)	0.347	0.040	0.674
C(3)	0.472	0.041	0.679
C(4)	0.537	0.069	0.584
C(5)	0.480	0.095	0.464
C(6)	0.350	0.093	0.447
N	0.669	0.073	0.598
O(1)	0.689	0.091	0.834
O(2)	0.854	0.113	0.672
C(7)	0.671	0.179	0.685
C(8)	0.727	0.218	0.603
C(9)	0.676	0.271	0.565
C(10)	0.580	0.287	0.664
C(11)	0.523	0.251	0.755
C(12)	0.572	0.196	0.777

They result in the structure shown in Figure XVB,105.

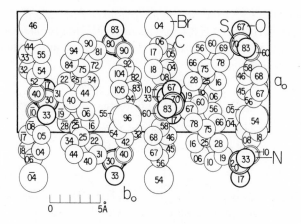

Fig. XVB,105. The orthorhombic structure of $N(p$-bromophenyl)benzene-sulfonamide projected along its c_0 axis.

TABLE XVB,73

Parameters of the Atoms in $N(p$-Bromophenyl)-p-chlorobenzenesulfonamide

Atom	x	y	z
Br	0.827	0.067	0.093
S	0.427	0.322	0.238
Cl	0.045	0.138	0.247
O(1)	0.533	0.342	0.284
O(2)	0.333	0.363	0.235
N	0.508	0.291	0.178
C(1)	0.719	0.137	0.121
C(2)	0.745	0.179	0.176
C(3)	0.674	0.230	0.195
C(4)	0.585	0.237	0.159
C(5)	0.558	0.194	0.103
C(6)	0.630	0.144	0.084
C(7)	0.319	0.271	0.241
C(8)	0.223	0.236	0.195
C(9)	0.135	0.194	0.197
C(10)	0.155	0.189	0.245
C(11)	0.236	0.223	0.291
C(12)	0.324	0.265	0.289

XV,bIII,27.   Crystals of *N(p-bromophenyl)-p-chlorobenzenesulfona-mide*, $BrC_6H_4NHSO_2 \cdot C_6H_4Cl$, are hexagonal, rhombohedral. According to a preliminary announcement, the hexagonal unit, which contains 18 molecules, has the edge lengths:

$$a_0 = 27.09(5) \text{ A.}; \quad c_0 = 9.40(5) \text{ A.}$$

The space group is C_{3i}^2 ($R\bar{3}$) with all atoms in the general positions:

$$(18f) \quad \pm (xyz; \bar{y},x-y,z; y-x,\bar{x},z); \ rh$$

Preliminary values for the parameters have been given as those of Table XVB,73.

XV,bIII,28.   *Deoxyanisoin*, $(CH_3OC_6H_4)_2CH_2CO$, forms monoclinic crystals whose tetramolecular unit has the dimensions:

$$a_0 = 15.13(2) \text{ A.}; \quad b_0 = 5.506(17) \text{ A.}; \quad c_0 = 21.94(6) \text{ A.}$$
$$\beta = 133°50(1)'$$

The space group is C_{2h}^5 ($P2_1/c$) with atoms in the positions:

$$(4e) \quad \pm (xyz; x,\tfrac{1}{2}-y,z+\tfrac{1}{2})$$

The determined parameters are listed in Table XVB,74; estimated values for hydrogen are included in the orginal paper.

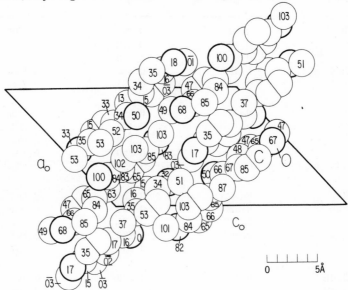

Fig. XVB,106.   The monoclinic structure of deoxyanisoin projected along its b_0 axis.

TABLE XVB,74

Parameters of the Atoms in Deoxyanisoin

Atom	x	y	z
C(1)	0.2278	0.8377	0.0806
C(2)	0.3077	0.8343	0.1744
C(3)	0.3897	0.0253	0.2229
C(4)	0.4641	0.0324	0.3100
C(5)	0.4544	0.8450	0.3481
C(6)	0.3739	0.6515	0.3007
C(7)	0.2998	0.6482	0.2138
C(8)	0.6008	0.0337	0.4841
C(9)	0.1375	0.6306	0.0314
C(10)	0.0514	0.6522	0.9357
C(11)	0.9673	0.8409	0.8902
C(12)	0.8848	0.8469	0.8030
C(13)	0.8863	0.6617	0.7599
C(14)	0.9683	0.4738	0.8034
C(15)	0.0519	0.4712	0.8919
C(16)	0.7913	0.4880	0.6269
O(1)	0.5215	0.8331	0.4326
O(2)	0.7989	0.6838	0.6734
O(3)	0.2384	0.9994	0.0481

The resulting structure is shown in Figure XVB,106. The bond dimensions of its molecules, similar to those in the related p,p'-dimethoxybenzophenone (**XV,bII,10**), are given in Figure XVB,107. The two benzene rings and their attached oxygen atoms are planar but the terminal methyl radicals are ca. 0.11 A. outside these planes; the C(1) and O(3) atoms are almost coplanar with the C(2)-to-C(7) ring.

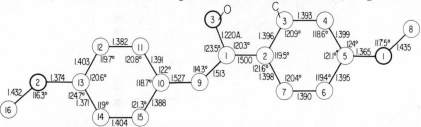

Fig. XVB,107.   Bond dimensions in the molecule of deoxyanisoin.

XV,bIII,29.   Crystals of the meso form of *1,2-dimethyl-1,2-diphenyl-diphosphine disulfide*, [(CH$_3$)(C$_6$H$_5$)PS]$_2$, are orthorhombic with a tetramolecular unit of the edge lengths:

$$a_0 = 17.104 \text{ A.}; \quad b_0 = 10.629 \text{ A.}; \quad c_0 = 8.592 \text{ A.}$$

The space group V_h^{15} (*Pbca*) places all atoms in the positions:

$$(8c) \quad \pm (xyz;\ x+1/2, 1/2-y, \bar{z};\ x, 1/2-y,\ z+1/2;\ x+1/2, y, 1/2-z)$$

The determined parameters are those of Table XVB,75.

TABLE XVB,75

Parameters of the Atoms in Meso[(CH$_3$)(C$_6$H$_5$)PS]$_2$

Atom	x	y	z
S	0.0625	0.222	0.019
P	0.005	0.072	0.092
C(1)	0.050	−0.014	0.251
C(2)	−0.099	0.105	0.149
C(3)	−0.138	0.023	0.250
C(4)	−0.215	0.046	0.298
C(5)	−0.252	0.151	0.229
C(6)	−0.215	0.236	0.127
C(7)	−0.138	0.209	0.084

The structure (Fig. XVB,108) is built up of dimeric molecules with a center of symmetry and the bond lengths of Figure XVB,109. In them

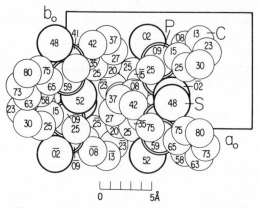

Fig. XVB,108.   The orthorhombic structure of 1,2-dimethyl-1,2-diphenyldiphosphine disulfide projected along its c_0 axis.

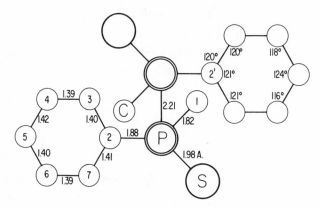

Fig. XVB,109.   Bond dimensions in the molecule of 1,2-dimethyl-1,2-diphenyl-diphosphine disulfide.

the bonds about the phosphorus atoms are tetrahedrally disposed, with angles ranging between 103 and 116°.

XV,bIII,30.   Crystals of *di-p-tolyl disulfide*, $(CH_3C_6H_4S)_2$, are monoclinic with a bimolecular unit of the dimensions:

$$a_0 = 14.86 \text{ A.}; \quad b_0 = 5.77 \text{ A.}; \quad c_0 = 7.69 \text{ A.}; \quad \beta = 94°$$

Atoms are in the positions

$$(2a) \quad xyz; \ \bar{x},y+{}^1/_2,\bar{z}$$

of the space group $C_2{}^2$ ($P2_1$). The parameters, as obtained in the recent refinement, are given in Table XVB,76.

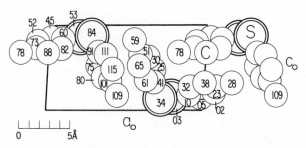

Fig. XVB,110.   The monoclinic structure of di-p-tolyl disulfide projected along its b_0 axis.

TABLE XVB,76

Parameters of the Atoms in $(CH_3C_6H_4S)_2$

Atom	x	y	z
S(1)	0.204	0.028	0.136
S(2)	0.271	0.342	0.124
C(1)	0.092	0.104	0.190
C(2)	0.082	0.321	0.274
C(3)	−0.005	0.375	0.325
C(4)	−0.079	0.233	0.290
C(5)	−0.175	0.276	0.335
C(6)	−0.065	0.019	0.212
C(7)	0.020	−0.047	0.159
C(8)	0.321	0.408	0.336
C(9)	0.370	0.611	0.354
C(10)	0.416	0.650	0.514
C(11)	0.410	0.512	0.665
C(12)	0.455	0.591	0.838
C(13)	0.363	0.297	0.632
C(14)	0.318	0.248	0.472

The resulting structure is shown in Figure XVB,110. Bond dimensions in the molecules that compose it are given in Figure XVB,111. The tolyl group C(1) through C(7) is planar but the methyl C(12) is considerably out of the plane of its C(8)-to-C(14) ring. Between molecules the shortest C–C = 3.65 A. and C–S = 3.98 A.

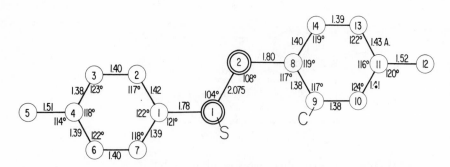

Fig. XVB,111.   Bond dimensions in the molecule of di-*p*-tolyl disulfide.

IV. Compounds in Which the Rings Are Separated by a Chain of Three Atoms

XV,bIV,1. Crystals of *dibenzoyl methane*, $C_6H_5C(O)CHC(OH)C_6H_5$, are orthorhombic with a unit that contains eight molecules and has

TABLE XVB,77

Parameters of the Atoms in Dibenzoyl Methane

Atom	x	y	z
C(1)	−0.05777	−0.01876	0.19003
C(2)	−0.12134	−0.05139	0.08528
C(3)	0.07535	−0.10199	0.04373
C(4)	−0.03396[a]	−0.11988	0.10034
C(5)	0.09997	−0.08812	0.20079
C(6)	0.05514	−0.03712	0.24353
C(7)	−0.03335	0.16173	0.47127
C(8)	−0.10279	0.20505	0.52414
C(9)	−0.05261	0.24220	0.62532
C(10)	0.06996	0.23780	0.66669
C(11)	0.13972	0.19476	0.61361
C(12)	0.08721	0.15682	0.51788
C(13)	−0.10661	0.03597	0.23060
C(14)	−0.04550	0.07085	0.33305
C(15)	−0.09208	0.12200	0.36467
O(16)	−0.20853	0.05034	0.16984
O(17)	−0.19330	0.13911	0.30089
H(18)	−0.20260	−0.03852	0.04833
H(19)	−0.12941	−0.12440	−0.02644
H(20)	0.07494	−0.15404	0.06953
H(21)	0.18910	−0.10372	0.23951
H(22)	0.10070	−0.02112	0.31957
H(23)	−0.18275	0.20689	0.48970
H(24)	−0.10456	0.27170	0.66327
H(25)	0.10809	0.26473	0.73244
H(26)	0.23072	0.19140	0.64943
H(27)	0.14430	0.13024	0.48462
H(28)	0.03194	0.06452	0.38663
H(29)	−0.21973	0.10171	0.22218

[a] It is presumed that this parameter should be minus rather than the plus of the original article.

the edge lengths:

$$a_0 = 10.857(2) \text{ A.}; \quad b_0 = 24.446(5) \text{ A.}; \quad c_0 = 8.756(2) \text{ A.}$$

All atoms are in the general positions of V_h^{15} (*Pbca*):

(8c) $\pm(xyz;\ x+\frac{1}{2},\frac{1}{2}-y,\bar{z};\ \bar{x},y+\frac{1}{2},\frac{1}{2}-z;\ \frac{1}{2}-x,\bar{y},z+\frac{1}{2})$

The atomic parameters, including those for hydrogen, are given in Table XVB,77.

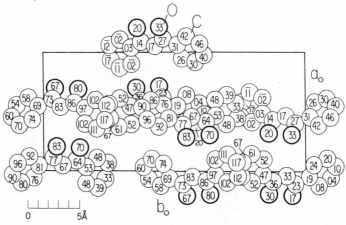

Fig. XVB,112a.   The orthorhombic structure of dibenzoyl methane projected along its c_0 axis.

The structure, illustrated in Figure XVB,112, contains molecules with the bond dimensions of Figure XVB,113. Each molecule consists of three planar sections: the two end benzene rings and a central enolic ring (labeled III in the figure). The angle between III and benzene ring I is $-3.8°$ and between III and II, $16.9°$.

XV,bIV,2. The orthorhombic crystals of *bis(m-chlorobenzoyl) methane*, $[ClC_6H_4C(O)]_2CH_2$, possess a tetramolecular unit of the edge lengths:

$$a_0 = 30.082(2) \text{ A.}; \quad b_0 = 3.850(5) \text{ A.}; \quad c_0 = 11.123(2) \text{ A.}$$

The space group is C_{2v}^5 (*Pca2₁*) with atoms in the positions:

(4a) $xyz;\ \bar{x},\bar{y},z+\frac{1}{2};\ \frac{1}{2}-x,y,z+\frac{1}{2};\ x+\frac{1}{2},\bar{y},z$

The determined parameters are stated in Table XVB,78.

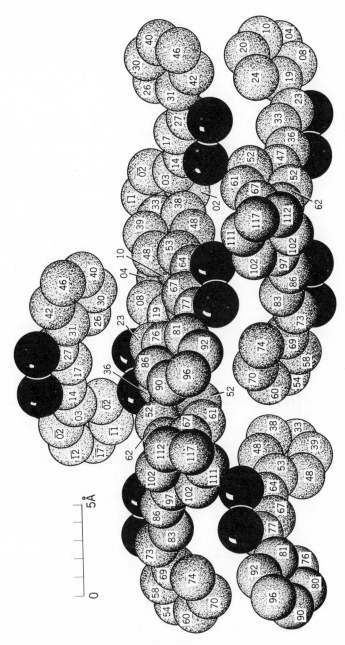

Fig. XVB,112b.   A packing drawing of the orthorhombic structure of dibenzoyl methane viewed along its c_0 axis. The oxygen atoms are black; those of carbon are dotted.

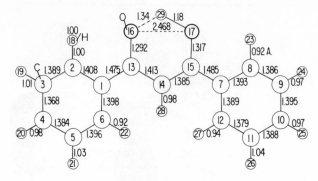

Fig. XVB,113a.   Bond lengths in the molecule of dibenzoyl methane.

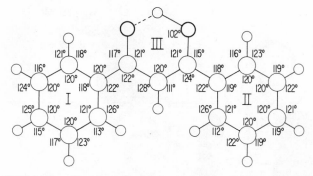

Fig. XVB,113b.   Bond angles in the molecule of dibenzoyl methane.

The molecules in this structure (Fig. XVB,114) are nearly planar, the largest departure from the mean molecular plane being a Cl(2) excursion of 0.07 A.

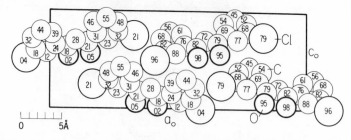

Fig. XVB,114.   The orthorhombic structure of bis(*m*-chlorobenzoyl) methane projected along its b_0 axis.

TABLE XVB,78

Parameters of the Atoms in Bis(m-chlorobenzoyl) Methane

Atom	x	y	z
Cl(1)	0.40885(7)	0.0408(7)	0.0000
Cl(2)	0.83345(6)	0.2093(9)	0.2196(4)
O(1)	0.58081(17)	0.0236(20)	0.0195(5)
O(2)	0.66147(19)	0.0476(21)	0.0627(5)
C(1)	0.4494(3)	0.1808(27)	0.0993(8)
C(2)	0.4936(3)	0.1269(23)	0.0718(8)
C(3)	0.5258(3)	0.2377(25)	0.1539(8)
C(4)	0.5133(3)	0.3918(24)	0.2621(8)
C(5)	0.4683(3)	0.4404(27)	0.2882(9)
C(6)	0.4352(2)	0.3243(25)	0.2075(9)
C(7)	0.5733(3)	0.1785(26)	0.1213(8)
C(8)	0.6084(2)	0.2781(26)	0.1969(8)
C(9)	0.6524(2)	0.2109(23)	0.1638(7)
C(10)	0.7656(3)	0.4844(28)	0.3765(9)
C(11)	0.7224(3)	0.5548(28)	0.4187(8)
C(12)	0.6856(3)	0.4625(26)	0.3486(8)
C(13)	0.6126(2)[a]	0.3073(23)	0.2384(7)
C(14)	0.7330(2)	0.2284(28)	0.2000(9)
C(15)	0.7700(3)	0.3187(27)	0.2701(9)

[a] This value is obviously in error. In preparing Figure XVB,114 it was assumed it should be 0.6926.

XV,bIV,3. Crystals of *bis(m-bromobenzoyl) methane*, $[BrC_6H_4C(O)]_2$-CH_2, are orthorhombic with a tetramolecular unit of the edge lengths:

$$a_0 = 26.48 \text{ A.}; \quad b_0 = 4.054 \text{ A.}; \quad c_0 = 12.79 \text{ A.}$$

The space group is V_h^{14} (*Pnca*) with atoms in the positions:

(4c) $\pm (1/4 \ 0 \ u; \ 1/4, 1/2, u + 1/2)$
(8d) $\pm (xyz; \ 1/2 - x, y + 1/2, z + 1/2; \ x, 1/2 - y, z + 1/2; \ x + 1/2, y, \bar{z})$

The chosen positions and parameters are listed in Table XVB,79.

The resulting structure, shown in Figure XVB,115, has molecules with the bond dimensions of Figure XVB,116. The entire molecule is roughly planar but individual atoms are as much as 0.10 A. away from the best plane through them. The position of the enolic hydrogen atom, which provides a very short hydrogen bond between the two oxygen

Positions and Parameters of the Atoms in Bis(*m*-bromobenzoyl) Methane

Atom	Position	x	y	z
Br	(8*d*)	0.03607(3)	0.2306(3)	0.15544(7)
O	(8*d*)	0.2097(2)	0.848(2)	0.2068(4)
C(1)	(8*d*)	0.0883(2)	0.399(2)	0.0684(6)
C(2)	(8*d*)	0.1291(2)	0.553(2)	0.1166(5)
C(3)	(8*d*)	0.1678(2)	0.683(2)	0.0546(5)
C(4)	(8*d*)	0.1648(2)	0.651(2)	0.9450(5)
C(5)	(8*d*)	0.1234(3)	0.494(2)	0.9005(5)
C(6)	(8*d*)	0.0843(3)	0.366(2)	0.9618(5)
C(7)	(8*d*)	0.2101(3)	0.850(2)	0.1047(5)
C(8)	(4*c*)	$^1/_4$	0	0.0520(7)
H(2)	(8*d*)	0.131	0.574	0.201
H(4)	(8*d*)	0.194	0.749	0.896
H(5)	(8*d*)	0.121	0.471	0.816
H(6)	(8*d*)	0.052	0.245	0.926
H(8)	(4c)	$^1/_4$	0	0.968

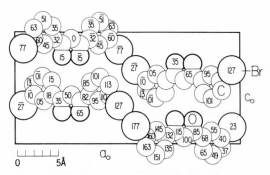

Fig. XVB,115.   The orthorhombic structure of bis(*m*-bromobenzoyl) methane projected along its b_0 axis.

atoms, is presumably not fixed, but throughout the crystal is sometimes closer to one and sometimes to the other of the oxygen atoms.

XV,bIV,4.   The monoclinic crystals of *bis(dibenzoylmethyl) palladium*, {[(C₆H₅)C(O)]₂CH}₂Pd, possess a tetramolecular unit of the dimensions:

$$a_0 = 17.37(2) \text{ A.;} \quad b_0 = 5.78(1) \text{ A.;} \quad c_0 = 24.23(4) \text{ A.;} \quad \beta = 103(^1/_2)°$$

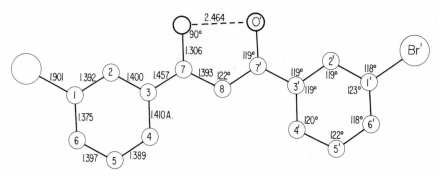

Fig. XVB,116. Bond dimensions in the molecule of bis(*m*-bromobenzoyl) methane.

The space group is C_{2h}^6 in the orientation $I2/c$. Atoms therefore are in the positions:

$$\text{Pd} : (4c) \quad \pm ({}^1/_4\, {}^1/_4\, {}^1/_4;\, {}^1/_4\, {}^3/_4\, {}^3/_4)$$

The other atoms are in the general positions:

$$(8f) \quad \pm (xyz;\, x,\bar{y},z+{}^1/_2;\, x+{}^1/_2,y+{}^1/_2,z+{}^1/_2;\, x+{}^1/_2,{}^1/_2-y,z)$$

TABLE XVB,80

Parameters of the Atoms in Bis(dibenzoylmethyl) Palladium

Atom	x	y	z
O(1)	0.3055(6)	0.519(2)	0.2898(4)
O(2)	0.2866(6)	0.310(2)	0.1808(4)
C(1)	0.3501(9)	0.648(3)	0.2671(7)
C(2)	0.3651(9)	0.632(4)	0.2143(7)
C(3)	0.3303(9)	0.480(4)	0.1725(7)
C(4)	0.3849(9)	0.840(3)	0.3068(7)
C(5)	0.3839(9)	0.822(3)	0.3642(7)
C(6)	0.4158(9)	0.996(4)	0.4029(7)
C(7)	0.4585(9)	0.172(3)	0.3848(7)
C(8)	0.4634(9)	0.188(3)	0.3280(6)
C(9)	0.4305(9)	0.017(4)	0.2914(7)
C(10)	0.3459(9)	0.495—	0.1144(7)
C(11)	0.3920(9)	0.665(3)	0.0967(7)
C(12)	0.3996(9)	0.672(3)	0.0409(7)
C(13)	0.3676(9)	0.504(4)	0.0029(6)
C(14)	0.3236(9)	0.328(3)	0.0176(7)
C(15)	0.3143(9)	0.321(3)	0.0744(7)

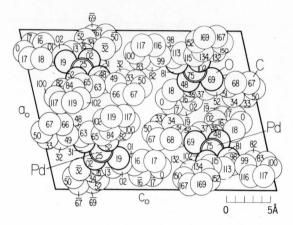

Fig. XVB,117.   The monoclinic structure of bis(dibenzoylmethyl) palladium projected along its b_0 axis.

Parameters are stated in Table XVB,80, values for hydrogen being listed in the original article.

The structure (Fig. XVB,117) is built of molecules with the bond dimensions of Figure XVB,118. In them coordination about the atom of palladium is square. They are stacked one above another in the crystal with C–O = 3.34 A. and a shortest Pd–C = 3.35 A.

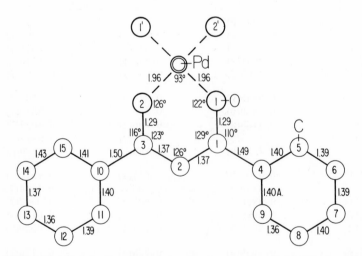

Fig. XVB,118.   Bond dimensions in the molecule of bis(dibenzoylmethyl) palladium.

XV,bIV,5. A determination has been made of the approximate structure of *p-bromobenzoic anhydride*, $(BrC_6H_4CO)_2O$. Its monoclinic crystals, with two molecules per cell, have the dimensions:

$$a_0 = 28.04 \text{ A.}; \quad b_0 = 5.91 \text{ A.}; \quad c_0 = 3.97 \text{ A.}; \quad \beta = 94°48'$$

The space group is C_2^3 ($C2$) with atoms in the positions:

$$O(10) : (2a) \quad 0u0; \; 1/2, u + 1/2, 0 \qquad \text{with } u = 0.522$$

All other atoms are in the general positions:

$$(4c) \quad xyz; \; \bar{x}y\bar{z}; \; x + 1/2, y + 1/2, z; \; 1/2 - x, y + 1/2, \bar{z}$$

The approximately determined parameters are stated in Table XVB,81. The x parameter for C(7), as given in the original, is clearly in error. It would appear that it should be ca. 0.13.

TABLE XVB,81

Parameters of the Atoms in *p*-Bromobenzoic Anhydride

Atom	x	y	z
Br(1)	0.209	0.000	0.585
C(2)	0.162	0.173	0.715
C(3)	0.167	0.401	0.843
C(4)	0.128	0.540	0.949
C(5)	0.084	0.458	0.934
C(6)	0.079	0.229	0.817
C(7)	0.166	0.089	0.700
C(8)	0.048	0.591	0.036
O(9)	0.049	0.784	0.197

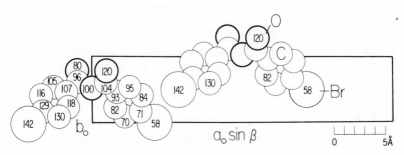

Fig. XVB,119. The contents of the elongated monoclinic cell of *o*-bromobenzoic anhydride projected along the c_0 axis.

The structure is shown in Figure XVB,119. The benzene rings of its molecules are turned 8° from those of their carboxyl groups and the planes they define make 55° with respect to one another.

XV,bIV,6. *Benzyl sulfide* forms an addition compound with *iodine* which has the composition $(C_6H_5CH_2)_2S \cdot I_2$. Its crystals are orthorhombic with a tetramolecular unit of the edge lengths:

$$a_0 = 9.86 \text{ A.}; \quad b_0 = 16.86 \text{ A.}; \quad c_0 = 9.36 \text{ A.}$$

The space group is V_h^{16} (*Pnma*) with atoms in the positions:

(4c)　$\pm (u\ ^1/_4\ v;\ u+^1/_2,^1/_4,^1/_2-v)$
(8d)　$\pm (xyz;\ ^1/_2-x,y+^1/_2,z+^1/_2;\ x,^1/_2-y,z;\ x+^1/_2,y,^1/_2-z)$

Positions and parameters are listed in Table XVB,82.

TABLE XVB,82

Positions and Parameters of the Atoms in $(C_6H_5CH_2)_2S \cdot I_2$

Atom	Position	x	y	z
I(1)	(4c)	0.3428	$^1/_4$	0.4849
I(2)	(4c)	0.5893	$^1/_4$	0.6374
S	(4c)	0.1025	$^1/_4$	0.3298
CH$_3$	(8d)	0.009	0.328	0.422
C(1)	(8d)	0.062	0.426	0.369
C(2)	(8d)	0.002	0.459	0.237
C(3)	(8d)	0.074	0.524	0.196
C(4)	(8d)	0.164	0.554	0.270
C(5)	(8d)	0.223	0.526	0.409
C(6)	(8d)	0.151	0.458	0.454

The structure (Fig. XVB,120) is built up of benzyl sulfide and I_2 molecules so arranged that the iodine atoms are in line with the atom of sulfur. In the I_2 molecule I–I = 2.819 A. and the I–S distance is 2.78 A. In the sulfide portion C–S–C = 93°.

XV,bIV,7. Crystals of *dichlorodiphenoxy titanium(IV)*, TiCl$_2$-(OC$_6$H$_5$)$_2$, are monoclinic with a tetramolecular unit of the dimensions:

$$a_0 = 9.821(3) \text{ A.}; \quad b_0 = 14.006(4) \text{ A.}; \quad c_0 = 9.836(3) \text{ A.}$$
$$\beta = 94°50(10)'$$

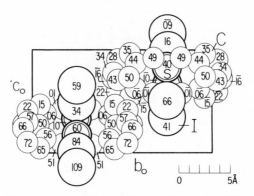

Fig. XVB,120. The orthorhombic structure of benzyl sulfide-iodine projected along its a_0 axis.

Atoms are in the positions

$$(4e) \quad \pm (xyz; x+\tfrac{1}{2}, \tfrac{1}{2}-y, z+\tfrac{1}{2})$$

of the space group C_{2h}^5 ($P2_1/n$). The parameters are those of Table XVB,83.

TABLE XVB,83

Parameters of the Atoms in $TiCl_2(OC_6H_5)_2$

Atom	x	y	z
Ti(1)	0.5031(2)	0.4714(1)	0.3389(2)
Cl(2)	0.3207(3)	0.5353(3)	0.2275(3)
Cl(3)	0.6958(3)	0.5269(3)	0.2663(3)
O(4)	0.4927(7)	0.4166(5)	0.5156(6)
O(5)	0.4972(9)	0.3628(5)	0.2518(7)
C(6)	0.4807(10)	0.3167(7)	0.5381(9)
C(7)	0.5958(13)	0.2656(8)	0.5736(11)
C(8)	0.5827(15)	0.1675(8)	0.5929(12)
C(9)	0.4563(16)	0.1256(8)	0.5801(12)
C(10)	0.3388(15)	0.1779(9)	0.5510(12)
C(11)	0.3510(12)	0.2769(9)	0.5272(11)
C(12)	0.4816(11)	0.2919(7)	0.1575(11)
C(13)	0.4564(12)	0.3191(8)	0.0218(11)
C(14)	0.4349(15)	0.2486(10)	−0.0768(11)
C(15)	0.4444(13)	0.1504(10)	−0.0392(14)
C(16)	0.4707(13)	0.1282(8)	0.0974(13)
C(17)	0.4908(13)	0.1983(8)	0.1991(11)

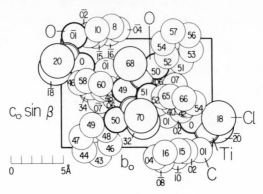

Fig. XVB,121. The monoclinic structure of dichlorodiphenoxy titanium projected along its a_0 axis.

The structure is shown in Figure XVB,121. It is built up of dimeric, centrosymmetric molecules that have the bond dimensions of Figure XVB,122. The titanium atoms have a fivefold coordination, the metal atom being at the centre of a distorted trigonal bipyramid with O(4) and O(5′) atoms at the apices. Titanium and all the oxygen atoms are approximately coplanar. The benzene rings have their expected dimensions.

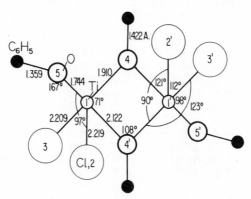

Fig. XVB,122.   Bond dimensions in the molecule of dichlorodiphenoxy titanium.

XV,bIV,8. Crystals of *(1,3-diphenyl urea)·di(iron tricarbonyl)*, $(C_6H_5N)_2CO \cdot Fe_2(CO)_6$, are monoclinic with a tetramolecular unit of the dimensions:

$$a_0 = 8.537(10)\ A.; \quad b_0 = 13.246(7)\ A.; \quad c_0 = 19.490(14)\ A.$$
$$\beta = 106.6(2)°$$

The space group is C_{2h}^5 ($P2_1/c$) with all atoms in the positions:

$$(4e) \quad \pm (xyz; \; x,{}^1/_2 - y, z + {}^1/_2)$$

The parameters are stated in Table XVB,84.

TABLE XVB,84

Parameters of the Atoms in $(C_6H_5N)_2CO \cdot Fe_2(CO)_6$

Atom	x	y	z
Fe(1)	0.1504(2)	0.1467(2)	0.2660(1)
Fe(2)	0.3514(2)	0.2638(2)	0.2474(1)
N(1)	0.3857(10)	0.1550(6)	0.3214(4)
N(2)	0.3085(9)	0.1247(6)	0.2087(4)
C(1)	−0.0242(16)	0.1674(11)	0.1917(7)
C(2)	0.2474(14)	0.3281(9)	0.1716(6)
C(3)	0.0843(16)	0.0250(11)	0.2845(7)
C(4)	0.5521(15)	0.2916(10)	0.2404(6)
C(5)	0.0680(15)	0.2188(10)	0.3215(6)
C(6)	0.3367(15)	0.3658(10)	0.3037(7)
C(7)	0.4240(13)	0.0884(9)	0.2723(5)
O(1)	−0.1319(14)	0.1832(9)	0.1403(6)
O(2)	0.1719(13)	0.3665(8)	0.1167(5)
O(3)	0.0487(13)	−0.0547(9)	0.2982(6)
O(4)	0.6823(12)	0.3155(8)	0.2367(5)
O(5)	0.0089(14)	0.2648(9)	0.3610(6)
O(6)	0.3176(12)	0.4309(8)	0.3396(5)
O(7)	0.5250(10)	0.0231(7)	0.2806(4)
C(8)	0.4745(15)	0.1495(9)	0.3979(6)
C(9)	0.3957(17)	0.1310(11)	0.4471(7)
C(10)	0.4819(26)	0.1255(16)	0.5199(11)
C(11)	0.6460(21)	0.1445(13)	0.5402(9)
C(12)	0.7261(28)	0.1638(16)	0.4917(16)
C(13)	0.6408(23)	0.1678(15)	0.4155(10)
C(14)	0.2988(12)	0.0798(8)	0.1420(5)
C(15)	0.2105(17)	−0.0079(12)	0.1235(7)
C(16)	0.1940(18)	−0.0534(12)	0.0538(8)
C(17)	0.2708(19)	−0.0047(12)	0.0087(8)
C(18)	0.3568(17)	0.0801(11)	0.0279(8)
C(19)	0.3671(17)	0.1252(11)	0.0955(8)

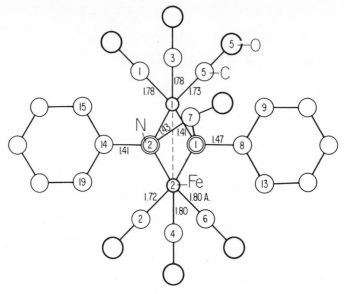

Fig. XVB,123. Some bond lengths in the molecule of (1,3-diphenyl urea) · di-(iron tricarbonyl).

Some of the bond dimensions of the molecules thus defined are shown in Figure XVB,123. The iron atoms have a sixfold coordination, with Fe–N = 2.00 A. and Fe–Fe = 2.416 A.

XV,bIV,9. *Diazoaminobenzene-copper(I)*, $(C_6H_5N\!\!=\!\!N\!\!-\!\!NC_6H_5)$-Cu(I), forms monoclinic crystals. The cell, containing eight of these formula units, has the dimensions:

$$a_0 = 16.00(5) \text{ A.}; \quad b_0 = 5.61(5) \text{ A.}; \quad c_0 = 24.01(5) \text{ A.}$$
$$\beta = 99°20(11)'$$

The space group is C_{2h}^6 in the orientation $I2/a$. Atoms accordingly are in the positions:

$$(8f) \quad +(xyz; x+{}^1\!/_2,\bar{y},z); \quad \text{B.C.}$$

The parameters are those of Table XVB,85.

The molecules in the resulting structure (Fig. XVB,124) are centrosymmetric dimers with the bond dimensions of Figure XVB,125. They are roughly planar, though rotations of the benzene rings about the C–N bonds bring carbon atoms as much as 0.1 A. out of the best plane

TABLE XVB,85

Parameters of the Atoms in Diazoaminobenzene-copper(I)

Atom	x	y	z
Cu	0.0097	0.0844	0.0477
N(1)	0.0788	0.3169	0.0200
N(2)	0.0991	0.3290	−0.0291
N(3)	0.0673	0.1618	−0.0648
C(1,1)	0.1194	0.5130	0.0559
C(1,2)	0.1642	0.6944	0.0343
C(1,3)	0.1992	0.8826	0.0702
C(1,4)	0.1894	0.8824	0.1268
C(1,5)	0.1418	0.7019	0.1473
C(1,6)	0.1051	0.5085	0.1121
C(2,1)	0.0884	0.1897	−0.1185
C(2,2)	0.0600	0.0169	−0.1588
C(2,3)	0.0741	0.0288	−0.2156
C(2,4)	0.1240	0.2156	−0.2299
C(2,5)	0.1560	0.3852	−0.1901
C(2,6)	0.1369	0.3782	−0.1340

through the entire dimer. The coordination of the copper atoms is linear. Between molecules the shortest atomic separations are Cu–C = 3.20 A., N–N = 3.27 A., C–N = 3.36 A. and C–C = 3.42 A.

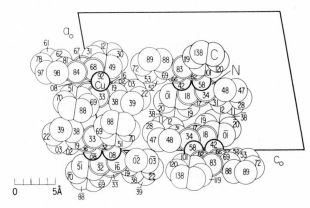

Fig. XVB,124.   The monoclinic structure of diazoaminobenzene-copper projected along its b_0 axis.

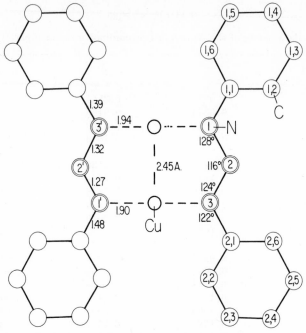

Fig. XVB,125.   Bond dimensions in the molecule of diazoaminobenzene-copper.

XV,bIV,10.   The β-modification of *p-bromodiazoaniline*, $BrC_6H_4NH \cdot N{=}N{-}C_6H_5$, is monoclinic with a tetramolecular unit of the dimensions:

$$a_0 = 11.49(1) \text{ A.}; \quad b_0 = 4.70(1) \text{ A.}; \quad c_0 = 22.35(2) \text{ A.}$$
$$\beta = 74°0(10)'$$

The space group C_{2h}^5 ($P2_1/c$) places all atoms in the positions:

$$(4e) \quad \pm(xyz; x, {}^1/_2 - y, z + {}^1/_2)$$

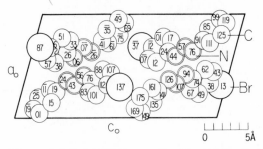

Fig. XVB,126.   The monoclinic structure of *p*-bromodiazoaniline projected along its b_0 axis.

Parameters are listed in Table XVB,86.

TABLE XVB,86

Parameters of the Atoms in p-Bromodiazoaniline

Atom	x	y	z
C(1)	0.319	0.121	0.418
C(2)	0.223	0.013	0.373
C(3)	0.247	0.828	0.330
C(4)	0.363	0.764	0.331
C(5)	0.459	0.876	0.376
C(6)	0.437	0.068	0.418
C(7)	0.247	0.092	0.172
C(8)	0.126	0.149	0.163
C(9)	0.040	0.007	0.117
C(10)	0.077	0.808	0.082
C(11)	0.199	0.750	0.091
C(12)	0.282	0.891	0.138
N(1)	0.397	0.556	0.290
N(2)	0.310	0.431	0.252
N(3)	0.340	0.243	0.218
Br	0.284	0.371	0.478

The structure, illustrated in Figure XVB,126, consists of molecules with the bond dimensions of Figure XVB,127. The C_6H_5 ring and its attached N(3) are planar and so is the C_6H_4Br group; the N(1) atom is ca. 0.15 A. outside the latter plane. Viewed edge-on these two planes make an angle of 6° with respect to one another. Between the N(1) and N(3) atoms of molecules across the screw axis there are considered to be

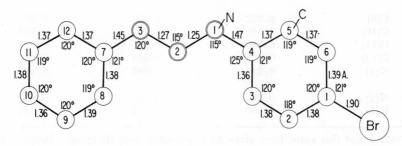

Fig. XVB,127. Bond dimensions in the molecule of p-bromodiazoaniline.

hydrogen bonds of length 3.20 A. The shortest separation between adjacent molecules is a Br–C = 3.45 A.

A partial structure has been described for the orthorhombic α-form of this compound.

XV,bIV,11. Crystals of *2,4-dibromodiazoaminobenzene*, $Br_2C_6H_3$-$NHN{=}NC_6H_5$, are monoclinic with a tetramolecular unit that has been given the edge lengths:

$$a_0 = 12.05(1) \text{ A.}; \quad b_0 = 23.88(2) \text{ A.}; \quad c_0 = 4.48(1) \text{ A.}$$
$$\beta = 84°13(15)'$$

All atoms are in the general positions of C_{2h}^5 $(P2_1/n)$:

$$(4e) \quad \pm (xyz; x+1/2, 1/2-y, z+1/2)$$

Parameters are stated in Table XVB,87.

TABLE XVB,87

Parameters of the Atoms in $Br_2C_6H_3NHN{=}NC_6H_5$

Atom	x	y	z
Br(1)	0.662	0.983	0.781[a]
Br(2)	0.078	0.082	0.286
C(1)	0.800	0.131	0.857
C(2)	0.903	0.132	0.967
C(3)	0.935	0.087	0.139
C(4)	0.865	0.043	0.210
C(5)	0.761	0.043	0.103
C(6)	0.729	0.087	0.930
C(7)	0.468	0.162	0.360
C(8)	0.360	0.157	0.280
C(9)	0.322	0.194	0.077
C(10)	0.390	0.237	0.954
C(11)	0.499	0.210	0.037
C(12)	0.536	0.204	0.245
N(1)	0.620	0.083	0.826
N(2)	0.595	0.126	0.618
N(3)	0.500	0.117	0.537

[a] Choice of this value for z gives an improbably long Br(1)–C(5) distance. It would appear to be in error.

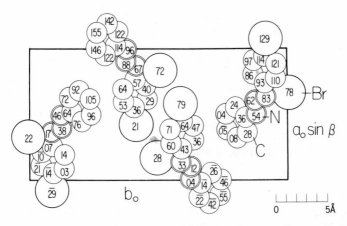

Fig. XVB,128. The monoclinic structure of 2,4-dibromodiazoaminobenzene projected along its short c_0 axis.

The structure, illustrated in Figure XVB,128, has molecules with the bond dimensions of Figure XVB,129. The planes of the benzene rings make 9° with one another, expressing a twist that puts N(2) and N(3) out of either plane.

XV,bIV,12.   The monoclinic crystals of *p-dibromodiazoaminobenzene*, p-BrC_6H_4N=N—NHC_6H_4Br, possess a tetramolecular unit of the dimensions:

$$a_0 = 16.75(1) \text{ A.}; \quad b_0 = 4.73(1) \text{ A.}; \quad c_0 = 15.78(1) \text{ A.}$$
$$\beta = 84°58(10)'$$

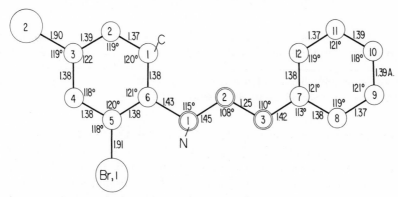

Fig. XVB,129.   Bond dimensions in the molecule of 2,4-dibromodiazoamino-benzene.

The space group, found to be C_{2h}^5 in the orientation $P2_1/n$, puts all atoms in the positions:

$$(4e) \quad \pm (xyz; x + {}^1/_2, {}^1/_2 - y, z + {}^1/_2)$$

The parameters are listed in Table XVB,88.

TABLE XVB,88

Parameters of the Atoms in p-Dibromodiazoaminobenzene
and Isostructural Compounds

Atom	x	y	z
Br(CH₃),1	0.317 (0.319)ᵃ [0.312]ᵇ	0.940	0.662 (0.653)ᵃ [0.648]ᵇ
Br(CH₃),2	0.606 (0.584) —	0.130	0.096 (0.107) —
C(1)	0.316 (0.316) [0.310]	0.753	0.556 (0.574) [0.573]
C(2)	0.376 (0.379) [0.377]	0.563	0.537 (0.536) [0.533]
C(3)	0.378 (0.376) [0.375]	0.414	0.462 (0.466) [0.458]
C(4)	0.316 (0.311) [0.305]	0.451	0.408 (0.421) [0.424]
C(5)	0.258 (0.248) [0.237]	0.641	0.425 (0.455) [0.466]
C(6)	0.258 (0.251) [0.238]	0.784	0.504 (0.529) [0.540]
C(7)	0.529 (0.522) [0.546]	0.381	0.145 (0.148) [0.148]
C(8)	0.460 (0.450) [0.470]	0.436	0.103 (0.120) [0.115]
C(9)	0.402 (0.392) [0.408]	0.627	0.138 (0.158) [0.150]
C(10)	0.412 (0.404) [0.420]	0.758	0.217 (0.228) [0.218]
C(11)	0.483 (0.478) [0.493]	0.712	0.257 (0.260) [0.251]
C(12)	0.538 (0.535) [0.558]	0.524	0.218 (0.220) [0.216]
N(1)	0.315 (0.302) [0.293]	0.283	0.332 (0.350) [0.350]
N(2)	0.366 (0.360) [0.361]	0.101	0.312 (0.332) [0.327]
N(3)	0.352 (0.345) [0.354]	0.970	0.245 (0.265) [0.254]

ᵃ For p-dimethyldiazoaminobenzene.

ᵇ For p-monomethyldiazoaminobenzene.

The structure is shown in Figure XVB,130, the bond dimensions of its molecules in Figure XVB,131. Each benzene ring and its attached bromine atom is planar and there is an angle of 6° between the planes. They are joined by the zigzag chain of the nitrogen atoms. Between molecules there is a shortest interatomic distance of N–N = 3.26 A. but, as the figure shows, the bromine atoms are also in contact.

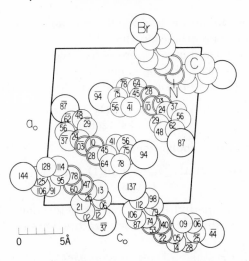

Fig. XVB,130. The monoclinic structure of *p*-dibromodiazoaminobenzene projected along its b_0 axis.

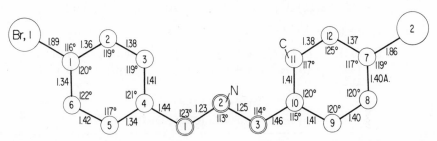

Fig. XVB,131. Bond dimensions in the molecule of *p*-dibromodiazoaminobenzene.

The corresponding chlorine compound is isostructural. Its unit has the dimensions:

$$a_0 = 16.94(1) \text{ A.}; \quad b_0 = 4.68(1) \text{ A.}; \quad c_0 = 15.21(1) \text{ A.}$$
$$\beta = 86°7(10)'$$

Parameters were not determined.

The *p-dimethyldiazoaminobenzene*, $p\text{-}CH_3C_6H_4N{=}N\text{—}NHC_6H_4CH_3$, also has this structure. For it:

$$a_0 = 17.83(2) \text{ A.}; \quad b_0 = 4.83(1) \text{ A.}; \quad c_0 = 14.40(2) \text{ A.}$$
$$\beta = 88°40(10)'$$

Parameters were found for the x and z but not for the y coordinates; they are given in parentheses in the table.

The monomethyl and monochloro derivatives have been described as being isostructural with the foregoing.

For *p-methyldiazoaminobenzene*, $CH_3C_6H_4N{=}N{-}NHC_6H_5$:

$$a_0 = 16.93(2) \text{ A.;} \quad b_0 = 4.83(1) \text{ A.;} \quad c_0 = 13.89(2) \text{ A.}$$
$$\beta = 84°42(10)'$$

For *p*-chlorodiazoaminobenzene, $ClC_6H_4N{=}N{-}NHC_6H_5$:

$$a_0 = 16.74(2) \text{ A.;} \quad b_0 = 4.76(1) \text{ A.;} \quad c_0 = 14.26(2) \text{ A.}$$
$$\beta = 84°40(10)'$$

Parameters of the methyl compound, as determined for the x and z but not the y coordinates, are in square brackets in the table.

XV,bIV,13. Crystals of *di-p-chlorophenyl hydrogen phosphate*, $(ClC_6H_4O)_2PO(OH)$, are orthorhombic with a tetramolecular unit that has the edges:

$$a_0 = 12.433(12) \text{ A.;} \quad b_0 = 4.611(5) \text{ A.;} \quad c_0 = 23.784(20) \text{ A.}$$

The space group is V_h^{10} (*Pnaa*) with atoms in the positions:

$$P : (4c) \quad \pm (u\ {}^1/_4\ {}^1/_4;\ u+{}^1/_2,{}^1/_4,{}^1/_4) \quad \text{with } u = 0.6198$$
$$H(O,2) : (4d) \quad \pm (u\ {}^1/_4\ {}^3/_4;\ u+{}^1/_2,{}^1/_4,{}^3/_4) \quad \text{with } u = 0.3170$$

TABLE XVB,89

Parameters of Atoms in $(ClC_6H_4O)_2PO(OH)$

Atom	x	y	z
Cl	0.3382	0.9248	0.0378
O(1)	0.5452	0.1433	0.2004
O(2)	0.6834	0.5084	0.2313
C(1)	0.3999	0.6981	0.0862
C(2)	0.5034	0.6019	0.0747
C(3)	0.5513	0.4113	0.1146
C(4)	0.4990	0.3436	0.1620
C(5)	0.3986	0.4443	0.1726
C(6)	0.3499	0.6350	0.1346
H(C,2)	0.545	0.661	0.037
H(C,3)	0.625	0.327	0.107
H(C,5)	0.359	0.387	0.209
H(C,6)	0.275	0.721	0.143

All other atoms are in the positions:

$(8e)$ $\pm(xyz;\, x,{}^1/_2-y,{}^1/_2-z;\, {}^1/_2-x,y+{}^1/_2,\bar{z};\, {}^1/_2-x,\bar{y},z+{}^1/_2)$

The parameters are listed in Table XVB,89.

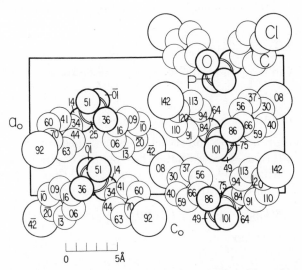

Fig. XVB,132.   The orthorhombic structure of di-p-chlorophenyl hydrogen phosphate projected along its b_0 axis.

The structure (Fig. XVB,132) contains molecules with the bond dimensions of Figure XVB,133. The phosphate groups are tetrahedral but with O(2) sensibly closer to phosphorus than is O(1). The O(2) atoms of different molecules are tied together along the b_0 axis by hydrogen bonds with the short O–H–O = 2.398 A. In this case the bond appears symmetrical, its hydrogen atom lying in the twofold axis. The

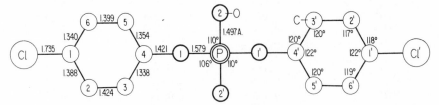

Fig. XVB,133.   Bond dimensions in the molecule of di-p-chlorophenyl hydrogen phosphate.

shortest van der Waals approach between molecules is a Cl–Cl = 3.655 A.; there is also a Cl–C = 3.80 A.

XV,bIV,14. The sulfonyl sulfide *bis(phenylsulfonyl)* *sulfide*, $(C_6H_5SO_2)_2S$, and the corresponding *selenide*, $(C_6H_5SO_2)_2Se$, are isostructural. Their monoclinic units, containing four molecules, have the dimensions:

For $(C_6H_5SO_2)_2S$:

$a_0 = 15.88$ A.; $\quad b_0 = 5.52$ A.; $\quad c_0 = 15.88$ A.; $\quad \beta = 112°54'$

For $(C_6H_5SO_2)_2Se$:

$a_0 = 16.14$ A.; $\quad b_0 = 5.60$ A.; $\quad c_0 = 15.84$ A.; $\quad \beta = 112°$

The space group is C_{2h}^6. The structure established for the sulfide was described in terms of coordinates having an origin obtained by a translation through $0\,^1/_4\,^1/_4$ with respect to that conventionally employed. In this orientation $(A2/a)$ the atomic positions are:

$$S(1)[Se] : (4e) \quad \pm(^1/_4\,u\,^1/_4;\quad ^1/_4,u+^1/_2,^3/_4)$$

All other atoms are in:

$$(8f) \quad \pm(xyz;\, x+^1/_2,\bar{y},z+^1/_2;\, x,y+^1/_2,z+^1/_2;\, x+^1/_2,^1/_2-y,z)$$

Determined atomic parameters are those of Table XVB,90 for the sulfide and Table XVB,91 for the selenide.

TABLE XVB,90

Parameters of the Atoms in Bis(phenylsulfonyl) Sulfide

Atom	x	y	z
C(1)	0.130	0.150	0.325
CH(2)	0.084	−0.025	0.355
CH(3)	0.079	0.004	0.440
CH(4)	0.116	0.205	0.492
CH(5)	0.161	0.382	0.463
CH(6)	0.168	0.355	0.378
S(1)	0.250	−0.109	0.250
S(2)	0.137	0.116	0.218
O(1)	0.067	−0.045	0.166
O(2)	0.147	0.333	0.175

TABLE XVB,91

Parameters of the Atoms in Bis(phenylsulfonyl) Selenide

Atom	x	y	z
C(1)	0.130	0.158	0.327
C(2)	0.090	0.980	0.358
C(3)	0.083	0.008	0.443
C(4)	0.120	0.209	0.495
C(5)	0.162	0.388	0.466
C(6)	0.168	0.360	0.382
Se	$1/4$	0.871	$1/4$
S	0.134	0.110	0.219
O(1)	0.062	0.959	0.167
O(2)	0.153	0.323	0.183

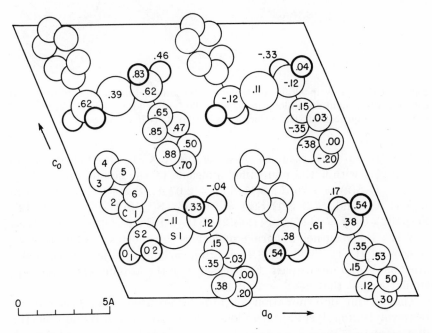

Fig. XVB,134a. The monoclinic structure of bis(phenylsulfonyl) sulfide projected along its b_0 axis. The two kinds of sulfur atom are represented by circles of large, but different, sizes. Left-hand axes.

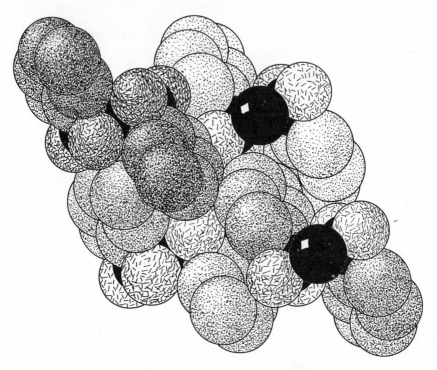

Fig. XVB,134b. A packing drawing of the monoclinic bis(phenylsulfonyl) sulfide arrangement seen along its b_0 axis. The sulfur atoms are black, those of oxygen are line shaded. The carbons are dotted. Left-hand axes.

The structure that results is shown in Figure XVB,134. Bond distances and angles within the nonplanar molecule of the sulfide are: C–C = 1.39 A., C–S(2) = 1.76 A., S(1)–S(2) = 2.07 A., S(2)–O(1) = S(2)–O(2) = 1.41 A., and S(2)–S(1)–S(2) = 106°30′, S(1)–S(2)–O(1) = 104°12′, S(1)–S(2)–O(2) = 111°54′, S(1)–S(2)–C(1) = 101°42′, O(2)–S(2)–C(1) = 115°54′. Between neighbouring molecules the shortest atomic separations are CH–O = 3.33 and 3.43 A.; ring carbons of adjacent molecules are 3.68 A. and more apart. For the selenide the atomic separations are similar except that Se–S = 2.20 A.

Though no determination of atomic positions has been made, bis(p-tolylsulfonyl) sulfide is undoubtedly isostructural. Its unit has the dimensions:

$$a_0 = 16.50 \text{ A.;} \quad b_0 = 5.85 \text{ A.;} \quad c_0 = 18.88 \text{ A.;} \quad \beta = 119°54′$$

V. Compounds in Which the Rings Are Separated by a Chain of Four or More Atoms

XV,bV,1. *Diphenyldiacetylene*, C_6H_5—$C{\equiv}C$—$C{\equiv}C$—C_6H_5, has an elongated bimolecular monoclinic unit of the dimensions:

$$a_0 = 6.61 \text{ A.}; \quad b_0 = 6.04 \text{ A.}; \quad c_0 = 14.92 \text{ A.}; \quad \beta = 105°$$

The space group is C_{2h}^5 ($P2_1/c$), this centering on the b_0c_0 plane being responsible for its long c_0 axis. All atoms are in the positions:

$$(4e) \quad \pm (xyz; x,{}^1/_2-y,z+{}^1/_2)$$

with the parameters of Table XVB,92.

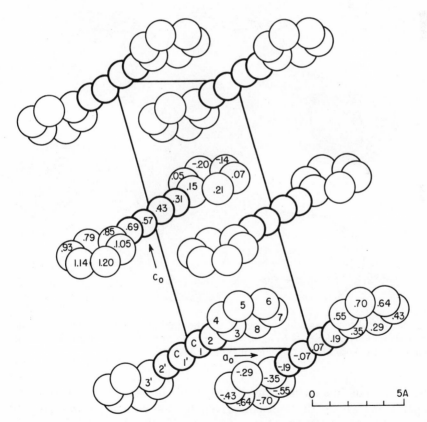

Fig. XVB,135a. The monoclinic structure of diphenyldiacetylene projected along its b_0 axis. Left-hand axes.

The center of symmetry of the molecule, in the origin, lies midway along the central single bond. As Figure XVB,135 shows, the planar molecules are in sheets with their long axes only moderately tilted from

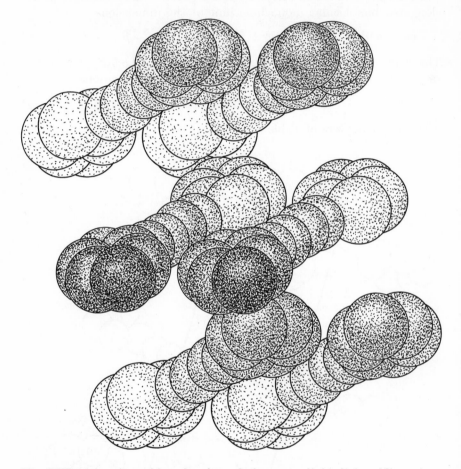

Fig. XVB,135b. A packing drawing of the monoclinic diphenyldiacetylene arrangement seen along its b_0 axis. The non-ring atoms of carbon are somewhat smaller than those of the benzene rings. Left-hand axes.

the a_0b_0 plane but inclined more steeply to the a_0c_0 plane. The atomic separations within the molecule given by the chosen parameters are those of Figure XVB,136; between molecules the shortest interatomic distances lie between 3.6 and 3.7 A.

TABLE XVB,92

Parameters of the Atoms in Diphenyldiacetylene

Atom	x	y	z
C(1)	0.080	0.072	0.021
C(2)	0.219	0.195	0.055
C(3)	0.384	0.346	0.094
C(4)	0.336	0.551	0.126
C(5)	0.497	0.697	0.165
C(6)	0.707	0.638	0.173
C(7)	0.757	0.433	0.142
C(8)	0.594	0.287	0.102

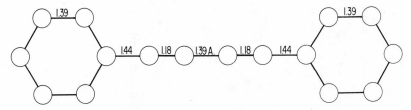

Fig. XVB,136.   Bond lengths in the molecule of diphenyldiacetylene.

XV,bV,2.   Crystals of *1,2-diphenoxyethane*, $(C_6H_5OCH_2)_2$, are ortho-rhombic. Their large unit, containing eight molecules, has the edge lengths:

$$a_0 = 34.74 \text{ A.}; \quad b_0 = 12.04 \text{ A.}; \quad c_0 = 5.58 \text{ A.}$$

TABLE XVB,93

Parameters of the Atoms in 1,2-Diphenoxyethane

Atom	x	y	z
O	−0.0311	0.0782	−0.003
C(1)	−0.0923	0.3731	0.935
C(2)	−0.0953	0.2905	0.762
C(3)	−0.0737	0.1932	0.786
C(4)	−0.0499	0.1783	0.985
C(5)	−0.0456	0.2630	0.154
C(6)	−0.0678	0.3602	0.129
C(7)	−0.0064	0.0607	0.203

The space group is C_{2v}^{19} ($Fdd2$) with all atoms in the positions:

(16b) xyz; $\bar{x}\bar{y}z$; $^1/_4-x,y+^1/_4,z+^1/_4$; $x+^1/_4,^1/_4-y,z+^1/_4$;   F.C.

The parameters are listed in Table XVB,93.

The resulting structure is shown in Figure XVB,137. It is built up of molecules with the bond dimensions of Figure XVB,138. They have twofold axes which pass through the midpoint of the central C(7)–C(7′)

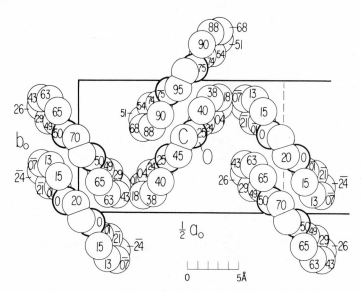

Fig. XVB,137.   The contents of half the large orthorhombic unit of 1,2-diphenoxyethane projected along its c_0 axis.

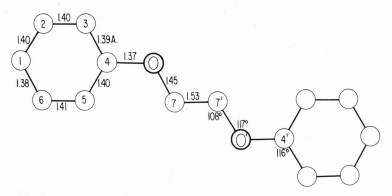

Fig. XVB,138.   Bond lengths in the molecule of 1,2-diphenoxyethane.

bonds. The C(7) atoms are 0.11 A. outside the planes of the benzene rings, and around the central bond the two halves of the molecule are twisted through 67°.

XV,bV,3. The monoclinic crystals of *p,p'-dichlorodiphenoxy-1,2-ethane*, $(ClC_6H_4OCH_2)_2$, possess a tetramolecular unit of the dimensions:

$$a_0 = 12.79 \text{ A.}; \quad b_0 = 9.89 \text{ A.}; \quad c_0 = 10.37 \text{ A.}; \quad \beta = 98°12'$$

All atoms are in the positions:

$$(4g) \quad \pm(xyz; x,\bar{y},z+{}^1/_2)$$

of the space group C_{2h}^4 $(P2/c)$. The parameters are given in Table XVB,94.

TABLE XVB,94

Parameters of the Atoms in $[ClC_6H_4OCH_2]_2$

Atom	x	y	z
Cl(1)	0.0876	0.0866	0.1455
Cl(2)	0.4169	0.3385	0.3635
O(1)	0.4438	0.1536	−0.1364
O(2)	0.0591	0.3980	0.6464
C(1)	0.1911	0.1151	0.0575
C(2)	0.2898	0.0645	0.1160
C(3)	0.3725	0.0848	0.0462
C(4)	0.3555	0.1477	−0.0780
C(5)	0.2582	0.1914	−0.1328
C(6)	0.1750	0.1745	−0.0620
C(7)	0.4482	0.2620	−0.2331
C(8)	0.3117	0.3614	0.4495
C(9)	0.2358	0.2579	0.4416
C(10)	0.1531	0.2739	0.5099
C(11)	0.1459	0.3909	0.5864
C(12)	0.2219	0.4923	0.5959
C(13)	0.3061	0.4751	0.5223
C(14)	0.0524	0.5145	0.7289

In the structure, as shown in Figure XVB,139, two crystallographically independent molecules have approximately the same dimensions as the molecule of 1,2-diphenoxyethane (**XV,bV,2**). Between molecules are short intermolecular separations of Cl–Cl = 3.33 and 3.39 A. and C–O = 3.37 A.

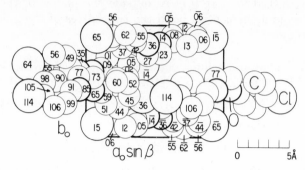

Fig. XVB,139. The monoclinic structure of p-p'-dichlorodiphenoxy-1,2-ethane projected along its c_0 axis.

XV,bV,4. *Dibenzoyl peroxide*, $[C_6H_5C(O)]_2O_2$, forms orthorhombic crystals. Its tetramolecular cell has the edge lengths:

$$a_0 = 8.95(1) \text{ A.}; \quad b_0 = 14.24(1) \text{ A.}; \quad c_0 = 9.40(2) \text{ A.}$$

TABLE XVB,95

Parameters of the Atoms in Dibenzoyl Peroxide

Atom	x	y	z
C(1)	0.1227(12)	−0.2761(6)	0.3772(13)
C(2)	0.2133(13)	−0.2311(6)	0.4703(13)
C(3)	0.2279(13)	−0.1350(7)	0.4749(12)
C(4)	0.1472(10)	−0.0828(6)	0.3788(9)
C(5)	0.0537(12)	−0.1284(7)	0.2808(12)
C(6)	0.0460(12)	−0.2231(6)	0.2808(12)
C(7)	0.1568(10)	0.0209(6)	0.3695(11)
C(8)	0.3585(11)	0.1893(6)	0.3937(11)
C(9)	0.4565(13)	0.4339(8)	0.2917(15)
C(10)	0.4507(12)	0.3370(6)	0.2982(14)
C(11)	0.3585(9)	0.2921(6)	0.3918(10)
C(12)	0.2664(12)	0.3453(8)	0.4844(13)
C(13)	0.2767(14)	0.4427(7)	0.4741(15)
C(14)	0.3654(12)	0.4849(6)	0.3789(13)
O(15)	0.2454(9)	0.0535(4)	0.4780(8)
O(16)	0.0973(8)	0.0698(4)	0.2868(8)
O(17)	0.2444(9)	0.1553(4)	0.4750(9)
O(18)	0.4394(8)	0.1391(5)	0.3289(10)

The space group is V^4 ($P2_12_12_1$) with atoms in the positions:

(4a) xyz; $1/2-x,\bar{y},z+1/2$; $x+1/2,1/2-y,\bar{z}$; $\bar{x},y+1/2,1/2-z$

The parameters are those of Table XVB,95; hydrogen positions are stated in the original article.

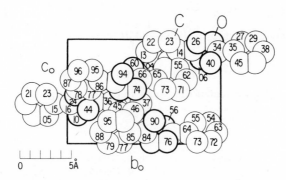

Fig. XVB,140.   The orthorhombic structure of dibenzoyl peroxide projected along its a_0 axis.

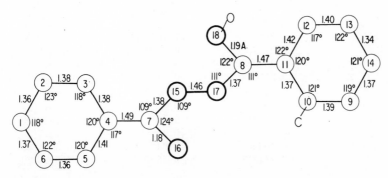

Fig. XVB,141.   Bond dimensions in the molecule of dibenzoyl peroxide.

The structure is illustrated in Figure XVB,140. Its molecules have the bond dimensions of Figure XVB,141. The two phenylcarboxyl ends are planar; their planes intersect at an angle of 84° in a line nearly parallel to the peroxide bond. Neither C(7), O(16), O(15), O(17) nor C(8), O(18), O(17), O(15) lie exactly in a plane; instead, the bond involving O(17) is twisted by 2.5° out of the plane of C(7)–O(15)–O(16) and that of O(15), by 4.2° out of the C(8)–O(17)–O(18) plane.

XV,bV,5. Crystals of *4,4'-dichlorodibenzoyl peroxide*, $[C_6H_4ClC(O)]_2O_2$, are monoclinic with a tetramolecular unit of the dimensions:

$$a_0 = 25.47(5) \text{ A.;} \quad b_0 = 7.80(2) \text{ A.;} \quad c_0 = 6.85(2) \text{ A.}$$
$$\beta = 98.60(17)°$$

Atoms are in the following positions of C_{2h}^5 $(P2_1/a)$:

$$(4e) \quad \pm (xyz; x+\tfrac{1}{2}, \tfrac{1}{2}-y, z)$$

The parameters are listed in Table XVB,96.

TABLE XVB,96

Parameters of the Atoms in $[C_6H_4ClC(O)]_2O_2$

Atom	x	y	z
Cl(1)	0.4036(1)	0.5704(6)	0.7736(7)
Cl(2)	−0.0233(2)	0.7084(7)	−0.8509(9)
O(1)	0.2315(4)	0.5346(16)	−0.0477(14)
O(2)	0.1851(4)	0.5238(14)	−0.2045(19)
O(3)	0.1839(3)	0.3827(12)	0.1342(14)
O(4)	0.1501(4)	0.7278(15)	−0.0296(18)
C(1)	0.2698(5)	0.4901(17)	0.2733(27)
C(2)	0.3147(5)	0.5865(17)	0.2468(27)
C(3)	0.3539(5)	0.6105(18)	0.4018(32)
C(4)	0.3532(5)	0.5361(19)	0.5894(28)
C(5)	0.3095(5)	0.4399(20)	0.6248(24)
C(6)	0.2684(5)	0.4081(20)	0.4682(30)
C(7)	0.2200(6)	0.4589(21)	0.1017(35)
C(8)	0.1062(4)	0.6459(18)	−0.3484(29)
C(9)	0.1113(5)	0.5655(22)	−0.5295(20)
C(10)	0.0708(5)	0.5884(20)	−0.6869(24)
C(11)	0.0287(5)	0.6786(22)	−0.6551(24)
C(12)	0.0221(5)	0.7548(22)	−0.4656(31)
C(13)	0.0615(5)	0.7465(22)	−0.3145(22)
C(14)	0.1483(5)	0.6363(21)	−0.1685(25)

The structure (Fig. XVB,142) is composed of molecules with the bond lengths of Figure XVB,143. Between molecules the shortest atomic separation is O(4)–C(5) = 3.26 A.

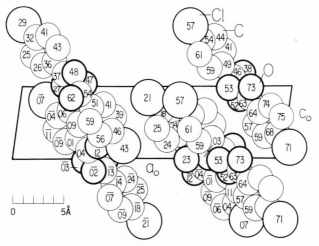

Fig. XVB,142.   The monoclinic structure of 4,4′-dichlorodibenzoyl peroxide projected along its b_0 axis.

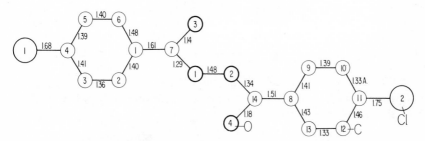

Fig. XVB,143.   Bond lengths in the molecule of 4,4′-dichlorodibenzoyl peroxide.

The corresponding bromine derivative $[C_6H_4BrC(O)]_2O_2$ is isostructural with a unit of the dimensions:

$$a_0 = 25.94(5) \text{ A.}; \quad b_0 = 7.91(2) \text{ A.}; \quad c_0 = 6.83(2) \text{ A.}$$
$$\beta = 96.83(17)°$$

Parameters have not been established.

XV,bV,6.   According to a preliminary announcement, crystals of *benzalazine*, $(C_6H_5CH{=}N)_2$, are orthorhombic with a tetramolecular unit of the edge lengths:

$$a_0 = 13.09 \text{ A.}; \quad b_0 = 11.76 \text{ A.}; \quad c_0 = 7.62 \text{ A.}$$

There is confusion in the statement of the space group, but it appears to have been chosen as V_h^{14} (*Pbcn*), in which case atoms are in the positions:

$$(8d) \quad \pm (xyz; x+\tfrac{1}{2}, y+\tfrac{1}{2}, \tfrac{1}{2}-z; \tfrac{1}{2}-x, y+\tfrac{1}{2}, z; x, \bar{y}, z+\tfrac{1}{2})$$

The parameters were given the values listed in Table XVB,97. A further refinement seems not to have been published.

TABLE XVB,97

Parameters of the Atoms in Benzalazine

Atom	x	y	z
N	0.053	0.011	0.009
C(1)	0.095	0.085	−0.043
C(2)	0.205	0.108	−0.022
C(3)	0.250	0.205	−0.087
C(4)	0.352	0.227	−0.067
C(5)	0.413	0.150	0.019
C(6)	0.370	0.053	0.083
C(7)	0.267	0.031	0.064

XV,bV,7. Crystals of *N-phenyl-N'-benzoylselenourea*, $C_6H_5NH \cdot C(Se) \cdot NH \cdot C(O) \cdot C_6H_5$, are monoclinic. Their tetramolecular unit has the dimensions:

$$a_0 = 13.16(1) \text{ A.}; \quad b_0 = 5.064(3) \text{ A.}; \quad c_0 = 19.94(1) \text{ A.}$$
$$\beta = 103.65(3)°$$

The space group is C_{2h}^5 (*P2₁/c*) with atoms in the positions:

$$(4e) \quad \pm (xyz; x, \tfrac{1}{2}-y, z+\tfrac{1}{2})$$

The parameters, including those for hydrogen, are given in Table XVB,98.

The structure is illustrated in Figure XVB,144. Its molecules, with the bond dimensions of Figure XVB,145, have been described in terms of three planar sections. One involves atoms Se, C(2), N(3) and the ring C(4)-to-C(9), another N(10), C(11), C(12) and O(18), and the third the benzene ring C(12)-to-C(17). The atom C(11) is 0.04 A. outside the last ring which is twisted by 31° about the C(11)–C(12) bond. Within the molecule the O(18)–N(3) distance is a notably short 2.59 A. Between

TABLE XVB,98

Parameters of the Atoms in N-Phenyl-N'-benzoylselenourea

Atom	x	y	z
Se(1)	0.1077(1)	0.2808(2)	0.5847(5)
C(2)	0.2096(7)	0.3771(17)	0.5416(5)
N(3)	0.3059(6)	0.2864(16)	0.5513(4)
C(4)	0.3630(7)	0.0872(19)	0.5952(5)
C(5)	0.3250(9)	−0.0692(23)	0.6406(6)
C(6)	0.3916(10)	−0.2584(25)	0.6793(6)
C(7)	0.4901(9)	−0.2926(24)	0.6721(6)
C(8)	0.5279(11)	−0.1365(28)	0.6253(7)
C(9)	0.4632(9)	0.0529(24)	0.5876(6)
N(10)	0.1818(6)	0.5676(15)	0.4897(4)
C(11)	0.2410(9)	0.6517(19)	0.4443(5)
C(12)	0.1880(7)	0.8354(19)	0.3893(5)
C(13)	0.2188(9)	0.8229(23)	0.3270(6)
C(14)	0.1765(9)	0.9909(24)	0.2741(6)
C(15)	0.1011(9)	0.1731(24)	0.2814(6)
C(16)	0.0688(9)	0.1817(22)	0.3427(5)
C(17)	0.1127(7)	0.0136(20)	0.3966(5)
O(18)	0.3295(6)	0.5724(17)	0.4485(4)
H(3)	0.358	0.403	0.526
H(5)	0.240	−0.040	0.643
H(6)	0.357	−0.383	0.720
H(7)	0.529	−0.420	0.715
H(8)	0.602	−0.153	0.598
H(9)[a]	(0.492)	(0.176)	(0.552)
H(10)	0.111	0.670	0.492
H(13)	0.277	0.711	0.323
H(14)	0.203	0.939	0.227
H(15)	0.083	0.294	0.240
H(16)	0.011	0.254	0.339
H(17)	0.090	0.041	0.437

[a] Values in parentheses are assumed.

molecules the comparatively short Se–N(10) = 3.83 A. plus an available hydrogen is thought to point to a hydrogen bond. The C–H distances range between 0.83 and 1.24 A., with an average of 1.06 A.

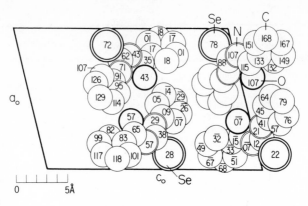

Fig. XVB,144. The monoclinic structure of N-phenyl-N'-benzoylselenourea projected along its b_0 axis.

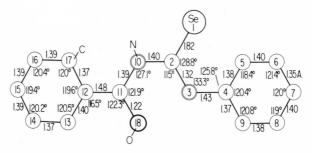

Fig. XVB,145. Bond dimensions in the molecule of N-phenyl-N'-benzoyl-selenourea.

XV,bV,8. Crystals   of   *α,4-dibromo-α-(4-methyl-2-nitrophenylazo)-acetanilide*, $C_{15}H_{12}N_4O_3Br_2$, are monoclinic with a tetramolecular unit of the dimensions:

$$a_0 = 19.54 \text{ A.}; \quad b_0 = 15.19 \text{ A.}; \quad c_0 = 5.61 \text{ A.}; \quad \beta = 100°30'$$

The space group is C_{2h}^5 ($P2_1/a$) with atoms in the positions:

$$(4e) \quad \pm(xyz; \ x+\tfrac{1}{2}, \tfrac{1}{2}-y, z)$$

The parameters are those of Table XVB,99.

The structure (Fig. XVB,146) is built up of molecules that are nearly planar, the maximum departure of any atom from the plane covering the entire molecule being less than 0.20 A. Their bond dimensions are those of Figure XVB,147.

TABLE XVB,99

Parameters of the Atoms in $C_{15}H_{12}N_4O_3Br_2$

Atom	x	y	z
C(1)	0.4229	0.3555	0.2963
C(2)	0.4571	0.3714	0.0911
C(3)	0.4828	0.3014	−0.0200
C(4)	0.4715	0.2184	0.0387
C(5)	0.4391	0.2000	0.2422
C(6)	0.4119	0.2706	0.3610
C(7)	0.4071	0.5095	0.3973
C(8)	0.3701	0.5693	0.5573
C(9)	0.2590	0.5615	0.9928
C(10)	0.2521	0.4659	0.0306
C(11)	0.2082	0.4375	0.1875
C(12)	0.1710	0.4911	0.3186
C(13)	0.1811	0.5826	0.2910
C(14)	0.2226	0.6133	0.1361
C(15)	0.1252	0.4597	0.4967
N(1)	0.3951	0.4236	0.4219
N(2)	0.3340	0.5365	0.7046
N(3)	0.3018	0.5927	0.8399
N(4)	0.2265	0.7126	0.1086
O(1)	0.4403	0.5469	0.2596
O(2)	0.2570	0.7406	0.9611
O(3)	0.1979	0.7572	0.2474
Br(1)	0.5116	0.1242	−0.1040
Br(2)	0.3799	0.6909	0.5111

XV,bV,9. *Dibenzyl phosphoric acid*, $(C_6H_5CH_2)_2PO_4H$, forms monoclinic crystals whose tetramolecular unit has the dimensions:

$$a_0 = 20.287(40) \text{ A.}; \quad b_0 = 5.709(10) \text{ A.}; \quad c_0 = 12.648(20) \text{ A.}$$
$$\beta = 103°13(10)'$$

All atoms are in the positions

$$(4e) \quad \pm(xyz; x+{}^1/_2, {}^1/_2-y, z)$$

of the space group C_{2h}^5 ($P2_1/a$). The determined parameters are stated in Table XVB,100. Assumed values for the hydrogen atoms are given in the original article.

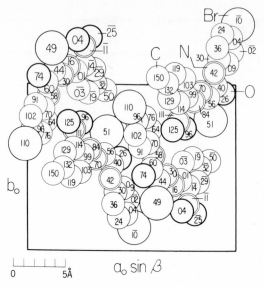

Fig. XVB,146. The monoclinic structure of α,4-dibromo-α-(4-methyl-2-nitro-
phenylazo)-acetanilide projected along its c_0 axis.

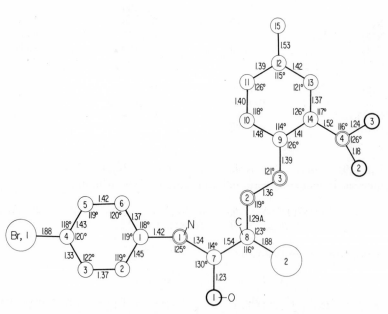

Fig. XVB,147. Bond dimensions in the molecule of α,4-dibromo-α-(4-methyl-
2-nitrophenylazo)-acetanilide.

TABLE XVB,100

Parameters of the Atoms in $(C_6H_5CH_2)_2PO_4H$

Atom	x	y	z
P	0.2896	0.3560	0.3739
O(1)	0.2517	0.1653	0.4108
O(2)	0.3165	0.5535	0.4560
O(3)	0.2449	0.4620	0.2695
O(4)	0.3539	0.2674	0.3371
C(1)	0.2646	0.6725	0.2180
C(2)	0.2072	0.7357	0.1248
C(3)	0.1702	0.9356	0.1298
C(4)	0.1164	0.9897	0.0412
C(5)	0.1006	0.8436	−0.0487
C(6)	0.1392	0.6543	−0.0516
C(7)	0.1908	0.5953	0.0329
C(8)	0.4040	0.1274	0.4128
C(9)	0.4620	0.0814	0.3610
C(10)	0.5148	0.2420	0.3697
C(11)	0.5671	0.1978	0.3180
C(12)	0.5689	−0.0020	0.2625
C(13)	0.5172	−0.1633	0.2542
C(14)	0.4641	−0.1227	0.3043

The structure is shown in Figure XVB,148. Its molecules have the bond dimensions of Figure XVB,149.

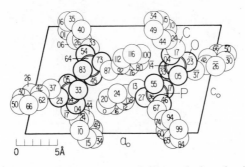

Fig. XVB,148.   The monoclinic structure of dibenzyl phosphoric acid projected along its b_0 axis.

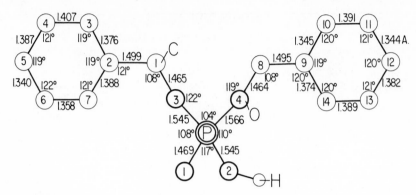

Fig. XVB,149. Bond dimensions in the molecule of dibenzyl phosphoric acid.

XV,bV,10. *Tellurium catecholate*, $Te(C_6H_4O_2)_2$, is monoclinic with a unit that contains eight molecules and has the dimensions:

$$a_0 = 22.678(3) \text{ A.}; \quad b_0 = 6.9355(8) \text{ A.}; \quad c_0 = 16.5274(15) \text{ A.}$$
$$\beta = 123.154(5)°$$

The space group is C_{2h}^6 ($C2/c$) with all atoms in the positions:

$$(8f) \quad \pm(xyz; \; x,\bar{y},z+{}^1/_2; \; x+{}^1/_2,y+{}^1/_2,z; \; x+{}^1/_2,{}^1/_2-y,z+{}^1/_2)$$

The determined parameters are listed in Table XVB,101; calculated values for the hydrogen atoms are stated in the original paper.

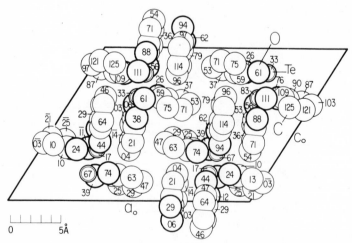

Fig. XVB,150. The monoclinic structure of tellurium catecholate projected along its b_0 axis.

TABLE XVB,101

Parameters of the Atoms in Te($C_6H_4O_2$)$_2$

Atom	x	y	z
Te	0.2389(1)	0.1688(3)	0.1762(1)
O(1)	0.2333(12)	0.441(4)	0.1315(18)
O(2)	0.1890(10)	0.122(4)	0.0353(13)
O(3)	0.1564(11)	0.241(3)	0.1807(14)
O(4)	0.3011(10)	0.390(3)	0.3319(13)
C(1)	0.3283(13)	0.208(5)	0.0184(18)
C(2)	0.3655(17)	0.207(6)	0.1168(23)
C(3)	0.3813(17)	0.035(7)	0.1630(24)
C(4)	0.1424(21)	0.356(7)	0.3874(28)
C(5)	0.1812(19)	0.358(7)	0.4873(25)
C(6)	0.1948(13)	0.467(5)	0.0351(18)
C(7)	0.1095(15)	0.095(6)	0.1553(21)
C(8)	0.0439(17)	0.133(6)	0.1406(23)
C(9)	0.4984(19)	0.469(7)	0.1135(24)
C(10)	0.4826(17)	0.293(6)	0.3959(23)
C(11)	0.4145(17)	0.254(5)	0.3817(22)
C(12)	0.3686(15)	0.410(5)	0.3507(19)

The structure, shown in Figure XVB,150, has molecules with the bond dimensions of Figure XVB,151. Both catecholate groups are planar, their planes making 99° with one another. The tellurium atom lies in the plane containing the C(1) atom but is 0.37 A. from the plane containing C(7). Besides the four oxygens of the molecule coordinated with tellurium, there is a fifth oxygen [O(4) from another molecule] that is 2.64 A. away.

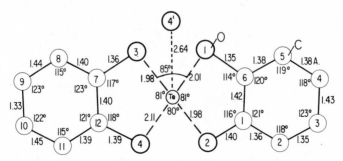

Fig. XVB,151.   Bond dimensions in the molecule of tellurium catecholate.

XV,bV,11. Crystals of *tellurium dibenzenethiosulfonate*, $Te(C_6H_5-S_2O_2)_2$, are orthorhombic with a tetramolecular unit of the edge lengths:

$$a_0 = 14.46 \text{ A.}; \quad b_0 = 11.18 \text{ A.}; \quad c_0 = 10.48 \text{ A.}$$

The tellurium atoms are in

$$(4c) \quad \pm(0\ u\ ^1/_4;\ ^1/_2, u + ^1/_2, ^1/_4) \qquad \text{with } u = 0.043$$

of the space group V_h^{14} (*Pbcn*). All the other atoms are in

$$(8d) \quad \pm(xyz;\ x + ^1/_2, y + ^1/_2, ^1/_2 - z;\ ^1/_2 - x, y + ^1/_2, z;\ x, \bar{y}, z + ^1/_2)$$

with the parameters of Table XVB,102.

TABLE XVB,102

Parameters of the Atoms in $Te(C_6H_5S_2O_2)_2$

Atom	x	y	z
S(2)	−0.068	0.186	0.395
S(1)	0.041	0.233	0.514
O(1)	−0.006	0.316	0.598
O(2)	0.083	0.139	0.566
C(1)	0.117	0.312	0.413
C(2)	0.207	0.271	0.396
C(3)	0.268	0.316	0.316
C(4)	0.238	0.435	0.253
C(5)	0.148	0.476	0.270
C(6)	0.087	0.414	0.349

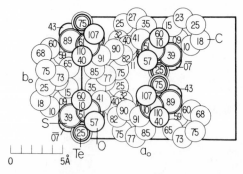

Fig. XVB,152.   The orthorhombic structure of tellurium dibenzenethiosulfonate projected along its c_0 axis.

The structure is shown in Figure XVB,152. In its molecules Te–S(2) = 2.41 A., S(2)–S(1) = 2.08 A., S–O = 1.44 A., and S–C = 1.77 A. The angles S(2)–Te–S(2) = 97° and Te–S(2)–S(1) = 104°. Positions given the carbon atoms were based on the assumption of a regular benzene ring (C–C = 1.39 A.).

XV,bV,12. Crystals of *tellurium di-p-tolylthiosulfonate*, Te-$(CH_3C_6H_4S_2O_2)_2$, are tetragonal with a unit that contains eight molecules and has the edge lengths:

$$a_0 = 10.93 \text{ A.,} \quad c_0 = 29.88 \text{ A.}$$

The space group is D_4^4 in the centered setting $C4_122_1$. Atoms in this atypical cell have been given the approximate parameters of Table XVB,103.

TABLE XVB,103

Parameters of the Atoms in $Te(CH_3C_6H_4S_2O_2)_2$

Atom	x	y	z
Te	0	0.136	0
S(1)	0.145	0.283	−0.028
S(2)	0.253	0.322	0.029
O(1)	0.365	0.361	0.009
O(2)	0.278	0.203	0.047
C(1)	0.154	0.413	0.062
C(2)	0.130	0.533	0.050
C(3)	0.052	0.604	0.075
C(4)	−0.003	0.555	0.113
C(5)	0.021	0.435	0.126
C(6)	0.099	0.363	0.100
C(7)	−0.089	0.634	0.105

XV,bV,13. The monoclinic crystals of *1,6-di-p-chlorophenyl-3,4-dimethyl hexatriene*, $C_{20}H_{18}Cl_2$, have a tetramolecular unit of the dimensions:

$$a_0 = 36.40(1) \text{ A.;} \quad b_0 = 4.184(2) \text{ A.;} \quad c_0 = 11.454(4) \text{ A.}$$
$$\beta = 107°57(1)'$$

Atoms are in the positions

(8*f*) $\pm (xyz; x,\bar{y},z+1/2; x+1/2,y+1/2,z; x+1/2,1/2-y,z+1/2)$

of the space group C_{2h}^6 ($C2/c$). The determined parameters are listed in Table XVB,104.

TABLE XVB,104

Parameters of the Atoms in 1,6-di-p-Chlorophenyl-3,4-dimethyl Hexatriene

Atom	x	y	z
Cl	0.0305	0.3823	0.1650
C(1)	0.1315	0.2092	0.0219
C(2)	0.0951	0.0877	−0.0399
C(3)	0.0633	0.1406	0.0025
C(4)	0.0693	0.3128	0.1096
C(5)	0.1050	0.4388	0.1746
C(6)	0.1359	0.3830	0.1295
C(7)	0.1636	0.1443	−0.0280
C(8)	0.1997	0.2544	0.0189
C(9)	0.2318	0.1946	−0.0289
C(10)	0.2220	0.0101	−0.1508

The resulting structure, shown in Figure XVB,153, is built of centrosymmetric and nearly planar molecules possessing the dimensions of Figure XVB,154. The shortest intermolecular separation is a Cl–Cl = 3.37 A.

This structure may be compared with that of the related ortho compound described in the next paragraph.

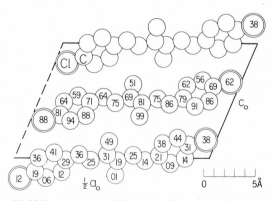

Fig. XVB,153a. Half the contents of the monoclinic unit of 1,6-di-p-chlorophenyl-3,4-dimethyl hexatriene projected along its b_0 axis.

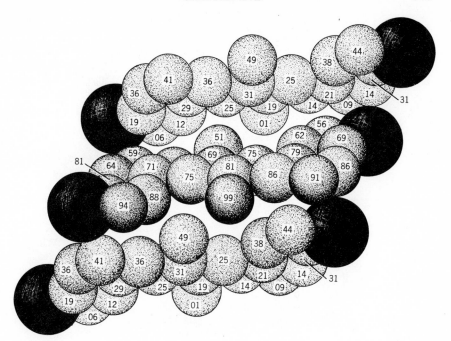

Fig. XVB,153b.   A packing drawing of the molecules in half the unit of the monoclinic 1,6-di-p-chlorophenyl-3,4-dimethyl hexatriene viewed along its b_0 axis. The chlorine atoms are black, the carbons dotted.

XV,bV,14.   Crystals of *1,6-di-o-chlorophenyl-3,4-dimethyl hexatriene*, $C_{20}H_{18}Cl_2$, are, like the para compound, **(XV,bV,13),** monoclinic but with a differently shaped cell. The bimolecular unit has the dimensions:

$$a_0 = 17.723(3) \text{ A.}; \quad b_0 = 6.264(1) \text{ A.}; \quad c_0 = 7.757(4) \text{ A.}$$
$$\beta = 102°58(2)'$$

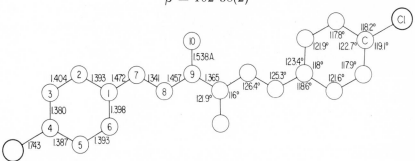

Fig. XVB,154.   Bond dimensions in the molecule of 1,6-di-p-chlorophenyl-3,4-dimethyl hexatriene.

All atoms are in the positions

$$(4e) \quad \pm (xyz; x + {}^1/_2, {}^1/_2 - y, z)$$

of C_{2h}^5 $(P2_1/a)$. The determined parameters are listed in Table XVB,105, positions for hydrogen not having been established.

TABLE XVB,105

Parameters of the Atoms in 1,6-di-o-Chlorophenyl-3,4-dimethyl Hexatriene

Atom	x	y	z
Cl	0.2022	0.7657	0.0971
C(1)	0.1251	0.5337	0.3009
C(2)	0.1702	0.7161	0.2892
C(3)	0.1900	0.8586	0.4257
C(4)	0.1699	0.8235	0.5841
C(5)	0.1247	0.6448	0.6038
C(6)	0.1032	0.5033	0.4662
C(7)	0.0995	0.3898	0.1486
C(8)	0.0512	0.2226	0.1466
C(9)	0.0238	0.0860	−0.0042
C(10)	0.0499	0.1435	−0.1721

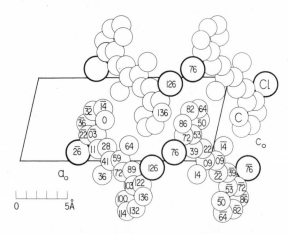

Fig. XVB,155a.   The monoclinic structure of 1,6-di-o-chlorophenyl-3,4-dimethyl hexatriene projected along its b_0 axis.

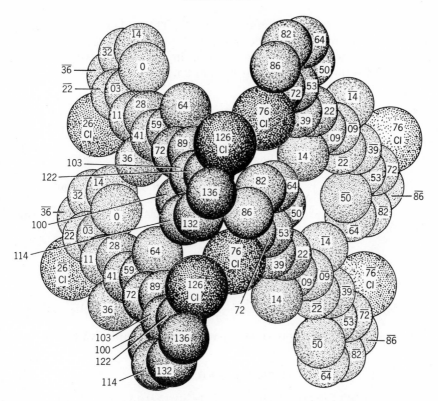

Fig XVB,155b. A packing drawing of the monoclinic 1,6-di-*o*-chlorophenyl-3,4-dimethyl hexatriene arrangement viewed along its b_0 axis. The larger, more coarsely dotted circles are chlorine, the more finely shaded circles are carbon.

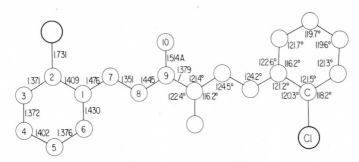

Fig. XVB,156.   Bond dimensions in the molecule of 1,6-di-*o*-chlorophenyl-3,4-dimethyl hexatriene.

The structure is that of Figure XVB,155. The molecules it contains, like those of the para compound, are almost planar and have a center of symmetry; bond dimensions are given in Figure XVB,156. Intermolecular distances range upwards from 3.65 A.

XV,bV,15. According to a preliminary announcement, *p-N, N-dimethylaminophenyldiazonium chlorozincate*, $[(CH_3)_2C_6H_3(NH_2)N_2]_2$-$ZnCl_4$, has monoclinic symmetry. The unit, using γ rather than the usual β as principal axis, has the dimensions:

$$a_0 = 14.25 \text{ A.;} \quad b_0 = 15.30 \text{ A.;} \quad c_0 = 10.65 \text{ A.;} \quad \gamma = 90°$$

The space group C_{2h}^6 in the orientation $B2/b$ has been assigned to this tetramolecular cell. Atoms of zinc, then, are in the positions:

$$(4e) \quad \pm(0 \; {}^1/_4 \; u; \; {}^1/_2, {}^1/_4, u + {}^1/_2) \qquad \text{with } u = 0.112$$

All other atoms, with the coordinates

$$(8e) \quad \pm(xyz; \; x,y + {}^1/_2, \bar{z}; \; x + {}^1/_2, y, z + {}^1/_2; \; x + {}^1/_2, y + {}^1/_2, {}^1/_2 - z)$$

have been given the approximate parameters of Table XVB,106.

TABLE XVB,106

Parameters of Atoms in $[(CH_3)_2C_6H_3(NH_2)N_2]_2ZnCl_4$

Atom	x	y	z
Cl(1)	0.45	0.36	0.48
Cl(2)	0.12	0.30	0.24
NH₂	0.39	0.43	0.05
N(1)	0.31	0.15	0.35
N(2)	0.29	0.20	0.43
CH₃(1)	0.34	0.44	0.18
CH₃(2)	0.45	0.35	0.03
C(1)	0.30	0.06	0.03
C(2)	0.29	0.12	0.13
C(3)	0.33	0.10	0.24
C(4)	0.39	0.02	0.28
C(5)	0.09	0.03	0.33
C(6)	0.13	0.02	0.46

XV,bV,16. The orthorhombic crystals of *1,8-diphenyl-1,3,5,7-octatetraene*, $C_{20}H_{18}$, possess a tetramolecular unit of the dimensions:

$$a_0 = 10.062(10) \text{ A.}; \quad b_0 = 7.625(10) \text{ A.}; \quad c_0 = 19.925(10) \text{ A.}$$
$$\text{(at } 18°\text{C.)}$$
$$a_0 = 10.196(10) \text{ A.}; \quad b_0 = 7.504(10) \text{ A.}; \quad c_0 = 19.579(5) \text{ A.}$$
$$\text{(at } -100°\text{C.)}$$

The space group is V_h^{15} (*Pcab*) with atoms in the positions:

$$(8c) \quad \pm(xyz; x+1/2, 1/2-y, z; 1/2-x, y, z+1/2; x, y+1/2, 1/2-z)$$

Parameters are listed in Table XVB,107.

TABLE XVB,107

Parameters of the Atoms in Diphenyl Octatetraene

Atom	x	y	z
C(1)	0.6939	−0.0861	0.2232
C(2)	0.5677	−0.1517	0.2329
C(3)	0.4695	−0.1148	0.1859
C(4)	0.7226	0.0145	0.1650
C(5)	0.6237	0.0523	0.1181
C(6)	0.4954	−0.0106	0.1279
C(7)	0.3936	0.0358	0.0778
C(8)	0.2683	−0.0221	0.0772
C(9)	0.1731	0.0301	0.0269
C(10)	0.0476	−0.0271	0.0251
H(1)	0.766	−0.112	0.257
H(2)	0.543	−0.233	0.279
H(3)	0.382	−0.165	0.197
H(4)	0.811	0.070	0.153
H(5)	0.654	0.130	0.077
H(7)	0.423	0.134	0.036
H(8)	0.235	−0.112	0.115
H(9)	0.210	0.122	−0.010
H(10)	0.013	−0.119	0.064

The structure is shown in Figure XVB,157. Bond dimensions in its molecules are those of Figure XVB,158. The terminal benzene rings are planar, and there is a plane through all the carbon atoms of the chain.

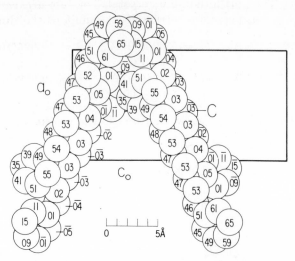

Fig. XVB,157.   The orthorhombic structure of 1,8-diphenyl-1,3,5,7-octatetraene projected along its b_0 axis.

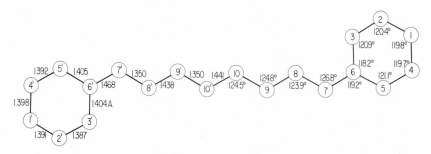

Fig. XVB,158.   Bond dimensions in the molecule of 1,8-diphenyl-1,3,5,7-octatetraene.

These planes make the small angle of 5.4° with one another. As the figure indicates, the molecules lie in layers normal to the b_0 axis of the crystal, with a shortest intermolecular distance of C–C = 3.57 A.

XV,bV,17.   The orthorhombic crystals of *N,N'-ethylene bis-(l-ephedrine) copper dihydrate*, $C_{22}H_{30}O_2N_2Cu \cdot 2H_2O$, have a tetramolecular unit of the edge lengths:

$$a_0 = 8.431(5) \text{ A.}; \quad b_0 = 15.92(3) \text{ A.}; \quad c_0 = 16.70(3) \text{ A.}$$

Atoms are in the positions

(4a) xyz; $1/2-x,\bar{y},z+1/2$; $x+1/2,1/2-y,\bar{z}$; $\bar{x},y+1/2,1/2-z$

of the space group V^4 ($P2_12_12_1$). The determined parameters are those of Table XVB,108.

TABLE XVB,108

Parameters of the Atoms in N,N'-Ethylene Bis(l-ephedrine) Copper Dihydrate

Atom	x	y	z
Cu	0.2219	0.0719	0.0119
O(1)	0.0145	0.0255	−0.0054
O(2)	0.1952	0.1182	0.1150
O(1,H$_2$O)	0.1389	0.6088	0.3841
O(2,H$_2$O)	0.1157	0.9589	0.1616
N(1)	0.2286	0.9488	0.3951
N(2)	0.4542	0.0856	0.0249
C(1)	0.3452	0.0192	0.3761
C(2)	0.2767	0.1063	0.4071
C(3)	0.5068	0.9945	0.4113
C(4)	0.2703	0.8729	0.3459
C(5)	0.3762	0.5659	0.0954
C(6)	0.2936	0.5736	0.1679
C(7)	0.1872	0.6421	0.1749
C(8)	0.1591	0.6965	0.1145
C(9)	0.2362	0.6880	0.0414
C(10)	0.3480	0.6195	0.0311
C(11)	0.0565	0.9646	0.3889
C(12)	0.0329	0.4015	0.0554
C(13)	0.4648	0.1629	0.0814
C(14)	0.4168	0.2448	0.0343
C(15)	0.3414	0.1441	0.1494
C(16)	0.5247	0.0099	0.0707
C(17)	0.3233	0.2183	0.2081
C(18)	0.1722	0.2570	0.2131
C(19)	0.1609	0.3260	0.2667
C(20)	0.2925	0.3519	0.3122
C(21)	0.4318	0.3109	0.3051
C(22)	0.4481	0.2448	0.2529

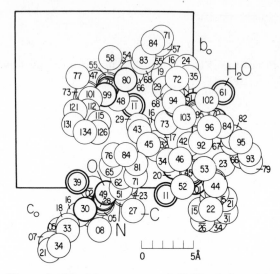

Fig. XVB,159.   The orthorhombic structure of N,N'-ethylene bis(l-ephedrine) copper dihydrate projected along its a_0 axis. Only one of the two water molecules appears. The atoms of copper are doubly ringed, the outer ring being the heavier.

The structure (Fig. XVB,159) is composed of molecules with the bond dimensions of Figure XVB,160. Coordination about the copper atom is approximately square. The water molecules, occurring in pairs, have

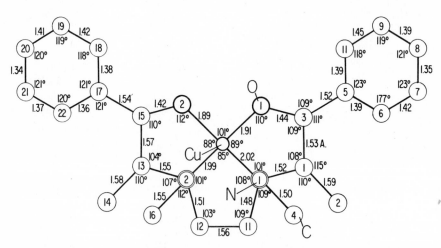

Fig. XVB,160.   Bond dimensions in the molecule of N,N'-ethylene bis(l-ephedrine) copper.

their oxygen atoms 3.30 A. apart. Each of these is also 2.74–3.10 A. away from two oxygen atoms of the molecule.

XV,bV,18. According to a preliminary announcement, *4,4'-dibromocinnamaldazine*, $C_{18}H_{14}N_2Br_2$, is triclinic with one molecule in a unit of the dimensions:

$$a_0 = 5.97(2) \text{ A.}; \quad b_0 = 16.94(3) \text{ A.}; \quad c_0 = 4.01(2) \text{ A.}$$
$$\alpha = 93(1)°; \quad \beta = 97(1)°; \quad \gamma = 100(1)°$$

All atoms are in the positions $(2i) \pm (xyz)$ of C_i^1 $(P\bar{1})$. The parameters are listed in Table XVB,109.

TABLE XVB,109

Parameters of the Atoms in $C_{18}H_{14}N_2Br_2$

Atom	x	y	z
Br	0.2376(10)	0.0567(3)	−0.3561(15)
C(1)	0.772(8)	0.286(3)	−0.285(12)
C(2)	0.573(9)	0.290(3)	−0.446(13)
C(3)	0.417(8)	0.223(3)	−0.470(13)
C(4)	0.454(8)	0.150(3)	−0.333(12)
C(5)	0.652(9)	0.143(3)	−0.172(13)
C(6)	0.808(8)	0.213(3)	−0.150(13)
C(7)	0.927(9)	0.362(3)	−0.266(13)
C(8)	0.122(8)	0.370(3)	−0.111(13)
C(9)	0.256(8)	0.449(3)	−0.122(12)
N	0.445(7)	0.460(2)	0.021(11)

This is a structure (Fig. XVB,161) made up of molecules that are essentially planar, the maximum departure from this plane being 0.05 A. for the atoms of bromine.

VI. Miscellaneous Compounds Containing Two Benzene Rings

The compounds described here contain two not directly connected benzene rings. Some are chelates that could more appropriately have been included in the first part of Chapter XV. They are, however, given in this volume for the sake of completeness.

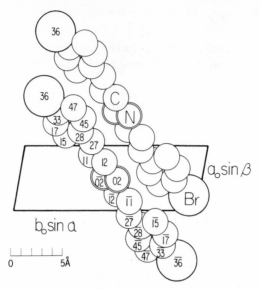

Fig. XVB,161. The triclinic structure of 4,4'-dibromocinnamaldazine projected along its c_0 axis.

TABLE XVB,110

Parameters of the Atoms in $C_6H_2(CN)_4 \cdot C_6(CH_3)_6$

Atom	x	y	z
N(1)	0.0788(12)	0.4064(18)	−0.0960(28)
N(2)	0.1908(10)	−0.2847(20)	0.2645(22)
C(1)	0.0547(11)	0.2925(22)	−0.0734(25)
C(2)	0.1349(12)	−0.2055(21)	0.1905(24)
C(3)	0.0924(11)	0.0457(19)	0.0596(21)
C(4)	0.0675(11)	−0.0984(20)	0.0949(25)
C(5)	0.0255(12)	0.1431(19)	−0.0396(22)
C(6)	0.1062(16)	0.2738(24)	0.4706(35)
C(7)	0.1006(17)	−0.2719(23)	0.6672(33)
C(8)	0.2020(13)	0.0032(36)	0.6394(34)
C(9)	0.0512(11)	0.1298(21)	0.4879(25)
C(10)	0.0461(13)	−0.1291(21)	0.5787(24)
C(11)	0.0953(11)	0.0001(22)	0.5655(26)

XV,bVI,1.   The complex *1,2,4,5-tetracyanobenzene-hexamethylben-zene*, $C_6H_2(CN)_4 \cdot C_6(CH_3)_6$, forms monoclinic crystals. Their bimolecular cells have the dimensions:

$$a_0 = 15.07(1) \text{ A.}; \quad b_0 = 8.92(1) \text{ A.}; \quad c_0 = 7.41(4) \text{ A.}; \quad \beta = 104.9(4)°$$

The space group C_{2h}^5 in the orientation $P2_1/n$ places all atoms in the positions:

$$(4e) \quad \pm (xyz; \, x + {}^1/_2, {}^1/_2 - y, z + {}^1/_2)$$

The determined parameters are recorded in Table XVB,110.

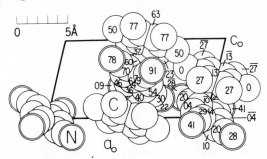

Fig. XVB,162.   The monoclinic structure of 1,2,4,5-tetracyanobenzene-hexa-methylbenzene projected along its b_0 axis.

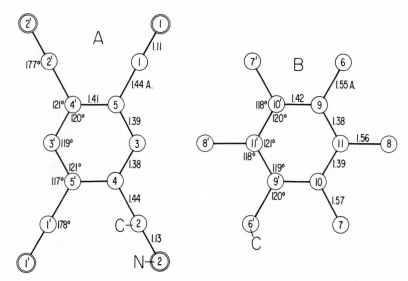

Fig. XVB,163.   Bond dimensions in the molecules of tetracyanobenzene (A) and hexamethylbenzene (B) in the complex crystals they form.

The structure is shown in Figure XVB,162. The two types of planar molecule that make it up have twofold axes of symmetry and the bond dimensions of Figure XVB,163. As Figure XVB,162 indicates, the molecules are nearly parallel to one another along the c_0 direction, with an angle between the planes of 2° and an interplanar distance of 3.54 A.

XV,bVI,2. Crystals of the complex *N,N,N',N'-tetramethyl-p-diaminobenzene-chloranil*, $C_6H_4[N(CH_3)_2]_2 \cdot C_6Cl_4O_2$, are monoclinic with a bimolecular unit of the dimensions:

$$a_0 = 16.320(15) \text{ A.}; \quad b_0 = 6.568(10) \text{ A.}; \quad c_0 = 8.811(10) \text{ A.}$$
$$\beta = 111.91(3)°$$

All atoms are in the special positions

$$(4i) \quad \pm (u0v; u+{}^1/_2, {}^1/_2, v)$$

of the space group C_{2h}^3 ($C2/m$). The values of u and v are listed in Table XVB,111.

TABLE XVB,111

Parameters of the Atoms in $C_6H_4[N(CH_3)_2]_2 \cdot C_6Cl_4O_2$

Atom	$u = x$	$v = z$
Cl(1)	$-0.06383(15)$	$0.2921(2)$
Cl(2)	$0.20180(13)$	$0.0717(3)$
O	$0.1184(3)$	$0.3110(5)$
N	$0.6309(4)$	$0.3100(7)$
C(1)	$-0.0309(4)$	$0.1288(7)$
C(2)	$0.0643(5)$	$0.1695(8)$
C(3)	$0.0899(4)$	$0.0267(8)$
C(4)	$0.4769(4)$	$0.1365(7)$
C(5)	$0.5667(4)$	$0.1582(7)$
C(6)	$0.5875(4)$	$0.0178(8)$
C(7)	$0.6101(6)$	$0.4573(8)$
C(8)	$0.7248(5)$	$0.3326(9)$
H(1)	0.459	0.243
H(2)	0.656	0.029
H(3)	0.671	0.564
H(4)	0.765	0.462
H(5)[a]	0.572	0.458
H(6)[a]	0.739	0.277

[a] For H(5) and H(6) [in positions ($8f$)], $y = 0.134$.

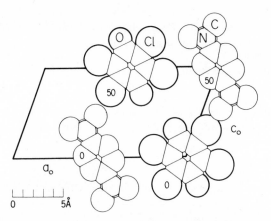

Fig. XVB,164. The monoclinic structure of the complex formed by N,N,N',N'-tetramethyl-p-diaminobenzene and chloranil projected along its b_0 axis.

The structure is shown in Figure XVB,164. Each molecular component is strictly planar [except for the H(5) and H(6) atoms] and has a twofold axis of symmetry. The bond dimensions are those of Figure XVB,165. The shortest distances between the two portions of the complex are Cl(2)–H(2) = 2.781 A., O–H(4) = 2.190 A., and H(5)–H(5) = 2.715 A.

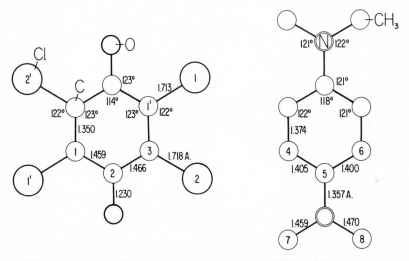

Fig. XVB,165. Bond dimensions of the molecules of N,N,N',N'-tetramethyl-p-diaminobenzene (at the left) and of chloranil (at the right) in their complex.

XV,bVI,3. Crystals of the complex between *N,N,N',N'-tetramethyl-p-phenylenediamine* and *1,2,4,5-tetracyanobenzene*, $C_6H_4[N(CH_3)_2]_2 \cdot C_6H_2(CN)_4$, are triclinic with a unimolecular cell of the dimensions:

$$a_0 = 7.654(12) \text{ A.}; \quad b_0 = 8.041(10) \text{ A.}; \quad c_0 = 7.462(13) \text{ A.}$$
$$\alpha = 96.7(1)°: \quad \beta = 85.9(1)°; \quad \gamma = 101.3(1)°$$

Atoms are in the positions $(2i) \pm (xyz)$ of the space group C_i^1 $(P\bar{1})$. The determined parameters are to be found in Table XVB,112.

TABLE XVB,112

Parameters of the Atoms in $C_6H_4[N(CH_3)_2]_2 \cdot C_6H_2(CN)_4$

Atom	x	y	z
Molecule $C_6H_4[N(CH)_3)_2]_2$			
C(1)	0.1238(13)	0.1386(11)	0.0773(11)
C(2)	0.1577(12)	−0.0242(11)	0.0555(12)
C(3)	0.0371(13)	−0.1607(11)	−0.0180(12)
N(1)	0.2472(11)	0.2775(10)	0.1617(11)
C(4)	0.4310(15)	0.2477(16)	0.1846(16)
C(5)	0.2273(16)	0.4503(12)	0.1284(16)
H(1)	0.238(14)	−0.066(13)	0.100(14)
H(2)	0.073(14)	−0.279(13)	−0.015(14)
Molecule $C_6H_2(CN)_4$			
C(6)	0.0689(11)	0.1725(9)	0.5358(10)
C(7)	0.1669(10)	0.0517(10)	0.5662(10)
C(8)	0.1006(11)	−0.1179(9)	0.5317(11)
C(9)	0.1375(12)	0.3522(10)	0.5747(12)
C(10)	0.3454(12)	0.1105(10)	0.6353(12)
N(2)	0.1913(12)	0.4911(9)	0.6101(12)
N(3)	0.4815(11)	0.1585(11)	0.6893(11)
H(3)	0.171(14)	−0.218(12)	0.554(14)

The structure is shown in Figure XVB,166. Its component centrosymmetric molecules have the bond dimensions of Figure XVB,167. The tetracyanobenzene molecule is planar and so is the other except for its methyl groups. In the crystal they are stacked alternately and nearly parallel to one another along the c_0 axis. Between them the shortest atomic separations are C–N = 3.15 A.

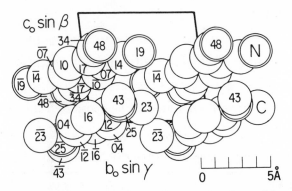

Fig. XVB,166.   The triclinic structure of the complex formed by N,N,N',N'-tetramethyl-p-phenylenediamine and 1,2,4,5-tetracyanobenzene projected along its a_0 axis.

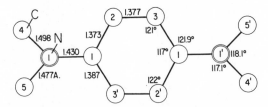

Fig. XVB,167a.   Bond dimensions in the molecule of N,N,N',N'-tetramethyl-p-phenylenediamine in its complex with 1,2,4,5-tetracyanobenzene.

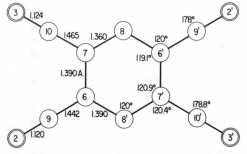

Fig. XVB,167b.   Bond dimensions in the molecule of 1,2,4,5-tetracyanobenzene in its complex with N,N,N',N'-tetramethyl-p-phenylenediamine.

XV,bVI,4.   According to a preliminary announcement, crystals of ($+$)-*methyl-n-propylphenylbenzyl-phosphonium bromide*, $[(CH_3)(C_3H_7)$-$(C_6H_5CH_2)(C_6H_5)]PBr$, are orthorhombic with a tetramolecular unit of the edge lengths:

$$a_0 = 12.85 \text{ A.}; \quad b_0 = 8.97 \text{ A.}; \quad c_0 = 14.66 \text{ A.}$$

The space group is V^4 $(P2_12_12_1)$ with atoms in the positions:

(4a) xyz; $^1/_2 - x, \bar{y}, z + ^1/_2$; $x + ^1/_2, ^1/_2 - y, \bar{z}$; $\bar{x}, y + ^1/_2, ^1/_2 - z$

The parameters have been stated to be those of Table XVB,113.

TABLE XVB,113

Parameters of the Atoms in $[(CH_3)(C_3H_7)(C_6H_5CH_2)(C_6H_5)]PBr$

Atom	x	y	z
Br	0.0776	0.2468	0.0440
P	0.1883	−0.0652	0.3322
$C(C_6H_6,1)$	0.3005	0.0083	0.2805
$C(C_6H_6,2)$	0.4004	−0.0442	0.3104
$C(C_6H_6,3)$	0.4929	0.0125	0.2749
$C(C_6H_6,4)$	0.4836	0.0987	0.1949
$C(C_6H_6,5)$	0.3877	0.1443	0.1655
$C(C_6H_6,6)$	0.2978	0.0980	0.2025
$C(C_3H_7,1)$	0.1921	−0.2671	0.3322
$C(C_3H_7,2)$	0.2122	−0.3417	0.2422
$C(C_3H_7,3)$	0.2151	−0.5106	0.2446
$C(CH_2C_6H_5,1)$	0.1802	−0.0189	0.4527
$C(CH_2C_6H_5,2)$	0.1821	0.1499	0.4703
$C(CH_2C_6H_5,3)$	0.2723	0.2302	0.4776
$C(CH_2C_6H_5,4)$	0.2720	0.3817	0.4971
$C(CH_2C_6H_5,5)$	0.1750	0.4453	0.5024
$C(CH_2C_6H_5,6)$	0.0837	0.3731	0.4909
$C(CH_2C_6H_5,7)$	0.0834	0.2184	0.4760
$C(CH_3)$	0.0764	−0.0118	0.2723

In this structure the four organic complexes are tetrahedrally distributed about phosphorus to form the cation.

XV,bVI,5. Crystals of *diphenyliodonium chloride*, $(C_6H_5)_2ICl$, are monoclinic with a large unit containing eight molecules and having the dimensions:

$$a_0 = 20.81(6) \text{ A.}; \quad b_0 = 5.82(2) \text{ A.}; \quad c_0 = 20.26(6) \text{ A.}$$
$$\beta = 102°34'$$

The space group is C_{2h}^6 ($C2/c$) with atoms in the positions:

(8f) $\pm(xyz; x,\bar{y},z+1/2; x+1/2,y+1/2,z; x+1/2,1/2-y,z+1/2)$

The parameters for the two crystallographically different molecules are listed in Table XVB,114; calculated values for hydrogen are given in the original paper.

TABLE XVB,114

Parameters of the Atoms in $(C_6H_5)_2ICl$

Atom	x	y	z
C(1)	0.183	−0.187	0.356
C(2)	0.231	−0.357	0.370
C(3)	0.232	−0.521	0.318
C(4)	0.187	−0.517	0.257
C(5)	0.144	−0.330	0.243
C(6)	0.139	−0.176	0.294
I	0.1802	0.0430	0.4344
C(1′)	0.077	0.058	0.420
C(2′)	0.039	0.238	0.386
C(3′)	−0.029	0.214	0.370
C(4′)	−0.058	−0.003	0.388
C(5′)	−0.018	−0.174	0.420
C(6′)	0.050	−0.151	0.437
Cl	0.3316	0.0636	0.4585

In this structure the true molecule appears to be a dimer with iodine atoms at the center of an approximate rectangle for which C–I = 2.08 A. and I–Cl = 3.08 and 3.20 A. The angles C(1)–I–C(1′) = 98° and C(1)–I–Cl = 87°; Cl–I–C(1′) = 174°. The benzene rings are rotated outside the central plane about the C–I bonds by 41 and 71°.

The iodide, $(C_6H_5)_2I_2$, is isostructural, with:

$$a_0 = 22.08(10) \text{ A.}; \quad b_0 = 6.27(3) \text{ A.}; \quad c_0 = 20.42(8) \text{ A.}$$
$$\beta = 101°$$

XV,bVI,6. Crystals of *trans-oxotrichloro bis(diethylphenylphosphine)-rhenium(V)*, $ReOCl_3[P(C_2H_5)_2C_6H_5]_2$, have a monoclinic, tetramolecular unit of the dimensions:

$$a_0 = 13.05(2) \text{ A.}; \quad b_0 = 7.83(1) \text{ A.}; \quad c_0 = 23.90(3) \text{ A.}$$
$$\beta = 91.5(1)°$$

The space group is C_{2h}^5 ($P2_1/c$) with atoms in the positions:

$$(4e) \quad \pm (xyz; x,{}^1\!/_2 - y, z + {}^1\!/_2)$$

The determined parameters are listed in Table XVB,115.

TABLE XVB,115

Parameters of the Atoms in $ReOCl_3[P(C_2H_5)_2C_6H_5]_2$

Atom	x	y	z
Re	0.2587	0.1097	0.3885
Cl(1)	0.1185	−0.073	0.4129
Cl(2)	0.3786	−0.102	0.4297
Cl(3)	0.4111	0.272	0.3663
P(1)	0.2780	−0.070	0.3055
P(2)	0.2646	0.212	0.4868
O	0.1820	0.249	0.3614
C(1)	0.1933	0.000	0.2515
C(2)	0.2210	0.128	0.2168
C(3)	0.1400	0.184	0.1765
C(4)	0.0490	0.122	0.1680
C(5)	0.0270	−0.019	0.2075
C(6)	0.0936	−0.067	0.2480
C(7)	0.4086	−0.074	0.2700
C(8)	0.4970	−0.124	0.3034
C(9)	0.2860	−0.301	0.3108
C(10)	0.2650	−0.413	0.2645
C(11)	0.2427	0.048	0.5380
C(12)	0.1445	0.005	0.5530
C(13)	0.1320	−0.123	0.5900
C(14)	0.2100	−0.200	0.6170
C(15)	0.3200	−0.170	0.6000
C(16)	0.3339	−0.041	0.5650
C(17)	0.3803	0.330	0.5093
C(18)	0.3930	0.420	0.5655
C(19)	0.1580	0.358	0.4936
C(20)	0.1910	0.537	0.4700

Some of the bond dimensions in the molecule of this compound are shown in Figure XVB,168. The structure is illustrated in Figure XVB,169. As this suggests, coordination about the rhenium is octahedral (with bond angles between 80 and 100°).

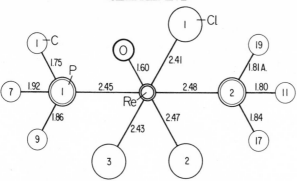

Fig. XVB,168.   Some bond lengths in the molecule of *trans*-oxotrichloro bis(diethylphenylphosphine)rhenium.

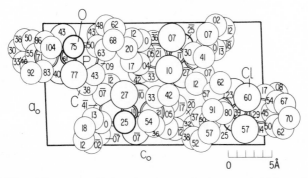

Fig. XVB,169.   The monoclinic structure of *trans*-oxotrichloro bis(diethylphenylphosphine)rhenium projected along its b_0 axis. The rhenium atoms are the small, doubly ringed circles.

XV,bVI,7.   Crystals of the complex *bis(l-ephedrine) copper(II)-benzene*, $(C_{10}H_4ON)_2Cu \cdot 2/3C_6H_6$, are hexagonal. They possess a unit that contains three of the foregoing formula weights and has the edge lengths:

$$a_0 = 11.89 \text{ A.}, \quad c_0 = 15.32 \text{ A.}$$

The space group is D_3^2 (*P*321) with copper atoms in the positions:

$$(3e) \quad u00; \; 0u0; \; \bar{u}\bar{u}0 \qquad \text{with } u = 0.1977$$

All other atoms are in the general positions:

$$(6g) \quad xyz; \; \bar{y}, x-y, z; \; y-x, \bar{x}, z; \; yx\bar{z}; \; \bar{x}, y-x, \bar{z}; \; x-y, \bar{y}, \bar{z}$$

Parameters for atoms of the ephedrine portion are listed in Table XVB,116. Though it could not be established with certainty, it seems probable that the benzene ring is either rotating or that its distribution is statistical about a threefold axis; possible atomic parameters for its atoms are given in the original article.

TABLE XVB,116

Parameters of Atoms in the Ephedrine–Copper–Benzene Complex

Atom	x	y	z
O	−0.1084	0.1260	0.0980
N	0.1446	0.2910	0.0906
C(1)	−0.2444	0.1517	0.2403
C(2)	−0.3169	0.1552	0.3141
C(3)	−0.2766	0.1541	0.4025
C(4)	−0.1637	0.1497	0.4096
C(5)	−0.0855	0.1552	0.3368
C(6)	−0.1184	0.1504	0.2501
C(7)	−0.0355	0.1498	0.1761
C(8)	0.0782	0.2982	0.1707
C(9)	0.0337	0.3986	0.1593
C(10)	0.2561	0.4170	0.0633

Fig. XVB,170.   Bond dimensions in the molecule of bis(l-ephedrine) copper in its complex with benzene.

The two ephedrine molecules chelating the copper atom are distributed about it in such a way that the metallic atom has its usual planar, square coordination. The bond dimensions are those of Figure XVB,170. Three of these chelate molecules are so related to one another about a threefold axis that they can enclose a benzene molecule; this then appears as the "guest" in a clathrate structure for which the trimeric chelate group provides the "host"

TABLE XVB,117

Parameters of the Atoms in $[(CH_3)_3(C_7H_5O_2)Pt]_2$

Atom	x	y	z
Molecule I			
Pt(1)	0.3917	0.5279	−0.1955
O(1)	0.403	0.445	0.058
O(2)	0.333	0.703	−0.157
C(1)	0.307	0.474	0.109
C(2)	0.268	0.395	0.221
C(3)	0.165	0.420	0.263
C(4)	0.108	0.536	0.212
C(5)	0.150	0.631	0.103
C(6)	0.248	0.609	0.047
C(7)	0.275	0.702	−0.063
C(8)	0.439	0.371	−0.237
C(9)	0.199	0.502	−0.306
C(10)	0.400	0.600	−0.409
Molecule II			
Pt(2)	0.0116	0.0501	−0.1869
O(2,1)	0.129	−0.058	0.062
O(2,2)	0.109	0.212	−0.109
C(2,1)	0.266	−0.029	0.137
C(2,2)	0.348	−0.128	0.243
C(2,3)	0.476	−0.104	0.326
C(2,4)	0.522	0.021	0.290
C(2,5)	0.428	0.100	0.184
C(2,6)	0.299	0.093	0.097
C(2,7)	0.223	0.200	−0.013
C(2,8)	−0.076	−0.098	−0.271
C(2,9)	0.148	0.045	−0.281
C(2,10)	−0.104	0.168	−0.412

XV,bVI,8. The triclinic crystals of *trimethyl(salicylaldehydato)-platinum(IV)*, $[(CH_3)_3(C_7H_5O_2)Pt]_2$, have a bimolecular cell of the dimensions:

$$a_0 = 10.828(5) \text{ A.}; \quad b_0 = 11.455(5) \text{ A.}; \quad c_0 = 9.147(5) \text{ A.}$$
$$\alpha = 74.7(5)°; \quad \beta = 110.8(5)°; \quad \gamma = 92.5(5)°$$

The space group is C_i^1 ($P\bar{1}$) with atoms in the positions $(2i) \pm (xyz)$. The determined parameters are listed in Table XVB,117.

The resulting structure (Fig. XVB,171) is built up of two dimeric molecules per cell, the components of each dimer being held together through the sharing of two oxygen atoms by two atoms of platinum. The Pt–O distances lie between 2.18 and 2.27 A. Coordination about the platinum atoms is that of a slightly distorted octahedron. Each of these atoms is surrounded by three oxygens and three methyl groups, with Pt–C = 1.95–2.13 A. The atoms of oxygen are equidistant from the two platinum atoms of a dimer and Pt–Pt = 3.41 A.

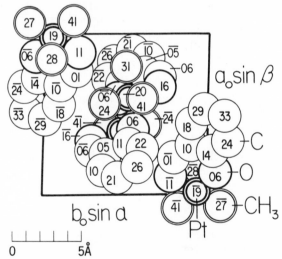

Fig. XVB,171. The triclinic structure of trimethyl(salicylaldehydato)platinum projected along its c_0 axis.

XV,bVI,9. Crystals of *chloro[N,N'-bis(salicylidene)ethylenediamine] iron(III)*, $[C_6H_4(O)C_2H_3N]_2FeCl$, are monoclinic with a tetramolecular unit of the dimensions:

$$a_0 = 11.37(2) \text{ A.}; \quad b_0 = 6.91(1) \text{ A.}; \quad c_0 = 19.16(2) \text{ A.}$$
$$\beta = 90.5(2)°$$

Atoms are in the positions

$$(4e) \quad \pm (xyz; x, \tfrac{1}{2} - y, z + \tfrac{1}{2})$$

of the space group C_{2h}^5 ($P2_1/c$). The determined parameters are listed in Table XVB,118; values for the hydrogen atoms are stated in the original.

TABLE XVB,118

Parameters of the Atoms in
Chloro[N,N'-bis(salicylidene)ethylenediamine] Iron(III)

Atom	x	y	z
Fe	0.13352	−0.00106	0.03377
Cl	0.21862	0.10889	0.13530
O(1)	0.2116	0.1724	−0.0270
O(2)	−0.0252	0.1020	0.0532
N(1)	0.2660	−0.2033	0.0121
N(2)	0.0713	−0.2495	0.0838
C(1)	0.3080	0.1569	−0.0663
C(2)	0.3379	0.3111	−0.1093
C(3)	0.4336	0.3078	−0.1531
C(4)	0.5047	0.1388	−0.1545
C(5)	0.4788	−0.0160	−0.1112
C(6)	0.3783	−0.0119	−0.0669
C(7)	0.3545	−0.1809	−0.0253
C(8)	0.2593	−0.3858	0.0559
C(9)	0.1318	−0.4328	0.0636
C(10)	−0.0044	−0.2503	0.1353
C(11)	−0.0658	−0.0777	0.1572
C(12)	−0.1264	−0.0879	0.2209
C(13)	−0.1888	0.0671	0.2441
C(14)	−0.1889	0.2388	0.2083
C(15)	−0.1337	0.2497	0.1441
C(16)	−0.0712	0.0940	0.1180

The structure is shown in Figure XVB,172. Its molecules actually are dimers of the formula given above, oxygen atoms being shared by the two irons of the dimer. The coordination about iron thus is octahedral, with Fe–O = 1.90 or 1.98 A., Fe–N = 2.09 or 2.10 A., and Fe–Cl = 2.29 A.

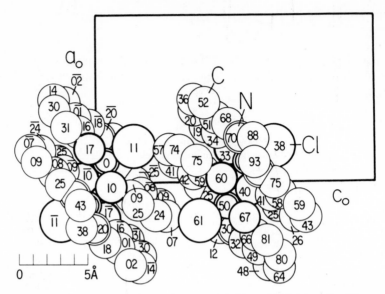

Fig. XVB,172.   The monoclinic structure of chloro[N,N'-bis(salicylidene)ethyl-
enediamine] iron projected along its b_0 axis. The oxygen atoms are the small,
heavily outlined circles, the irons are smaller and doubly, and heavily, ringed.

XV,bVI,10.   Crystals of   *N,N'-disalicylidene-ethylenediamine   zinc
monohydrate*, $Zn(C_8H_7NO)_2 \cdot H_2O$, are orthorhombic with a tetramolecular unit of the edge lengths:

$$a_0 = 25.03(3) \text{ A.}; \quad b_0 = 8.42(2) \text{ A.}; \quad c_0 = 7.36(2) \text{ A.}$$

The space group has been chosen as C_{2v}^9 ($P2_1nb$) with atoms in the
positions:

$$(4a) \quad xyz; \; x+1/2,\bar{y},\bar{z}; \; x+1/2,1/2-y,z+1/2; \; x,y+1/2,1/2-z$$

The parameters are listed in Table XVB,119.

Because of limited data the bond dimensions are not very precise but
show no abnormalities. The atom of zinc has two oxygen and two
nitrogen neighbors as well as a close water molecule, with H_2O–Zn =
2.0 A. The planes of the salicylidene groups include the zinc atom but
are turned through 23° with respect to one another.

A satisfactory drawing cannot be made of the structure as here
described. Either parameters or the space group attribution would
appear to be in error.

TABLE XVB,119

Parameters of the Atoms in $Zn(C_8H_7NO)_2 \cdot H_2O$

Atom	x	y	z
Zn	0.2500	0.085	0.078
O(1)	0.1973	0.248	0.093
O(2)	0.3107	0.241	0.107
O(3)	0.2574	−0.025	0.319
N(1)	0.2008	−0.059	−0.064
N(2)	0.3105	−0.056	−0.066
C(1)	0.1461	0.234	0.049
C(2)	0.1145	−0.130	0.099
C(3)	0.0563	−0.133	0.045
C(4)	0.0362	0.239	0.011
C(5)	0.0687	0.108	−0.067
C(6)	0.1210	0.111	−0.040
C(7)	0.1493	−0.039	−0.090
C(8)	0.2328	−0.182	−0.189
C(9)	0.2819	−0.212	−0.097
C(10)	0.3595	−0.032	−0.080
C(11)	0.3833	0.106	−0.028
C(12)	0.4426	0.120	−0.073
C(13)	0.4667	0.248	−0.048
C(14)	0.4492	0.128	0.076
C(15)	0.3934	0.137	0.104
C(16)	0.3603	0.241	0.057

C. COMPOUNDS CONTAINING THREE BENZENE RINGS

XV,c1. Crystals of *p-diphenylbenzene* (terphenyl), $C_6H_4(C_6H_5)_2$, have a bimolecular monoclinic unit of the dimensions:

$$a_0 = 8.08 \text{ A.}; \quad b_0 = 5.60 \text{ A.}; \quad c_0 = 13.59 \text{ A.}; \quad \beta = 91° 55'$$

The space group is $C_{2h}^5 (P2_1/a)$ with all atoms in the positions:

$$(4e) \quad \pm (xyz; x + {}^1/_2, {}^1/_2 - y, z)$$

The parameters, determined many years ago, are listed in Table XVC,1.

TABLE XVC,1

Parameters of the Atoms in p-Diphenylbenzene

Atom	x	y	z
C(1)	0·059	0·182	0·064
C(2)	−0.046	0.000	0.100
C(3)	−0.105	−0.182	0.036
C(4)	−0.094	0.000	0.204
C(5)	−0.036	0.182	0.268
C(6)	−0.082	0.182	0.368
C(7)	−0.187	0.000	0.402
C(8)	−0.246	−0.182	0.339
C(9)	−0.200	−0.182	0.239

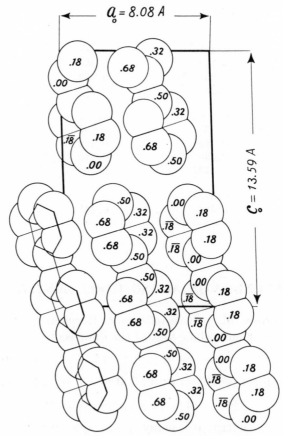

Fig. XVC,1a. The monoclinic structure of p-diphenylbenzene projected along its b_0 axis. Left-hand axes.

The structure is illustrated in Figure XVC,1. Comparison with Figure XVF,1 brings out its similarity to the naphthalene arrangement (**XV,f1**).

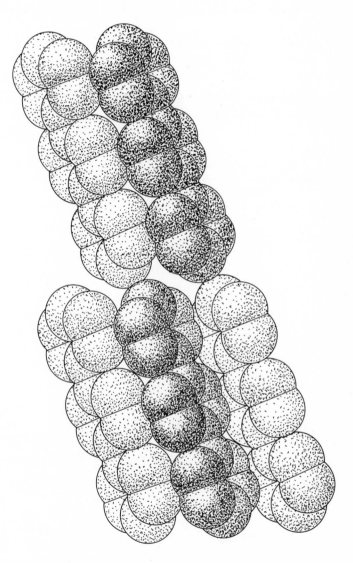

Fig. XVC,1b.   A packing drawing of the monoclinic structure of p-diphenyl-benzene viewed along its b_0 axis. Left-hand axes.

XV,c2. The hexagonal crystals of *triphenyl bromomethane*, $(C_6H_5)_3$-CBr, have a unit that contains six molecules and has the edge lengths:

$$a_0 = 13.85 \text{ A.}, \quad c_0 = 13.42 \text{ A.}$$

The space group has been chosen as C_{3i}^1 $(P\bar{3})$ with atoms in the positions:

$$
\begin{aligned}
&(2c) && \pm (00u) \\
&(2d) && \pm (^1/_3\ ^2/_3\ u) \\
&(6g) && \pm (xyz;\ \bar{y}, x-y, z;\ y-x, \bar{x}, z)
\end{aligned}
$$

The atomic distribution is pseudorhombohedral, with parameters that seemingly are those of Table XVC,2.

TABLE XVC,2

Positions and Parameters of the Atoms in $(C_6H_5)_3$CBr

Atom	Position	x	y	z
Br	(2c)	0	0	0.1292
C(7)	(2c)	0	0	0.281
C(1)	(6g)	0.120	0.076	0.314
C(2)	(6g)	0.172	0.049	0.384
C(3)	(6g)	0.281	0.122	0.410
C(4)	(6g)	0.337	0.227	0.378
C(5)	(6g)	0.284	0.265	0.301
C(6)	(6g)	0.175	0.193	0.273
Br'	(2d)	$^1/_3$	$^2/_3$	−0.4822
C(7')	(2d)	$^1/_3$	$^2/_3$	−0.618
C(1')	(6g)	0.777	0.427	0.663
C(2')	(6g)	0.877	0.439	0.624
C(3')	(6g)	0.979	0.527	0.654
C(4')	(6g)	0.983	0.597	0.740
C(5')	(6g)	0.884	0.579	0.775
C(6')	(6g)	0.784	0.494	0.740
Br''	(2d)	$^1/_3$	$^2/_3$	−0.2451
C(7'')	(2d)	$^1/_3$	$^2/_3$	−0.094
C(1'')	(6g)	0.553	0.247	0.065
C(2'')	(6g)	0.457	0.233	0.104
C(3'')	(6g)	0.355	0.156	0.076
C(4'')	(6g)	0.346	0.074	−0.002
C(5'')	(6g)	0.442	0.086	−0.045
C(6'')	(6g)	0.548	0.166	−0.013

XV,c3. *Triphenylmethyl   perchlorate*, $C(C_6H_5)_3ClO_4$, is cubic at 85°C. Its large unit cube, containing 16 molecules, has the edge length:

$$a_0 = 18.91(2) \text{ A.}$$

The space group has been found to be $O^4(F4_132)$. The methyl carbon atom has been placed in:

C(m): (16c) $1/8\ 1/8\ 1/8$; $1/8\ 7/8\ 7/8$; $7/8\ 1/8\ 7/8$; $7/8\ 7/8\ 1/8$;   F.C.

Chlorine atoms are of two sorts:

Cl(1): (8a) 000; $1/4\ 1/4\ 1/4$;   F.C.
Cl(2): (8b) $1/2\ 1/2\ 1/2$; $3/4\ 3/4\ 3/4$;   F.C.

The perchlorate ion which has Cl(1) at its center is disordered, best results being obtained with Cl–O = 1.45 A. Atoms of oxygen, tetrahedrally distributed about Cl(2), are in the fixed positions:

(32e) uuu; $1/4-u,1/4-u,1/4-u$; $u\bar{u}\bar{u}$; $1/4-u,u+1/4,u+1/4$;   tr
  with $u = 0.5419(9)$

Two sets of the phenyl carbon atoms are in the positions:

(48g) $1/8,u,1/4-u$; $1/8,\bar{u},u+3/4$; $7/8,u,u+3/4$; $7/8,\bar{u},1/4-u$;   tr;   F.C.
        C(1): (48g)   with $u = 0.0709(7)$
        C(4): (48g)   with $u = -0.0316(19)$

The others are in the general positions:

(96h) xyz; $\bar{x}y\bar{z}$; $1/4-x,1/4-z,1/4-y$; $x+1/4,1/4-z,y+1/4$;
        $x\bar{y}\bar{z}$; $z\bar{x}\bar{y}$; $1/4-x,z+1/4,y+1/4$; $y+1/4,x+1/4,1/4-z$;   tr;   F.C.
  C(2): (96h)   with $x = 0.2482(6)$, $y = 0.1012(6)$, $z = 0.0838(6)$
  C(3): (96h)   with $x = 0.2997(6)$, $y = 0.0993(7)$, $z = 0.0335(6)$

The three planar phenyl rings of the carbonium cation are arranged propellor-wise about the central carbon atom C(m), adjacent rings making 54° with one another. The distance between C(m) and the ring carbon to which it is bound is 1.454(18) A.; the ring bonds are normal, with C(1)–C(2) = 1.408(14) A., C(2)–C(3) = 1.364(16) A., and C(3)–C(4) = 1.372(28) A. The included angles within the rings are 115, 123, 116, and 125° for C(1)-to-C(4). In the tetrahedral, ordered ClO_4^- ion, Cl–O = 1.37 A.

Above 110°C. crystals of the compound triphenylmethyl tetrafluoroborate, $C(C_6H_5)_3BF_4$, are isostructural. As is true of the perchlorate, transition from the disorder that prevails at lower temperatures is gradual. For the fluoroborate: $a_0 = 18.87(3)$ A. at 110°C.

XV,c4. The red tautomer of *3,5-dibromo-p-hydroxytriphenyl methane carbinol*, $(C_6H_5)_2(C_6H_2Br_2OH)COH$, forms monoclinic crystals whose unit, containing eight molecules, has the dimensions:

$$a_0 = 8.63(2) \text{ A.}; \quad b_0 = 22.14(7) \text{ A.}; \quad c_0 = 17.89(8) \text{ A.}; \quad \beta = 101°$$

The space group is $C_{2h}^6(C2/c)$ with atoms in the positions:

$$(8f) \quad \pm(xyz; x,\bar{y},z+{}^1/_2; x+{}^1/_2,y+{}^1/_2,z; x+{}^1/_2,{}^1/_2-y,z+{}^1/_2)$$

According to a preliminary announcement, the atoms have the approximate parameters of Table XVC,3.

TABLE XVC,3

Parameters of the Atoms in $(C_6H_5)_2(C_6H_2Br_2OH)COH$

Atom	x	y	z
C(1)	0.18	0.50	0.37
C(2)	0.20	0.49	0.30
C(3)	0.25	0.43	0.29
C(4)	0.30	0.38	0.35
C(5)	0.28	0.38	0.43
C(6)	0.22	0.45	0.43
C(1')	0.13	0.76	0.37
C(2')	0.18	0.82	0.39
C(3')	0.19	0.00	0.38
C(4')	0.14	0.09	0.35
C(5')	0.08	0.86	0.35
C(6')	0.08	0.86	0.33
C(1″)	0.10	0.54	0.46
C(2″)	0.08	0.40	0.46
C(3″)	0.07	0.36	0.54
C(4″)	0.07	0.48	0.59
C(5″)	0.09	0.62	0.58
C(6″)	0.10	0.67	0.51
Br(1)	0.044	0.158	0.549
Br(2)	0.089	0.770	0.660
O(1)	0.05	0.43	0.66
C(7)	0.12	0.59	0.38
O(2)	0.07	0.53	0.32

XV,c5. The free radical *tri-p-nitrophenylmethyl*, $(NO_2C_6H_4)_3C$, forms orthorhombic crystals whose tetramolecular unit has the edge

lengths:

$$a_0 = 7.80 \text{ A.}; \quad b_0 = 11.50 \text{ A.}; \quad c_0 = 18.60 \text{ A.}$$

Atoms are in the following positions of the space group V_h^{14} (*Pbcn*):

(4c) $\pm (0 \ u \ 1/4; \ 1/2, u + 1/2, 1/4)$

(8d) $\pm (xyz; \ x + 1/2, y + 1/2, 1/2 - z; \ 1/2 - x, y + 1/2, z; \ x, \bar{y}, z + 1/2)$

The determined parameters are those of Table XVC,4. Values for hydrogen are stated in the original paper.

TABLE XVC,4

Positions and Parameters of the Atoms in $(NO_2C_6H_4)_3C$

Atom	Position	x	y	z
C(0)	(4c)	0	0.1002(21)	$1/4$
C(1)	(4c)	0	0.2291(22)	$1/4$
C(2)	(8d)	0.0503(24)	0.2893(14)	0.3107(10)
C(3)	(8d)	0.0581(24)	0.4113(15)	0.3106(10)
C(4)	(4c)	0	0.4683(21)	$1/4$
C(1′)	(8d)	−0.0573(24)	0.0408(14)	0.3148(9)
C(2′)	(8d)	−0.1725(26)	0.0869(16)	0.3638(9)
C(3′)	(8d)	−0.2260(25)	0.0262(16)	0.4242(9)
C(4′)	(8d)	−0.1521(26)	−0.0777(17)	0.4377(9)
C(5′)	(8d)	−0.0384(25)	−0.1296(14)	0.3942(9)
C(6′)	(8d)	0.0134(26)	−0.0721(14)	0.3316(9)
N(1)	(4c)	0	0.6012(23)	$1/4$
N(1′)	(8d)	−0.2181(29)	−0.1373(17)	0.5040(9)
O(1)	(8d)	0.0915(18)	0.6398(10)	0.2956(7)
O(1′)	(8d)	−0.3062(25)	−0.0858(13)	0.5440(9)
O(2′)	(8d)	−0.1481(22)	−0.2307(14)	0.5152(7)

The resulting structure is shown in Figure XVC,2. The molecule, which has a twofold axis of symmetry, has the bond dimensions of Figure XVC,3. About the central C(0) the three bonds are coplanar. The benzene ring that contains the twofold axis is twisted through 40° about C(0)–C(1) and the other two phenyl groups through 30° about their bonds to C(0). The NO_2 groups of these two rings lie in the ring plane but that belonging to the ring with an axis is turned through an additional 20°. Between molecules the shortest interatomic distances are O–O = 3.15 A., C–O = 3.29 A., C–N = 3.32 A. and C–C = 3.37 A.

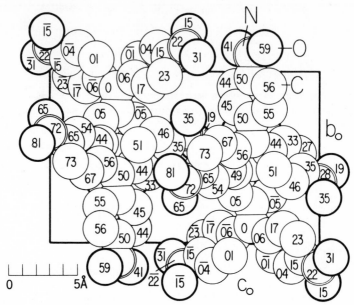

Fig. XVC,2.   The orthorhombic structure of the free radical tri-*p*-nitrophenyl-
methyl projected along its a_0 axis.

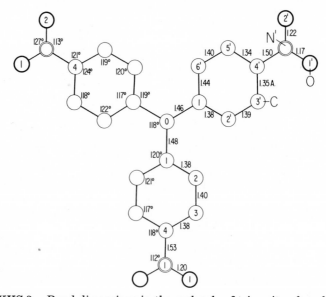

Fig. XVC,3.   Bond dimensions in the molecule of tri-*p*-nitrophenylmethyl.

XV,c6. Crystals of *benzil-α-monoxime-p-bromobenzoate*, $C_6H_5C(O)$-$C(C_6H_5)NOC(O)C_6H_4Br$, are triclinic with a bimolecular unit of the dimensions:

$$a_0 = 8.78 \text{ A.}; \quad b_0 = 17.01 \text{ A.}; \quad c_0 = 6.27 \text{ A.}$$
$$\alpha = 95.1°; \quad \beta = 91.1°; \quad \gamma = 102.3°$$

All atoms are in the positions $(2i) \pm (xyz)$ of the space group $C_i^1(P\bar{1})$. The parameters are listed in Table XVC,5.

TABLE XVC,5

Parameters of the Atoms in Benzil-α-monoxime-*p*-bromobenzoate

Atom	x	y	z
C(1)	0.229(3)	0.9993(14)	0.150(4)
C(2)	0.301(3)	0.9921(16)	0.946(4)
C(3)	0.234(3)	0.9308(17)	0.789(5)
C(4)	0.097(3)	0.8760(14)	0.856(4)
C(5)	0.020(3)	0.8895(17)	0.054(4)
C(6)	0.083(3)	0 9450(14)	0.205(4)
C(7)	0.011(3)	0.8108(14)	0.682(4)
C(8)	0.137(2)	0.7120(13)	0.229(3)
C(9)	0.060(3)	0.6580(14)	0.046(3)
C(10)	−0.080(3)	0.5971(13)	0.084(4)
C(11)	−0.102(3)	0.5632(12)	0.283(4)
C(12)	−0.227(4)	0.4956(16)	0.318(5)
C(13)	−0.326(3)	0.4661(18)	0.143(5)
C(14)	−0.304(4)	0.4981(20)	−0.051(5)
C(15)	−0.170(3)	0.5670(17)	−0.089(4)
C(16)	0.311(2)	0.7255(14)	0.250(4)
C(17)	0.406(2)	0.7541(14)	0.078(4)
C(18)	0.564(3)	0.7544(16)	0.095(4)
C(19)	0.624(3)	0.7358(16)	0.273(4)
C(20)	0.524(3)	0.7093(15)	0.451(4)
C(21)	0.367(3)	0.7091(15)	0.440(4)
N(22)	0.037(2)	0.7424(11)	0.357(3)
O(23)	0.113(2)	0.6744(10)	−0.126(3)
O(24)	0.115(2)	0.7971(9)	0.529(2)
O(25)	−0.113(2)	0.7691(10)	0.723(3)
Br(26)	0.3197(4)	0.08090(18)	0.357(5)

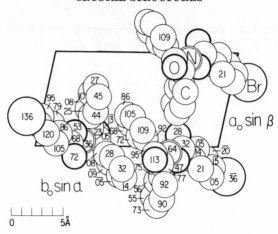

Fig. XVC,4.   The triclinic structure of benzil-α-monoxime-*p*-bromobenzoate
projected along its c_0 axis.

The structure, shown in Figure XVC,4 has molecules with the bond
dimensions of Figure XVC,5. In these molecules the *p*-bromobenzoate
group is *trans* to the carbonyl group. Between molecules the interatomic distances range upward from an O(23)–C(7) = 2.99 A. and an
O(23)–O(25) = 3.00 A.

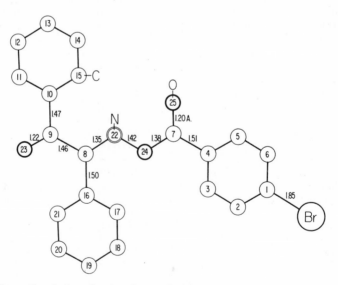

Fig. XVC,5.   Bond lengths in the molecule of benzil-α-monoxime-*p*-bromo-
benzoate.

XV,c7. *Phenoquinone*, $C_6H_4O_2 \cdot 2C_6H_5OH$, crystallizes in the mono-clinic system with a bimolecular unit of the dimensions:

$$a_0 = 11.152(4) \text{ A.}; \quad b_0 = 5.970(4) \text{ A.}; \quad c_0 = 11.499(9) \text{ A.}$$
$$\beta = 100.0(3)°$$

All its atoms are in the general positions of the space group $C_{2h}^5(P2_1/c)$:

$$(4e) \quad \pm (xyz; x, 1/2 - y, z + 1/2)$$

Recently refined parameters are listed in Table XVC,6; values for hydrogen are given in the same article.

TABLE XVC,6

Parameters of the Atoms in Phenoquinone

Atom	x	y	z
Quinone molecules			
O	0.906	−0.3540	0.1253
C(1)	0.0492	−0.1897	0.0674
C(2)	0.0138	0.0111	−0.1205
C(3)	0.0619	−0.1660	−0.0580
Phenol molecules			
OH	0.2238	0.3285	0.0259
C(4)	0.2790	0.1607	0.0975
C(5)	0.2728	0.1495	0.2167
C(6)	0.3278	−0.0291	0.2835
C(7)	0.3888	−0.1935	0.2308
C(8)	0.3959	−0.1805	0.1138
C(9)	0.3406	−0.0018	0.0456

In the resulting structure (Fig. XVC,6) the phenol and quinone molecules have the bond dimensions of Figure XVC,7. Two phenols are centered about a quinone molecule in such a way that the quinone ring is sandwiched between the two phenol rings. These molecules, stacked in columns, are held together by hydrogen bonds linking phenolic hydroxyls and quinonic oxygen, with O–O = 2.774 A.

XV,c8. Crystals of *acetyl triphenylgermane*, $(C_6H_5)_3GeCOCH_3$, are monoclinic with a tetramolecular unit of the dimensions:

$$a_0 = 15.30(2) \text{ A.}; \quad b_0 = 14.53(2) \text{ A.}; \quad c_0 = 7.68(2) \text{ A.}; \quad \beta = 94.8(3)°$$

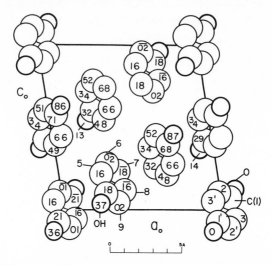

Fig. XVC,6a.   The monoclinic structure of phenoquinone projected along its
b_0 axis. Left-hand axes.

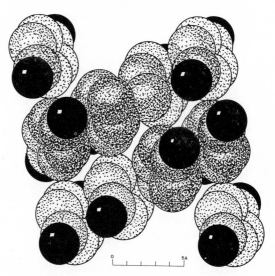

Fig. XVC,6b.   A packing drawing of the monoclinic phenoquinone structure
seen along its b_0 axis. The oxygens are black, the carbons dotted. Left-hand
axes.

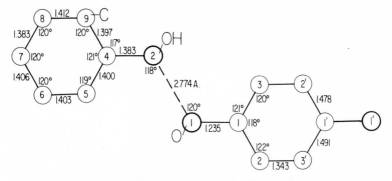

Fig. XVC,7.   Bond dimensions in phenoquinone.

The space group is $C_{2h}^5(P2_1/c)$ with atoms in the positions:

$$(4e) \quad \pm (xyz; \, x, 1/2 - y, z + 1/2)$$

Parameters are listed in Table XVC,7; values for the hydrogen atoms are stated in the original paper.

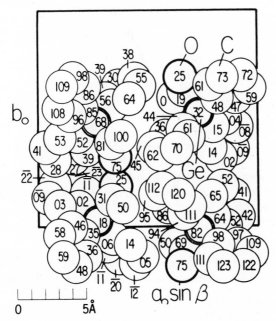

Fig. XVC,8.   The monoclinic structure of acetyl triphenylgermane projected along its c_0 axis.

TABLE XVC,7

Parameters of Atoms in $(C_6H_5)_3GeCOCH_3$

Atom	x	y	z
Ge(1)	0.7354	0.5264	0.3163
O(2)	0.6499	0.6951	0.2546
C(3)	0.6615	0.6227	0.1862
CH₃(4)	0.6200	0.6011	0.0052
C(5)	0.8176	0.5900	0.4806
C(6)	0.9082	0.5855	0.4723
C(7)	0.9629	0.6324	0.5891
C(8)	0.9307	0.6852	0.7191
C(9)	0.8419	0.6933	0.7268
C(10)	0.7854	0.6451	0.6095
C(11)	0.6584	0.4484	0.4411
C(12)	0.5776	0.4199	0.3651
C(13)	0.5260	0.3600	0.4554
C(14)	0.5520	0.3306	0.6208
C(15)	0.6315	0.3596	0.6976
C(16)	0.6845	0.4177	0.6072
C(17)	0.7961	0.4539	0.1501
C(18)	0.7867	0.3602	0.1383
C(19)	0.8311	0.3108	0.0183
C(20)	0.8855	0.3539	−0.0853
C(21)	0.8959	0.4478	−0.0778
C(22)	0.8508	0.4983	0.0384

In this structure (Fig. XVC,8) distribution around the germanium atom is that of a very nearly regular tetrahedron, with Ge–C (phenyl) = 1.945 A. and Ge–C(acetyl) = 2.011 A.; the angles C–Ge–C = 107.4–111.4°.

XV,c9. Crystals of *tribenzyl tin acetate*, $(C_6H_5CH_2)_3SnOC(O)CH_3$, are monoclinic with a tetramolecular unit expressed in the little-used orientation that makes c_0 the unique axis. For it:

$$a_0 = 11.254(8) \text{ A.}; \quad b_0 = 16.716(12) \text{ A.}; \quad c_0 = 10.943(9) \text{ A.}$$
$$\gamma = 105°11(4)'$$

The space group C_{2h}^5 is thus in the orientation $P2_1/b$. All atoms therefore are in the positions:

$$(4e) \quad \pm (xyz\,;\ x,y+^1/_2,^1/_2-z)$$

The parameters are listed in Table XVC,8.

TABLE XVC,8

Parameters of the Atoms in Tribenzyl Tin Acetate

Atom	x	y	z
Sn	−0.0108(2)	0.1979(1)	0.7191(2)
C(1)	0.147(3)	0.303(2)	0.727(4)
C(2)	0.262(3)	0.285(2)	0.675(4)
C(3)	0.306(3)	0.311(2)	0.564(4)
C(4)	0.414(3)	0.294(2)	0.514(4)
C(5)	0.476(3)	0.250(3)	0.585(6)
C(6)	0.434(4)	0.227(2)	0.698(6)
C(7)	0.328(3)	0.241(2)	0.748(4)
C(1,1)	−0.033(3)	0.088(2)	0.833(4)
C(1,2)	0.090(3)	0.074(1)	0.872(3)
C(1,3)	0.132(3)	0.095(2)	0.989(3)
C(1,4)	0.247(4)	0.080(2)	0.028(4)
C(1,5)	0.312(3)	0.046(2)	0.953(5)
C(1,6)	0.270(3)	0.023(2)	0.824(4)
C(1,7)	0.159(3)	0.038(2)	0.774(4)
C(2,1)	−0.191(3)	0.198(2)	0.648(4)
C(2,2)	−0.272(3)	0.110(2)	0.637(4)
C(2,3)	−0.345(3)	0.070(2)	0.741(3)
C(2,4)	−0.412(3)	−0.013(3)	0.732(4)
C(2,5)	−0.419(3)	−0.054(3)	0.624(5)
C(2,6)	−0.346(4)	−0.016(3)	0.524(5)
C(2,7)	−0.272(3)	0.067(2)	0.538(5)
C(3,1)	0.062(2)	0.168(2)	0.456(3)
C(3,2)	0.108(4)	0.116(2)	0.364(5)
O(1)	0.042(2)	0.136(1)	0.566(2)
O(2)	0.040(2)	0.234(1)	0.431(2)

In the resulting structure tin has the unusual coordination of five due to close approach of the acetate oxygen of an adjacent molecule. Within a molecule Sn–O(1) = 2.14 A. and Sn–C separations lie between 2.15 and 2.18 A. The close intermolecular distance is Sn–O(2) = 2.65 A.

XV,c10.   The monoclinic crystals of *triphenyl phosphorus*, $P(C_6H_5)_3$, possess a tetramolecular unit of the dimensions:

$$a_0 = 11.413 \text{ A.}; \quad b_0 = 15.032 \text{ A.}; \quad c_0 = 8.500 \text{ A.}; \quad \beta = 92°53'$$

The space group $C_{2h}^5(P2_1/a)$ places atoms in the positions:

$$(4e) \quad \pm(xyz; x+1/2, 1/2-y, z)$$

The determined parameters are those of Table XVC,9.

TABLE XVC,9

Parameters of the Atoms in $P(C_6H_5)_3$

Atom	x	y	z
P	0.2126	0.0374	0.1412
C(1)	0.3124	0.1316	0.1659
C(2)	0.2950	0.1865	0.2967
C(3)	0.3671	0.2595	0.3267
C(4)	0.4546	0.2807	0.2271
C(5)	0.4739	0.2261	0.0971
C(6)	0.4017	0.1522	0.0653
C(7)	0.3012	−0.0573	0.2120
C(8)	0.4120	−0.0494	0.2881
C(9)	0.4677	−0.1251	0.3527
C(10)	0.4146	−0.2088	0.3390
C(11)	0.3046	−0.2169	0.2635
C(12)	0.2475	−0.1418	0.1982
C(13)	0.2043	0.0192	−0.0720
C(14)	0.2904	−0.0281	−0.1518
C(15)	0.2777	−0.0359	−0.3160
C(16)	0.1810	0.0003	−0.4001
C(17)	0.0946	0.0461	−0.3197
C(18)	0.1079	0.0556	−0.1544

The structure that results is shown in Figure XVC,9. Its molecules depart considerably from a threefold symmetry owing to a twist of one of the rings through 30° from the position that would give this symmetry. The benzene rings are planar, with the phosphorus atom 0.18 A. outside the plane of the twisted ring. The P–C bonds have lengths of 1.822 and 1.831(twice) A.

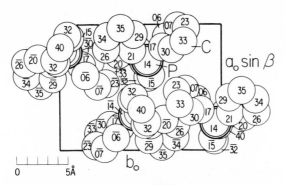

Fig. XVC,9.   The monoclinic structure of triphenyl phosphorus projected along
its c_0 axis.

XV,c11.   Crystals of *tris(phenylethynyl)phosphine*, $(C_6H_5C_2)_3P$, are
hexagonal with a trimolecular unit of the edge lengths:

$$a_0 = 18.540(5) \text{ A.,}   c_0 = 4.630(5) \text{ A.}$$

The atom of phosphorus is in the position

$$(3a)   00u; \text{rh}   \text{with } u = 0 \text{ (arbitrary)}$$

of the space group $C_3^4(R3)$. All other atoms are in

$$(9b)   xyz; \bar{y},x-y,z; y-x,\bar{x},z;   \text{rh}$$

with the assigned parameters of Table XVC,10.

TABLE XVC,10

Parameters of Atoms in $(C_6H_5C_2)_3P$

Atom	x	y	z
C(1)	0.2468(6)	0.1259(7)	0.602(3)
C(2)	0.2647(7)	0.0803(8)	0.399(3)
C(3)	0.3415(9)	0.1145(9)	0.273(3)
C(4)	0.4028(8)	0.1964(11)	0.336(3)
C(5)	0.3868(8)	0.2420(10)	0.540(3)
C(6)	0.3088(7)	0.2063(7)	0.667(3)
C(7)	0.1668(6)	0.0878(5)	0.729(2)
C(8)	0.0977(6)	0.0521(7)	0.826(3)

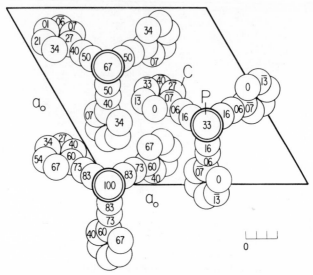

Fig. XVC,10.   The hexagonal structure of tris(phenylethynyl)phosphine projected along its c_0 axis.

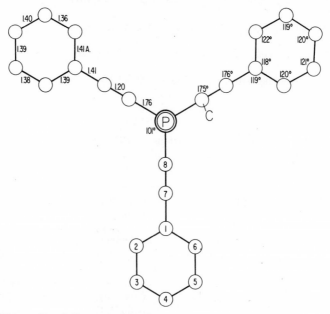

Fig. XVC,11.   Bond dimensions in the molecule of tris(phenylethynyl)phosphine.

The structure, shown in Figure XVC,10, is composed of molecules with the bond dimensions of Figure XVC,11. The coordination of phosphorus is that of a flat trigonal pyramid. Between molecules all atomic separations exceed 3.5 A.

XV,c12. According to a brief preliminary announcement, crystals of *p-bromophenyldiphenyl phosphine*, $BrC_6H_4(C_6H_5)_2P$, are orthorhombic with a unit that contains eight molecules and has the edge lengths:

$$a_0 = 18.92 \text{ A.}; \quad b_0 = 19.66 \text{ A.}; \quad c_0 = 8.66 \text{ A.}$$

All atoms are in the general positions of V_h^{15} (*Pcab*):

$$(8c) \quad \pm (xyz; \ x+\tfrac{1}{2}, \tfrac{1}{2}-y, z; \ \tfrac{1}{2}-x, y, z+\tfrac{1}{2}; \ x, y+\tfrac{1}{2}, \tfrac{1}{2}-z)$$

The parameters are given in Table XVC,11.

TABLE XVC,11

Parameters of the Atoms in $BrC_6H_4(C_6H_5)_2P$

Atom	x	y	z
Br	0.1978	0.7082	0.2759
P	0.1280	0.3852	0.3406
C(1)	0.148	0.476	0.322
C(2)	0.125	0.523	0.435
C(3)	0.140	0.593	0.422
C(4)	0.178	0.616	0.294
C(5)	0.202	0.571	0.181
C(6)	0.186	0.501	0.195
C(7)	0.040	0.381	0.260
C(8)	0.007	0.319	0.246
C(9)	−0.061	0.315	0.184
C(10)	−0.096	0.374	0.137
C(11)	−0.064	0.438	0.152
C(12)	0.004	0.440	0.213
C(13)	0.115	0.373	0.550
C(14)	0.172	0.358	0.645
C(15)	0.163	0.349	0.804
C(16)	0.095	0.356	0.868
C(17)	0.038	0.372	0.773
C(18)	0.048	0.380	0.614

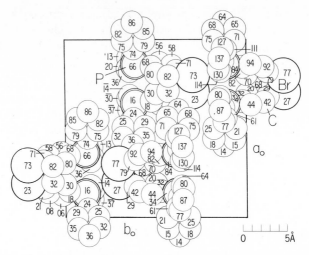

Fig. XVC,12.   The orthorhombic structure of p-bromophenyldiphenyl phosphine projected along its c_0 axis.

The structure is shown in Figure XVC,12. As in other phosphines, the distribution about phosphorus is that of a trigonal pyramid with phosphorus at the apex. The average P–C distance is 1.83 A. and the angle about phosphorus is 103°.

XV,c13. *Triphenyl   phosphoranylideneketen,* $(C_6H_5)_3P{=}C{=}C{=}O$, forms monoclinic crystals. Their tetramolecular unit has the dimensions:

$$a_0 = 14.527(16) \text{ A.}; \quad b_0 = 10.299(11) \text{ A.}; \quad c_0 = 10.794(12) \text{ A.}$$
$$\beta = 94°10(15)'$$

The space group is C_{2h}^5 in the orientation $P2_1/n$ with all atoms in the positions:

$$(4e) \quad \pm(xyz;\, x + {}^1\!/_2, {}^1\!/_2 - y, z + {}^1\!/_2)$$

The parameters are listed in Table XVC,12.

The structure is built up of molecules with the bond dimensions of Figure XVC,13. The planes of the benzene rings are turned by different amounts with respect to one another about the bonds connecting them with the central phosphorus atom; this atom is 0.09 and 0.07 A. outside the planes of two of these rings. As usual its coordination is tetrahedral. The distances in the cumulene chain, P=C and C=C, are notably short. The C–H separations range between 0.85 and 1.22 A.

TABLE XVC,12

Parameters of the Atoms in $(C_6H_5)_3P{=}C{=}C{=}O$

Atom	x	y	z
P	0.2351	0.2218	0.0940
O	0.1292	−0.0931	−0.0544
C(1)	0.1601	0.2204	0.2214
C(2)	0.0687	0.1869	0.1954
C(3)	0.0093	0.1842	0.2904
C(4)	0.0408	0.2138	0.4094
C(5)	0.1315	0.2497	0.4358
C(6)	0.1920	0.2514	0.3417
C(7)	0.2161	0.3727	0.0134
C(8)	0.1524	0.4635	0.0459
C(9)	0.1368	0.5749	−0.0227
C(10)	0.1848	0.5953	−0.1252
C(11)	0.2490	0.5071	−0.1587
C(12)	0.2641	0.3963	−0.0912
C(13)	0.3537	0.2251	0.1578
C(14)	0.4056	0.1103	0.1531
C(15)	0.4961	0.1087	0.2054
C(16)	0.5319	0.2183	0.2598
C(17)	0.4819	0.3293	0.2641
C(18)	0.3921	0.3348	0.2113
C(19)	0.2169	0.0983	−0.0017
C(20)	0.1721	0.0010	−0.0242
H(2)	0.049	0.160	0.087
H(3)	−0.054	0.154	0.264
H(4)	0.004	0.204	0.474
H(5)	0.155	0.290	0.530
H(6)	0.257	0.282	0.367
H(8)	0.122	0.449	0.126
H(9)	0.089	0.645	−0.006
H(10)	0.176	0.685	−0.164
H(11)	0.287	0.518	−0.238
H(12)	0.307	0.333	−0.108
H(14)	0.377	0.037	0.098
H(15)	0.524	0.037	0.196
H(16)	0.592	0.209	0.301
H(17)	0.512	0.412	0.306
H(18)	0.350	0.420	0.226

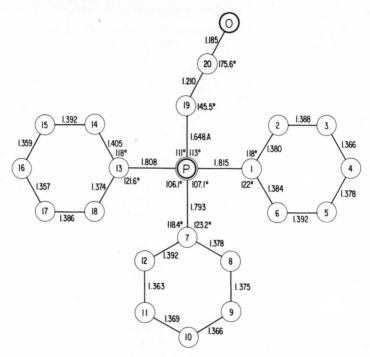

Fig. XVC,13. Bond dimensions in the molecule of triphenyl phosphoranylidene-keten.

XV,c14. Crystals of *triphenyl phosphoranylidenethioketen*, $(C_6H_5)_3P{=}C{=}C{=}S$, like those of the corresponding oxygen compound (**XV,c13**), are monoclinic but the two crystals are not isostructural. For the thioketen there are four molecules in a unit of the dimensions:

$$a_0 = 8.722(7) \text{ A.}; \quad b_0 = 10.819(9) \text{ A.}; \quad c_0 = 17.664(15) \text{ A.}$$
$$\beta = 98°8(15)'$$

Atoms are in the positions

$$(4e) \quad \pm(xyz; \; x,{}^1\!/_2-y,z+{}^1\!/_2)$$

of the space group $C_{2h}^5(P2_1/c)$. The parameters are stated in Table XVC,13; values for the hydrogen atoms are to be found in the original paper.

The structure (Fig. XVC,14) is composed of molecules with bond lengths of Figure XVC,15. The cumulene chain in this compound, as

TABLE XVC,13

Parameters of the Atoms in $(C_6H_5)_3P{=}C{=}C{=}S$

Atom	x	y	z
S	0.3827	−0.5168	0.1390
P	0.1020	−0.1961	0.1991
C(1)	0.1269	−0.1886	0.3009
C(2)	0.2664	−0.2224	0.3412
C(3)	0.2924	−0.2103	0.4182
C(4)	0.1798	−0.1691	0.4576
C(5)	0.0387	−0.1380	0.4200
C(6)	0.0129	−0.1461	0.3412
C(7)	0.1809	−0.0576	0.1636
C(8)	0.2558	0.0310	0.2112
C(9)	0.3128	0.1363	0.1805
C(10)	0.2953	0.1508	0.1033
C(11)	0.2226	0.0631	0.0552
C(12)	0.1651	−0.0406	0.0856
C(13)	−0.1131	−0.1954	0.1617
C(14)	−0.1902	−0.0929	0.1708
C(15)	−0.3476	−0.0974	0.1428
C(16)	−0.4111	−0.2019	0.1070
C(17)	−0.3218	−0.3016	0.0974
C(18)	−0.1666	−0.2995	0.1241
C(19)	0.1878	−0.3231	0.1704
C(20)	0.2738	−0.4055	0.1572

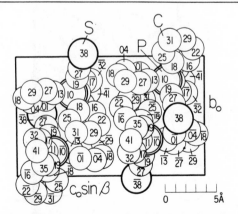

Fig. XVC,14. The monoclinic structure of triphenyl phosphoranylidenethioketen projected along its a_0 axis.

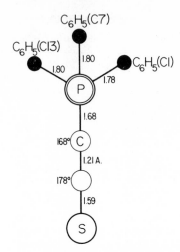

Fig. XVC,15. Some bond lengths in the molecule of triphenyl phosphoranyli-denethioketen.

in the oxygen keten, has short atomic separations, and the average C–C = 1.37 A. is considered to be 0.02 A. less than usual. Distribution about the phosphorus atom is tetrahedral.

XV,c15. Crystals of the addition compound *di(triphenylphosphine-sulfide)-tri(iodine)*, $2(C_6H_5)_3PS \cdot 3I_2$, are triclinic with a unimolecular

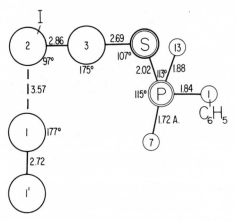

Fig. XVC,16. Some bond dimensions in the molecule of di(triphenylphosphine-sulfide)-tri(iodine).

cell of the dimensions:

$$a_0 = 9.70(2) \text{ A.}; \quad b_0 = 13.28(3) \text{ A.}; \quad c_0 = 8.94(2) \text{ A.}$$
$$\alpha = 65.2(5)°; \quad \beta = 81.5(5)°; \quad \gamma = 90.6(5)°$$

All atoms are in the positions $(2i) \pm (xyz)$ of the space group C_i^1 ($P\bar{1}$). The parameters are listed in Table XVC,14.

TABLE XVC,14

Parameters of the Atoms in $2(C_6H_5)_3PS \cdot 3I_2$

Atom	x	y	z
I(1)	−0.1385(8)	−0.0274(5)	0.0653(9)
I(2)	−0.4974(7)	−0.0990(4)	0.2581(8)
I(3)	−0.5549(6)	0.1116(4)	0.2588(7)
S	−0.5874(27)	0.3086(16)	0.2724(27)
P	−0.7549(21)	0.3709(13)	0.1658(22)
C(1)	−0.790(7)	0.502(4)	0.184(7)
C(2)	−0.924(8)	0.508(5)	0.270(9)
C(3)	−0.937(8)	0.609(5)	0.289(9)
C(4)	−0.832(7)	0.694(4)	0.237(7)
C(5)	−0.695(8)	0.677(5)	0.178(8)
C(6)	−0.667(7)	0.581(5)	0.152(8)
C(7)	−0.737(8)	0.404(5)	−0.044(8)
C(8)	−0.716(7)	0.520(4)	−0.168(8)
C(9)	−0.692(6)	0.540(4)	−0.346(7)
C(10)	−0.663(9)	0.459(6)	−0.404(10)
C(11)	−0.687(6)	0.359(4)	−0.288(6)
C(12)	−0.707(7)	0.328(5)	−0.113(8)
C(13)	−0.914(8)	0.268(5)	0.252(8)
C(14)	−0.016(11)	0.284(8)	0.168(13)
C(15)	−0.144(6)	0.207(4)	0.247(7)
C(16)	−0.152(9)	0.136(7)	0.403(11)
C(17)	−0.055(7)	0.138(4)	0.516(7)
C(18)	−0.902(22)	0.210(16)	0.459(27)

Bond dimensions are shown in Figure XVC,16, which represents half the molecule, the other half being an inversion of that shown here.

XV,c16. Crystals of *triphenyl phosphate*, $(C_6H_5)_3PO_4$, are monoclinic with a tetramolecular unit of the dimensions:

$$a_0 = 17.124(48) \text{ A.}; \quad b_0 = 5.833(36) \text{ A.}; \quad c_0 = 16.970(42) \text{ A.}$$
$$\beta = 105°21(15)'$$

The space group is $C_{2h}^5(P2_1/a)$ with all atoms in the general positions:

$$(4e) \quad \pm(xyz; x+^1/_2, ^1/_2-y, z)$$

The parameters determined in the more recent study are those of Table XVC,15.

TABLE XVC,15

Parameters of the Atoms in $(C_6H_5)_3PO_4$

Atom	x	y	z
P(1)	0.3888(2)	0.3454(7)	0.2539(2)
O(2)	0.4388(5)	0.5459(16)	0.2253(5)
O(3)	0.3229(5)	0.2812(18)	0.1752(6)
O(4)	0.3317(5)	0.4742(18)	0.2966(5)
O(5)	0.4396(5)	0.1687(17)	0.2989(5)
C(6)	0.5246(7)	0.5460(30)	0.2358(8)
C(7)	0.5652(7)	0.3702(24)	0.2114(8)
C(8)	0.6479(9)	0.3800(33)	0.2189(9)
C(9)	0.6902(9)	0.5835(34)	0.2557(9)
C(10)	0.6444(9)	0.7644(25)	0.2789(9)
C(11)	0.5622(9)	0.7391(31)	0.2689(9)
C(12)	0.3420(8)	0.2102(27)	0.1022(8)
C(13)	0.3794(9)	0.0086(25)	0.1017(9)
C(14)	0.4013(9)	0.9551(34)	0.0289(12)
C(15)	0.3846(14)	0.0917(55)	0.9629(12)
C(16)	0.3453(14)	0.2890(48)	0.9694(12)
C(17)	0.3244(8)	0.3680(28)	0.0411(9)
C(18)	0.3516(7)	0.5172(29)	0.3794(8)
C(19)	0.3234(10)	0.3638(34)	0.4261(9)
C(20)	0.3383(10)	0.4037(47)	0.5062(11)
C(21)	0.3840(11)	0.5970(42)	0.5453(11)
C(22)	0.4105(10)	0.7457(30)	0.4944(13)
C(23)	0.3954(9)	0.7099(31)	0.4133(11)

The structure is shown in Figure XVC,17. Its molecules have the bond lengths of Figure XVC,18. In the PO_4 group the angles range between 97 and 119° and the O–P–C angles are 123°. Within the phenyl groups the angles range from 112 to 127°. The dihedral angles between the benzene rings and the planes through one of its carbon atoms and the adjacent oxygen and phosphorus vary greatly, equaling 50°41', 69°5', and 94°52'.

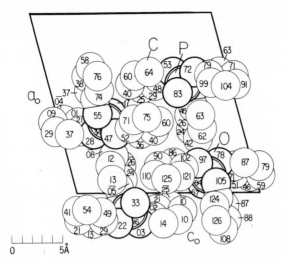

Fig. XVC,17. The monoclinic structure of triphenyl phosphate projected along its b_0 axis.

XV,c17. The hexagonal (rhombohedral) crystals of *tri-p-tolylarsine*, $(CH_3C_6H_4)_3As$, have a bimolecular rhombohedral unit of the dimensions:

$$a_0 = 9.84 \text{ A.,} \quad \alpha = 80°2'$$

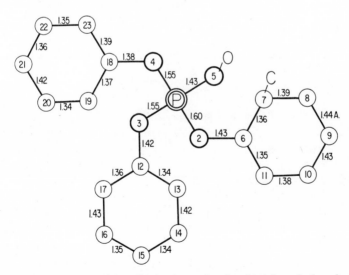

Fig. XVC,18. Bond lengths in the molecule of triphenyl phosphate.

The space group appears to be $C_{3i}^2(R\bar{3})$ with arsenic atoms in

$$(2c) \quad \pm(uuu) \qquad \text{with } u = 0.1867$$

and other atoms in

$$(6f) \quad \pm(xyz; zxy; yzx)$$

with the parameters stated in Table XVC,16.

TABLE XVC,16

Parameters of Atoms in $(CH_3C_6H_4)_3As$

Atom	x	y	z
C(1)	0.251	−0.014	0.192
C(2)	0.392	−0.065	0.151
C(3)	0.439	−0.207	0.156
C(4)	0.343	−0.301	0.201
C(5)	0.201	−0.251	0.244
C(6)	0.155	−0.108	0.238
C(7)	0.393	−0.454	0.208

In this structure the arsenic atom is surrounded by three planar tolyl groups, with the plane of each group rotated through 36° from

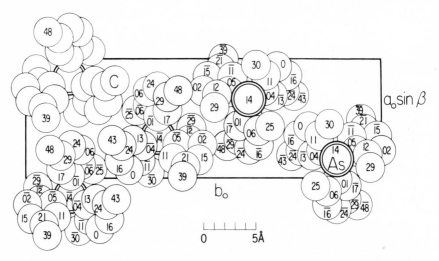

Fig. XVC,19. The monoclinic structure of tri-p-xylylarsine projected along its c_0 axis.

the plane of arsenic and its three attached carbon atoms. The As–C separation is 1.96 A.

XV,c18. Crystals of *tri-p-xylylarsine*, $(C_8H_9)_3As$, are monoclinic with a tetramolecular unit of the dimensions:

$$a_0 = 10.81 \text{ A.}; \quad b_0 = 33.4 \text{ A.}; \quad c_0 = 5.72 \text{ A.}; \quad \beta = 96°28'$$

Atoms are in the positions

$$(4e) \quad \pm (xyz; x + 1/2, 1/2 - y, z)$$

of the space group $C_{2h}^5(P2_1/a)$. The parameters are stated to be those of Table XVC,17.

TABLE XVC,17

Parameters of the Atoms in $(C_8H_9)_3As$

Atom	x	y	z
As	0.1467	0.1250	0.1893
C(1)	0.233	0.169	0.042
C(2)	0.355	0.175	0.109
C(3)	0.416	0.207	0.006
C(4)	0.355	0.232	−0.165
C(5)	0.233	0.225	−0.236
C(6)	0.172	0.194	−0.132
C(7)	0.421	0.148	0.302
C(8)	0.168	0.252	−0.427
C(9)	0.224	0.078	0.049
C(10)	0.187	0.040	0.115
C(11)	0.243	0.007	0.020
C(12)	0.335	0.013	−0.146
C(13)	0.371	0.051	−0.212
C(14)	0.316	0.084	−0.114
C(15)	0.086	0.033	0.295
C(16)	0.473	0.057	−0.391
C(17)	−0.024	0.126	0.011
C(18)	−0.103	0.155	0.057
C(19)	−0.223	0.157	−0.059
C(20)	−0.261	0.129	−0.236
C(21)	−0.180	0.099	−0.290
C(22)	−0.063	0.098	−0.167
C(23)	−0.061	0.186	0.255
C(24)	−0.222	0.069	−0.477

The structure is shown in Figure XVC,19. Each xylyl group is planar and includes the central arsenic atom; they are rotated about the C–As bonds in "propellor" fashion. The bond C–As = 1.98–2.00 A.

TABLE XVC,18

Parameters of the Atoms in $As(NO_2C_6H_4C_2)_3$

Atom	x	y	z
As	0.1114(1)	0.2529(6)	0.2239(1)
C(1,1)	0.2581(12)	0.5684(44)	0.0964(7)
C(1,2)	0.3127(11)	0.7977(40)	0.1019(7)
C(1,3)	0.3535(12)	0.8885(45)	0.0604(7)
C(1,4)	0.3339(11)	0.7828(45)	0.0161(7)
C(1,5)	0.2836(13)	0.5602(45)	0.0103(7)
C(1,6)	0.2404(11)	0.4617(39)	0.0495(7)
C(1,7)	0.2154(12)	0.4623(42)	0.1381(7)
C(1,8)	0.1760(11)	0.3673(37)	0.1712(7)
N(1)	0.3799(12)	0.8929(43)	−0.0275(7)
O(1,1)	0.4204(11)	0.0901(43)	−0.0225(7)
O(1,2)	0.3642(10)	0.7887(39)	−0.0664(7)
C(2,1)	−0.0281(10)	−0.3571(34)	0.1326(6)
C(2,2)	0.0117(12)	−0.4429(42)	0.0907(7)
C(2,3)	−0.0268(12)	−0.6469(39)	0.0614(7)
C(2,4)	−0.0997(10)	−0.7300(43)	0.0759(6)
C(2,5)	−0.1432(12)	−0.6518(39)	0.1162(7)
C(2,6)	−0.1054(11)	−0.4563(36)	0.1464(7)
C(2,7)	0.0110(11)	−0.1577(35)	0.1635(7)
C(2,8)	0.0505(11)	−0.0151(40)	0.1857(7)
N(2)	−0.1416(11)	−0.9658(36)	0.0449(6)
O(2,1)	−0.1097(10)	−0.9950(33)	0.0049(6)
O(2,2)	−0.1997(10)	−0.0618(34)	0.0618(5)
C(3,1)	0.3078(11)	−0.2773(46)	0.2902(7)
C(3,2)	0.3358(13)	−0.2029(50)	0.3383(8)
C(3,3)	0.4027(13)	−0.3392(41)	0.3582(8)
C(3,4)	0.4355(13)	−0.5365(46)	0.3302(8)
C(3,5)	0.4094(13)	−0.6173(52)	0.2836(8)
C(3,6)	0.3464(14)	−0.4742(47)	0.2624(8)
C(3,7)	0.2420(12)	−0.1152(44)	0.2691(7)
C(3,8)	0.1887(12)	0.0102(40)	0.2509(7)
N(3)	0.5120(12)	−0.6770(44)	0.3517(8)
O(3,1)	0.5334(11)	−0.6132(40)	0.3904(6)
O(3,2)	0.5347(11)	−0.8563(40)	0.3257(7)

XV,c19. The crystals of *tris(p-nitrophenylethynyl) arsine,* As($NO_2C_6H_4C_2$)$_3$, are monoclinic with a tetramolecular unit of the dimensions:

$$a_0 = 16.562(6) \text{ A.}; \quad b_0 = 4.974(10) \text{ A.}; \quad c_0 = 27.549(6) \text{ A.}$$
$$\beta = 89.75(5)°$$

The space group has been given as C_{2h}^4 (*P2/c*) with atoms in the positions:

$$(4g) \quad \pm(xyz; x,\bar{y},z+{}^1/_2)$$

Chosen parameters are those of Table XVC,18.

The structure is shown in Figure XVC,20. In the molecule the arsenic atom is at the apex of a trigonal pyramid in which the C–As–C angles are all close to 96°. The NO_2 radicals are turned through 3 or 11° with respect to the plane of their benzene rings. In the As–C≡C–C

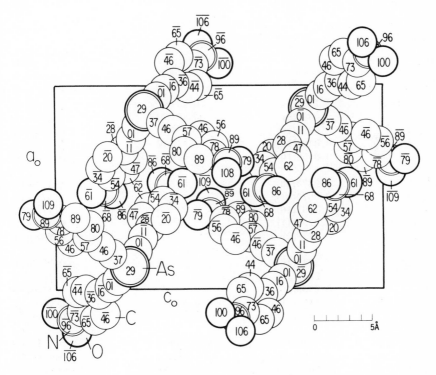

Fig. XVC,20. The monoclinic structure of tris(*p*-nitrophenylethynyl) arsine projected along its b_0 axis.

groups of atoms the average bond lengths are As–C = 1.93 A., C≡C = 1.18 A., and C–C = 1.46 A. but these lines seem slightly bent, the arsenic lying 0.16–0.45 A. off the line of the rest. Other bond lengths are normal.

XV,c20. *Triphenyl bismuth,* $Bi(C_6H_5)_3$, forms monoclinic crystals that possess a unit containing eight molecules and having the dimensions:

$$a_0 = 27.70(3) \text{ A.;} \quad b_0 = 5.82(2) \text{ A.;} \quad c_0 = 20.45(3) \text{ A.}$$
$$\beta = 114.48°$$

The space group is $C_{2h}^6 (C2/c)$ with all atoms in the general positions:

$$(8f) \quad \pm(xyz; x,\bar{y},z+1/2; x+1/2,y+1/2,z; x+1/2,1/2-y,z+1/2)$$

The parameters are stated in Table XVC,19.

TABLE XVC,19

Parameters of the Atoms in $Bi(C_6H_5)_3$

Atom	x	y	z
Bi	0.11448(4)	0.2718(2)	0.03017(5)
C(1)	0.0845(9)	0.477(5)	0.0965(12)
C(2)	0.0916(12)	0.401(6)	0.1164(16)
C(3)	0.0729(13)	0.518(6)	0.2089(18)
C(4)	0.0468(10)	0.716(5)	0.1850(14)
C(5)	0.0376(12)	0.823(6)	0.1160(17)
C(6)	0.0575(11)	0.692(6)	0.0736(15)
C(7)	0.1991(10)	0.315(5)	0.1105(13)
C(8)	0.2155(11)	0.517(6)	0.1530(15)
C(9)	0.2678(9)	0.525(5)	0.2049(13)
C(10)	0.3041(10)	0.348(5)	0.2125(14)
C(11)	0.2882(13)	0.154(6)	0.1651(17)
C(12)	0.2339(12)	0.135(5)	0.1143(15)
C(13)	0.1177(11)	0.567(5)	−0.0392(14)
C(14)	0.1650(11)	0.665(6)	−0.0313(16)
C(15)	0.1619(13)	0.855(6)	−0.0780(17)
C(16)	0.1162(12)	0.931(6)	−0.1311(17)
C(17)	0.0699(11)	0.820(6)	−0.1360(15)
C(18)	0.0695(11)	0.655(5)	−0.0901(15)

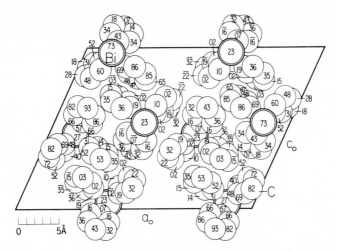

Fig. XVC,21.   The monoclinic structure of triphenyl bismuth projected along
its b_0 axis.

The structure is shown in Figure XVC,21. In its molecules the bismuth atom is at the apex of a trigonal pyramid, with C–Bi–C angles between 92 and 96° and with C–Bi = 2.21 or 2.25 A.

XV,c21.   The orthorhombic crystals of *triphenyl stibine dichloride*, $(C_6H_5)_3SbCl_2$, have a tetramolecular unit of the dimensions:

$$a_0 = 13.17(3) \text{ A.}; \quad b_0 = 11.08(3) \text{ A.}; \quad c_0 = 12.39(3) \text{ A.}$$

Atoms are in the positions

(4a) $xyz; \frac{1}{2}-x,\bar{y},z+\frac{1}{2}; x+\frac{1}{2},\frac{1}{2}-y,\bar{z}; \bar{x},y+\frac{1}{2},\frac{1}{2}-z$

of the space group V^4 ($P2_12_12_1$). The parameters are given in Table XVC,20. Calculated values for the hydrogen atoms are to be found in the original.

The structure is built up of molecules which have the general configuration of those in SbCl$_5$ (**V,e4**) and $(CH_3)_3SbCl_2$ (**XIV,a65**). The coordination of the antimony atoms is that of a trigonal bipyramid with the three phenyl groups in the equatorial plane. In this molecule Sb–Cl = 2.458 and 2.509 A. and Sb–C = 2.14, 2.16, and 2.18 A.

XV,c22.   The unit in the orthorhombic crystals of *triphenyl bismuth dichloride*, $(C_6H_5)_3BiCl_2$, contains eight molecules and has the edge

TABLE XVC,20

Parameters of the Atoms in $(C_6H_5)_3SbCl_2$

Atom	x	y	z
Sb	0.2292	0.3691	0.2302
Cl(1)	0.223	0.171	0.319
Cl(2)	0.226	0.575	0.145
C(1)	0.225	0.282	0.072
C(2)	0.305	0.198	0.040
C(3)	0.306	0.159	−0.060
C(4)	0.231	0.188	−0.131
C(5)	0.160	0.264	−0.105
C(6)	0.155	0.320	0.001
C(7)	0.374	0.398	0.304
C(8)	0.405	0.347	0.395
C(9)	0.508	0.367	0.435
C(10)	0.582	0.420	0.373
C(11)	0.545	0.467	0.275
C(12)	0.443	0.452	0.238
C(13)	0.095	0.339	0.327
C(14)	0.087	0.513	0.381
C(15)	0.001	0.551	0.441
C(16)	−0.088	0.483	0.437
C(17)	−0.085	0.379	0.362
C(18)	0.009	0.337	0.314

TABLE XVC,21

Parameters of the Atoms in Triphenyl Bismuth Dichloride

Atom	x	y	z
		Molecule A	
Bi(1)	0.08946(28)	0.04135(13)	0.01371(9)
Cl(1)	−0.0039(20)	0.1669(8)	0.0684(6)
Cl(2)	0.1994(23)	−0.0769(10)	−0.0371(7)
C(1)	−0.032(9)	−0.033(4)	0.081(3)
C(2)	−0.060(14)	−0.006(7)	0.135(5)
C(3)	−0.118(11)	−0.062(5)	0.173(4)

(*continued on next page*)

TABLE XVC,21 (*continued*)

Parameters of the Atoms in Triphenyl Bismuth Dichloride

Atom	x	y	z
C(4)	−0.151(13)	−0.135(6)	0.161(4)
C(5)	−0.128(12)	−0.165(5)	0.100(4)
C(6)	−0.078(11)	−0.109(5)	0.059(3)
C(7)	−0.030(11)	0.075(5)	−0.071(4)
C(8)	0.009(12)	0.043(6)	−0.119(5)
C(9)	−0.057(12)	0.072(5)	−0.171(4)
C(10)	−0.181(13)	0.119(6)	−0.165(4)
C(11)	−0.248(14)	0.147(6)	−0.118(5)
C(12)	−0.132(12)	0.128(5)	−0.064(4)
C(13)	0.277(7)	0.061(3)	0.037(2)
C(14)	0.358(10)	0.137(4)	0.021(3)
C(15)	0.496(12)	0.165(6)	0.042(4)
C(16)	0.587(10)	0.109(4)	0.074(3)
C(17)	0.559(12)	0.039(6)	0.091(4)
C(18)	0.388(9)	0.025(4)	0.070(3)
	Molecule B		
Bi(1′)	0.64814(29)	0.22579(13)	0.30119(9)
Cl(1′)	0.7413(22)	0.1098(9)	0.2381(7)
Cl(2′)	0.5318(24)	0.3357(11)	0.3626(8)
C(1′)	0.758(8)	0.191(3)	0.382(2)
C(2′)	0.795(10)	0.109(5)	0.387(3)
C(3′)	0.871(12)	0.067(5)	0.434(4)
C(4′)	0.905(13)	0.112(6)	0.480(4)
C(5′)	0.873(15)	0.191(6)	0.481(5)
C(6′)	0.795(10)	0.234(5)	0.428(3)
C(7′)	0.767(12)	0.305(5)	0.238(4)
C(8′)	0.787(11)	0.385(5)	0.262(4)
C(9′)	0.858(13)	0.444(6)	0.224(4)
C(10′)	0.894(13)	0.415(6)	0.168(4)
C(11′)	0.863(10)	0.342(4)	0.150(3)
C(12′)	0.817(12)	0.282(6)	0.189(4)
C(13′)	0.445(10)	0.192(4)	0.282(3)
C(14′)	0.394(10)	0.183(4)	0.226(3)
C(15′)	0.217(11)	0.166(5)	0.206(4)
C(16′)	0.137(11)	0.151(4)	0.250(3)
C(17′)	0.179(12)	0.166(6)	0.312(4)
C(18′)	0.322(9)	0.180(4)	0.324(3)

lengths:

$$a_0 = 9.18(2) \text{ A.}; \quad b_0 = 17.11(3) \text{ A.}; \quad c_0 = 22.30(3) \text{ A.}$$

The space group is V^4 $(P2_12_12_1)$ with atoms in the positions:

$$(4a) \quad xyz; \; {}^1/_2 - x, \bar{y}, z + {}^1/_2; \; x + {}^1/_2, {}^1/_2 - y, \bar{z}; \; \bar{x}, y + {}^1/_2, {}^1/_2 - z$$

The parameters are listed in Table XVC,21.

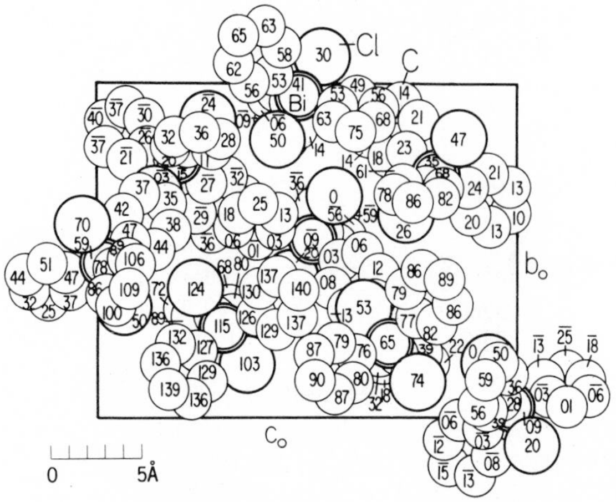

Fig. XVC,22.   The orthorhombic structure of triphenyl bismuth dichloride projected along its a_0 axis.

This structure (Fig. XVC,22) contains two crystallographically different but very similar molecules. They have the form of trigonal bipyramids, with chlorine atoms at the apices and the carbon atoms that connect the phenyl groups to bismuth coplanar with it. The angles between the phenyl groups and this plane are not the same, being turned 65(68)°, 66(69)°, and 35(33)° with reference to it (values in parentheses refer to the second molecule). It is noteworthy that the groups containing C(1) and C(7) are rotated in opposite senses with respect to the plane. The angle Cl–Bi–Cl is 175(176)°. The bond distances Bi–Cl = 2.53–2.62 A. and Bi–C = 2.14–2.25 A. for the C(1) and C(7) groups but are much shorter (1.83 and 2.00 A.) for the phenyl group associated with C(13).

XV,c23. Crystals of *dimethoxytriphenyl antimony*, $(C_6H_5)_3$-Sb(OCH$_3$)$_2$, are monoclinic with a tetramolecular unit of the

dimensions:

$$a_0 = 11.51(2) \text{ A.}; \quad b_0 = 9.40(2) \text{ A.}; \quad c_0 = 7.30(3) \text{ A.}; \quad \beta = 101.75(1)°$$

Atoms are in the positions

$$(4e) \quad \pm (xyz; x, {}^1\!/_2 - y, z + {}^1\!/_2)$$

of the space group $C_{2h}^5 (P2_1/c)$. Parameters are listed in Table XVC,22.

TABLE XVC,22

Parameters of the Atoms in $(C_6H_5)_3Sb(OCH_3)_2$

Atom	x	y	z
Sb	−0.24451(6)	0.21039(9)	0.18937(4)
O(1)	−0.0660(6)	0.2455(7)	0.2114(4)
O(2)	−0.4192(6)	0.1599(7)	0.1708(4)
C(1)	−0.0076(10)	0.3108(12)	0.1552(6)
C(2)	−0.4913(10)	0.2119(13)	0.2214(7)
C(3)	−0.2150(10)	0.1271(15)	0.3056(6)
C(4)	−0.1295(12)	0.1818(14)	0.3646(7)
C(5)	−0.1180(14)	0.1280(20)	0.4394(8)
C(6)	−0.1876(14)	0.0198(17)	0.4563(8)
C(7)	−0.2678(11)	−0.0384(13)	0.3982(10)
C(8)	−0.2814(10)	0.0168(15)	0.3221(7)
C(9)	−0.2416(11)	0.0863(11)	0.0874(6)
C(10)	−0.3386(11)	0.0875(14)	0.0249(9)
C(11)	−0.3320(14)	0.0112(16)	−0.0429(9)
C(12)	−0.2346(16)	−0.0646(16)	−0.0475(9)
C(13)	−0.1430(12)	−0.0671(14)	0.0131(11)
C(14)	−0.1437(11)	0.0063(14)	0.0804(8)
C(15)	−0.2823(11)	0.4288(12)	0.1666(6)
C(16)	−0.3869(10)	0.4676(14)	0.1149(7)
C(17)	−0.4107(11)	0.6082(16)	0.0957(7)
C(18)	−0.3339(14)	0.7134(15)	0.1254(8)
C(19)	−0.2286(14)	0.6735(16)	0.1766(8)
C(20)	−0.2034(11)	0.5322(15)	0.1981(7)

The structure, shown in Figure XVC,23, has molecules with the bond dimensions of Figure XVC,24. As is true for the tetraphenyl compound (**XV,d34**), the coordination of the antimony atom is five-fold, with the two oxygen atoms at the apices of a trigonal bipyramid.

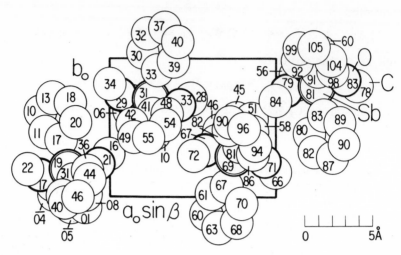

Fig. XVC,23. The monoclinic structure of dimethoxytriphenyl antimony
projected along its c_0 axis.

XV,c24. *Cesium tetracyanoquinodimethanide,* which has the composition $Cs_2[C_6H_4C_2(CN)_4]_3$, forms monoclinic crystals having a bimolecular cell of the dimensions:

$$a_0 = 7.34(1) \text{ A.}; \quad b_0 = 10.40(2) \text{ A.}; \quad c_0 = 21.98(4) \text{ A.}; \quad \beta = 97°11(3)'$$

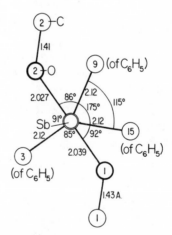

Fig. XVC,24. Some bond dimensions in the molecule of dimethoxytriphenyl
antimony.

All atoms are in the general positions

$$(4e) \quad \pm (xyz;\, x, {}^{1}/_{2} - y, z + {}^{1}/_{2})$$

of $C_{2h}{}^{5}(P2_{1}/c)$. Parameters, refined in a recent study, are stated in Table XVC,23, those for the hydrogen atoms being approximate only.

TABLE XVC,23

Parameters of the Atoms in $Cs_2[C_6H_4C_2(CN)_4]_3$

Atom	x	y	z
Cs(1)	0.51750(3)	0.23071(3)	0.26651(1)
N(2)	−0.38808(49)	0.00335(37)	0.16494(16)
N(3)	0.20271(51)	0.00096(39)	0.23681(15)
N(4)	0.47584(49)	0.30093(42)	−0.08661(17)
N(5)	−0.08822(51)	0.29856(36)	−0.16944(16)
N(6)	−0.41626(53)	0.34015(46)	0.12713(17)
N(7)	0.16208(55)	0.36427(41)	0.20226(16)
C(8)	−0.23755(52)	0.01926(36)	0.15706(16)
C(9)	−0.05195(50)	0.04119(33)	0.14743(16)
C(10)	0.08598(53)	0.01881(36)	0.19739(16)
C(11)	−0.14012(43)	0.11155(32)	0.03929(16)
C(12)	−0.00554(46)	0.09049(31)	0.09115(15)
C(13)	0.18004(48)	0.12070(33)	0.08434(16)
C(14)	0.22664(45)	0.16927(34)	0.03107(16)
C(15)	0.09230(45)	0.19037(31)	−0.02082(15)
C(16)	−0.09342(45)	0.15877(32)	−0.01400(15)
C(17)	0.32597(54)	0.27421(38)	−0.08142(17)
C(18)	0.14142(47)	0.24091(31)	−0.07562(15)
C(19)	0.01321(51)	0.27053(35)	−0.12741(16)
C(20)	−0.26902(56)	0.36950(39)	0.12017(17)
C(21)	−0.08562(50)	0.40318(33)	0.11133(15)
C(22)	0.05244(54)	0.38327(38)	0.16243(16)
C(23)	−0.18286(48)	0.46627(32)	0.00467(15)
C(24)	−0.04397(48)	0.45104(31)	0.05659(15)
C(25)	0.14255(48)	0.48715(33)	0.04894(15)
H(26)	−0.267	0.088	0.046
H(27)	0.274	0.112	0.121
H(28)	0.355	0.188	0.026
H(29)	−0.192	0.172	−0.049
H(30)	−0.311	0.442	0.010
H(31)	0.238	0.477	0.085

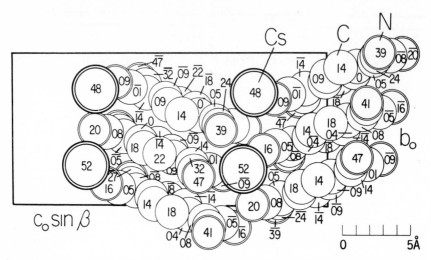

Fig. XVC,25. The monoclinic structure of cesium tetracyanoquinodimethanide projected along its a_0 axis.

The structure is shown in Figure XVC,25. Its three complex anions are of two sorts, one having a center of symmetry, while the other two are asymmetric. Bond dimensions in these anions are shown in Figure XVC,26. Each cesium atom has about it an approximate cube of eight nitrogen atoms at distances between 3.07 and 3.49 A. The anions are nearly but not exactly planar. Adjacent asymmetric anions are only 3.26 A. apart.

XV,c25. *Tris(1-phenyl-1,3-butanedionato)aquoyttrium(III)*, $Y(C_6-H_5COCHCOCH_3)_3 \cdot H_2O$, forms triclinic crystals whose bimolecular

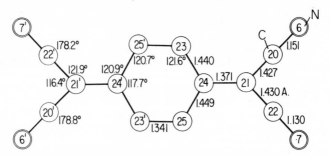

Fig. XVC,26a. Bond dimensions in the centrosymmetric anion of cesium tetracyanoquinodimethanide.

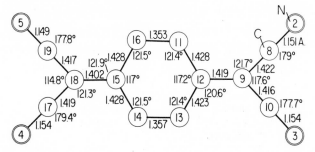

Fig. XVC,26b.   Bond dimensions in the asymmetric anion of cesium tetracyanoquinodimethanide.

unit has the dimensions:

$$a_0 = 6.214(5) \text{ A.};\quad b_0 = 12.462(8) \text{ A.};\quad c_0 = 19.299(8) \text{ A.}$$
$$\alpha = 95°37(3)';\qquad \beta = 104°53(3)';\qquad \gamma = 96°20(3)'$$

The space group is C_i^1 $(P\bar{1})$ with all atoms in the positions $(2i) \pm (xyz)$.
The parameters of the atoms heavier than hydrogen are listed in Table
XVC,24; values for hydrogen are to be found in the original article.

The structure is shown in Figure XVC,27. Its yttrium atoms have a
sevenfold coordination, each of the three anionic groups supplying
two oxygens, the seventh being the water oxygen. The Y–O distances,
nearly equal, range between 2.248 and 2.341 A., the latter applying
to the oxygen of water. These water molecules appear to be involved

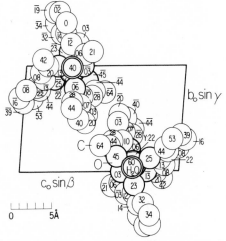

Fig. XVC,27.   The triclinic structure of tris(1-phenyl-1,3-butanedionato)aquoyt-
trium projected along its a_0 axis.

TABLE XVC,24

Parameters of the Atoms in $Y(C_6H_5COCHCOCH_3)_3 \cdot H_2O$

Atom	x	y	z
Y	0.22040(5)	0.06298(3)	0.28734(2)
O(1)	0.4515(6)	0.1484(3)	0.3939(2)
O(2)	0.0645(5)	0.2123(2)	0.3129(2)
C(1)	0.6417(11)	0.2526(5)	0.5047(3)
C(2)	0.4475(8)	0.2296(4)	0.4380(2)
C(3)	0.2820(8)	0.2972(4)	0.4279(2)
C(4)	0.0997(7)	0.2865(3)	0.3658(2)
C(5)	−0.0701(7)	0.3654(3)	0.3591(2)
C(6)	−0.0276(9)	0.4643(4)	0.4029(3)
C(7)	−0.1982(10)	0.5313(5)	0.3947(3)
C(8)	−0.4009(11)	0.5020(5)	0.3451(3)
C(9)	−0.4429(11)	0.4020(5)	0.3005(3)
C(10)	−0.2755(9)	0.3355(4)	0.3076(3)
O(3)	−0.1271(5)	0.0098(2)	0.2034(2)
O(4)	0.2473(5)	0.1311(2)	0.1848(2)
C(11)	−0.4230(10)	−0.0592(5)	0.0998(3)
C(12)	−0.1952(8)	0.0059(4)	0.1355(2)
C(13)	−0.0793(9)	0.0598(4)	0.0924(3)
C(14)	0.1275(7)	0.1266(3)	0.1206(2)
C(15)	0.2160(8)	0.1996(4)	0.0733(2)
C(16)	0.0783(11)	0.2282(5)	0.0113(3)
C(17)	0.1648(13)	0.3018(6)	−0.0285(4)
C(18)	0.3860(13)	0.3442(6)	−0.0068(4)
C(19)	0.5296(11)	0.3120(5)	0.0531(4)
C(20)	0.4428(9)	0.2413(4)	0.0939(3)
O(5)	0.0256(5)	−0.0126(3)	0.3609(2)
O(6)	0.2307(6)	−0.1169(3)	0.2669(2)
C(21)	−0.2128(10)	−0.1217(5)	0.4129(3)
C(22)	−0.0613(8)	−0.1089(4)	0.3627(2)
C(23)	−0.0274(8)	−0.2021(4)	0.3227(3)
C(24)	0.1156(7)	−0.2022(3)	0.2771(2)
C(25)	0.1402(8)	−0.3059(4)	0.2371(2)
C(26)	0.3212(10)	−0.3077(4)	0.2057(3)
C(27)	0.3447(12)	−0.4032(5)	0.1670(4)
C(28)	0.1915(12)	−0.4973(6)	0.1573(4)
C(29)	0.0214(11)	−0.4971(5)	0.1896(4)
C(30)	−0.0057(10)	−0.4025(5)	0.2292(3)
O(7,H_2O)	0.5955(5)	0.0545(2)	0.2882(2)

in hydrogen bonds that tie the large organic molecules together along the a_0 direction.

XV,c26. The triclinic crystals of *triphenylmethylphosphonium bis(1,2-dicyanoethylene-1,2-dithiolato)nickelate(III)*, $[PCH_3(C_6H_5)_3]^+Ni[S_2C_2(CN)_2]_2^-$, possess a bimolecular unit of the dimensions:

$$a_0 = 12.815(3) \text{ A.}; \quad b_0 = 8.409(2) \text{ A.}; \quad c_0 = 14.276(3) \text{ A.}$$
$$\alpha = 108°47(1)'; \quad \beta = 102°8(1)'; \quad \gamma = 93°56(1)'$$

The space group is C_i^1 ($P\bar{1}$) with all atoms in the positions $(2i) \pm (xyz)$. Determined parameters are listed in Table XVC,25.

TABLE XVC,25

Parameters of the Atoms in $[PCH_3(C_6H_5)_3] \cdot Ni[S_2C_2(CN)_2]_2$

Atom	x	y	z
Ni(1)	0.0362(1)	0.2358(1)	−0.0301(1)
S(2)	0.1910(1)	0.3817(2)	−0.0013(1)
S(3)	−0.1211(1)	0.0930(2)	−0.0653(1)
S(4)	0.0083(1)	0.3929(2)	0.1120(1)
S(5)	0.0696(1)	0.0775(2)	−0.1681(1)
P(6)	0.3182(1)	−0.1974(2)	−0.3607(1)
N(7)	0.3822(6)	0.7368(11)	0.1840(5)
N(8)	0.1421(7)	0.7586(11)	0.3447(5)
N(9)	−0.0460(7)	−0.2993(12)	−0.3951(6)
N(10)	−0.2916(5)	−0.2981(9)	−0.2609(5)
C(11)	0.3029(6)	0.6478(10)	0.1556(5)
C(12)	0.2033(6)	0.5336(8)	0.1165(5)
C(13)	0.1238(6)	0.5384(8)	0.1658(5)
C(14)	0.1342(7)	0.6616(10)	0.2647(5)
C(15)	−0.0466(6)	−0.1982(10)	−0.3185(6)
C(16)	−0.0444(5)	−0.0689(8)	−0.2245(5)
C(17)	−0.1281(5)	−0.0632(8)	−0.1794(5)
C(18)	−0.2206(5)	−0.1933(9)	−0.2247(5)
C(19)	0.2086(6)	−0.3638(10)	−0.4335(6)
C(20)	0.3697(5)	−0.2186(8)	−0.2385(5)
C(21)	0.4715(5)	−0.2604(9)	−0.2123(5)
C(22)	0.5092(7)	−0.2674(12)	−0.1182(6)
C(23)	0.4453(7)	−0.2507(12)	−0.0502(6)
C(24)	0.3436(8)	−0.2116(14)	−0.0759(7)
C(25)	0.3042(7)	−0.1955(12)	−0.1706(6)

(*continued on next page*)

TABLE XVC,25 (*continued*)

Parameters of the Atoms in $[PCH_3(C_6H_5)_3] \cdot Ni[S_2C_2(CN)_2]_2$

Atom	x	y	z
C(26)	0.4222(5)	−0.2094(8)	−0.4288(5)
C(27)	0.4091(7)	−0.3327(9)	−0.5232(5)
C(28)	0.4911(8)	−0.3370(11)	−0.5724(6)
C(29)	0.5866(7)	−0.2252(11)	−0.5282(7)
C(30)	0.5977(6)	−0.1001(11)	−0.4343(6)
C(31)	0.5151(6)	−0.0924(9)	−0.3844(5)
C(32)	0.2696(5)	0.0043(8)	−0.3373(5)
C(33)	0.1774(6)	0.0215(10)	−0.4026(5)
C(34)	0.1428(7)	0.1774(12)	−0.3830(7)
C(35)	0.1976(8)	0.3157(11)	−0.2985(7)
C(36)	0.2883(8)	0.2995(10)	−0.2359(7)
C(37)	0.3255(6)	0.1439(9)	−0.2539(5)
H(38)	0.514 —	−0.277 —	−0.260 —
H(39)	0.579 —	−0.304 —	−0.100 —
H(40)	0.471 —	−0.262 —	0.014 —
H(41)	0.300 —	−0.192 —	−0.028 —
H(42)	0.236 —	−0.166 —	−0.188 —
H(43)	0.343 —	−0.415 —	−0.555 —
H(44)	0.484 —	−0.418 —	−0.637 —
H(45)	0.641 –	−0.228 —	−0.563 —
H(46)	0.665 —	−0.017 —	−0.403 —
H(47)	0.523 —	−0.014 —	−0.321 —
H(48)	0.138 —	−0.076 —	−0.462 —
H(49)	0.081 —	0.190 —	−0.427 —
H(50)	0.173 —	0.422 —	−0.287 —
H(51)	0.329 —	0.397 —	−0.175 —
H(52)	0.386 —	0.131 —	−0.210 —
H(53)	0.227 —	−0.471 —	−0.437 —
H(54)	0.180 —	−0.375 —	−0.503 —
H(55)	0.148 —	−0.342 —	−0.390 —

In this structure (Fig. XVC,28) the complex anions have the bond dimensions of Figure XVC,29. The anion, with the nickel atom at its center, is nearly planar, the greatest departure from the best plane through it being ca. 0.14 A.; as the figure indicates, the coordination of the metal atom is fourfold and nearly square. In the cation P–C lies between 1.786 and 1.800 A., and all angles about the tetrahedrally

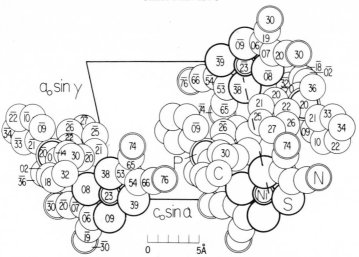

Fig. XVC,28.   The triclinic structure of triphenylmethylphosphonium bis(1,2-dicyanoethylene-1,2-dithiolato)nickelate projected along its b_0 axis.

coordinated phosphorus atom are close to 109°. In the structure the anions are stacked one above another along b_0, with the cations alternating in the large voids between.

XV,c27.   The   compound   *tris(phenyldiethylphosphine)nonachlorotri-rhenium(III)*, [C₆H₅(C₂H₅)₂P]₃Re₃Cl₉, forms cubic crystals with eight of these molecules in a unit of the edge length:

$$a_0 = 20.53(2) \text{ A.}$$

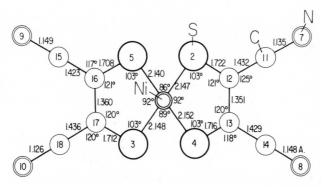

Fig. XVC,29.   Bond dimensions in the anion of triphenylmethylphosphonium bis(1,2-dicyanoethylene-1,2-dithiolato)nickelate.

The space group is $T_h^6 (Pa3)$ with all atoms in the general positions:

(24d) $\pm (xyz; x+\frac{1}{2}, \frac{1}{2}-y, \bar{z}; \bar{x}, y+\frac{1}{2}, \frac{1}{2}-z; \frac{1}{2}-x, \bar{y}, z+\frac{1}{2})$;   tr

The determined parameters are those of Table XVC,26.

TABLE XVC,26

Parameters of the Atoms in $[C_6H_5(C_2H_5)_2P]_3Re_3Cl_9$

Atom	x	y	z
Re	0.1358(1)	0.2182(1)	0.2247(1)
P	0.0294(9)	0.2685(9)	0.2834(9)
Cl(1)	0.1827(8)	0.2891(8)	0.2980(8)
Cl(2)	0.0547(8)	0.1621(7)	0.1692(8)
Cl(3)	0.1459(8)	0.1318(8)	0.3025(8)
C(1)	0.020(4)	0.239(4)	0.364(4)
C(2)	−0.024(3)	0.170(3)	0.370(3)
C(3)	−0.028(4)	0.146(4)	0.432(4)
C(4)	0.002(3)	0.177(3)	0.486(3)
C(5)	0.032(3)	0.230(3)	0.481(3)
C(6)	0.048(3)	0.260(3)	0.422(4)
C(7)	−0.044(3)	0.249(3)	0.236(3)
C(8)	−0.053(3)	0.295(3)	0.174(3)
C(9)	0.040(3)	0.359(3)	0.294(3)
C(10)	−0.029(3)	0.384(3)	0.331(3)

The structure appears to be fundamentally molecular with the Re_3Cl_9 part similar to that in other compounds involving three rhenium atoms in ring form. In this compound Re–Re = 2.493 A. and Re–Cl = 2.30–2.40 A. In the organic portion P–C = 1.77–1.88 A. and P–Re = 2.700 A.

XV,c28. The monoclinic crystals of *nitridodichloro tris(diethylphenylphosphine)rhenium(V)*, $[P(C_2H_5)_2C_6H_5]_3ReNCl_2$, have a tetramolecular unit of the dimensions:

$a_0 = 18.669(4)$ A.; $b_0 = 11.514(2)$ A.; $c_0 = 18.186(4)$ A.
$$\beta = 122°34(1)'$$

Atoms are in the positions

(4e) $\pm (xyz; x, \frac{1}{2}-y, z+\frac{1}{2})$

of the space group C_{2h}^5 ($P2_1/c$). Some parameters are listed in Table XVC,27; those for the individual carbon atoms of the phenyl groups are stated in the original article.

TABLE XVC,27

Parameters of the Atoms in $[P(C_2H_5)_2C_6H_5]_3ReNCl_2$

Atom	x	y	z
Re	0.22715(5)	0.04334(7)	0.14000(5)
Cl(1)	0.1985(3)	0.0389(4)	0.2568(3)
Cl(2)	0.3599(3)	0.1584(4)	0.2504(3)
P(1)	0.3071(3)	−0.1401(4)	0.2063(3)
P(2)	0.2853(3)	0.0812(4)	0.0500(3)
P(3)	0.1419(3)	0.2234(4)	0.0920(3)
N	0.1360(7)	−0.0347(10)	0.0579(7)
C(1)	0.4156(9)	−0.1354(14)	0.3058(10)
C(2)	0.4139(10)	−0.0857(16)	0.3856(11)
C(3)	0.2377(12)	−0.2279(18)	0.2332(13)
C(4)	0.2734(15)	−0.3394(23)	0.2642(16)
C(5)	0.3222(9)	0.2354(14)	0.0581(10)
C(6)	0.3570(10)	0.2615(15)	−0.0023(11)
C(7)	0.2100(9)	0.0716(13)	−0.0706(10)
C(8)	0.1597(10)	−0.0460(15)	−0.1052(11)
C(9)	0.1962(9)	0.3596(14)	0.1491(10)
C(10)	0.2180(10)	0.3590(15)	0.2457(11)
C(11)	0.0479(10)	0.2119(15)	0.1036(10)
C(12)	−0.0058(10)	0.3277(15)	0.0732(11)
Ph(1)	0.3199(5)	−0.3195(6)	0.0779(5)
Ph(2)	0.4554(4)	−0.0538(6)	0.0916(5)
Ph(3)	0.0515(4)	0.2729(7)	−0.1122(5)

Note: Ph refers to the center of a phenyl (C_6H_5) group.

The structure is molecular. Coordination about the central rhenium atom of the molecule is octahedral; bond dimensions involving this atom are shown in Figure XVC,30. The atoms Re, Cl(1), Cl(2), P(2), and N are coplanar. The $P(C_2H_5)_2C_6H_5$ groups have their expected sizes and shapes.

XV,c29. Crystals of *trichloro(p-methoxyphenylimino) bis(diethylphenylphosphine)rhenium(V)*, $[P(C_2H_5)_2(C_6H_5)]_2ReCl_3(NC_6H_4OCH_3)$,

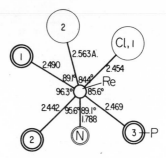

Fig. XVC,30. Dimensions of the bonds about the rhenium atom in nitridodichloro tris(diethylphenylphosphine)rhenium.

TABLE XVC,28

Parameters of Atoms in $[P(C_2H_5)_2(C_6H_5)]_2ReCl_3(NC_6H_4OCH_3)$

Atom	Position	x	y	z
Re	(4e)	0	0.02204(1)	$^1/_4$
Cl(1)	(4e)	0	−0.15171(10)	$^1/_4$
Cl(2)	(8f)	0.04583(10)	0.01249(9)	0.13695(8)
P	(8f)	−0.19339(9)	0.00325(7)	0.11448(8)
N	(4e)	0	0.14410(31)	$^1/_4$
O	(8f)	0.0332(8)	0.5345(5)	0.2500(8)
C(CH₃)	(8f)	−0.0284(12)	0.5959(8)	0.2657(9)
C(1,C₂H₅)	(8f)	−0.2581(4)	−0.0955(4)	0.1305(4)
C(2,C₂H₅)	(8f)	−0.3791(5)	−0.1043(5)	0.0517(5)
C(3,C₂H₅)	(8f)	−0.2262(4)	−0.0146(4)	−0.0069(3)
C(4,C₂H₅)	(8f)	−0.1947(4)	−0.1130(4)	0.0208(4)
C(1,C₆H₅)[a]	(8f)	−0.2725(9)	0.1070(5)	0.1022(5)
C(2,C₆H₅)	(8f)	−0.3158(4)	0.1706(3)	0.0246(2)
C(3,C₆H₅)	(8f)	−0.3718(6)	0.2510(4)	0.0200(4)
C(4,C₆H₅)	(8f)	−0.3846(10)	0.2677(5)	0.0930(6)
C(5,C₆H₅)	(8f)	−0.3414(4)	0.2041(3)	0.1706(3)
C(6,C₆H₅)	(8f)	−0.2853(6)	0.1238(4)	0.1752(4)
C(1′,C₆H₅)	(8f)	0.0078(9)	0.2419(2)	0.2525(9)
C(2′,C₆H₅)	(8f)	0.0832(7)	0.2851(5)	0.2443(8)
C(3′,C₆H₅)	(8f)	0.0891(5)	0.3844(5)	0.2432(5)
C(4′,C₆H₅)	(8f)	0.0195(5)	0.4404(2)	0.2503(5)
C(5′,C₆H₅)	(8f)	−0.0559(5)	0.3972(5)	0.2585(5)
C(6′,C₆H₅)	(8f)	−0.0618(7)	0.2979(5)	0.2596(7)

[a]Parameters assigned to the benzene carbon atoms involve the assumption that C–C = 1.392 A.

are monoclinic with a tetramolecular unit of the dimensions:

$$a_0 = 15.427(6) \text{ A.}; \quad b_0 = 13.998(6) \text{ A.}; \quad c_0 = 17.062(7) \text{ A.}$$
$$\beta = 126.55(2)°$$

The space group is C_{2h}^6 ($C2/c$) with atoms in the positions:

(4e) $\quad \pm (0 \; u \; 1/4; \; 1/2, u + 1/2, 1/4)$

(8f) $\quad \pm (xyz; \; x, \bar{y}, z + 1/2; \; x + 1/2, y + 1/2, z; \; x + 1/2, 1/2 - y, z + 1/2)$

The chosen positions and parameters are listed in Table XVC,28.

In the resulting structure the rhenium atoms have a sixfold coordination with three chlorine atoms at distances of 2.421 and 2.432 A., the nitrogen atom at a distance of 1.709 A., and the two phosphorus atoms 2.470 A. away. Choice of C_{2h}^6 as space group gives the molecules a twofold axis of symmetry, but this is not possible for the ($NC_6H_4OCH_3$) part unless disorder exists in the distribution about the Re–N line.

XV,c30. Crystals of *trichloro(p-acetylphenylimino) bis(diethylphenyl-phosphine)rhenium(V)*, $[P(C_2H_5)_2(C_6H_5)]_2ReCl_3(NC_6H_4COCH_3)$, are monoclinic. Their tetramolecular unit has the dimensions:

$$a_0 = 10.52(2) \text{ A.}; \quad b_0 = 16.06(3) \text{ A.}; \quad c_0 = 18.15(3) \text{ A.};$$
$$\beta = 91.65(5)°$$

Atoms are in the positions

(4e) $\quad \pm (xyz; \; x, 1/2 - y, z + 1/2)$

of C_{2h}^5 ($P2_1/c$). The parameters are listed in Table XVC,29.

The molecules in the resulting structure are similar to those in the corresponding methoxy compound (**XV,c29**) but in this case there is no evidence for disorder. Distribution about the rhenium atom is again sixfold, with Re–Cl = 2.399–2.433 A., Re–P = 2.457 or 2.461 A., and Re–N = 1.690 A.

XV,c31. *Tricarbonyl(triphenylphosphine)-σ-tetrafluoroethyl cobalt*, $CHF_2 \cdot CF_2 \cdot Co(CO)_3P(C_6H_5)_3$, forms monoclinic crystals whose tetramolecular unit has the dimensions:

$$a_0 = 9.675(10) \text{ A.}; \quad b_0 = 19.99(1) \text{ A.}; \quad c_0 = 11.975(10) \text{ A.}$$
$$\beta = 91°22(10)'$$

The space group is C_{2h}^5 ($P2_1/c$) with atoms in the positions:

(4e) $\quad \pm (xyz; \; x, 1/2 - y, z + 1/2)$

TABLE XVC,29

Parameters of Atoms in $[P(C_2H_5)_2(C_6H_5)]_2ReCl_3(NC_6H_4COCH_3)$

Atom	x	y	z
Re	0.21119(2)	0.18152(2)	0.16740(1)
Cl(1)	0.38606(16)	0.12483(11)	0.23996(10)
Cl(2)	0.11627(17)	0.21468(12)	0.28235(10)
Cl(3)	0.33544(18)	0.14328(12)	0.06162(10)
P(1)	0.11521(16)	0.04263(11)	0.17692(9)
P(2)	0.32831(17)	0.31408(11)	0.17329(10)
N	0.09053(48)	0.21447(31)	0.11126(28)
C(Ac)	−0.2945(7)	0.2870(5)	−0.1030(4)
O(Ac)	−0.2735(6)	0.2706(4)	−0.1669(3)
CH_3(Ac)	−0.4152(10)	0.3212(6)	−0.0818(6)
C(1,C_2H_5)	0.1314(7)	−0.0038(4)	0.2676(4)
C(2,C_2H_5)	0.0780(8)	−0.0920(5)	0.2733(5)
C(3,C_2H_5)	0.1715(7)	−0.0347(4)	0.1134(4)
C(4,C_2H_5)	0.3090(8)	−0.0608(5)	0.1273(4)
C(5,C_2H_5)	0.4965(7)	0.3062(4)	0.1560(4)
C(6,C_2H_5)	0.5634(8)	0.3885(5)	0.1506(5)
C(7,C_2H_5)	0.3201(7)	0.3724(4)	0.2588(4)
C(8,C_2H_5)	0.3886(8)	0.3310(5)	0.3236(4)
C(1,C_6H_5)[a]	−0.0562(3)	0.0468(5)	0.1595(2)
C(2,C_6H_5)	−0.1300(4)	0.0904(3)	0.2089(2)
C(3,C_6H_5)	−0.2607(4)	0.0972(3)	0.1963(2)
C(4,C_6H_5)	−0.3177(3)	0.0603(5)	0.1343(2)
C(5,C_6H_5)	−0.2439(4)	0.0167(3)	0.0849(2)
C(6,C_6H_5)	−0.1132(4)	0.0100(3)	0.0975(2)
C(1′,C_6H_5)	0.2684(7)	0.3840(4)	0.1019(2)
C(2′,C_6H_5)	0.2865(7)	0.3626(3)	0.0286(3)
C(3′,C_6H_5)	0.2405(5)	0.4141(4)	−0.2077(2)
C(4′,C_6H_5)	0.1764(8)	0.4871(4)	−0.0108(3)
C(5′,C_6H_5)	0.1582(7)	0.5086(3)	0.0624(3)
C(6′,C_6H_5)	0.2043(5)	0.4570(3)	0.1187(2)
C(1″,C_6H_5)	−0.0031(3)	0.2321(5)	0.0589(2)
C(2″,C_6H_5)	−0.1257(4)	0.2490(2)	0.0815(2)
C(3″,C_6H_5)	−0.2220(3)	0.2658(4)	0.0294(2)
C(4″,C_6H_5)	−0.1956(4)	0.2658(5)	−0.0453(2)
C(5″,C_6H_5)	−0.0730(4)	0.2489(3)	−0.0679(2)
C(6″,C_6H_5)	0.0232(3)	0.2321(4)	−0.0158(2)

[a]Parameters assigned to the benzene carbon atoms involve the assumption of rigid rings having C–C = 1.392 A.

Atoms have been assigned the parameters of Table XVC,30; calculated values for hydrogen are stated in the original article.

TABLE XVC,30

Parameters of the Atoms in $CHF_2 \cdot CF_2 \cdot Co(CO)_3P(C_6H_5)_3$

Atom	x	y	z
Co	0.1984(4)	0.3960(2)	0.6111(3)
P	0.3234(6)	0.4346(3)	0.7601(5)
F(1)	−0.0123(16)	0.3135(8)	0.5265(12)
F(2)	−0.0081(16)	0.4099(8)	0.4399(13)
F(3)	0.2346(20)	0.2810(10)	0.4154(15)
F(4)	0.0507(19)	0.3032(9)	0.3143(15)
O(1)	0.2911(20)	0.2576(12)	0.6655(16)
O(2)	0.3475(20)	0.4773(10)	0.4548(16)
O(3)	−0.0564(22)	0.4556(10)	0.6923(17)
C(1)	0.2533(29)	0.3074(16)	0.6460(24)
C(2)	0.2852(28)	0.4475(14)	0.5215(24)
C(3)	0.0382(29)	0.4309(13)	0.6636(22)
C(4)	0.0816(27)	0.3629(14)	0.4871(22)
C(5)	0.1541(34)	0.3295(16)	0.3825(27)
C(1,1)	0.2298(24)	0.4335(13)	0.8871(19)
C(1,2)	0.1528(27)	0.3794(13)	0.9133(21)
C(1,3)	0.0816(26)	0.3708(13)	0.0136(22)
C(1,4)	0.0940(27)	0.4246(14)	0.0908(22)
C(1,5)	0.1653(27)	0.4807(14)	0.0669(22)
C(1,6)	0.2439(27)	0.4868(13)	0.9675(21)
C(2,1)	0.3782(25)	0.5238(12)	0.7447(19)
C(2,2)	0.2853(29)	0.5667(16)	0.6924(23)
C(2,3)	0.3297(39)	0.6333(21)	0.6828(31)
C(2,4)	0.4520(29)	0.6553(14)	0.7379(23)
C(2,5)	0.5359(25)	0.6135(13)	0.7940(20)
C(2,6)	0.5040(26)	0.5452(12)	0.7990(21)
C(3,1)	0.4854(26)	0.3900(13)	0.7879(21)
C(3,2)	0.5739(32)	0.3768(15)	0.6965(25)
C(3,3)	0.6982(27)	0.3430(13)	0.7235(22)
C(3,4)	0.7299(27)	0.3234(13)	0.8273(23)
C(3,5)	0.6574(30)	0.3370(14)	0.9147(24)
C(3,6)	0.5265(28)	0.3664(14)	0.8942(23)

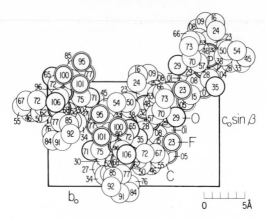

Fig. XVC,31.   The monoclinic structure of tricarbonyl(triphenylphosphine)-σ-tetrafluoroethyl cobalt projected along its a_0 axis. The cobalt atoms are the small, heavily and doubly ringed circles. Left-hand axes.

The structure is shown in Figure XVC,31. Some of the bond dimensions of its molecules are recorded in Figure XVC,32. In them the fivefold coordination about cobalt is that of a trigonal bipyramid.

XV,c32.   A complicated structure has been described for the complex *triphenyltin-pentacarbonylmanganese*, $Sn(C_6H_5)_3 \cdot Mn(CO)_5$. Its large

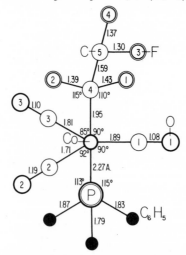

Fig. XVC,32.   Some bond dimensions in the molecule of tricarbonyl(triphenyl-phosphine)-σ-tetrafluoroethyl cobalt.

monoclinic unit containing eight molecules has the dimensions:

$$a_0 = 12.17(2) \text{ A.}; \quad b_0 = 32.22(5) \text{ A.}; \quad c_0 = 11.39(2) \text{ A.}$$
$$\beta = 90°33(5)'$$

The space group has been chosen as the low symmetry C_2^2 ($P2_1$) with all atoms in the positions:

$$(2a) \quad xyz; \, \bar{x}, y + {}^1/_2, \bar{z}$$

There are thus four crystallographically different molecules in the unit; they have been assigned the parameters of Table XVC,31.

TABLE XVC,31

Parameters of the Atoms in $Sn(C_6H_5)_3 \cdot Mn(CO)_5$

Atom	x	y	z
Sn(1)	−0.0602(1)	0.0003(0)	0.1501(2)
Sn(2)	0.4395(1)	−0.0003(0)	0.5974(1)
Sn(3)	0.1904(1)	0.2559(0)	0.3490(2)
Sn(4)	0.6904(1)	0.2560(0)	0.8978(2)
Mn(1)	−0.0459(4)	−0.0597(1)	−0.0112(5)
Mn(2)	0.4535(4)	−0.0601(1)	0.7602(5)
Mn(3)	0.2047(4)	0.3155(1)	0.5115(5)
Mn(4)	0.7042(4)	0.3158(1)	0.7362(5)
C(1,1)	−0.033(3)	−0.0957(12)	−0.126(4)
C(2,1)	0.459(4)	−0.0989(14)	0.878(4)
C(3,1)	0.215(4)	0.3545(14)	0.628(4)
C(4,1)	0.715(3)	0.3554(12)	0.630(4)
O(1,1)	−0.030(2)	−0.1254(8)	−0.186(2)
O(2,1)	0.476(3)	−0.1237(8)	0.941(3)
O(3,1)	0.226(3)	0.3806(8)	0.697(2)
O(4,1)	0.726(2)	0.3812(8)	0.554(3)
C(1,2)	−0.183(3)	−0.0668(12)	0.017(4)
C(2,2)	0.313(3)	−0.0681(11)	0.738(3)
C(3,2)	0.062(3)	0.3237(12)	0.484(4)
C(4,2)	0.567(3)	0.3239(10)	0.763(3)
O(1,2)	−0.279(3)	−0.0740(8)	0.039(3)
O(2,2)	0.220(3)	−0.0766(8)	0.709(3)
O(3,2)	−0.026(3)	0.3325(8)	0.457(3)
O(4,2)	0.469(3)	0.3319(9)	0.790(3)

(*continued*)

TABLE XVC,31 (continued)

Parameters of the Atoms in $Sn(C_6H_5)_3 \cdot Mn(CO)_5$

Atom	x	y	z
C(1,3)	0.093(3)	−0.0461(9)	−0.020(3)
C(2,3)	0.600(3)	−0.0472(11)	0.776(3)
C(3,3)	0.343(3)	0.3027(12)	0.531(3)
C(4,3)	0.847(3)	0.3024(11)	0.727(3)
O(1,3)	0.192(3)	−0.0420(8)	−0.025(3)
O(2,3)	0.690(2)	−0.0409(8)	0.775(2)
O(3,3)	0.441(2)	0.2968(8)	0.525(4)
O(4,3)	0.943(2)	0.2966(7)	0.724(2)
C(1,4)	−0.074(4)	−0.0193(14)	−0.120(4)
C(2,4)	0.426(3)	−0.0201(11)	0.864(3)
C(3,4)	0.176(4)	0.2765(12)	0.611(4)
C(4,4)	0.681(4)	0.2742(12)	0.637(4)
O(1,4)	−0.094(3)	0.0099(9)	−0.176(3)
O(2,4)	0.405(3)	0.0102(8)	0.920(2)
O(3,4)	0.157(2)	0.2467(8)	0.674(2)
O(4,4)	0.656(2)	0.2465(8)	0.574(2)
C(1,5)	−0.017(3)	−0.0935(10)	0.101(3)
C(2,5)	0.483(3)	−0.0959(10)	0.651(3)
C(3,5)	0.236(3)	0.3501(10)	0.405(3)
C(4,5)	0.729(3)	0.3517(11)	0.847(3)
O(1,5)	0.002(2)	−0.1169(7)	0.178(2)
O(2,5)	0.503(3)	−0.1176(8)	0.575(3)
O(3,5)	0.254(3)	0.3746(8)	0.328(3)
O(4,5)	0.754(2)	0.3740(8)	0.926(2)
C(1,11)	0.081(4)	0.0422(15)	0.142(4)
C(2,11)	0.574(4)	0.0415(12)	0.603(4)
C(3,11)	0.330(4)	0.2142(12)	0.357(4)
C(4,11)	0.828(4)	0.2135(11)	0.885(4)
C(1,12)	0.129(4)	0.0556(14)	0.245(4)
C(2,12)	0.630(4)	0.0574(14)	0.500(4)
C(3,12)	0.374(4)	0.1993(13)	0.250(4)
C(4,12)	0.878(5)	0.2000(16)	0.991(5)
C(1,13)	0.218(4)	0.0842(15)	0.240(4)
C(2,13)	0.720(4)	0.0830(14)	0.503(4)
C(3,13)	0.467(4)	0.1730(16)	0.258(4)
C(4,13)	0.969(4)	0.1722(13)	0.987(4)

(continued)

TABLE XVC,31 (*continued*)

Parameters of the Atoms in $Sn(C_6H_5)_3 \cdot Mn(CO)_5$

Atom	x	y	z
C(1,14)	0.258(5)	0.0958(16)	0.130(5)
C(2,14)	0.753(3)	0.0945(11)	0.614(3)
C(3,14)	0.505(4)	0.1609(16)	0.370(4)
C(4,14)	0.002(4)	0.1597(12)	0.888(4)
C(1,15)	0.207(4)	0.0804(14)	0.029(4)
C(2,15)	0.710(4)	0.0828(13)	0.720(4)
C(3,15)	0.458(4)	0.1735(14)	0.467(4)
C(4,15)	0.958(4)	0.1763(15)	0.775(4)
C(1,16)	0.116(4)	0.0532(14)	0.037(4)
C(2,16)	0.615(4)	0.0563(13)	0.712(4)
C(3,16)	0.369(4)	0.1999(13)	0.468(4)
C(4,16)	0.866(4)	0.2018(14)	0.781(4)
C(1,21)	−0.067(3)	−0.0234(10)	0.329(3)
C(2,21)	0.431(4)	−0.0229(14)	0.413(4)
C(3,21)	0.185(4)	0.2795(12)	0.174(4)
C(4,21)	0.686(3)	0.2793(11)	0.068(3)
C(1,22)	−0.157(3)	−0.0117(10)	0.395(3)
C(2,22)	0.345(4)	−0.0130(12)	0.351(4)
C(3,22)	0.093(4)	0.2707(12)	0.101(4)
C(4,22)	0.594(4)	0.2701(15)	0.139(4)
C(1,23)	−0.163(4)	−0.0278(15)	0.514(4)
C(2,23)	0.341(4)	−0.0300(14)	0.234(4)
C(3,23)	0.094(3)	0.2857(11)	−0.012(3)
C(4,23)	0.591(4)	0.2850(15)	0.261(4)
C(1,24)	−0.077(4)	−0.0513(15)	0.561(4)
C(2,24)	0.425(4)	−0.0527(14)	0.191(4)
C(3,24)	0.178(4)	0.3086(14)	−0.056(4)
C(4,24)	0.679(4)	0.3097(15)	0.299(4)
C(1,25)	0.020(4)	−0.0618(14)	0.488(4)
C(2,25)	0.519(4)	−0.0611(13)	0.263(4)
C(3,25)	0.267(4)	0.3173(14)	0.014(4)
C(4,25)	0.776(4)	0.3168(14)	0.233(4)
C(1,26)	0.021(4)	−0.0459(13)	0.372(4)
C(2,26)	0.522(3)	−0.0479(12)	0.378(3)
C(3,26)	0.275(4)	0.3046(13)	0.129(4)
C(4,26)	0.772(4)	0.3034(13)	0.118(4)

(*continued*)

TABLE XVC,31 (continued)

Parameters of the Atoms in $Sn(C_6H_5)_3 \cdot Mn(CO)_5$

Atom	x	y	z
C(1,31)	−0.206(3)	0.0382(11)	0.134(4)
C(2,31)	0.296(4)	0.0372(12)	0.619(4)
C(3,31)	0.044(3)	0.2197(11)	0.363(3)
C(4,31)	0.544(3)	0.2180(11)	0.878(4)
C(1,32)	−0.197(4)	0.0785(15)	0.102(4)
C(2,32)	0.309(4)	0.0811(13)	0.632(4)
C(3,32)	0.057(5)	0.1781(16)	0.377(5)
C(4,32)	0.558(4)	0.1768(14)	0.847(4)
C(1,33)	−0.291(5)	0.1026(16) ·	0.090(5)
C(2,33)	0.208(4)	0.1043(15)	0.634(5)
C(3,33)	−0.046(5)	0.1482(17)	0.390(5)
C(4,33)	0.455(4)	0.1542(13)	0.840(4)
C(1,34)	−0.395(4)	0.0851(15)	0.102(5)
C(2,34)	0.104(4)	0.0851(16)	0.630(5)
C(3,34)	−0.144(4)	0.1718(15)	0.381(5)
C(4,34)	0.356(4)	0.1702(14)	0.849(4)
C(1,35)	−0.404(4)	0.0452(14)	0.136(4)
C(2,35)	0.096(4)	0.0438(15)	0.609(5)
C(3,35)	−0.155(4)	0.2104(15)	0.358(5)
C(4,35)	0.346(4)	0.2104(13)	0.877(4)
C(1,36)	−0.311(4)	0.0211(12)	0.146(4)
C(2,36)	0.187(4)	0.0213(13)	0.607(4)
C(3,36)	−0.064(3)	0.2325(11)	0.362(3)
C(4,36)	0.444(4)	0.2347(13)	0.899(4)

These molecules (Fig. XVC,33) are nearly identical and thus a considerable measure of pseudosymmetry is introduced into the unit. Their averaged bond dimensions are shown in the figure.

XV,c33. Crystals of *triphenylphosphine-ironcarbonyl*, $P(C_6H_5)_3 \cdot Fe_3(CO)_{11}$, are monoclinic with a large unit containing 16 molecules and having the dimensions:

$$a_0 = 37.14(3) \text{ A.}; \quad b_0 = 12.26(1) \text{ A.}; \quad c_0 = 26.05(2) \text{ A.}$$
$$\beta = 93.96(15)°$$

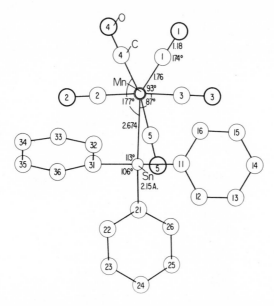

Fig. XVC,33.   Average bond dimensions in the four crystallographically different molecules present in crystals of triphenyltin-pentacarbonylmanganese.

The space group has been chosen as $C_{2h}^6(C2/c)$ with atoms in the positions:

$$(8f) \quad \pm (xyz; x,\bar{y},z+\tfrac{1}{2}; x+\tfrac{1}{2},y+\tfrac{1}{2},z; x+\tfrac{1}{2},\tfrac{1}{2}-y,z+\tfrac{1}{2})$$

Parameters assigned the atoms, except those of the phenyl groups, are listed in Table XVC,32; values for the phenyl carbons were given in the original.

TABLE XVC,32

Parameters of the Atoms in $P(C_6H_5)_3 \cdot Fe_3(CO)_{11}$

Atom	x	y	z
Fe(A,1)	0.3958(2)	0.2132(5)	0.3506(2)
Fe(A,2)	0.4235(2)	0.4164(5)	0.3617(2)
Fe(A,3)	0.4665(2)	0.2527(5)	0.3595(2)
Fe(B,1)	0.2993(2)	0.0319(5)	0.1424(2)
Fe(B,2)	0.2362(2)	0.0979(6)	0.0944(2)

(continued)

TABLE XVC,32 (*continued*)

Parameters of the Atoms in $P(C_6H_5)_3 \cdot Fe_3(CO)_{11}$

Atom	x	y	z
Fe(B,3)	0.2536(2)	0.1740(6)	0.1842(3)
P(A)	0.4507(3)	0.5789(9)	0.3705(4)
P(B)	0.3332(3)	−0.0769(10)	0.0957(4)
C(A,1)	0.399	0.220	0.284
O(A,1)	0.399	0.218	0.239
C(A,2)	0.398	0.216	0.417
O(A,2)	0.396	0.255	0.463
C(A,3)	0.395	0.075	0.351
O(A,3)	0.400	−0.025	0.354
C(A,4)	0.348	0.227	0.344
O(A,4)	0.319	0.242	0.344
C(A,5)	0.390	0.443	0.400
O(A,5)	0.364	0.462	0.426
C(A,6)	0.397	0.449	0.306
O(A,6)	0.379	0.470	0.268
C(A,7)	0.470	0.155	0.312
O(A,7)	0.471	0.083	0.282
C(A,8)	0.472	0.164	0.411
O(A,8)	0.476	0.104	0.447
C(A,9)	0.511	0.287	0.362
O(A,9)	0.543	0.300	0.363
C(B,1)	0.319	0.150	0.120?
O(B,1)	0.337	0.212	0.101
C(B,2)	0.272	−0.071	0.161
O(B,2)	0.256	−0.147	0.175
C(B,3)	0.326	0.026	0.197
O(B,3)	0.342	0.024	0.240
C(B,4)	0.225	−0.039	0.075
O(B,4)	0.216	−0.123	0.059
C(B,5)	0.262	0.129	0.044
O(B,5)	0.277	0.157	0.008
C(B,6)	0.198	0.149	0.070
O(B,6)	0.171	0.194	0.051
C(B,7)	0.289	0.252	0.208
O(B,7)	0.314	0.308	0.222

(*continued*)

Parameters of the Atoms in $P(C_6H_5)_3 \cdot Fe_3(CO)_{11}$

Atom	x	y	z
C(B,8)	0.255	0.089	0.240
O(B,8)	0.254	0.034	0.276
C(B,9)	0.221	0.258	0.200
O(B,9)	0.200	0.325	0.217
C(AB,1)	0.453	0.358	0.415
O(AB,1)	0.466	0.376	0.458
C(AB,2)	0.455	0.361	0.309
O(AB,2)	0.462	0.382	0.265
C(BB,1)	0.214	0.071	0.155
O(BB,1)	0.189	0.024	0.172
C(BB,2)	0.255	0.248	0.122
O(BB,2)	0.261	0.338	0.102

In the structure as thus defined there are two different molecules, designated A and B. In the A molecule the $P(C_6H_5)_3$ portion is bonded to an iron atom that has (CO) bridges to another iron atom; in the B molecule it is attached to the unbridged iron. Bond dimensions in these molecules are shown in Figure XVC,34; in both, the unbridged Fe–C = 1.65–1.79 A.

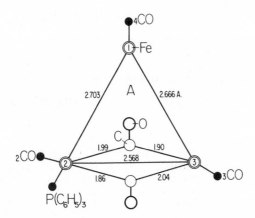

Fig. XVC,34a. Bond dimensions in the A type molecules present in triphenyl-phosphine-ironcarbonyl.

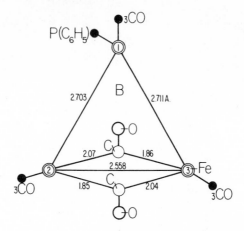

Fig. XVC,34b.   Bond dimensions in the B type molecules of triphenylphosphine-
ironcarbonyl.

XV,c34. A structure has been described for the *phenylacetylene-
ironcarbonyl* complex $(C_6H_5C_2H)_3 \cdot Fe_2(CO)_6$. It is monoclinic with
a tetramolecular unit of the dimensions:

$$a_0 = 11.963(3) \text{ A.}; \quad b_0 = 20.442(5) \text{ A.}; \quad c_0 = 10.326(3) \text{ A.}$$
$$\beta = 93°24(5)'$$

The space group is C_{2h}^5 $(P2_1/n)$ with atoms in the positions:

$$(4e) \quad \pm (xyz; x + \tfrac{1}{2}, \tfrac{1}{2} - y, z + \tfrac{1}{2})$$

The parameters are listed in Table XVC,33.

TABLE XVC,33

Parameters of the Atoms in $(C_6H_5C_2H)_3 \cdot Fe_2(CO)_6$

Atom	x	y	z
Fe(1)	0.5410	0.3893	0.7315
Fe(2)	0.3858	0.3243	0.8202
O(1)	0.6923	0.2314	0.8467
O(2)	0.2594	0.4159	0.9705
O(3)	0.2590	0.2180	0.9307

(continued)

TABLE XVC,33 (*continued*)

Parameters of the Atoms in $(C_6H_5C_2H)_3 \cdot Fe_2(CO)_6$

Atom	x	y	z
O(4)	0.4994	0.5236	0.8205
O(5)	0.5693	0.4118	0.4536
O(6)	0.7786	0.3967	0.8070
C(1)	0.3781	0.3728	0.6545
C(2)	0.3647	0.3066	0.6192
C(3)	0.4507	0.2630	0.6662
C(4)	0.5663	0.2897	0.6942
C(5)	0.6069	0.2647	0.8242
C(6)	0.5408	0.2897	0.9292
C(7)	0.5221	0.3578	0.9130
C(8)	0.3083	0.3811	0.9056
C(9)	0.3083	0.2608	0.8860
C(10)	0.5150	0.4722	0.7820
C(11)	0.5586	0.4046	0.5627
C(12)	0.6848	0.3958	0.7784
C(13)	0.3086	0.4233	0.5852
C(14)	0.2572	0.4754	0.6480
C(15)	0.1872	0.5171	0.5775
C(16)	0.1703	0.5088	0.4477
C(17)	0.2154	0.4582	0.3818
C(18)	0.2844	0.4148	0.4513
C(19)	0.4305	0.1909	0.6515
C(20)	0.3221	0.1653	0.6391
C(21)	0.3105	0.0972	0.6120
C(22)	0.3997	0.0576	0.6053
C(23)	0.5074	0.0830	0.6202
C(24)	0.5248	0.1477	0.6445
C(25)	0.5323	0.2538	0.0519
C(26)	0.5050	0.2867	0.1668
C(27)	0.4970	0.2530	0.2793
C(28)	0.5144	0.1876	0.2892
C(29)	0.5404	0.1531	0.1768
C(30)	0.5518	0.1885	0.0600

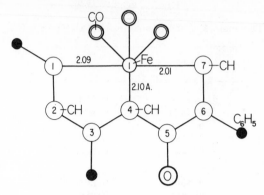

Fig. XVC,35a. Coordination and bond lengths about the Fe(1) atom in phenyl-acetylene-ironcarbonyl.

In this complicated arrangement the principal bond lengths and the atomic associations that prevail are indicated in Figure XVC,35. The coordination about Fe(1) is octahedral; Fe(2) is coordinated with five carbon atoms of the organic chain, two carbonyl carbons, and Fe(1) [Fe(1)–Fe(2) = 2.50 A.]. The intermolecular distances are said to be normal.

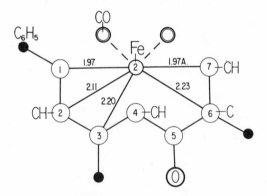

Fig. XVC,35b. Coordination and bond lengths about the Fe(2) atom in phenyl-acetylene-ironcarbonyl.

D. COMPOUNDS CONTAINING FOUR BENZENE RINGS

XV,d1. Crystals of *1,3,5-triphenylbenzene*, $(C_6H_5)_3C_6H_3$, are orthorhombic with a unit which, containing four molecules, has the dimensions:

$$a_0 = 7.47 \text{ A.}; \quad b_0 = 19.66 \text{ A.}; \quad c_0 = 11.19 \text{ A.}$$

All atoms are in general positions of C_{2v}^9 (*Pna*)

(4a) $xyz; \bar{x}, \bar{y}, z + 1/2; 1/2 - x, y + 1/2, z + 1/2; x + 1/2, 1/2 - y, z$

with the parameters of Table XVD,1.

TABLE XVD,1

Parameters of the Atoms in 1,3,5-Triphenylbenzene

Atom	x	y	z
C(1)	0.216	0.169	0.109
C(2)	0.218	0.229	0.046
C(3)	0.250	0.229	−0.076
C(4)	0.285	0.169	−0.135
C(5)	0.280	0.108	−0.071
C(6)	0.253	0.109	0.050
C(7)	0.179	0.169	0.242
C(8)	0.087	0.113	0.294
C(9)	0.048	0.113	0.416
C(10)	0.111	0.169	0.484
C(11)	0.216	0.221	0.434
C(12)	0.241	0.221	0.312
C(13)	0.250	0.295	−0.144
C(14)	0.301	0.356	−0.086
C(15)	0.293	0.416	−0.148
C(16)	0.254	0.417	−0.270
C(17)	0.217	0.356	−0.328
C(18)	0.212	0.296	−0.265
C(19)	0.320	0.043	−0.134
C(20)	0.289	0.038	−0.256
C(21)	0.317	−0.023	−0.315
C(22)	0.387	−0.078	−0.252
C(23)	0.415	−0.073	−0.128
C(24)	0.379	−0.012	−0.071

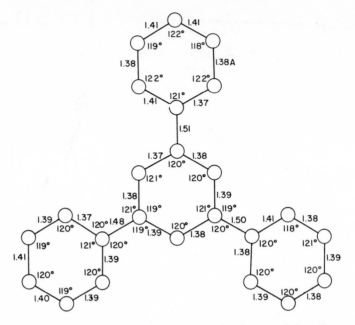

Fig. XVD,1.   Bond dimensions in the molecule of 1,3,5-triphenylbenzene.

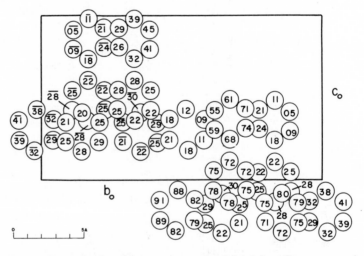

Fig. XVD,2.   The orthorhombic structure of 1,3,5-triphenylbenzene projected along its a_0 axis. Left-hand axes.

The dimensions of the resulting molecule are indicated in Figure XVD,1. The molecule, however, is not planar; instead, the plane of each phenyl group is rotated about the bond joining it with the central benzene ring. The angles of twist are 34, 27, and 24°, the direction of the second being in the opposite sense to those of the other two. Figure XVD,2 gives an idea of the way these molecules pack in the crystal.

XV,d2. *Tetraphenylmethane*, $(C_6H_5)_4C$, forms tetragonal crystals. Their bimolecular cell has the edges:

$$a_0 = 10.87 \text{ A.}, \quad c_0 = 7.23 \text{ A.}$$

The space group is V_d^4 ($P\bar{4}2_1c$) with the central carbon atoms in the body-centered positions

$$(2a) \quad 000; \, {}^1/_2, {}^1/_2, {}^1/_2$$

and all other atoms in general positions:

$$(8e) \quad xyz; \, {}^1/_2-x, y+{}^1/_2, {}^1/_2-z; \, \bar{x}\bar{y}z; \, x+{}^1/_2, {}^1/_2-y, {}^1/_2-z;$$
$$\bar{y}x\bar{z}; \, y+{}^1/_2, x+{}^1/_2, z+{}^1/_2; \, y\bar{x}\bar{z}; \, {}^1/_2-y, {}^1/_2-x, z+{}^1/_2$$

The assigned parameters of the benzene carbon atoms are those of Table XVD,2.

TABLE XVD,2

Parameters of Atoms in Tetraphenylmethane

Atom	x	y	z
C(2)	0.112	−0.016	0.119
CH(3)	0.104	−0.079	0.311
CH(4)	0.207	−0.093	0.423
CH(5)	0.223	0.034	0.037
CH(6)	0.327	0.020	0.149
CH(7)	0.319	−0.043	0.341

The resulting structure is shown in Figure XVD,3. Within the rings the C–C distance is 1.39 A.; ring carbon atoms are separated from the central carbon atom by 1.47 A.

Tetraphenyl lead, $Pb(C_6H_5)_4$, has recently been shown to have the same structure but with different parameters for the benzene carbon atoms which give these rings a different orientation with respect to the central lead atom. These parameters are stated in Table XVD,3.

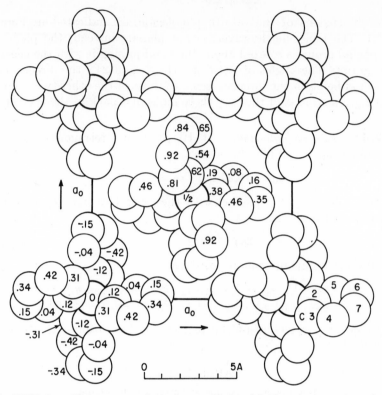

Fig. XVD,3a.   The tetragonal structure of tetraphenylmethane projected along
its c_0 axis.

TABLE XVD,3

Parameters of Atoms in $Pb(C_6H_5)_4$

Atom	x	y	z
C(1)	0.0195	0.1483	0.1868
C(2)	0.1068	0.1556	0.3247
C(3)	0.1192	0.2503	0.4439
C(4)	0.0444	0.3375	0.4251
C(5)	−0.0428	0.3301	0.2871
C(6)	−0.0553	0.2355	0.1679

The following compounds are tetragonal with bimolecular units of
the same shapes as the foregoing. It is to be presumed that they are
isostructural with them.

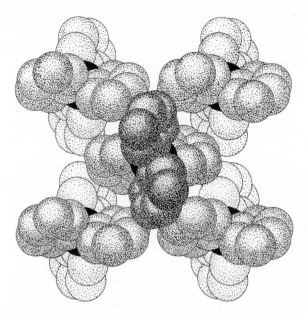

Fig. XVD,3b. A packing drawing of the tetragonal tetraphenylmethane structure seen along its c_0 axis. The central methyl carbon is black, all the others are dotted.

Compound	a_0, A.	c_0, A.
$Si(C_6H_5)_4$	11.30	7.05
$Ge(C_6H_5)_4$	11.60	6.85
$Sn(C_6H_5)_4$	11.85	6.65
Tetra(p-tolyl)tin	13.45	6.35
Tetra(p-methoxyphenyl)tin	14.30	6.50

Tetraphenylarsonium iodide (**XV,d25**) also is tetragonal with a similar unit, but the structure assigned it makes use of the different space group $S_4{}^2$.

XV,d3. Crystals of *cis,cis-1,2,3,4-tetraphenylbutadiene*, $C_{28}H_{22}$, are monoclinic with a bimolecular unit of the dimensions:

$$a_0 = 5.87(2) \text{ A.}; \quad b_0 = 21.31(2) \text{ A.}; \quad c_0 = 8.13(2) \text{ A.}$$
$$\beta = 97°5(10)'$$

The space group is C_{2h}^5 $(P2_1/c)$ with all atoms in the positions:

$$(4e) \quad \pm (xyz; x,{}^1/_2-y,z+{}^1/_2)$$

The assigned parameters are listed in Table XVD,4.

TABLE XVD,4

Parameters of the Atoms in $C_{28}H_{22}$

Atom	x	y	z
C(1)	0.2372(13)	0.0117(3)	0.6706(9)
C(2)	0.0725	0.0252	0.5434
C(3)	0.0309	0.0918	0.4804
C(4)	0.1955	0.1188	0.3914
C(5)	0.1603	0.1828	0.3379
C(6)	−0.0276	0.2160	0.3769
C(7)	−0.1864	0.1889	0.4657
C(8)	−0.1614	0.1251	0.5181
C(9)	0.3908	0.0547	0.7746
C(10)	0.3254	0.1164	0.8168
C(11)	0.4766	0.1522	0.9254
C(12)	0.6947	0.1304	0.9906
C(13)	0.7616	0.0701	0.9465
C(14)	0.6071	0.0328	0.8416

The values of $\sigma(x)$, $\sigma(y)$ and $\sigma(z)$ are the same for all atoms.

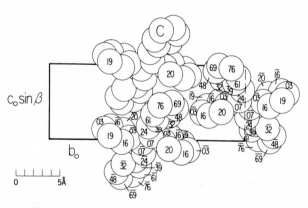

Fig. XVD,4.   The monoclinic structure of *cis,cis*-1,2,3,4-tetraphenylbutadiene projected along its a_0 axis.

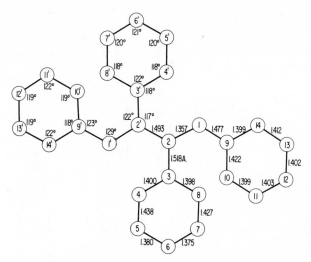

Fig. XVD,5.   Bond dimensions in the molecule of *cis,cis*-1,2,3,4-tetraphenyl-butadiene.

The resulting atomic arrangement is shown in Figure XVD,4. Each of the two kinds of benzene ring is planar to within ca. 0.10 A.; the dihedral angle between them is 69°. The butadiene chain is also planar; it makes an angle of 75° with that of ring I and 34° with that of ring II. Bond dimensions within this molecule are those of Figure XVD,5. The shortest distance between molecules is a C—C = 3.31 A.

XV,d4.   The more stable, black form of *bis(diphenylacetylene)-ironoctacarbonyl*, $(C_6H_5C_2C_6H_5)_2 \cdot Fe_3(CO)_8$, is monoclinic, like the violet modification. Its tetramolecular unit has the dimensions:

$$a_0 = 9.39 \text{ A.}; \quad b_0 = 18.45 \text{ A.}; \quad c_0 = 18.29 \text{ A.}; \quad \beta = 96.8°$$

Atoms are in the positions

$$(4e) \quad \pm (xyz; x, 1/2 - y, z + 1/2)$$

of C_{2h}^5 $(P2_1/c)$. Parameters have been determined to have the values listed in Table XVD,5.

The molecular structure of this form is different from that described in the following paragraph. Each iron atom has two carbonyl groups attached exclusively to it and shares an additional group with another metallic atom. Distances Fe–C are longer (1.77–1.99 A.) for the shared carbon than for the others (1.70–1.77 A.). The four non-benzene carbons

TABLE XVD,5

Parameters of the Atoms in Black Bis(diphenylacetylene)-ironoctacarbonyl

Atom	x	y	z
Fe(1)	0.0843(3)	0.1747(1)	0.2939(1)
Fe(2)	0.3026(3)	0.0891(1)	0.1887(1)
Fe(3)	0.3421(3)	0.1574(1)	0.3032(1)
O(1)	0.3687(15)	−0.0604(8)	0.1546(7)
O(2)	0.3912(13)	0.1170(7)	0.0451(7)
O(3)	0.6030(15)	0.1054(7)	0.2439(7)
O(4)	0.5332(15)	0.2793(8)	0.3228(7)
O(5)	0.5103(14)	0.0884(8)	0.4292(7)
O(6)	0.2326(12)	0.2405(7)	0.4249(7)
O(7)	−0.0845(13)	0.3057(7)	0.2889(6)
O(8)	−0.1292(13)	0.1278(7)	0.3858(7)
C(1)	0.3385(22)	0.0004(12)	0.1679(11)
C(2)	0.3532(19)	0.1047(10)	0.1022(11)
C(3)	0.4707(23)	0.1105(12)	0.2384(11)
C(4)	0.4543(22)	0.2296(12)	0.3147(11)
C(5)	0.4378(19)	0.1167(10)	0.3820(10)
C(6)	0.2180(19)	0.2081(10)	0.3680(11)
C(7)	−0.0148(18)	0.2551(10)	0.2876(9)
C(8)	−0.0406(20)	0.1454(10)	0.3497(10)
C(9)	0.0998(16)	0.1489(8)	0.1798(8)
C(10)	0.2204(16)	0.1913(8)	0.2103(8)
C(11)	0.1911(16)	0.0762(8)	0.2852(8)
C(12)	0.0874(16)	0.0811(9)	0.2194(8)
C(13)	−0.0112(17)	0.1671(9)	0.1167(8)
C(14)	0.2523(17)	0.2624(9)	0.1719(9)
C(15)	0.1924(16)	0.0133(9)	0.3372(9)
C(16)	−0.0206(16)	0.0239(9)	0.1936(9)
C(17)	−0.1540(19)	0.1689(10)	0.1281(9)
C(18)	−0.2585(18)	0.1832(10)	0.0645(10)
C(19)	−0.2125(19)	0.1968(10)	−0.0034(10)
C(20)	−0.0705(20)	0.1982(10)	−0.0134(9)
C(21)	0.0386(18)	0.1845(9)	0.0484(9)
C(22)	0.1495(18)	0.3164(10)	0.1507(9)
C(23)	0.1826(19)	0.3774(10)	0.1127(9)
C(24)	0.3185(19)	0.3858(10)	0.0935(9)

(*continued*)

TABLE XVD,5 (*continued*)

Parameters of the Atoms in Black Bis(diphenylacetylene)-ironoctacarbonyl

Atom	x	y	z
C(25)	0.4226(19)	0.3351(11)	0.1094(9)
C(26)	0.3940(18)	0.2722(9)	0.1524(9)
C(27)	0.2074(17)	−0.0595(8)	0.3118(9)
C(28)	0.2034(17)	−0.1147(9)	0.3634(10)
C(29)	0.1949(17)	−0.1034(10)	0.4369(9)
C(30)	0.1819(17)	−0.0318(10)	0.4609(9)
C(31)	0.1826(17)	0.0270(10)	0.4123(9)
C(32)	−0.1191(18)	0.0006(9)	0.2398(9)
C(33)	−0.2288(19)	−0.0482(10)	0.2112(10)
C(34)	−0.2290(20)	−0.0741(11)	0.1408(11)
C(35)	−0.1319(20)	−0.0519(10)	0.0948(10)
C(36)	−0.0235(19)	−0.0015(10)	0.1218(10)

are bound to one another, with C–C = 1.43–1.46 A. Each of these atoms is about equally distant from Fe(1) and Fe(2); the two end atoms, C(10) and C(11), of the bent chain they form are at about the same distance from Fe(3). For all these, Fe–C = 2.03–2.20 A. The Fe–Fe separations are only slightly greater (2.43 A.).

XV,d5. The violet, less stable form of *bis(diphenylacetylene)-ironoctacarbonyl*, $(C_6H_5C_2C_6H_5)_2 \cdot Fe_3(CO)_8$, is monoclinic with a unit that contains eight molecules and has the dimensions:

$$a_0 = 38.49 \text{ A.}; \quad b_0 = 8.31 \text{ A.}; \quad c_0 = 21.75 \text{ A.}; \quad \beta = 115.1°$$

The space group has been chosen as C_{2h}^6 (*C2/c*) with all atoms in the positions:

$$(8f) \quad \pm (xyz; x,\bar{y},z+1/2; x+1/2,y+1/2,z; x+1/2,1/2-y,z+1/2)$$

The assigned parameters are stated in Table XVD,6.

In the molecule of this structure two of the iron atoms have three associated carboxyl groups, with Fe–C = 1.73–1.80 A.; the third, Fe(1), has two carboxyl neighbors with Fe–C = 1.70 or 1.72 A. The four carbon atoms C(9) through C(12) and their attached benzene rings are situated between the Fe(1) on the one side and the Fe(2) and Fe(3) atoms on the other, the rings being twisted and somewhat bent for steric accommodation. Among these non-ring carbons C(9)–C(10) = 1.39 A. and C(11)–C(12) = 1.37 A.

TABLE XVD,6

Parameters of the Atoms in
Violet Bis(diphenylacetylene)-ironoctacarbonyl

Atom	x	y	z
Fe(1)	0.1308(1)	0.2598(4)	0.1568(2)
Fe(2)	0.1551(1)	−0.0176(5)	0.1641(2)
Fe(3)	0.0845(1)	0.0422(5)	0.1405(2)
O(1)	0.0905(5)	0.5554(25)	0.1458(9)
O(2)	0.1983(6)	0.4406(25)	0.1785(9)
O(3)	0.1369(5)	−0.2749(26)	0.0601(9)
O(4)	0.2360(5)	0.0489(23)	0.1921(8)
O(5)	0.1609(5)	−0.2731(25)	0.2619(9)
O(6)	0.0782(5)	−0.1723(21)	0.2407(8)
O(7)	0.0126(5)	0.1946(22)	0.1163(8)
O(8)	0.0501(6)	−0.1927(27)	0.0302(11)
C(1)	0.1068(7)	0.4385(35)	0.1496(12)
C(2)	0.1710(8)	0.3679(33)	0.1715(13)
C(3)	0.1459(7)	−0.1680(35)	0.0994(14)
C(4)	0.2042(8)	0.0113(34)	0.1805(12)
C(5)	0.1639(6)	−0.1689(30)	0.2254(12)
C(6)	0.0827(8)	−0.0901(36)	0.2017(15)
C(7)	0.0417(9)	0.1393(39)	0.1254(15)
C(8)	0.0639(8)	−0.0997(39)	0.0720(15)
C(9)	0.1568(6)	0.1256(25)	0.2389(10)
C(10)	0.1192(5)	0.1592(24)	0.2285(9)
C(11)	0.0960(5)	0.1636(24)	0.0693(9)
C(12)	0.1339(6)	0.1345(25)	0.0833(10)
C(13)	0.1902(6)	0.1632(26)	0.3039(10)
C(14)	0.1053(6)	0.2134(27)	0.2782(10)
C(15)	0.0674(6)	0.2151(28)	0.0022(10)
C(16)	0.1525(5)	0.1787(25)	0.0377(9)
C(17)	0.2208(6)	0.0499(28)	0.3357(10)
C(18)	0.2520(7)	0.0866(30)	0.3974(12)
C(19)	0.2552(6)	0.2363(32)	0.4254(10)
C(20)	0.2264(8)	0.3516(35)	0.3970(13)
C(21)	0.1948(6)	0.3152(27)	0.3366(10)
C(22)	0.1222(6)	0.1386(28)	0.3434(11)
C(23)	0.1103(6)	0.1900(28)	0.3926(11)
C(24)	0.0825(7)	0.3124(31)	0.3811(12)

(continued)

TABLE XVD,6 (*continued*)

Parameters of the Atoms in
Violet Bis(diphenylacetylene)-ironoctacarbonyl

Atom	x	y	z
C(25)	0.0643(8)	0.3921(34)	0.3170(13)
C(26)	0.0767(7)	0.3350(30)	0.2648(12)
C(27)	0.0427(6)	0.3483(28)	−0.0032(11)
C(28)	0.0156(7)	0.4037(30)	−0.0709(13)
C(29)	0.0088(8)	0.3120(39)	−0.1264(14)
C(30)	0.0303(7)	0.1849(33)	−0.1206(12)
C(31)	0.0598(7)	0.1226(30)	−0.0571(12)
C(32)	0.1421(6)	0.3297(26)	0.0018(10)
C(33)	0.1583(7)	0.3709(30)	−0.0427(12)
C(34)	0.1829(6)	0.2593(30)	−0.0548(10)
C(35)	0.1929(6)	0.1120(27)	−0.0213(11)
C(36)	0.1777(6)	0.0713(25)	0.0219(10)

XV,d6. Crystals of *dichlorotetrabenzoato dirhenium(III)-di-chloroform*, $Re_2Cl_2(C_6H_5CO_2)_4 \cdot 2CHCl_3$, are monoclinic with a bimolecular unit of the dimensions:

$$a_0 = 11.029(2) \text{ A.}; \quad b_0 = 10.760(2) \text{ A.}; \quad c_0 = 18.383(5) \text{ A.}$$
$$\beta = 119.02(3)°$$

All atoms are in the positions

$$(4e) \quad \pm (xyz; x, {}^1/_2 - y, z + {}^1/_2)$$

of the space group C_{2h}^5 ($P2_1/c$). Parameters are listed in Table XVD,7.

In the dimeric molecules that result each rhenium atom is octahedrally surrounded by four oxygens from four different $C_6H_5CO_2$ radicals, one chlorine atom, and the second rhenium. The oxygen atoms are at the corners of an essentially perfect square, with Re–O = 2.00–2.03 A. and O–Re–O = 88.6–91.8°. The Re–Cl and Re–Re bonds are normal to this plane, with Re–Cl = 2.489 A. and Re–Re = 2.235 A. The chloroform molecules occupy tetrahedral voids between four phenyl groups; a short 3.65 A. separates their carbon atoms from chlorine atoms attached to rhenium.

XV,d7. *Cerium tetrakis-dibenzoylmethane*, $Ce\{[C_6H_5C(O)]_2CH_2\}_4$, forms orthorhombic crystals whose tetramolecular unit has the edge

CRYSTAL STRUCTURES

TABLE XVD,7

Parameters of the Atoms in $Re_2Cl_2(C_6H_5CO_2)_4 \cdot 2CHCl_3$

Atom	x	y	z
Re	0.07356	0.08017	0.01969
Cl(1)	0.2296	0.2645	0.0655
O(1,1)	0.0888	−0.1972	0.0339
O(2,1)	0.2353	−0.0352	0.0739
O(1,2)	−0.0752	−0.0720	0.0900
O(2,2)	0.0707	0.0893	0.1290
C(0,1)	0.2107	−0.1583	0.0713
C(1,1)	0.3284	−0.2386	0.1155
C(2,1)	0.3069	−0.3657	0.1114
C(3,1)	0.4190	−0.4456	0.1601
C(4,1)	0.5513	−0.4002	0.2085
C(5,1)	0.5705	−0.2707	0.2060
C(6,1)	0.4619	−0.1903	0.1625
C(0,2)	−0.0021	0.0093	0.1437
C(1,2)	0.0037	0.0114	0.2251
C(2,2)	−0.0745	−0.0689	0.2408
C(3,2)	−0.0656	−0.0674	0.3223
C(4,2)	0.0248	0.0140	0.3790
C(5,2)	0.1009	0.0946	0.3637
C(6,2)	0.0908	0.0961	0.2844
C(1)	0.5880	0.2380	0.1076
Cl(2)	0.7176	0.2599	0.2046
Cl(3)	0.6100	0.1023	0.0695
Cl(4)	0.5759	0.3538	0.0448
H(2,1)	0.203	−0.402	0.075
H(3,1)	0.400	−0.545	0.158
H(4,1)	0.638	−0.462	0.243
H(5,1)	0.675	−0.234	0.246
H(6,1)	0.478	−0.091	0.161
H(2,2)	−0.143	−0.131	0.192
H(3,2)	−0.127	−0.132	0.335
H(4,2)	0.033	0.016	0.440
H(5,2)	0.170	0.157	0.413
H(6,2)	0.154	0.158	0.269

TABLE XVD,8

Parameters of the Atoms in Ce{[C$_6$H$_5$C(O)]$_2$CH$_2$}$_4$

Atom	x	y	z
C(1)	0.422	0.057	0.255
C(2)	0.470	0.012	0.293
C(3)	0.465	−0.055	0.283
C(4)	0.412	−0.077	0.232
C(5)	0.368	−0.035	0.191
C(6)	0.373	0.032	0.204
C(7)	0.320	0.080	0.162
C(8)	0.310	0.070	0.104
C(9)	0.291	0.112	0.057
C(10)	0.291	0.094	−0.004
C(11)	0.359	0.040	−0.026
C(12)	0.359	0.025	−0.083
C(13)	0.291	0.065	−0.119
C(14)	0.223	0.119	−0.098
C(15)	0.223	0.134	−0.040
O(1)	0.305	0.141	0.181
O(2)	0.271	0.174	0.064
O(3)	0.044	0.211	0.108
O(4)	0.111	0.236	0.225
C(16)	−0.121	0.157	0.030
C(17)	−0.203	0.127	−0.009
C(18)	−0.325	0.107	0.009
C(19)	−0.363	0.117	0.064
C(20)	−0.281	0.147	0.102
C(21)	−0.160	0.167	0.085
C(22)	−0.073	0.199	0.128
C(23)	−0.087	0.206	0.185
C(24)	−0.010	0.219	0.232
C(25)	−0.039	0.221	0.296
C(26)	−0.126	0.177	0.319
C(27)	−0.155	0.179	0.376
C(28)	−0.097	0.226	0.410
C(29)	−0.010	0.271	0.387
C(30)	0.019	0.269	0.330

lengths:

$$a_0 = 10.320(2) \text{ A.}; \quad b_0 = 20.109(3) \text{ A.}; \quad c_0 = 23.514(5) \text{ A.}$$

The space group has been chosen as V_h^{10} (*Pccn*). The cerium atoms are in:

(4c) $\pm (^{1}/_4 \, ^{1}/_4 \, u; \, ^{1}/_4, ^{1}/_4, u + ^{1}/_2)$ with $u = 0.142$

The other atoms are in:

(8d) $\pm (xyz; \, x + ^{1}/_2, y + ^{1}/_2, \bar{z}; \, ^{1}/_2 - x, y, z + ^{1}/_2; \, x, ^{1}/_2 - y, z + ^{1}/_2)$

Assigned parameters are given in Table XVD,8.

According to the brief description that has been published, the cerium atom has about it eight chelating oxygen atoms. Bond dimensions have not been stated.

The corresponding thorium and uranium compounds are isostructural.

For $Th\{[C_6H_5C(O)]_2CH_2\}_4$:

$$a_0 = 10.398(2) \text{ A.}; \quad b_0 = 20.298(3) \text{ A.}; \quad c_0 = 23.334(5) \text{ A.}$$

For $U\{[C_6H_5C(O)]_2CH_2\}_4$:

$$a_0 = 10.303(2) \text{ A.}; \quad b_0 = 20.136(3) \text{ A.}; \quad c_0 = 23.613(5) \text{ A.}$$

XV,d8. The monoclinic crystals of *bis(1,3-diphenyl-1,3-propane-dionato)copper(II)*, $Cu(C_6H_5COCH \cdot COC_6H_5)_2$, have a tetramolecular unit of the dimensions:

$$a_0 = 26.18 \text{ A.}; \quad b_0 = 6.07 \text{ A.}; \quad c_0 = 16.64 \text{ A.}; \quad \beta = 115.3°$$

The copper atoms are in

(4a) $000; \, ^{1}/_2 \, ^{1}/_2 \, 0; \, 0 \, 0 \, ^{1}/_2; \, ^{1}/_2 \, ^{1}/_2 \, ^{1}/_2$

of the space group C_{2h}^6 (*C2/c*). All other atoms are in the general positions:

(8f) $\pm (xyz; \, x, \bar{y}, z + ^{1}/_2; \, x + ^{1}/_2, y + ^{1}/_2, z; \, x + ^{1}/_2, ^{1}/_2 - y, z + ^{1}/_2)$

The determined parameters are listed in Table XVD,9.

The structure, shown in Figure XVD,6, is built up of molecules possessing the bond dimensions of Figure XVD,7. They are nearly planar, with the major atomic excursion of 0.16 A. from the mean plane through all atoms. Coordination about Cu(II) is the usual planar square.

TABLE XVD,9

Parameters of the Atoms in $Cu(C_6H_5COCH \cdot COC_6H_5)_2$

Atom	x	y	z
O(1)	−0.0359	0.2609	0.0142
O(2)	0.0677	0.0544	0.1052
C(1)	−0.1117	0.5722	0.0127
C(2)	−0.1501	0.7399	0.0089
C(3)	−0.1333	0.9135	0.0687
C(4)	−0.0775	0.9294	0.1301
C(5)	−0.0385	0.7679	0.1340
C(6)	−0.0554	0.5828	0.0703[a]
C(7)	−0.0164	0.4006	0.0808
C(8)	0.0373	0.3803	0.1514
C(9)	0.0771	0.2190	0.1583
C(10)	0.1359	0.2291	0.2324
C(11)	0.1543	0.3957	0.2962
C(12)	0.2097	0.3961	0.3638
C(13)	0.2468	0.2326	0.3674
C(14)	0.2297	0.0618	0.3055
C(15)	0.1738	0.0599	0.2376

[a] This parameter as stated in the original article is 0.7703. Such a value is unreasonable and it is presumed that perhaps it should be 0.0703.

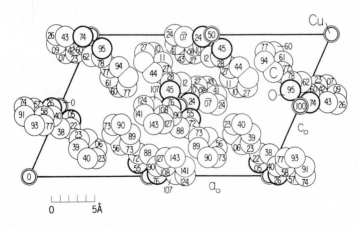

Fig. XVD,6.   The monoclinic structure of bis(1,3-diphenyl-1,3-propanedionato) copper projected along its b_0 axis.

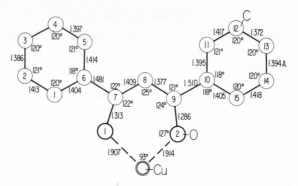

Fig. XVD,7. Bond dimensions in the molecule of bis(1,3-diphenyl-1,3-pro-panedionato)copper.

The palladium compound is isostructural with a cell of the dimensions:

$$a_0 = 26.4 \text{ A.}; \quad b_0 = 5.89 \text{ A.}; \quad c_0 = 17.75 \text{ A.}; \quad \beta = 118°$$

XV,d9. Crystals of *bis(N-phenylsalicylaldiminato)copper(II)*, Cu-$(OC_6H_4CHNC_6H_5)_2$, are monoclinic with a bimolecular unit of the dimensions:

$$a_0 = 12.145(10) \text{ A.}; \quad b_0 = 7.956(10) \text{ A.}; \quad c_0 = 11.935(10) \text{ A.}$$
$$\beta = 111°44'$$

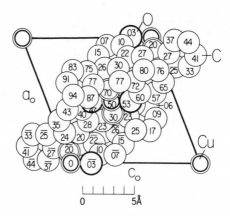

Fig. XVD,8. The monoclinic structure of bis(N-phenylsalicylaldiminato)copper projected along its b_0 axis.

The space group is C_{2h}^5 ($P2_1/n$) with atoms in the positions:

$$\text{Cu} : (2a) \quad 000; \; {}^1/_2 \, {}^1/_2 \, {}^1/_2$$

All the other atoms are in the positions:

$$(4e) \quad \pm (xyz; \, x + {}^1/_2, {}^1/_2 - y, z + {}^1/_2)$$

Determined parameters are those of Table XVD,10.

TABLE XVD,10

Parameters of Atoms in $Cu(OC_6H_4CHNC_6H_5)_2$

Atom	x	y	z
O	−0.0036(3)	−0.0256(4)	0.1550(3)
N	0.1008(3)	−0.2042(5)	0.0211(3)
C(1)	0.0248(5)	−0.3703(8)	−0.1628(5)
C(2)	0.0384(7)	−0.4377(10)	−0.2656(6)
C(3)	0.1368(7)	−0.4139(10)	−0.2871(6)
C(4)	0.2268(7)	−0.3268(10)	−0.2071(6)
C(5)	0.2183(5)	−0.2547(8)	−0.1010(5)
C(6)	0.1170(4)	−0.2775(6)	−0.0819(3)
C(7)	0.1607(4)	−0.2706(6)	0.1257(3)
C(8)	0.1531(4)	−0.1478(7)	0.4662(3)
C(9)	0.2338(4)	−0.2630(8)	0.4576(4)
C(10)	0.2351(4)	−0.2985(8)	0.3457(4)
C(11)	0.1558(3)	−0.2204(6)	0.2393(3)
C(12)	0.0735(3)	−0.1032(6)	0.2482(3)
C(13)	0.0722(4)	−0.0689(7)	0.3646(3)
H(1)	−0.054	−0.390	−0.147
H(2)	−0.029	−0.515	−0.325
H(3)	0.145	−0.457	−0.368
H(4)	0.308	−0.313	−0.220
H(5)	0.289	−0.183	−0.040
H(7)	0.219	−0.370	−0.128
H(8)	0.151	−0.119	0.552
H(9)	0.293	−0.324	0.535
H(10)	0.298	−0.385	0.338
H(13)	0.010	0.016	0.374

The structure (Fig. XVD,8) is composed of centrosymmetric molecules possessing the bond dimensions indicated in Figure XVD,9. The coordination of Cu(II) is fourfold and planar, the departure from

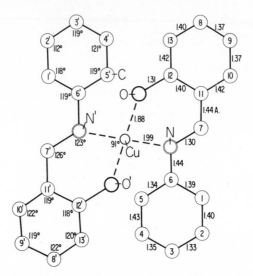

Fig. XVD,9. Bond dimensions in the molecule of bis(N-phenylsalicylaldiminato) copper.

a square being set by the difference between Cu–N = 1.99 A. and Cu–O = 1.88 A. The salicylaldimine groups are nearly planar; the distance apart of their planes is 0.89 A. and they make angles of 18° with the CuN_2O_2 plane. The planar benzene rings make angles of 65° with their salicylaldimine planes. The shortest intermolecular distance is a C–C = 3.33 A.

The corresponding nickel compound appears to be isostructural, though a complete study has not been made. Its cell has the dimensions:

$$a_0 = 12.64 \text{ A.}; \quad b_0 = 7.63 \text{ A.}; \quad c_0 = 11.81 \text{ A.}; \quad \beta = 112°13'$$

XV,d10. A brief statement, unsupported by experimental evidence, has been made of the atomic arrangement in *p-bromobenzeneazotri-benzoylmethane*, $(C_6H_5CO)_3CN{=}NC_6H_4Br$. Its crystals are monoclinic with a tetramolecular unit of the dimensions:

$$a_0 = 9.84(3) \text{ A.}; \quad b_0 = 14.19(4) \text{ A.}; \quad c_0 = 18.56(5) \text{ A.}$$
$$\beta = 111°50(10)'$$

The space group is C_{2h}^5 ($P2_1/c$) with all atoms in the positions:

$$(4e) \quad \pm(xyz; x,{}^1/_2-y,z+{}^1/_2)$$

The parameters are stated to be those of Table XVD,11.

TABLE XVD,11

Parameters of the Atoms in $(C_6H_5CO)_3CN{=}NC_6H_4Br$

Atom	x	y	z
C(1)	0.5079	0.6154	0.2672
C(2)	0.5442	0.5211	0.2757
C(3)	0.4271	0.4548	0.2586
C(4)	0.2831	0.4897	0.2266
C(5)	0.2514	0.5840	0.2162
C(6)	0.3635	0.6506	0.2335
C(7)	−0.0605	0.3634	0.1681
C(8)	−0.0622	0.3019	0.0992
C(9)	−0.2128	0.4137	0.1510
C(10)	−0.0152	0.3067	0.2447
C(11)	−0.1952	0.2433	0.0547
C(12)	−0.2119	0.2231	−0.0221
C(13)	−0.3352	0.1636	−0.0667
C(14)	−0.4255	0.1295	−0.0329
C(15)	−0.4065	0.1454	0.0459
C(16)	−0.2882	0.2062	0.0893
C(17)	−0.2589	0.4879	0.0943
C(18)	−0.2006	0.4965	0.0343
C(19)	−0.2604	0.5683	−0.0208
C(20)	−0.3792	0.6331	−0.0192
C(21)	−0.4289	0.6206	0.0414
C(22)	−0.3743	0.5475	0.0980
C(23)	0.0285	0.3534	0.3211
C(24)	0.0095	0.4513	0.3269
C(25)	0.0557	0.4929	0.4031
C(26)	0.1236	0.4354	0.4664
C(27)	0.1453	0.3390	0.4596
C(28)	0.0971	0.2948	0.3856
N(1)	0.1764	0.4138	0.2153
N(2)	0.0448	0.4452	0.1764
O(1)	0.0440	0.2990	0.0825
O(2)	−0.2829	0.3858	0.1877
O(3)	−0.0106	0.2230	0.2389
Br	0.6654	0.7012	0.2960

XV,d11. The monoclinic crystals of the free radical *2,2-diphenyl-1-picrylhydrazyl·benzene*, $C_{18}H_{12}N_2(NO_2)_3 \cdot C_6H_6$, have a bimolecular unit of the dimensions:

$$a_0 = 7.764(2) \text{ A.}; \quad b_0 = 10.648(2) \text{ A.}; \quad c_0 = 14.780(3) \text{ A.}$$
$$\beta = 109.02(2)°$$

Atoms are in the positions

$$(2a) \quad xyz; \, x,\bar{y},z + \tfrac{1}{2}$$

of the space group C_s^2 (Pc). The parameters for the heavier atoms are those of Table XVD,12; values for hydrogen are given in the original article.

TABLE XVD,12

Parameters of Atoms in 2,2-Diphenyl-1-picrylhydrazil·benzene

Atom	x	y	z
C(1)	0.0039(9)	0.2304(5)	0.3062(4)
C(2)	0.8616(8)	0.3161(5)	0.3025(4)
C(3)	0.6925(8)	0.3095(5)	0.2351(5)
C(4)	0.6524(9)	0.2110(6)	0.1700(4)
C(5)	0.7812(9)	0.1203(5)	0.1717(4)
C(6)	0.9518(8)	0.1336(5)	0.2373(4)
C(7)	0.2438(8)	0.4385(5)	0.3463(5)
C(8)	0.2319(9)	0.5548(6)	0.3839(5)
C(9)	0.2097(9)	0.6602(6)	0.3268(6)
C(10)	0.2017(10)	0.6504(6)	0.2331(6)
C(11)	0.2155(10)	0.5340(6)	0.1953(5)
C(12)	0.2359(9)	0.4266(6)	0.2503(5)
C(13)	0.4113(9)	0.3156(6)	0.4913(4)
C(14)	0.4043(10)	0.2217(7)	0.5533(5)
C(15)	0.5456(11)	0.2080(7)	0.6386(5)
C(16)	0.6917(11)	0.2888(9)	0.6628(5)
C(17)	0.6985(9)	0.3832(8)	0.6002(6)
C(18)	0.5578(10)	0.3985(6)	0.5127(5)
N(19)	0.1779(7)	0.2219(4)	0.3690(4)
N(20)	0.2677(—)	0.3278(4)	0.4029(—)
N(21)	0.8804(9)	0.4073(5)	0.3792(5)
O(22)	0.9801(7)	0.3809(4)	0.4606(4)
O(23)	0.7879(8)	0.5018(5)	0.3589(4)
N(24)	0.4710(8)	0.2005(6)	0.1004(4)
O(25)	0.3696(8)	0.2926(5)	0.0897(4)

(continued)

TABLE XVD,12 (*continued*)

Parameters of Atoms in 2,2-Diphenyl-1-picrylhydrazil·benzene

Atom	x	y	z
O(26)	0.4295(7)	0.1025(5)	0.0562(4)
N(27)	0.0909(9)	0.0436(6)	0.2307(4)
O(28)	0.0544(9)	−0.0664(5)	0.2308(5)
O(29)	0.2298(8)	0.0856(5)	0.2214(4)
C(30)	0.9582(11)	0.9510(16)	0.4799(7)
C(31)	0.8464(18)	0.0513(9)	0.4632(6)
C(32)	0.6624(15)	0.0354(10)	0.4154(6)
C(33)	0.6009(10)	0.9188(11)	0.3906(6)
C(34)	0.7100(18)	0.8228(9)	0.4091(7)
C(35)	0.8890(17)	0.8367(13)	0.4557(7)

The general shape of the molecule and some of its more important bond lengths are shown in Figure XVD,10. The arrangement of these molecules and of the single benzene of crystallization are given in Figure XVD,11. All the benzene rings are planar; their angles with respect to one another are determined by the distribution of the bonds around the nitrogen atoms.

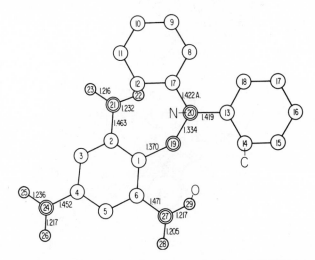

Fig. XVD,10.ʻ Bond dimensions in the molecule of 2,2-diphenyl-1-picryl-hydrazyl.

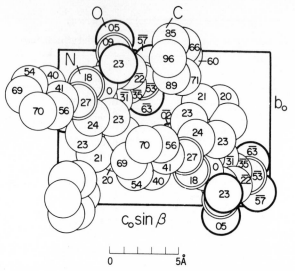

Fig. XVD,11. The monoclinic structure of 2,2-diphenyl-1-picrylhydrazyl ·
benzene projected along its a_0 axis.

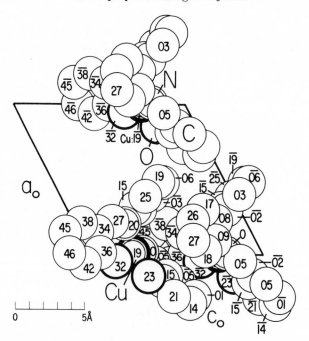

Fig. XVD,12. The monoclinic structure of N,N'-(2,2'-biphenyl)bis(salicylaldiminato)copper projected along its b_0 axis.

XV,d12. Crystals of *N,N'-(2,2'-biphenyl)bis(salicylaldiminato)-copper(II)*, $CuC_{26}H_{18}N_2O_2$, are monoclinic with a bimolecular unit of the dimensions:

$$a_0 = 11.30(2) \text{ A.}; \quad b_0 = 10.00(2) \text{ A.}; \quad c_0 = 12.04(2) \text{ A.}$$
$$\beta = 118°6'$$

The space group is C_s^2 (Pc) with atoms in the positions:

$$(2a) \quad xyz; x,\bar{y},z + 1/2$$

The determined parameters are those of Table XVD,13.

The structure is shown in Figure XVD,12. The bond dimensions of the copper ligands are those of Figure XVD,13. Among these the biphenyl rings make an angle of 57° with respect to one another. There is some departure from a planar distribution of the chelating radicals but

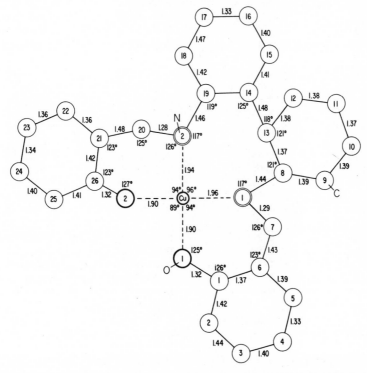

Fig. XVD,13. Bond dimensions in the molecule of *N,N'-(2,2'-biphenyl)bis (salicylaldiminato)copper.*

TABLE XVD,13

Parameters of the Atoms in $CuC_{26}H_{18}N_2O_2$

Atom	x	y	z
C(1)	−0.1931(11)	0.1529(13)	0.3392(12)
C(2)	−0.2923(16)	0.2140(17)	0.3631(17)
C(3)	−0.3409(17)	0.1381(21)	0.4351(18)
C(4)	−0.2969(12)	0.0076(16)	0.4744(14)
C(5)	−0.2085(13)	−0.0460(17)	0.4435(13)
C(6)	−0.1543(11)	0.0239(13)	0.3780(10)
C(7)	−0.0618(12)	−0.0487(13)	0.3505(11)
C(8)	0.0860(13)	−0.0887(13)	0.2692(12)
C(9)	0.0170(18)	−0.1838(13)	0.1778(15)
C(10)	0.0864(27)	−0.2734(17)	0.1419(19)
C(11)	0.2230(22)	−0.2656(20)	0.1967(20)
C(12)	0.2938(21)	−0.1739(19)	0.2903(19)
C(13)	0.2235(14)	−0.0838(13)	0.3246(12)
C(14)	0.3006(13)	0.0161(16)	0.4227(12)
C(15)	0.4014(16)	−0.0325(21)	0.5392(14)
C(16)	0.4873(18)	0.0574(25)	0.6316(17)
C(17)	0.4748(15)	0.1888(22)	0.6165(17)
C(18)	0.3704(16)	0.2453(24)	0.4987(15)
C(19)	0.2874(12)	0.1509(15)	0.4070(12)
C(20)	0.2282(15)	0.2660(16)	0.2195(12)
C(21)	0.1424(16)	0.3401(15)	0.1026(13)
C(22)	0.1999(19)	0.3836(16)	0.0320(15)
C(23)	0.1324(21)	0.4486(19)	−0.0797(16)
C(24)	0.0007(25)	0.4659(17)	−0.1212(15)
C(25)	−0.0706(21)	0.4224(14)	−0.0589(14)
C(26)	0.0031(15)	0.3598(14)	0.0590(12)
N(1)	0.0120(9)	0.0022(10)	0.3061(9)
N(2)	0.1855(12)	0.2038(11)	0.2868(10)
O(1)	−0.1516(9)	0.2266(9)	0.2738(9)
O(2)	−0.0604(10)	0.3190(9)	0.1212(9)
Cu	−0.0002(3)	0.1878(1)	0.2500(3)

the largest distortion is about the central copper atom. Its coordination is fourfold with a distribution that is nearer the square than the tetrahedral.

XV,d13. Crystals of *tetra-p-anisylethylene dichloroiodide*, $[(CH_3\text{-}OC_6H_4)_4C_2](ICl_2)_2$, are monoclinic with a bimolecular unit of the dimensions:

$$a_0 = 9.798(3) \text{ A.}; \quad b_0 = 10.956(4) \text{ A.}; \quad c_0 = 16.865(9) \text{ A.}$$
$$\beta = 114.54(5)°$$

Atoms are in the positions

$$(4g) \quad \pm(xyz; x,\bar{y},z+1/2)$$

of the space group C_{2h}^4 (*P2/c*). Determined parameters are listed in Table XVD,14; values for the atoms of hydrogen are stated in the original article.

TABLE XVD,14

Parameters of the Atoms in $[(CH_3OC_6H_4)_4C_2](ICl_2)_2$

Atom	x	y	z
I	0.2494(2)	0.1125(1)	0.9530(1)
Cl(1)	0.1867(8)	0.9012(5)	0.9847(4)
Cl(2)	0.3353(8)	0.3280(6)	0.9348(5)
O(1)	0.822(2)	0.285(1)	0.200(1)
O(1')	0.928(2)	0.443(1)	0.396(1)
C(e)	0.579(2)	0.856(1)	0.261(1)
C(1)	0.641(2)	0.967(2)	0.245(1)
C(2)	0.794(2)	0.003(2)	0.296(1)
C(3)	0.846(3)	0.112(2)	0.280(1)
C(4)	0.761(2)	0.181(2)	0.211(2)
C(5)	0.613(2)	0.149(2)	0.156(1)
C(6)	0.557(2)	0.038(2)	0.173(2)
C(1')	0.673(2)	0.753(1)	0.303(1)
C(2')	0.627(2)	0.667(2)	0.346(1)
C(3')	0.710(2)	0.559(2)	0.380(1)
C(4')	0.838(3)	0.545(2)	0.372(2)
C(5')	0.888(3)	0.626(2)	0.326(2)
C(6')	0.813(2)	0.727(2)	0.289(2)
C(CH₃,1)	0.734(4)	0.368(2)	0.130(3)
C(CH₃,1')	0.886(3)	0.350(2)	0.439(2)

The structure is shown in Figure XVD,14. The ICl_2^- anions are nearly linear (Cl–I–Cl = 174°) but the two I–Cl distances are somewhat different (2.510 and 2.567 A.). In the complex cation, bond lengths

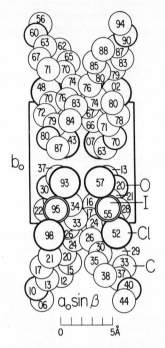

Fig. XVD,14.   The monoclinic structure of tetra-*p*-anisylethylene dichloroiodide
projected along its c_0 axis.

in the planar phenyl groups are normal, C(1)–C(ethylene) = 1.44 A.,
C(4)–O = C(methyl)–O = 1.44 A.

XV,d14.   The monoclinic crystals of *1,1,6,6-tetraphenylhexapentaene*,
$(C_6H_5)_4C_6$, possess a tetramolecular cell of the dimensions:

$$a_0 = 14.4 \text{ A.}; \quad b_0 = 7.8 \text{ A.}; \quad c_0 = 18.1 \text{ A.}; \quad \beta = 108°30'$$

All atoms are in general positions of C_{2h}^5 $(P2_1/c)$:

$$(4e) \quad \pm (xyz; x, 1/2 - y, z + 1/2)$$

Assigned atomic parameters are listed in Table XVD,15.

The general shape of the molecule that results is shown in Figure
XVD,15. Within the limited accuracy of the determination, the ring
C–C = 1.37 A., the double-bonded chain C–C = 1.31 A., and the
single-bonded C(1)–C(13) = 1.51 A. As the figure suggests, the benzene rings at the ends of the linear carbon chain are rotated about the

TABLE XVD,15

Parameters of the Atoms in 1,1,6,6-Tetraphenylhexapentaene

Atom	x	y	z
C(1)	0.112	0.205	0.247
C(2)	0.104	0.285	0.177
C(3)	0.019	0.285	0.112
C(4)	0.940	0.205	0.121
C(5)	0.948	0.125	0.193
C(6)	0.032	0.125	0.255
C(7)	0.138	0.125	0.406
C(8)	0.148	0.125	0.488
C(9)	0.223	0.205	0.542
C(10)	0.294	0.285	0.521
C(11)	0.286	0.285	0.443
C(12)	0.207	0.205	0.385
C(13)	0.209	0.205	0.304
C(14)	0.288	0.205	0.288
C(15)	0.366	0.205	0.272
C(16)	0.445	0.205	0.256
C(17)	0.522	0.205	0.239
C(18)	0.600	0.205	0.223
C(19)	0.605	0.205	0.146
C(20)	0.528	0.285	0.089
C(21)	0.520	0.285	0.010
C(22)	0.590	0.205	0.991
C(23)	0.665	0.125	0.046
C(24)	0.674	0.125	0.123
C(25)	0.771	0.125	0.271
C(26)	0.854	0.125	0.334
C(27)	0.863	0.205	0.406
C(28)	0.785	0.285	0.411
C(29)	0.698	0.285	0.348
C(30)	0.693	0.205	0.280

bonds connecting them with the terminal chain atoms. The planes defined by these chains and the axes of twist of the rings are parallel to the a_0c_0 plane in the crystal (Fig. XVD,16).

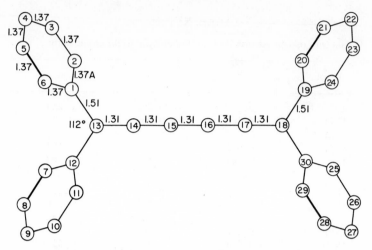

Fig. XVD,15.   Bond lengths in the molecule of 1,1,6,6-tetraphenylhexapentaene.

XV,d15.   Crystals of *di[3-methyl-1(or 5)-phenyl-5(or 1)-p-tolylfor-mazyl]nickel*, C₃₀H₃₀N₈Ni, are triclinic with a bimolecular cell which, for the purposes of structural analysis, was chosen to have the

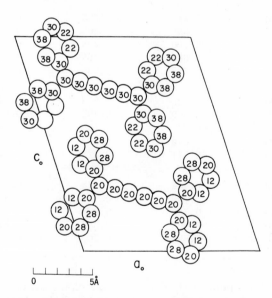

Fig. XVD,16.   Half the molecules in the monoclinic unit of 1,1,6,6-tetraphenyl-hexapentaene projected along the b_0 axis. Left-hand axes.

dimensions:

$$a_0 = 10.50 \text{ A.}; \quad b_0 = 9.26 \text{ A.}; \quad c_0 = 8.25 \text{ A.}$$
$$\alpha = 105°30'; \quad \beta = 75°54'; \quad \gamma = 105°48'$$

The space group is C_i^1 ($P\bar{1}$) with the nickel atoms in the origin (1a) 000 and all other atoms in (2i) $\pm(xyz)$. The determined parameters are listed in Table XVD,16. Probable hydrogen parameters are stated in the original paper.

TABLE XVD,16

Parameters of the Atoms in $C_{30}H_{30}N_8Ni$

Atom	x	y	z
N(1)	0.179(6)	0.110(6)	−0.026(6)
N(2)	0.282(6)	0.043(7)	−0.089(7)
C(3)	0.261(8)	−0.099(8)	−0.189(8)
N(4)	−0.160(6)	0.163(6)	0.274(6)
N(5)	−0.050(6)	0.108(5)	0.222(6)
C(6)	0.217(7)	0.271(8)	0.038(7)
C(7)	0.334(8)	0.336(8)	0.102(8)
C(8)	0.365(9)	0.494(10)	0.170(9)
C(9)	0.281(9)	0.587(8)	0.171(9)
C(10)	0.164(10)	0.520(9)	0.103(9)
C(11)	0.131(8)	0.362(8)	0.036(8)
C(13)	0.381(9)	−0.178(10)	−0.240(11)
C(14)	0.038(7)	0.161(7)	0.344(7)
C(15)	0.123(8)	0.067(8)	0.340(7)
C(16)	0.217(9)	0.123(9)	0.448(8)
C(17)	0.227(9)	0.262(9)	0.554(9)
C(18)	0.136(9)	0.351(9)	0.561(9)
C(19)	0.042(8)	0.301(8)	0.458(8)
$^1/_2$C(12′)	0.320(15)	0.756(11)	0.244(15)
$^1/_2$C(12″)	0.348(18)	0.333(20)	0.634(17)

The bond dimensions in half the molecule in this structure are those of Figure XVD,17. The nickel atom and its four coordinated nitrogen atoms are strictly coplanar but there is buckling of the rest of the chelating part of the molecule. The height of the C(12′) and C(12″) peaks indicates disorder in the sense that this methyl group appears sometimes on one and sometimes on the other of the benzene rings.

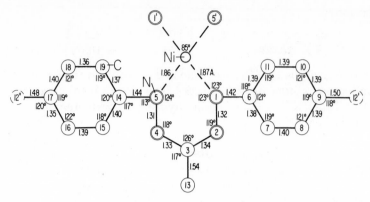

Fig. XVD,17. Bond dimensions in half the molecule of di[3-methyl-1(or 5)-phenyl-5(or 1)-*p*-tolylformazyl]nickel.

XV,d16. It might have been anticipated that the tetragonal structure of *tetra-m-tolylsilane*, $(m\text{-}CH_3C_6H_4)_4Si$, would resemble that described for tetraphenylmethane (**XV,d2**). It is, however, entirely different, though possessed of a bimolecular unit. The edges of this cell are:

$$a_0 = 10.0 \text{ A.,} \quad c_0 = 11.3 \text{ A.}$$

The chosen space group is V_d^2 $(P\bar{4}2c)$ with silicon atoms, at the centers of sphenoidally distributed tolyl groups, in:

$$(2f) \quad {}^1/_2\,{}^1/_2\,0;\, {}^1/_2\,{}^1/_2\,{}^1/_2$$

The carbon atoms are in the general positions

$$(8n) \quad xyz;\, \bar{x}\bar{y}z;\, \bar{x},y,{}^1/_2-z;\, x,\bar{y},{}^1/_2-z;$$
$$\bar{y}x\bar{z};\, y\bar{x}\bar{z};\, y,x,z+{}^1/_2;\, \bar{y},\bar{x},z+{}^1/_2$$

with the parameters of Table XVD,17.

TABLE XVD,17

Parameters of the Carbon Atoms in Tetra-*m*-tolylsilane

Atom	x	y	z
C(1)	0.556	0.663	0.067
C(2)	0.638	0.663	0.166
C(3)	0.678	0.783	0.216
C(4)	0.638	0.903	0.166
C(5)	0.556	0.903	0.067
C(6)	0.515	0.783	0.018
C(7)	0.762	0.783	0.315

This arrangement cannot be considered well established, both because of the limited data and because of the extraordinarily short C–C = 2.24 A. that prevails between adjacent molecules.

XV,d17. According to a brief note, crystals of *rubidium tetraphenylboronate*, $Rb[B(C_6H_5)_4]$, are tetragonal with a bimolecular unit of the edge lengths:

$$a_0 = 11.23 \text{ A.}; \quad c_0 = 8.08 \text{ A.}$$

Atoms have been placed in the following positions of V_d^{11} ($I\bar{4}2m$):

Rb : (2a) 000; $^1/_2$ $^1/_2$ $^1/_2$

  B : (2b) 0 0 $^1/_2$; $^1/_2$ $^1/_2$ 0

C(1) : (8i) uuv; $\bar{u}\bar{u}v$; $u\bar{u}\bar{v}$; $\bar{u}u\bar{v}$; B.C.
       with $u = 0.078$, $v = 0.392$

C(2) : (8i) with $u = 0.206$, $v = 0.143$

C(3) : (16j) xyz; $\bar{x}\bar{y}z$; $\bar{x}y\bar{z}$; $x\bar{y}\bar{z}$;
       $\bar{y}x\bar{z}$; $y\bar{x}\bar{z}$; yxz; $\bar{y}\bar{x}z$; B.C
       with $x = 0.040$, $y = 0.194$, $z = 0.342$

C(4) : (16j) with $x = 0.100$, $y = 0.249$, $z = 0.213$

The structure is shown in Figure XVD,18. In it B–C = 1.51 A. and Rb–C = 3.46 A. In the benzene rings C(1)–C(2) = 1.43 A., C(2)–C(3) = 1.38 A., and C(3)–C(4) = 1.40 A.

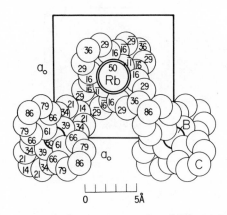

Fig. XVD,18. The tetragonal structure of rubidium tetraphenylboronate projected along its c_0 axis.

XV,d18. Crystals of *tetraphenylphosphonium iodide*, $(C_6H_5)_4PI$, are tetragonal with a bimolecular unit of the edge lengths:

$$a_0 = 11.88(6) \text{ A.}, \quad c_0 = 6.96(2) \text{ A.}$$

The space group is S_4^2 ($I\bar{4}$) with atoms in the positions:

$$\text{I} : (2a) \quad 000; \tfrac{1}{2}\tfrac{1}{2}\tfrac{1}{2}$$
$$\text{P} : (2c) \quad 0\,\tfrac{1}{2}\,\tfrac{1}{4}; \tfrac{1}{2}\,0\,\tfrac{3}{4}$$

The other atoms are in

$$(8g) \quad xyz; \ \bar{x}\bar{y}z; \ y\bar{x}\bar{z}; \ \bar{y}x\bar{z}; \quad \text{B.C.}$$

with the following parameters:

Atom	x	y	z
C(1)	0.067	0.402	0.091
C(2)	0.040	0.286	0.085
C(3)	0.096	0.207	−0.032
C(4)	0.173	0.251	−0.160
C(5)	0.200	0.369	−0.154
C(6)	0.143	0.447	−0.037

Calculated hydrogen positions are also stated in the original paper.

The structure, shown in Figure XVD,19, consists of tetrahedral $(C_6H_5)_4P^+$ cations and I^- anions. In the cations P–C = 1.80 A. and the dihedral angles between the planes of the phenyl groups are 92.5 and 118.5°.

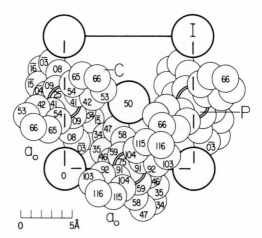

Fig. XVD,19. The tetragonal structure of tetraphenylphosphonium iodide projected along its c_0 axis.

XV,d19. *Benzoyl(triphenylphosphoranylidine)methyl chloride,* $(C_6H_5)_3PC(Cl)C(O)C_6H_5$, forms monoclinic crystals whose tetramolecular unit has the dimensions:

$$a_0 = 11.130 \text{ A.}; \quad b_0 = 12.641 \text{ A.}; \quad c_0 = 15.470 \text{ A.}; \quad \beta = 97°32'$$

The space group is C_{2h}^5 ($P2_1/a$) with atoms in the positions:

$$(4e) \quad \pm (xyz; x + {}^1/_2, {}^1/_2 - y, z)$$

The parameters are those of Table XVD,18.

TABLE XVD,18

Parameters of the Atoms in $(C_6H_5)_3PC(Cl)C(O)C_6H_5$

Atom	x	y	z
Cl	0.3726	0.4844	0.0776
P	0.3412	0.4394	0.2638
O	0.4199	0.7410	0.2913
C(1)	0.2070	0.4794	0.3115
C(2)	0.1502	0.4016	0.3591
C(3)	0.0463	0.4338	0.3970
C(4)	0.0042	0.5361	0.3846
C(5)	0.0544	0.6098	0.3357
C(6)	0.1613	0.5814	0.2972
C(7)	0.3042	0.3126	0.2146
C(8)	0.3770	0.2272	0.2297
C(9)	0.3487	0.1297	0.1915
C(10)	0.2382	0.1192	0.1340
C(11)	0.1633	0.2049	0.1189
C(12)	0.1927	0.3031	0.1568
C(13)	0.4634	0.4171	0.3501
C(14)	0.4426	0.3826	0.4317
C(15)	0.5485	0.3576	0.4956
C(16)	0.6636	0.3677	0.4742
C(17)	0.6819	0.4034	0.3939
C(18)	0.5814	0.4255	0.3295
C(19)	0.3802	0.5277	0.1862
C(20)	0.4255	0.6247	0.2109
C(21)	0.4709	0.7050	0.1527
C(22)	0.4180	0.8006	0.1445
C(23)	0.4620	0.8788	0.0902
C(24)	0.5597	0.8554	0.0476
C(25)	0.6135	0.7584	0.0564
C(26)	0.5686	0.6791	0.1091

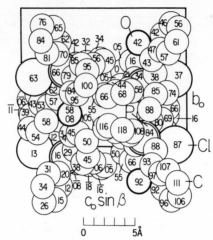

Fig. XVD,20.   The monoclinic structure of benzoyl(triphenylphosphoranylidene)-
methyl chloride projected along its a_0 axis.

The structure, as shown in Figure XVD,20, is composed of molecules
with the bond dimensions of Figure XVD,21. The shortest intermolecular separation is a C–O = 3.27 A.

The isostructural bromine compound has the cell dimensions:

$$a_0 = 11.149 \text{ A.}; \quad b_0 = 12.663 \text{ A.}; \quad c_0 = 15.686 \text{ A.}; \quad \beta = 97°27'$$

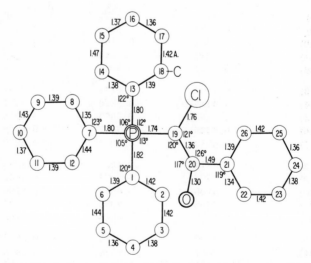

Fig. XVD,21.   Bond dimensions in the molecule of benzoyl(triphenylphosphoranylidene)methyl chloride.

Parameters have not been established for its atoms. A different structure has been found for the iodine derivative (**XV,d20**).

XV,d20. Crystals of *benzoyl(triphenylphosphoranylidene)methyl iodide*, $(C_6H_5)_3PC(I)C(O)C_6H_5$, though monoclinic, have a structure different from that of the chloride. The tetramolecular unit has the

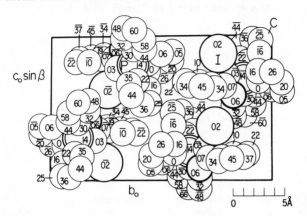

Fig. XVD,22. The monoclinic structure of benzoyl(triphenylphosphoranylidene) methyl iodide projected along its a_0 axis. The smaller, heavily ringed circles are oxygen.

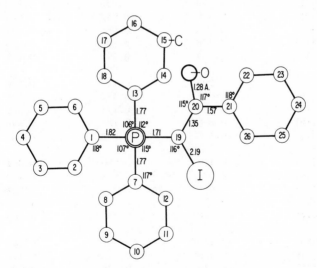

Fig. XVD,23. Bond dimensions in the molecule of benzoyl(triphenylphosphoranylidene)methyl iodide.

dimensions:

$$a_0 = 8.248(10) \text{ A.}; \quad b_0 = 20.544(40) \text{ A.}; \quad c_0 = 13.488(28) \text{ A.}$$
$$\beta = 101°21(15)'$$

The space group is C_{2h}^5 ($P2_1/c$) with atoms in the positions:

$$(4e) \quad \pm(xyz; x, 1/2 - y, z + 1/2)$$

Determined parameters are listed in Table XVD,19.

TABLE XVD,19

Parameters of the Atoms in $(C_6H_5)_3PC(I)C(O)C_6H_5$

Atom	x	y	z
I	−0.0187	0.2423	0.1265
P	0.1359	0.1260	0.2889
O	−0.061	0.174	0.417
C(1)	0.001	0.059	0.308
C(2)	−0.160	0.058	0.252
C(3)	−0.257	0.002	0.258
C(4)	−0.200	−0.045	0.320
C(5)	−0.046	−0.050	0.373
C(6)	0.062	0.006	0.362
C(7)	0.224	0.106	0.183
C(8)	0.159	0.056	0.118
C(9)	0.250	0.049	0.033
C(10)	0.365	0.086	0.024
C(11)	0.436	0.133	0.082
C(12)	0.355	0.143	0.167
C(13)	0.299	0.126	0.396
C(14)	0.321	0.172	0.480
C(15)	0.484	0.175	0.552
C(16)	0.603	0.124	0.538
C(17)	0.576	0.083	0.467
C(18)	0.440	0.080	0.394
C(19)	0.027	0.198	0.277
C(20)	−0.066	0.214	0.344
C(21)	−0.193	0.272	0.338
C(22)	−0.343	0.258	0.350
C(23)	−0.447	0.317	0.341
C(24)	−0.370	0.373	0.331
C(25)	−0.225	0.390	0.318
C(26)	−0.095	0.330	0.329

The structure is shown in Figure XVD,22. Significant bond dimensions in its molecules are given in Figure XVD,23. Between molecules there is a short O–I = 3.25 A.; all other separations are greater than 3.5 A.

TABLE XVD,20

Parameters of the Atoms in $(C_6H_5)_3P{=}CHSO_2C_6H_4(CH_3)$

Atom	x	y	z
S	0.0914	0.0548	0.1538
P	0.1855	0.1200	0.1054
O(1)	0.0600	0.1462	0.0929
O(2)	0.0911	0.1021	0.2203
C(1)	0.1577	0.0299	0.1578
C(2)	0.0589	−0.1222	0.1396
C(3)	0.0257	−0.1669	0.0710
C(4)	0.0004	−0.3011	0.0608
C(5)	0.0073	−0.3993	0.1164
C(6)	0.0398	−0.3563	0.1812
C(7)	0.0660	−0.2205	0.1939
C(8)	0.1430	0.0956	0.0129
C(9)	0.1433	0.2101	−0.0340
C(10)	0.1127	0.1844	−0.1043
C(11)	0.0848	0.0541	−0.1275
C(12)	0.0854	−0.0544	−0.0818
C(13)	0.1169	−0.0348	−0.0084
C(14)	0.1987	0.3141	0.1180
C(15)	0.2542	0.3720	0.1340
C(16)	0.2626	0.5211	0.1413
C(17)	0.2175	0.6224	0.1331
C(18)	0.1651	0.5722	0.1172
C(19)	0.1555	0.4150	0.1123
C(20)	0.2527	0.0304	0.1219
C(21)	0.2666	−0.0251	0.0672
C(22)	0.3171	−0.1024	0.0824
C(23)	0.3512	−0.1343	0.1505
C(24)	0.3375	−0.0754	0.2051
C(25)	0.2882	0.0023	0.1910
C(26)	−0.0226	−0.5505	0.1029

XV,d21. The Wittig reagent, *p-tolyl(triphenylphosphoranylidene)-methyl sulfone*, $(C_6H_5)_3P{=}CHSO_2C_6H_4(CH_3)$, forms monoclinic crystals whose unit contains eight molecules and has the cell dimensions:

$$a_0 = 25.633(87) \text{ A.}; \quad b_0 = 8.981(8) \text{ A.}; \quad c_0 = 20.733(71) \text{ A.}$$
$$\beta = 111°54'$$

Atoms are in the positions

$$(8f) \quad \pm(xyz;\ x,\bar{y},z+{}^1/_2;\ x+{}^1/_2,y+{}^1/_2,z;\ x+{}^1/_2,{}^1/_2-y,z+{}^1/_2)$$

of the space group C_{2h}^6 ($C2/c$). The determined parameters are listed in Table XVD,20.

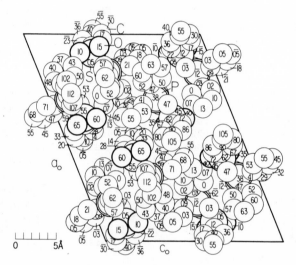

Fig. XVD,24. The monoclinic structure of *p*-tolyl(triphenylphosphoranylidene)-methyl sulfone projected along its b_0 axis.

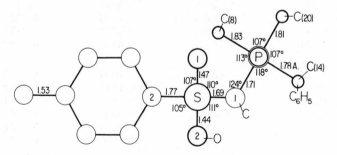

Fig. XVD,25. Bond dimensions in the molecule of *p*-tolyl(triphenylphosphoranylidene)methyl sulfone.

In the structure (Fig. XVD,24) the molecules have the bond dimensions shown in Figure XVD,25. As usual the planar benzene rings are twisted about the phosphorus atom to which they are attached. Distribution about both the phosphorus and the sulfur atoms is tetrahedral.

XV,d22. A reaction product of $(C_6H_5)_3P{=}N(C_6H_4Br)$ and $CH_3O_2C{\equiv}C{\cdot}CO_2CH_3$ is shown by x-ray analysis to be $(C_6H_5)_3P{=}C(CO_2CH_3){-}C(CO_2CH_3){=}N(C_6H_4Br)$. Its crystals have monoclinic symmetry with a tetramolecular unit of the dimensions:

$$a_0 = 11.83(2) \text{ A.}; \quad b_0 = 9.24(2) \text{ A.}; \quad c_0 = 25.43(4) \text{ A.}$$
$$\beta = 104°15(5)'$$

The space group is C_{2h}^5 ($P2_1/c$) with atoms in the positions:

$$(4e) \quad \pm(xyz; x,{}^1/_2-y,z+{}^1/_2)$$

The parameters are listed in Table XVD,21.

The resulting structure is shown in Figure XVD,26. The molecules thus defined have the bond dimensions of Figure XVD,27. The distance of the bromine atom from C(3) is 1.95 A.; atomic separations in the planar benzene rings are as expected.

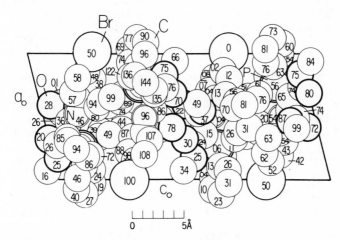

Fig. XVD,26.   The monoclinic structure of $(C_6H_5)_3P{=}C(CO_2CH_3){-}C(CO_2CH_3)$ $={N}(C_6H_4Br)$ projected along its b_0 axis.

TABLE XVD,21

Parameters of the Atoms in $(C_6H_5)_3P=C(CO_2CH_3)-C(CO_2CH_3)=N(C_6H_4Br)$

Atom	x	y	z
Br(1)	0.0105	0.5028	0.2402
P(2)	0.2520	0.4363	0.1156
C(3)	0.8472	0.4858	0.2005
C(4)	0.7851	0.3770	0.2159
C(5)	0.6642	0.3721	0.1863
C(6)	0.6203	0.4581	0.1446
C(7)	0.6867	0.5748	0.1300
C(8)	0.8038	0.5826	0.1610
C(9)	0.1015	0.3751	0.1054
C(10)	0.0741	0.2447	0.1314
C(11)	−0.0362	0.1915	0.1209
C(12)	−0.1248	0.2746	0.0834
C(13)	−0.0997	0.4000	0.0568
C(14)	0.0144	0.4605	0.0698
C(15)	0.3243	0.3916	0.1851
C(16)	0.3428	0.4925	0.2280
C(17)	0.3924	0.4352	0.2804
C(18)	0.4255	0.2965	0.2921
C(19)	0.4020	0.1920	0.2478
C(20)	0.3581	0.2451	0.1963
C(21)	0.2452	0.6376	0.1064
C(22)	0.1836	0.7210	0.1369
C(23)	0.1809	0.8640	0.1295
C(24)	0.2380	0.9398	0.0934
C(25)	0.2988	0.8493	0.0649
C(26)	0.3084	0.6973	0.0705
C(27)	0.4511	0.3595	0.0880
C(28)	0.3219	0.3496	0.0739
C(29)	0.5179	0.2621	0.0707
C(30)	0.2710	0.2623	0.0254
C(31)	0.5952	0.0123	0.0765
C(32)	0.0956	0.1622	−0.0323
N(33)	0.5004	0.4596	0.1231
O(34)	0.5229	0.1250	0.0931
O(35)	0.5817	0.2781	0.0364
O(36)	0.1533	0.2551	0.0152
O(37)	0.3261	0.2002	−0.0015

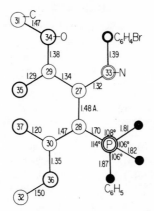

Fig. XVD,27.   Bond dimensions in the molecule of $(C_6H_5)_3P$=$C(CO_2CH_3)$—
$C(CO_2CH_3)$=$N(C_6H_4Br)$.

XV,d23.   Crystals of *hydridochloro bis(diphenylethylphosphine) platinum*, $PtHCl[P(C_6H_5)_2C_2H_5]_2$, are monoclinic with a tetramolecular unit of the dimensions:

$$a_0 = 11.80(2) \text{ A.}; \quad b_0 = 16.93(3) \text{ A.}; \quad c_0 = 14.31(2) \text{ A.}$$
$$\beta = 108.4(3)°$$

The space group is C_{2h}^5 ($P2_1/c$) with atoms in the positions:

$$(4e) \quad \pm (xyz; x, 1/2 - y, z + 1/2)$$

The parameters are those of Table XVD,22.

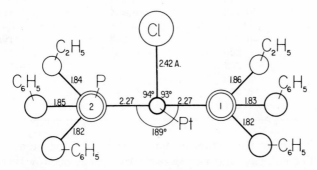

Fig. XVD,28.   Bond dimensions in the molecule of hydridochloro bis(diphenylethylphosphine) platinum. The hydrogen attached to platinum is not shown.

TABLE XVD,22

Parameters of the Atoms in $PtHCl[P(C_6H_5)_2C_2H_5]_2$

Atom	x	y	z
Pt	0.17910(9)	0.41664(7)	0.28903(8)
Cl	0.2588(7)	0.4506(6)	0.4619(5)
P(1)	−0.0015(6)	0.3837(5)	0.3022(5)
P(2)	0.3438(5)	0.4567(5)	0.2526(5)
C(1,1)	−0.010(2)	0.362(2)	0.427(2)
C(1,2)	0.060(3)	0.281(3)	0.459(3)
C(2,1)	0.485(2)	0.467(2)	0.354(2)
C(2,2)	0.527(3)	0.389(2)	0.406(2)
C(1)	−0.073	0.298	0.230
C(2)	−0.001	0.236	0.216
C(3)	−0.054	0.167	0.171
C(4)	−0.178	0.159	0.138
C(5)	−0.249	0.221	0.152
C(6)	−0.196	0.290	0.197
C(1′)	−0.111	0.462	0.260
C(2′)	−0.132	0.491	0.165
C(3′)	−0.209	0.554	0.132
C(4′)	−0.263	0.590	0.194
C(5′)	−0.242	0.562	0.290
C(6′)	−0.165	0.498	0.323
C(1″)	0.393	0.398	0.164
C(2″)	0.347	0.322	0.140
C(3″)	0.384	0.276	0.075
C(4″)	0.468	0.305	0.033
C(5″)	0.514	0.381	0.057
C(6″)	0.477	0.427	0.123
C(1‴)	0.324	0.553	0.194
C(2‴)	0.220	0.568	0.116
C(3‴)	0.202	0.641	0.070
C(4‴)	0.289	0.699	0.100
C(5‴)	0.393	0.685	0.177
C(6‴)	0.411	0.612	0.224

The molecules in this structure have the bond dimensions of Figure XVD,28. The chlorine and two phosphorus atoms occupy three of the four corners of a distorted square about platinum. In the analogous triethyl bromide (**XIV,b112**) it was inferred that the hydrogen atom occupied the fourth corner of this square; it may also be true in this case.

XV,d24. The triclinic crystals of *di-μ-diphenylphosphinatoacetyl-acetonato chromium(III)*, $[(CH_3CO)_2CH_2]_2Cr[OP(C_6H_5)_2O]_2Cr[(CH_3CO)_2CH_2]_2$, have been assigned a bimolecular unit of the dimensions:

$$a_0 = 12.64(5) \text{ A.}; \quad b_0 = 15.57(5) \text{ A.}; \quad c_0 = 13.35(1) \text{ A.}$$
$$\alpha = 112.4(2)°; \quad \beta = 112.5(2)°; \quad \gamma = 84.2(1)°$$

All atoms are in the positions $(2i) \pm (xyz)$ of C_i^1 ($P\bar{1}$). The parameters are listed in Table XVD,23.

TABLE XVD,23

Parameters of the Atoms in Di-μ-diphenylphosphinatoacetylacetonato Chromium (III)

Atom	x	y	z
Cr(1)	0.2458	0.2449	0.3490
Cr(2)	0.1722	0.2008	−0.0658
O(1)	0.0999	0.2654	0.2372
O(2)	0.3341	0.2919	0.2852
O(3)	0.2493	0.3708	0.4591
O(4)	0.2485	0.1165	0.2432
O(5)	0.3871	0.2290	0.4672
O(6)	0.1591	0.1972	0.4116
O(7)	0.0824	0.2526	0.0326
O(8)	0.1832	0.3223	−0.0747
O(9)	0.2570	0.1475	−0.1675
O(10)	0.0316	0.1765	−0.2052
O(11)	0.1577	0.0766	−0.0659
O(12)	0.3165	0.2284	0.0680
P(1)	0.0533	0.3005	0.1378
P(2)	0.3850	0.2785	0.1968
C(1)	0.091	0.424	0.195
C(2)	−0.101	0.291	0.086
C(3)	0.513	0.215	0.231
C(4)	0.427	0.393	0.222
C(5)	0.525	0.135	0.145
C(6)	0.626	0.084	0.173
C(7)	0.709	0.114	0.281
C(8)	0.697	0.194	0.364
C(9)	0.600	0.247	0.341

(*continued*)

TABLE XVD,23 (*continued*)

Parameters of the Atoms in Di-μ-diphenylphosphinatoacetylacetonato
Chromium (III)

Atom	x	y	z
C(10)	−0.164	0.280	−0.028
C(11)	−0.285	0.281	−0.068
C(12)	−0.341	0.285	0.005
C(13)	−0.276	0.294	0.121
C(14)	−0.156	0.298	0.161
C(15)	0.087	0.481	0.301
C(16)	0.112	0.577	0.342
C(17)	0.134	0.618	0.275
C(18)	0.137	0.560	0.169
C(19)	0.112	0.464	0.128
C(20)	0.440	0.465	0.329
C(21)	0.479	0.554	0.352
C(22)	0.501	0.572	0.267
C(23)	0.487	0.503	0.162
C(24)	0.452	0.410	0.137
C(25)	0.448	0.158	0.460
C(26)	0.557	0.171	0.567
C(27)	0.424	0.076	0.365
C(28)	0.327	0.063	0.265
C(29)	0.311	−0.032	0.163
C(30)	0.231	0.347	−0.129
C(31)	0.223	0.444	−0.124
C(32)	0.288	0.286	−0.195
C(33)	0.296	0.189	−0.214
C(34)	0.356	0.128	−0.294
C(35)	−0.035	0.104	−0.255
C(36)	−0.137	0.108	−0.357
C(37)	−0.021	0.028	−0.223
C(38)	0.074	0.018	−0.133
C(39)	0.080	−0.071	−0.108
C(40)	0.201	0.401	0.530
C(41)	0.215	0.507	0.602
C(42)	0.127	0.345	0.539
C(43)	0.117	0.247	0.487
C(44)	0.043	0.195	0.513

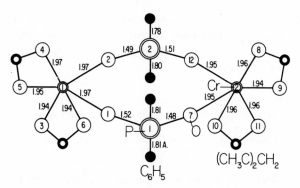

Fig. XVD,29. Some of the bond lengths in the molecule of di-μ-diphenyl-phosphinatoacetylacetonato chromium.

The distribution of the atoms in the central part of the trimeric molecule and the principal bond dimensions are those of Figure XVD,29. Chromium atoms are octahedrally and the phosphorus atoms tetrahedrally coordinated.

XV,d25. The tetragonal structure of *tetraphenylarsonium iodide*, $As(C_6H_5)_4I$, has a bimolecular unit of the dimensions:

$$a_0 = 12.19 \text{ A.}, \quad c_0 = 7.085 \text{ A.}$$

Atoms have been placed in the following special and general positions of S_4^2 ($I\bar{4}$):

$(2a)$ 000; $^1/_2$ $^1/_2$ $^1/_2$

$(2d)$ 0 $^1/_2$ $^3/_4$; $^1/_2$ 0 $^1/_4$

$(8g)$ xyz; $\bar{x}\bar{y}z$; $y\bar{x}\bar{z}$; $\bar{y}x\bar{z}$; B.C.

with the positions and parameters of Table XVD,24.

TABLE XVD,24

Positions and Parameters of the Atoms in Tetraphenylarsonium Iodide

Atom	Position	x	y	z
I	(2d)	0	$^1/_2$	$^3/_4$
As	(2a)	0	0	0
C(1)	(8g)	0.065	0.110	0.162
C(2)	(8g)	0.008	0.205	0.216
C(3)	(8g)	0.053	0.280	0.344
C(4)	(8g)	0.164	0.268	0.396
C(5)	(8g)	0.219	0.180	0.344
C(6)	(8g)	0.174	0.102	0.216

Bond lengths within the molecule are C–C = 1.39 A. and As–C = 1.95 A. The closest approaches of carbon atoms of different phenyl groups are 3.40 A. within the same phenylarsonium ion and 3.50 A. between neighboring ions.

XV,d26. *Tetraphenylarsonium triiodide*, $As(C_6H_5)_4I_3$, forms monoclinic crystals whose bimolecular unit has the dimensions:

$$a_0 = 15.34(1) \text{ A.}; \quad b_0 = 7.63(1) \text{ A.}; \quad c_0 = 10.63(1) \text{ A.}$$
$$\beta = 93°24'$$

The assigned space group is C_{2h}^4 (not the C_{2h}^2 stated in the article) in the non-standard orientation $P2/n$. The arsenic atom has been put in the positions

$$(2f) \quad \pm (1/4 \; u \; 1/4) \qquad \text{with } u = 0.000$$

and I(1) in:

$$(2e) \quad \pm (1/4 \; u \; 3/4) \qquad \text{with } u = 0.158$$

Other atoms have been placed in the general positions

$$(4g) \quad \pm (xyz; x + 1/2, \bar{y}, z + 1/2)$$

with the parameters of Table XVD,25, the values for the carbon atoms being calculated.

TABLE XVD,25

Parameters of the Atoms in $As(C_6H_5)_4I_3$

Atom	x	y	z
I(2)	0.062	0.170	0.703
C(1)	0.231	0.853	0.393
C(2)	0.146	0.800	0.421
C(3)	0.294	0.800	0.478
C(4)	0.131	0.695	0.518
C(5)	0.281	0.695	0.579
C(6)	0.197	0.642	0.600
C(7)	0.353	0.147	0.288
C(8)	0.367	0.200	0.410
C(9)	0.407	0.200	0.194
C(10)	0.439	0.305	0.438
C(11)	0.477	0.305	0.219
C(12)	0.500	0.358	0.346

In this structure the I_3^- anion is symmetrical and regular, with
I–I = 2.90 A.

XV,d27. Crystals of *bis(diphenylarsenic)oxide*, $[(C_6H_5)_2As]_2O$, are
monoclinic. Their tetramolecular cell has the dimensions:

$$a_0 = 11.48(3) \text{ A.}; \quad b_0 = 30.2(1) \text{ A.}; \quad c_0 = 5.97(2) \text{ A.}$$
$$\beta = 93.8(5)°$$

TABLE XVD,26

Parameters of the Atoms in $[(C_6H_5)_2As]_2O$

Atom	x	y	z
As(1)	0.1105	0.1034	0.1100
As(2)	0.3075	0.1688	0.0000
O	0.216	0.141	0.154
C(1)	0.153	0.056	0.284
C(2)	0.221	0.062	0.502
C(3)	0.246	0.028	0.668
C(4)	0.210	−0.011	0.602
C(5)	0.146	−0.017	0.390
C(6)	0.118	0.017	0.228
C(7)	0.426	0.123	−0.049
C(8)	0.484	0.128	−0.234
C(9)	0.565	0.095	−0.268
C(10)	0.592	0.058	−0.117
C(11)	0.532	0.053	0.065
C(12)	0.452	0.086	0.099
C(13)	−0.020	0.129	0.270
C(14)	−0.131	0.124	0.175
C(15)	−0.225	0.142	0.290
C(16)	−0.207	0.163	0.492
C(17)	−0.095	0.169	0.587
C(18)	−0.003	0.151	0.483
C(19)	0.394	0.203	0.200
C(20)	0.345	0.223	0.392
C(21)	0.413	0.251	0.555
C(22)	0.528	0.257	0.502
C(23)	0.575	0.236	0.320
C(24)	0.507	0.210	0.155

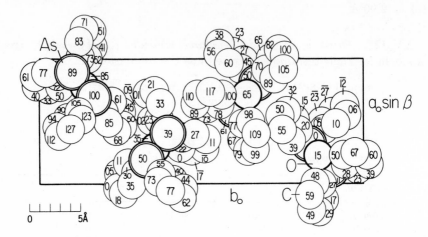

Fig. XVD,30. The monoclinic structure of bis(diphenylarsenic)oxide projected along its c_0 axis.

The space group is C_{2h}^5 ($P2_1/n$) with atoms in the positions:

$$(4e)\quad \pm (xyz;\ x+\tfrac{1}{2}, \tfrac{1}{2}-y, z+\tfrac{1}{2})$$

The parameters are listed in Table XVD,26.

The structure is illustrated in Figure XVD,30. The four benzene rings in its molecules are planar. In these molecules the As–C distances lie between 1.82 and 1.99 A.; the C–As–C angles are either 102 or 98°. The distance As–O = 1.67 A. and the angle As–O–As = 137°. The angle C–As–O ranges between 101 and 106°.

XV,d28. *Tetraphenylarsonium - 3 - fluoro - 1,1,4,5,5 - pentacyano - 2 - azapentadienide,* $[(C_6H_5)_4As]^+[(CN)_2C \cdot N \cdot C(F) \cdot C(CN) \cdot C(CN)_2]^-$, forms orthorhombic crystals. Their unit, containing eight molecules, has the edge lengths:

$$a_0 = 9.789(3)\ \text{A.};\quad b_0 = 24.601(6)\ \text{A.};\quad c_0 = 23.918(6)\ \text{A.}$$

The space group is V_h^{15} ($Pcab$) with atoms in the positions:

$$(8c)\quad \pm (xyz;\ x+\tfrac{1}{2}, \bar{y}, \tfrac{1}{2}-z;\ \bar{x}, \tfrac{1}{2}-y, z+\tfrac{1}{2};\ \tfrac{1}{2}-x, y+\tfrac{1}{2}, \bar{z})$$

The determined parameters are listed in Table XVD,27; suggested positions for the hydrogen atoms are to be found in the original article.

TABLE XVD,27

Parameters of the Atoms in $[(C_6H_5)_4As][(CN)_2C \cdot N \cdot C(F) \cdot C(CN) \cdot C(CN)_2]$

Atom	x	y	z
C(1)	0.2283(7)	0.0346(3)	0.2185(2)
C(2)	0.1605(7)	0.0694(3)	0.2543(2)
C(3)	0.1549(7)	0.0567(3)	0.3116(2)
C(4)	0.2172(7)	0.0108(3)	0.3313(2)
C(5)	0.2839(7)	−0.0236(3)	0.2957(2)
C(6)	0.2905(7)	−0.0114(3)	0.2380(2)
C(7)	0.3026(7)	−0.0097(3)	0.1029(2)
C(8)	0.2252(7)	−0.0574(3)	0.1059(2)
C(9)	0.2728(7)	−0.1038(3)	0.0790(2)
C(10)	0.3960(7)	−0.1030(3)	0.0503(2)
C(11)	0.4699(7)	−0.0560(3)	0.0478(2)
C(12)	0.4247(7)	−0.0090(3)	0.0741(2)
C(13)	0.0775(7)	0.0797(3)	0.1123(2)
C(14)	0.0002(7)	0.0484(3)	0.0756(2)
C(15)	−0.1205(7)	0.0690(3)	0.0536(2)
C(16)	−0.1626(7)	0.1214(3)	0.0680(2)
C(17)	−0.0865(7)	0.1522(3)	0.1048(2)
C(18)	0.0341(7)	0.1316(3)	0.1270(2)
C(19)	0.3745(7)	0.1094(3)	0.1345(2)
C(20)	0.4764(7)	0.1160(3)	0.1740(2)
C(21)	0.5683(7)	0.1588(3)	0.1699(2)
C(22)	0.5567(7)	0.1946(3)	0.1253(2)
C(23)	0.4582(7)	0.1883(3)	0.0849(2)
C(24)	0.3657(7)	0.1455(3)	0.0893(2)
C(25)	−0.1816(7)	0.2783(3)	0.2560(2)
C(26)	−0.0366(7)	0.3352(3)	0.3127(2)
C(27)	−0.0716(7)	0.1841(3)	0.3004(2)
C(28)	0.2873(7)	0.2267)3)	0.4525(2)
C(29)	0.2705(7)	0.1311(3)	0.4380(2)
C(30)	−0.0771(7)	0.2812(3)	0.2989(2)
C(31)	−0.0213(7)	0.2349(3)	0.3222(2)
C(32)	0.0782(7)	0.2319(3)	0.3634(2)
C(33)	0.2255(7)	0.1830(3)	0.4221(2)
F	0.1228(4)	0.2805(1)	0.3829(2)
N(1)	−0.2646(7)	0.2772(2)	0.2231(2)
N(2)	−0.0104(7)	0.3796(2)	0.3213(2)
N(3)	−0.1157(7)	0.1455(2)	0.2816(2)
N(4)	0.3421(7)	0.2595(2)	0.4780(2)
N(5)	0.3087(7)	0.0892(2)	0.4525(2)
N(6)	0.1298(7)	0.1869(2)	0.3823(2)
As	0.24400(6)	0.05314(2)	0.14186(2)

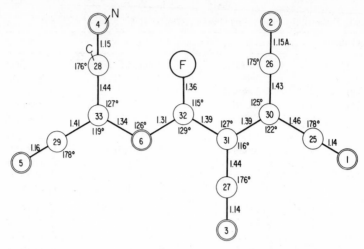

Fig. XVD,31a. Bond dimensions in the complex anion of tetraphenylarsonium-
3-fluoro-1,1,4,5,5-pentacyano-2-azapentadienide.

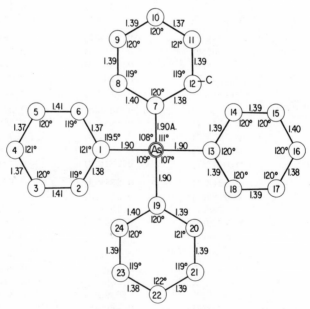

Fig. XVD,31b. Bond dimensions in the cation of tetraphenylarsonium-3-
fluoro-1,1,4,5,5-pentacyano-2-azapentadienide.

The bond dimensions of the ions of this structure are shown in Figure XVD,31. In the cation the four planar C_6H_5 rings are tetrahedrally distributed as a pinwheel around the central arsenic, which is up to 0.10 A. outside the planes of the phenyl groups.

XV,d29. Crystals of *tetraphenylarsonium cis-diaquotetrachlororuthenate monohydrate*, $[As(C_6H_5)_4][RuCl_4 \cdot 2H_2O] \cdot H_2O$, are monoclinic with a tetramolecular unit of the dimensions:

$$a_0 = 15.059(5) \text{ A.;} \quad b_0 = 16.711(5) \text{ A.;} \quad c_0 = 10.996(5) \text{ A.}$$
$$\beta = 99.88(1)°$$

Atoms are in the positions

$$(4e) \quad \pm (xyz; x+{}^1/_2, {}^1/_2-y, z+{}^1/_2)$$

of C_{2h}^5 ($P2_1/n$). The parameters are listed in Table XVD,28.

TABLE XVD,28

Parameters of the Atoms in $[As(C_6H_5)_4][RuCl_4 \cdot 2H_2O] \cdot H_2O$

Atom	x	y	z
Ru	0.4872	0.5164	0.7585
As	0.4169	0.0785	0.7332
Cl(1)	0.3829	0.5818	0.8566
Cl(2)	0.3813	0.4884	0.5838
Cl(3)	0.5308	0.6382	0.6796
Cl(4)	0.4614	0.3895	0.8391
O(1)	0.5865	0.5358	0.9173
O(2)	0.5884	0.4623	0.6748
$H_2O(3)$	0.0442	0.1822	0.0983
C(1,1)	0.3493	0.1309	0.5926
C(1,2)	0.3189	0.0852	0.4865
C(1,3)	0.2656	0.1249	0.3866
C(1,4)	0.2489	0.2063	0.3914
C(1,5)	0.2840	0.2497	0.4922
C(1,6)	0.3338	0.2124	0.5995
C(2,1)	0.3914	0.1309	0.8793
C(2,2)	0.3130	0.1121	0.9194
C(2,3)	0.2920	0.1577	0.0226
C(2,4)	0.3530	0.2161	0.0785
C(2,5)	0.4301	0.2313	0.0359

(continued)

TABLE XVD,28 (*continued*)

Parameters of the Atoms in $[As(C_6H_5)_4][RuCl_4 \cdot 2H_2O] \cdot H_2O$

Atom	x	y	z
C(2,6)	0.4502	0.1898	0.9339
C(3,1)	0.5421	0.0904	0.7312
C(3,2)	0.6046	0.0611	0.8340
C(3,3)	0.6969	0.0656	0.8253
C(3,4)	0.7249	0.0968	0.7267
C(3,5)	0.6651	0.1247	0.6252
C(3,6)	0.5707	0.1204	0.6283
C(4,1)	0.3893	0.9670	0.7260
C(4,2)	0.3014	−0.0579	0.7264
C(4,3)	0.2828	−0.1419	0.7187
C(4,4)	0.3528	−0.1957	0.7106
C(4,5)	0.4380	−0.1698	0.7054
C(4,6)	0.4599	−0.0871	0.7150

In this structure the arsenic atoms are tetrahedrally surrounded by the four planar benzene rings which, however, are turned so that the total cation is without symmetry; in it the average As–C = 1.91 A. In the octahedral (RuCl$_4$·2H$_2$O) anions Ru–Cl = 2.32–2.36 A. and Ru–O = 2.11 or 2.12 A. The water molecules of crystallization, O(3), are considered to be hydrogen-bonded to water molecules and chlorine atoms within the anion.

XV,d30. The tetragonal crystals of *tetraphenylarsonium tetrachloroferrate*, $As(C_6H_5)_4FeCl_4$, possess a bimolecular unit of the dimensions:

$$a_0 = 13.160(5) \text{ A.}, \quad c_0 = 7.15(2) \text{ A.}$$

The space group is $S_4{}^2$ ($I\overline{4}$) with atoms in the positions:

$$\text{As} : (2a) \quad 000; \, {}^1/_2 \, {}^1/_2 \, {}^1/_2$$
$$\text{Fe} : (2c) \quad 0 \, {}^1/_2 \, {}^1/_4; \, {}^1/_2 \, 0 \, {}^3/_4$$

The other atoms are in:

$$(8g) \quad xyz; \, \bar{x}\bar{y}z; \, y\bar{x}\bar{z}; \, \bar{y}x\bar{z}; \quad \text{B.C.}$$

The assigned parameters are stated in Table XVD,29, those for carbon being based on the assumption that the benzene rings are regular.

TABLE XVD,29

Parameters of Atoms in As(C₆H₅)₄FeCl₄

Atom	x	y	z
Cl	0.038	0.365	0.084
C(1)	0.047	0.112	0.841
C(2)	0.981	0.160	0.722
C(3)	0.007	0.235	0.608
C(4)	0.111	0.270	0.616
C(5)	0.178	0.222	0.736
C(6)	0.147	0.141	0.850

The structure is shown in Figure XVD,32. Its FeCl₄⁻ ion is a flattened tetrahedron, with Fe–Cl = 2.19 A. and Cl–Fe–Cl = 114 and 107°. The four phenyl groups are similarly distributed about arsenic.

XV,d31. *Tetraphenylarsonium oxotetrabromoaquomolybdate,* [As-(C₆H₅)₄][MoOBr₄·H₂O], is tetragonal with a bimolecular unit of the edge lengths:

$$a_0 = 13.14(1) \text{ A.}, \quad c_0 = 7.89(1) \text{ A.}$$

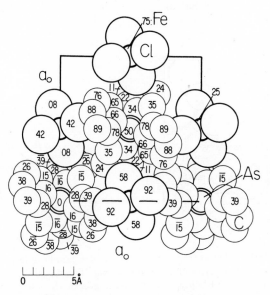

Fig. XVD,32. The tetragonal structure of tetraphenylarsonium tetrachloro-ferrate projected along its c_0 axis.

The space group is C_{4h}^3 $(P4/n)$ with atoms in the positions:

$$\text{As} : (2b) \quad {}^1/_4\,{}^3/_4\,{}^1/_2;\; {}^3/_4\,{}^1/_4\,{}^1/_2$$
$$\text{Mo} : (2c) \quad \pm({}^1/_4\,{}^1/_4\,u) \qquad \text{with } u = 0.2862(6)$$
$$\text{O} : (2c) \quad \text{with } u = 0.062(9)$$
$$\text{H}_2\text{O} : (2c) \quad \text{with } u = 0.589(3)$$

The other atoms are in

$$(8g) \quad \pm(xyz;\; x+{}^1/_2,y+{}^1/_2,\bar{z};\; y+{}^1/_2,\bar{x},\bar{z};\; y,{}^1/_2-x,z)$$

with the parameters of Table XVD,30.

TABLE XVD,30

Parameters of Atoms in $[\text{As}(\text{C}_6\text{H}_5)_4][\text{MoOBr}_4\cdot\text{H}_2\text{O}]$

Atom	x	y	z
Br	0.1474(3)	0.0902(3)	0.3267(6)
C(1)	0.866(2)	0.244(2)	0.646(5)
C(2)	0.940(2)	0.170(2)	0.634(5)
C(3)	0.018(3)	0.163(3)	0.745(6)
C(4)	0.019(3)	0.236(3)	0.876(6)
C(5)	0.947(3)	0.316(3)	0.903(5)
C(6)	0.867(2)	0.318(2)	0.775(5)

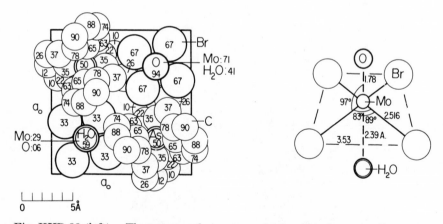

Fig. XVD,33 (left).   The tetragonal structure of tetraphenylarsonium oxotetra-bromoaquomolybdate projected along its c_0 axis. The molybdenum atoms, which do not show, are situated below either the oxygen or the water oxygen atoms.

Fig. XVD,34 (right).   Bond dimensions in the anion of tetraphenylarsonium oxotetrabromoaquomolybdate.

The ionic structure, shown in Figure XVD,33, is composed of $[As(C_6H_5)_4]^+$ cations and $[MoOBr_4]^-$ anions. In the tetrahedral cation As–C = 1.91 A. The anion has the shape and bond dimensions indicated in Figure XVD,34. The water in this crystal is presumably weakly bound and acquired when the substance stands in a moist atmosphere.

XV,d32. Crystals of *tetraphenylarsonium oxotetrabromoaceto-nitrilerhenate(V)*, $[As(C_6H_5)_4][ReBr_4O(CH_3CN)]$, are triclinic with a bimolecular unit of the dimensions:

$$a_0 = 8.47(2)\ A.;\qquad b_0 = 12.90(3)\ A.;\qquad c_0 = 13.82(3)\ A.$$
$$\alpha = 104°42(18)';\qquad \beta = 98°48(18)';\qquad \gamma = 90°6(18)'$$

All atoms are in the positions $(2i)\ \pm(xyz)$ of C_i^1 $(P\bar{1})$. The parameters are listed in Table XVD,31.

In this ionic structure the phenyl groups of its $As(C_6H_5)_4^+$ cations are tetrahedrally distributed pinwheel-fashion about the central arsenic. The bond As–C = 1.87–2.01 A. and the angle C–As–C = 105–118°. The distorted octahedral shape of the $ReBr_4O(CH_3CN)^-$ anion and its bond dimensions are indicated in Figure XVD,35.

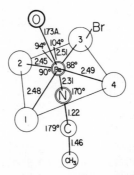

Fig. XVD,35.   Bond dimensions in the anion of tetraphenylarsonium oxo-tetrabromoacetonitrilerhenate.

XV,d33.   The ionic compound *palladium o-phenylene bis(o-dimethyl-arsinophenylmethylarsine)chloride perchlorate* crystallizes with one molecule of *benzene*. Its monoclinic crystals have the composition $Pd(C_{24}H_{30}As_4Cl)ClO_4·C_6H_6$ and a tetramolecular unit of the dimensions:

$$a_0 = 11.07(2)\ A.;\qquad b_0 = 20.34(2)\ A.;\qquad c_0 = 18.24(2)\ A.$$
$$\beta = 122.1(1)°$$

TABLE XVD,31

Parameters of the Atoms in [As(C₆H₅)₄][ReBr₄O(CH₃CN)]

Atom	x	y	z
Re	0.2876(6)	0.1711(1)	0.2559(10)
Br(1)	0.2567(4)	0.0267(5)	0.0960(4)
Br(2)	0.2281(8)	0.0413(6)	0.3472(7)
Br(3)	0.2335(3)	0.3169(1)	0.4055(1)
Br(4)	0.2693(1)	0.3071(6)	0.1553(3)
O	0.493(8)	0.156(5)	0.278(4)
N	0.015(7)	0.181(5)	0.215(4)
C(CN)	−0.128(8)	0.183(5)	0.180(5)
C(CH₃)	−0.297(10)	0.185(7)	0.134(6)
As	0.4773(3)	0.6915(3)	0.2759(2)

Phenyl carbons

Atom	x	y	z
C(1,1)	0.605(1)	0.569(3)	0.308(4)
C(2,1)	0.756(4)	0.553(1)	0.277(2)
C(3,1)	0.865(4)	0.477(3)	0.307(2)
C(4,1)	0.789(1)	0.398(6)	0.358(4)
C(5,1)	0.649(3)	0.424(4)	0.386(7)
C(6,1)	0.539(8)	0.507(6)	0.353(5)
C(1,2)	0.336(7)	0.646(5)	0.155(4)
C(2,2)	0.355(7)	0.549(5)	0.094(4)
C(3,2)	0.254(10)	0.523(7)	0.003(6)
C(4,2)	0.127(13)	0.586(9)	−0.029(8)
C(5,2)	0.102(9)	0.688(6)	0.044(5)
C(6,2)	0.203(9)	0.719(6)	0.141(5)
C(1,3)	0.613(8)	0.782(6)	0.254(5)
C(2,3)	0.606(11)	0.818(7)	0.150(6)
C(3,3)	0.732(12)	0.882(7)	0.148(6)
C(4,3)	0.843(9)	0.912(6)	0.220(5)
C(5,3)	0.857(8)	0.890(6)	0.313(5)
C(6,3)	0.748(11)	0.817(8)	0.338(7)
C(1,4)	0.339(9)	0.743(6)	0.383(5)
C(2,4)	0.397(8)	0.833(6)	0.464(5)
C(3,4)	0.320(9)	0.885(6)	0.543(5)
C(4,4)	0.183(12)	0.840(8)	0.526(6)
C(5,4)	0.120(9)	0.753(7)	0.460(6)
C(6,4)	0.208(8)	0.698(5)	0.376(4)

The space group is C_{2h}^5 ($P2_1/c$) with all atoms in the positions:

$$(4e) \quad \pm(xyz; \ x, 1/2 - y, z + 1/2)$$

The parameters are stated in Table XVD,32.

TABLE XVD,32

Parameters of the Atoms in $Pd(C_{24}H_{30}As_4Cl)ClO_4 \cdot C_6H_6$

Atom	x	y	z
Pd	0.0099(3)	0.1067(1)	0.2384(2)
As(1)	−0.0672(5)	0.0253(2)	0.3008(3)
As(2)	−0.1115(5)	0.0118(2)	0.1018(3)
As(3)	0.2198(5)	0.0530(2)	0.2730(3)
As(4)	0.0781(5)	0.1877(2)	0.1734(3)
Cl(1)	−0.188(1)	0.171(2)	0.2036(7)
C(1,1)	−0.136(4)	0.060(2)	0.370(2)
C(1,2)	0.055(5)	−0.048(2)	0.374(3)
C(12,1)	0.232(5)	−0.018(2)	0.213(3)
C(12,2)	−0.323(5)	−0.054(2)	0.225(3)
C(12,3)	−0.448(4)	−0.085(2)	0.160(3)
C(12,4)	−0.467(5)	−0.089(2)	0.076(3)
C(12,5)	−0.365(5)	−0.063(2)	0.060(3)
C(12,6)	−0.243(4)	−0.027(2)	0.130(3)
C(2,1)	−0.193(4)	−0.019(2)	−0.024(3)
C(23,1)	−0.044(5)	0.447(2)	0.341(3)
C(23,2)	0.018(5)	−0.119(2)	0.137(3)
C(23,3)	0.124(5)	−0.166(2)	0.173(3)
C(23,4)	0.256(5)	−0.146(2)	0.239(3)
C(23,5)	0.292(4)	−0.082(2)	0.268(3)
C(23,6)	0.179(4)	−0.039(2)	0.226(2)
C(3,1)	0.382(4)	0.041(2)	0.393(2)
C(34,1)	0.311(4)	0.095(2)	0.223(2)
C(34,2)	0.431(4)	0.070(2)	0.218(2)
C(34,3)	0.485(4)	0.110(2)	0.182(3)
C(34,4)	0.427(5)	0.170(2)	0.139(3)
C(34,5)	0.312(4)	0.195(2)	0.138(2)
C(34,6)	0.245(4)	0.158(2)	0.177(3)
C(4,1)	−0.067(5)	0.201(2)	0.052(3)
C(4,2)	0.132(5)	0.275(2)	0.230(3)
Cl(2)	−0.2656(6)	−0.1197(8)	0.444(1)

(continued)

TABLE XVD,32 (*continued*)

Parameters of the Atoms in $Pd(C_{24}H_{30}As_4Cl)ClO_4 \cdot C_6H_6$

Atom	x	y	z
O(1)	−0.392(2)	−0.151(2)	0.381(2)
O(2)	0.317(4)	0.447(2)	0.070(3)
O(3)	−0.230(4)	−0.146(2)	0.521(4)
O(4)	0.155(5)	0.367(2)	0.067(3)
C(5,1)	0.330(5)	−0.167(3)	0.476(3)
C(5,2)	0.289(6)	−0.167(2)	0.540(3)
C(5,3)	−0.366(5)	0.200(2)	0.382(3)
C(5,4)	0.473(5)	−0.239(3)	0.621(3)
C(5,5)	0.515(6)	−0.240(3)	0.562(3)
C(5,6)	0.438(5)	−0.204(3)	0.489(3)

Note: Carbon atoms designated C(12,*n*) belong to the benzene ring shared by As(1) and As(2), etc. The C(5,*n*) atoms belong to the isolated benzene ring.

The structure consists of the chelated palladium complex functioning as a univalent cation, a ClO_4^- anion, and isolated benzene molecules. The cation has the bond dimensions of Figure XVD,36. The fivefold coordination of the palladium is that of an elongated square pyramid with palladium at the center of the base and As(2) at its apex. In the tetrahedrally ClO_4^- ion, Cl–O = 1.36–1.42 A.

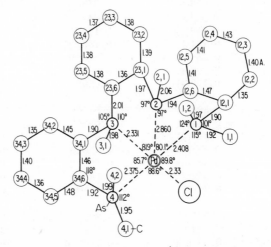

Fig. XVD,36. Bond dimensions in the cation of palladium *o*-phenylene bis(*o*-dimethylarsinophenylmethylarsine)chloride perchlorate.

XV,d34. Crystals of *methoxytetraphenyl antimony*, $(C_6H_5)_4SbOCH_3$, are orthorhombic with a unit that contains weight molecules and has the edge lengths:

$$a_0 = 14.81(2) \text{ A.}; \quad b_0 = 16.95(3) \text{ A.}; \quad c_0 = 16.74(2) \text{ A.}$$

Atoms are in the positions

$$(8c) \quad \pm (xyz; \, 1/2-x,y+1/2,z; \, x,1/2-y,z+1/2; \, x+1/2,y,1/2-z)$$

of the space group V_h^{15} (*Pbca*). The parameters are those of Table XVD,33.

TABLE XVD,33

Parameters of the Atoms in $(C_6H_5)_4SbOCH_3$

Atom	x	y	z
Sb	0.12070(6)	0.20264(5)	0.04221(4)
O	0.1933(5)	0.1417(4)	0.1277(4)
C(1)	0.1734(8)	0.1502(8)	0.2083(7)
C(2)	0.2184(10)	0.1575(8)	−0.0397(8)
C(3)	0.2943(12)	0.1179(8)	−0.0148(8)
C(4)	0.3572(9)	0.0891(8)	−0.0671(11)
C(5)	0.3446(11)	0.1047(8)	−0.1500(10)
C(6)	0.2713(12)	0.1458(9)	−0.1770(7)
C(7)	0.2095(10)	0.1730(7)	−0.1207(10)
C(8)	0.0056(7)	0.1373(6)	0.0805(6)
C(9)	0.0098(8)	0.0567(7)	0.0964(7)
C(10)	−0.0683(11)	0.0166(7)	0.1166(8)
C(11)	−0.1495(8)	0.0545(9)	0.1239(8)
C(12)	−0.1536(8)	0.1350(8)	0.1097(7)
C(13)	−0.0769(9)	0.1745(7)	0.0876(7)
C(14)	0.0415(10)	0.2639(9)	−0.0503(8)
C(15)	−0.0218(11)	0.2244(8)	−0.0968(10)
C(16)	−0.0657(10)	0.2594(14)	−0.1586(10)
C(17)	−0.0511(12)	0.3375(15)	−0.1764(8)
C(18)	0.0117(12)	0.3801(9)	−0.1317(10)
C(19)	0.0574(9)	0.3442(10)	−0.0695(8)
C(20)	0.1544(12)	0.3099(8)	0.0988(8)
C(21)	0.0951(10)	0.3489(12)	0.1457(11)
C(22)	0.1134(15)	0.4247(15)	0.1747(9)
C(23)	0.1937(20)	0.4611(9)	0.1552(12)
C(24)	0.2504(13)	0.4237(15)	0.1103(11)
C(25)	0.2339(14)	0.3456(13)	0.0814(8)

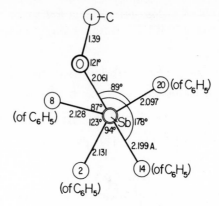

Fig. XVD,37. Some bond dimensions in the molecule of methoxytetraphenyl antimony.

The structure is made up of molecules with the bond dimensions of Figure XVD,37. Coordination about the antimony atom is fivefold and has been described as that of a trigonal bipyramid having the oxygen atom and C(14) of a benzene ring at the apices.

XV,d35. The monoclinic crystals of *bis(tetra-n-butylammonium)-di[bis(1,2,3,4-tetrachlorobenzene-5,6-dithiolato)cobaltate]*, $[(n\text{-}C_4H_9)_4N]_2\text{-}[Co_2(S_2C_6Cl_4)_4]$, have a bimolecular unit of the dimensions:

$$a_0 = 13.95(3) \text{ A.}; \quad b_0 = 18.98(4) \text{ A.}; \quad c_0 = 15.36(3) \text{ A.}$$
$$\beta = 114.8(4)°$$

The atoms are in the positions

$$(4e) \quad \pm(xyz; x, 1/2 - y, z + 1/2)$$

of C_{2h}^5 ($P2_1/c$). Parameters are stated in Table XVD,34.

TABLE XVD,34

Parameters of the Atoms in $[(n\text{-}C_4H_9)_4N]_2[Co_2(S_2C_6Cl_4)_4]$

Atom	x	y	z
Co	0.0651(3)	0.0571(2)	-0.0237(2)
S(1)	0.1932(6)	0.1294(3)	0.0581(4)
S(2)	0.1757(5)	-0.0114(3)	-0.0488(4)
S(3)	-0.0604(5)	-0.0114(3)	-0.1205(3)
S(4)	-0.0493(5)	0.1433(3)	-0.0472(4)

(*continued*)

TABLE XVD,34 (*continued*)

Parameters of the Atoms in $[(n\text{-}C_4H_9)_4N]_2[Co_2(S_2C_6Cl_4)_4]$

Atom	x	y	z
Cl(1)	−0.2644(5)	−0.0621(3)	−0.2969(4)
Cl(2)	−0.4697(6)	0.0255(3)	−0.3675(4)
Cl(3)	−0.4673(6)	0.1697(4)	−0.2723(5)
Cl(4)	−0.2581(6)	0.2317(3)	−0.1194(4)
Cl(5)	0.4142(6)	0.1815(3)	0.2099(4)
Cl(6)	0.6155(6)	0.0933(5)	0.2512(5)
Cl(7)	0.5982(6)	−0.0487(4)	0.1484(5)
Cl(8)	0.3796(6)	−0.1015(4)	−0.0017(4)
C(1)	−0.176(2)	0.042(1)	−0.170(1)
C(2)	−0.267(2)	0.016(1)	−0.243(1)
C(3)	−0.352(2)	0.057(1)	−0.277(1)
C(4)	−0.351(2)	0.122(1)	−0.233(2)
C(5)	−0.264(2)	0.150(1)	−0.167(1)
C(6)	−0.169(2)	0.111(1)	−0.134(1)
C(7)	0.306(2)	0.081(1)	0.082(1)
C(8)	0.409(2)	0.104(1)	0.151(1)
C(9)	0.495(2)	0.067(1)	0.170(1)
C(10)	0.486(2)	0.003(1)	0.122(2)
C(11)	0.386(2)	−0.023(1)	0.055(2)
C(12)	0.302(2)	0.016(1)	0.037(1)
C[Bu(1,1)]	0.171(2)	0.299(1)	0.326(1)
C[Bu(1,2)]	0.265(3)	0.251(2)	0.349(2)
C[Bu(1,3)]	0.356(2)	0.276(1)	0.449(2)
C[Bu(1,4)]	0.456(2)	0.231(1)	0·467(1)
C[Bu(2,1)]	0.114(2)	0.287(1)	0.144(2)
C[Bu(2,2)]	0.155(2)	0.361(1)	0.138(1)
C[Bu(2,3)]	0.210(3)	0.354(2)	0.065(2)
C[Bu(2,4)]	0.237(2)	0.424(1)	0.049(2)
C[Bu(3,1)]	−0.003(3)	0.341(1)	0.211(2)
C[Bu(3,2)]	−0.116(2)	0.327(1)	0.111(2)
C[Bu(3,3)]	−0.185(3)	0.390(1)	0.109(2)
C[Bu(3,4)]	−0.288(3)	0.379(2)	0.018(2)
C[Bu(4,1)]	0.038(2)	0.205(1)	0.219(2)
C[Bu(4,2)]	−0.010(2)	0.192(1)	0.293(2)
C[Bu(4,3)]	−0.049(3)	0.116(2)	0.277(2)
C[Bu(4,4)]	−0.094(3)	0.091(1)	0.354(2)
N	0.072(2)	0.282(1)	0.225(1)

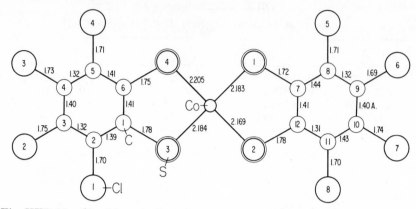

Fig. XVD,38.   Bond lengths in the anion of bis(tetra-*n*-butylammonium)di[bis (1,2,3,4-tetrachlorobenzene-5,6-dithiolato)cobaltate].

The structure is made up of tetrahedral cations and dimeric anions. In these anions each half is approximately planar, with the bond lengths of Figure XVD,38. In the half-anions the sulfur atoms are as much as 0.15 A. outside the best plane and the cobalt 0.26 A. from it. The two half-anions, related by a center of symmetry, are tied together by Co–S(3) bonds of length 2.404(7) A. Each cobalt atom thus is coordinated with five atoms of sulfur, the S–Co–S angles ranging from 87 to 104°. In the tetrahedral cations, N–C = 1.49–1.62 A. and C–C = 1.44–1.70 A.

E.   COMPOUNDS CONTAINING FIVE OR MORE BENZENE RINGS

XV,e1.   *Pentaphenyl phosphorus*, $P(C_6H_5)_5$, forms monoclinic crystals whose tetramolecular unit has the dimensions:

$$a_0 = 10.029(15) \text{ A.}; \quad b_0 = 17.215(18) \text{ A.}; \quad c_0 = 14.170(17) \text{ A.}$$
$$\beta = 112°3(30)'$$

The space group is the low symmetry C_s^4 (*Cc*) with atoms in the positions:

$$(4a) \quad xyz;\ x,\bar{y},z+1/2;\ x+1/2,y+1/2,z;\ x+1/2,1/2-y,z+1/2$$

The parameters are listed in Table XVE,1.

TABLE XVE,1

Parameters of the Atoms in $P(C_6H_5)_5$

Atom	x	y	z
C(1)	−0.0103	0.2230	−0.0417
C(2)	−0.0133	0.2355	−0.1391
C(3)	0.0007	0.3107	−0.1728
C(4)	0.0189	0.3743	−0.1083
C(5)	0.0221	0.3627	−0.0118
C(6)	0.0089	0.2875	0.0225
C(7)	−0.0124	0.1392	0.1263
C(8)	−0.1364	0.1718	0.1303
C(9)	−0.1456	0.1869	0.2246
C(10)	−0.0279	0.1745	0.3143
C(11)	0.0984	0.1431	0.3087
C(12)	0.1044	0.1249	0.2163
C(13)	0.0190	0.0034	0.0501
C(14)	−0.0954	−0.0288	0.0709
C(15)	−0.0878	−0.1033	0.1109
C(16)	0.0364	−0.1481	0.1277
C(17)	0.1459	−0.1185	0.1066
C(18)	0.1401	−0.0427	0.0693
C(19)	−0.1680	0.0884	−0.1079
C(20)	−0.2827	0.1417	−0.1393
C(21)	−0.4088	0.1218	−0.2195
C(22)	−0.4217	0.0524	−0.2703
C(23)	−0.3069	0.0009	−0.2411
C(24)	−0.1814	0.0189	−0.1598
C(25)	0.1694	0.1078	−0.0168
C(26)	0.1821	0.0657	−0.0974
C(27)	0.3126	0.0587	−0.1105
C(28)	0.4367	0.0929	−0.0374
C(29)	0.4232	0.1344	0.0398
C(30)	0.2913	0.1447	0.0498
P	−0.0013	0.1126	0.0019

In the molecules of this crystal the five benzene rings are distributed in triangular bipyramidal fashion about the central phosphorus atom. For the two apical rings C–P = 1.989 and 1.986 A.; for the three equatorial rings C–P = 1.830–1.865 A. The angles between the bonds linking the apical and equatorial rings to phosphorus lie between 86 and 92°; for the equatorial rings these bond angles are 118–123°. The rings are planar and twisted through various angles, presumably to avoid crowding.

TABLE XVE,2

Parameters of the Atoms in $Sb(C_6H_5)_5$

Atom	x	y	z
Sb	0.1640	0.0387	0.2529
C(1)	0.231	0.197	0.348
C(2)	0.250	0.157	0.444
C(3)	0.298	0.262	0.496
C(4)	0.324	0.407	0.459
C(5)	0.287	0.439	0.363
C(6)	0.247	0.344	0.267
C(7)	0.224	−0.094	0.349
C(8)	0.122	−0.215	0.439
C(9)	0.167	−0.301	0.504
C(10)	0.333	−0.237	0.484
C(11)	0.427	−0.110	0.400
C(12)	0.384	−0.052	0.343
C(13)	−0.018	−0.174	0.239
C(14)	−0.163	−0.208	0.230
C(15)	−0.261	−0.345	0.225
C(16)	−0.239	−0.461	0.225
C(17)	−0.103	−0.420	0.231
C(18)	0.007	−0.275	0.239
C(19)	0.362	0.159	0.112
C(20)	0.365	0.094	0.039
C(21)	0.481	0.171	−0.065
C(22)	0.609	0.312	−0.076
C(23)	0.611	0.379	−0.010
C(24)	0.480	0.298	0.096
C(25)	0.042	0.125	0.195
C(26)	−0.062	0.123	0.254
C(27)	−0.173	0.185	0.221
C(28)	−0.158	0.217	0.122
C(29)	−0.065	0.131	0.055
C(30)	0.049	0.161	0.089

XV,e2. Crystals of *pentaphenyl antimony*, $Sb(C_6H_5)_5$, are triclinic with a bimolecular unit of the dimensions:

$$a_0 = 10.28(5) \text{ A.}; \quad b_0 = 10.57(5) \text{ A.}; \quad c_0 = 13.59(7) \text{ A.}$$
$$\alpha = 79°0(30)'; \quad \beta = 79°34(30)'; \quad \gamma = 119°37(30)'$$

All atoms are in the positions $(2i)$ $\pm(xyz)$ of C_i^1 $(P\bar{1})$. The determined parameters are those of Table XVE,2.

In this structure planar phenyl groups are distributed in an approximately square pyramidal way about the central antimony atom. Within the limit of accuracy of the determination, the Sb–C bond to the apical phenyl group is a twofold axis. The lengths of the various Sb–C bonds lie between 2.05 and 2.23 A., and the angles C–Sb–C range from 84 to 104°.

XV,e3. The monoclinic crystals of *tri-μ-chlorochloropentakis(diethylphenylphosphine)diruthenium(II)*, $Ru_2Cl_4[(C_2H_5)_2(C_6H_5)P]_5$, possess a bimolecular unit of the dimensions:

$$a_0 = 15.882(13) \text{ A.}; \quad b_0 = 19.078(17) \text{ A.}; \quad c_0 = 10.418(18) \text{ A.}$$
$$\beta = 104.18(7)°$$

The space group is C_2^2 $(P2_1)$ with atoms in the positions:

$$(2a) \quad xyz; \quad \bar{x}, y + 1/2, \bar{z}$$

Determined parameters are stated in Table XVE,3.

TABLE XVE,3

Parameters of the Atoms in $Ru_2Cl_4[(C_2H_5)_2(C_6H_5)P]_5$

Atom	x	y	z
Ru(1)	0.39136(9)	0.00000(0)	0.29003(14)
Ru(2)	0.20727(9)	0.91750(11)	0.13407(15)
Cl(1)	0.32116(29)	0.98193(25)	0.05305(48)
Cl(2)	0.23330(28)	0.01518(24)	0.28785(46)
Cl(3)	0.34551(29)	0.87621(26)	0.29399(47)
Cl(4)	0.21585(39)	0.82346(34)	−0.01443(61)
P(1)	0.39704(33)	0.12104(29)	0.26527(54)
P(2)	0.43499(29)	0.99151(28)	0.52127(48)
P(3)	0.52193(30)	0.97189(27)	0.24478(49)
P(4)	0.11959(33)	0.86003(29)	0.23939(54)
P(5)	0.10058(38)	0.96954(35)	−0.02013(61)

(continued)

TABLE XVE,3 (*continued*)

Parameters of the Atoms in $Ru_2Cl_4[(C_2H_5)_2(C_6H_5)P]_5$

Atom	x	y	z
C(1,1)	0.5277(28)	0.1966(16)	0.1845(28)
C(1,2)	0.6050(19)	0.2347(16)	0.2078(29)
C(1,3)	0.6508(18)	0.2532(16)	0.3430(29)
C(1,4)	0.6161(18?)	0.2331(15)	0.4451(28)
C(1,5)	0.5370(15)	0.1929(13)	0.4220(24)
C(1,6)	0.4961(13)	0.1732(11)	0.2882(20)
C(1,7)	0.3347(20)	0.1649(17)	0.3725(31)
C(1,8)	0.3416(22)	0.2441(19)	0.3940(34)
C(1,9)	0.3388(18)	0.1471(16)	0.0967(28)
C(1,10)	0.3212(21)	0.2272(18)	0.0667(33)
C(4,1)	0.9353(13)	0.8560(11)	0.1931(22)
C(4,2)	0.8564(13)	0.8852(12)	0.2157(22)
C(4,3)	0.8541(13)	0.9451(12)	0.2699(22)
C(4,4)	0.9312(13)	0.9842(11)	0.3050(21)
C(4,5)	0.0072(14)	0.9601(12)	0.2970(22)
C(4,6)	0.0103(13)	0.8921(12)	0.2347(22)
C(4,7)	0.1641(14)	0.8619(12)	0.4167(22)
C(4,8)	0.8875(13)	0.3199(12)	0.4945(21)
C(4,9)	0.0973(13)	0.7681(12)	0.1987(22)
C(4,10)	0.1795(13)	0.7203(12)	0.2473(22)
C(3,1)	0.6709(13)	0.9343(11)	0.4417(21)
C(3,2)	0.7531(14)	0.9474(12)	0.5426(21)
C(3,3)	0.7829(13)	0.0145(11)	0.5461(21)
C(3,4)	0.7434(13)	0.0693(12)	0.4681(21)
C(3,5)	0.6630(13)	0.0533(12)	0.3778(22)
C(3,6)	0.6289(13)	0.9869(12)	0.3623(20)
C(3,7)	0.6203(12)	0.9950(13)	0.0504(20)
C(3,8)	0.5371(13)	0.0145(11)	0.0945(21)
C(3,9)	0.5326(14)	0.8771(12)	0.2066(22)
C(3,10)	0.4777(13)	0.8509(11)	0.0753(22)
C(2,1)	0.4112(13)	0.6672(11)	0.2097(21)
C(2,2)	0.3705(13)	0.6166(11)	0.2694(21)
C(2,3)	0.5039(13)	0.6604(11)	0.2329(22)
C(2,4)	0.5455(13)	0.6122(11)	0.3113(21)
C(2,5)	0.4160(13)	0.5633(12)	0.3461(21)
C(2,6)	0.5039(13)	0.5599(12)	0.3717(21)
C(2,7)	0.5033(12)	0.4110(13)	0.4262(19)
C(2,8)	0.4530(13)	0.4008(11)	0.2847(20)
C(2,9)	0.6597(13)	0.4824(11)	0.4091(21)
C(2,10)	0.6473(13)	0.4603(11)	0.2704(21)

(*continued*)

TABLE XVE,3 *(continued)*

Parameters of the Atoms in $Ru_2Cl_4[(C_2H_5)_2(C_6H_5)P]_5$

Atom	x	y	z
C(5,1)	0.1028(13)	0.3328(12)	0.1608(21)
C(5,2)	0.1733(13)	0.3751(12)	0.1453(21)
C(5,3)	0.1651(13)	0.4396(11)	0.0989(21)
C(5,4)	0.0163(13)	0.3655(11)	0.1119(21)
C(5,5)	0.0088(13)	0.4324(11)	0.0680(20)
C(5,6)	0.0828(13)	0.4718(11)	0.0540(21)
C(5,7)	0.0851(13)	0.0628(12)	0.0233(21)
C(5,8)	0.9572(13)	0.6113(11)	0.0916(21)
C(5,9)	0.8669(13)	0.4736(11)	0.1826(21)
C(5,10)	0.9432(13)	0.4839(11)	0.3115(20)

The central atoms of the molecules in this structure have the general distribution and bond dimensions of Figure XVE,1. Coordination about each ruthenium atom is roughly octahedral, and Ru–Ru = 3.367 A.

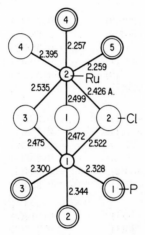

Fig. XVE,1. Bond lengths involving atoms in the central portion of the molecule of tri-μ-chloro-chloropentakis(diethylphenylphosphine)diruthenium.

XV,e4. Crystals of *triphenylmethylarsonium bis(toluene-3,4-dithiolato)cobaltate hemiethanolate*, $[(C_6H_5)_3CH_3As][Co(CH_3C_6H_3S_2)_2]\cdot{}^1/_2C_2H_5\text{-}$ OH, are triclinic with a bimolecular unit of the dimensions:

$$a_0 = 18.61(1) \text{ A.}; \quad b_0 = 10.848(6) \text{ A.}; \quad c_0 = 10.206(5) \text{ A.}$$
$$\alpha = 112.45(2); \quad \beta = 71.55(2)°; \quad \gamma = 115.90(2)°$$

The space group is C_i^1 ($P\bar{1}$) with cobalt atoms in the centrosymmetric positions:

$$(1a) \quad 000 \qquad (1f) \quad {}^1/_2 \ 0 \ {}^1/_2$$

All other atoms are in the general positions $(2i) \pm (xyz)$ with the parameters of Table XVE,4.

TABLE XVE,4

Parameters of Atoms in $[(C_6H_5)_3CH_3As][Co(CH_3C_6H_3S_2)_2]\cdot{}^1/_2C_2H_5OH$

Atom	x	y	z
As	0.2686(1)	−0.0431(2)	0.0115(2)
S(1)	0.1201(2)	0.1576(5)	−0.0402(4)
S(2)	−0.0243(2)	0.1310(6)	0.2182(4)
S(3)	0.5178(3)	0.2228(6)	0.5592(5)
S(4)	0.3727(3)	−0.0607(6)	0.4953(5)
C(1)	0.1299(9)	0.295(2)	0.125(2)
C(2)	0.0667(9)	0.282(2)	0.239(2)
C(3)	0.074(1)	0.388(2)	0.374(2)
C(4)	0.148(1)	0.505(2)	0.386(2)
C(5)	0.2110(9)	0.521(2)	0.274(2)
C(6)	0.203(1)	0.415(2)	0.136(2)
C(7)	0.423(1)	0.229(3)	0.572(2)
C(8)	0.356(1)	0.104(2)	0.546(2)
C(9)	0.277(1)	0.099(3)	0.552(2)
C(10)	0.275(1)	0.238(3)	0.585(2)
C(11)	0.338(2)	0.364(3)	0.612(2)
C(12)	0.412(1)	0.359(3)	0.600(2)
C(13,CH₃)	0.158(1)	0.618(2)	0.537(2)
C(14,CH₃)	0.193(1)	0.247(3)	0.597(2)
C(15,CH₃)	0.2248(9)	−0.109(2)	−0.164(2)
C(1,1)	0.1823(6)	−0.092(1)	0.162(1)
C(1,2)	0.1655(7)	−0.210(1)	0.207(1)
C(1,3)	0.0960(8)	−0.252(1)	0.307(1)
C(1,4)	0.0431(6)	−0.178(1)	0.363(1)
C(1,5)	0.0599(7)	−0.060(1)	0.318(1)
C(1,6)	0.1294(8)	−0.018(1)	0.217(1)

(continued)

TABLE XVE,4 (*continued*)

Parameters of Atoms in $[(C_6H_5)_3CH_3As][Co(CH_3C_6H_3S_2)_2]\cdot^1/_2C_2H_5OH$

Atom	x	y	z
C(2,1)	0.3188(6)	0.159(1)	0.063(1)
C(2,2)	0.3262(6)	0.238(1)	0.205(1)
C(2,3)	0.3663(6)	0.386(1)	0.239(1)
C(2,4)	0.3989(6)	0.454(1)	0.132(1)
C(2,5)	0.3915(7)	0.375(1)	−0.011(1)
C(2,6)	0.3514(7)	0.227(1)	−0.045(1)
C(3,1)	0.3469(6)	−0.121(1)	−0.014(1)
C(3,2)	0.3714(7)	−0.113(1)	0.105(1)
C(3,3)	0.4305(7)	−0.167(1)	0.087(1)
C(3,4)	0.4651(6)	−0.229(1)	−0.051(1)
C(3,5)	0.4406(7)	−0.238(1)	−0.171(1)
C(3,6)	0.3815(7)	−0.184(1)	−0.152(1)
X(1)	0.006(2)	0.395(4)	0.025(3)
X(2)	0.045(4)	0.485(7)	−0.049(6)
X(3)	0.113(3)	0.549(5)	−0.111(5)

Note: Carbon atoms C(1,1), C(2,1), C(3,1), etc. belong to the three phenyl groups around arsenic. Atoms X are the C or O atoms of the alcohol molecule.

The structure contains planar anions, pseudotetrahedral cations, and C_2H_5OH molecules trapped within the ionic complex. Coordination about the cobalt atoms is square, the two crystallographically different anions having the bond dimensions of Figure XVE,2.

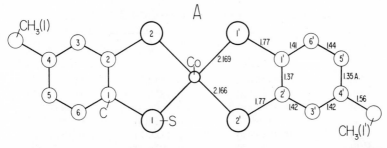

Fig. XVE,2a. Bond lengths in one (A) of the crystallographically different anions of triphenylmethylarsonium bis(toluene-3,4-dithiolato)cobaltate.

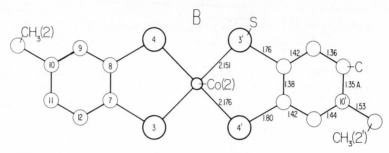

Fig. XVE,2b   Bond lengths in the second (B) of the crystallographically different anions of triphenylmethylarsonium bis(toluene-3,4-dithiolato)cobaltate.

XV,e5.   There are four molecules in the monoclinic unit of *tetraphenylphosphonium bis(tetracyanoquinodimethanide)*, $P(C_6H_5)_4\{C_6H_4[C(CN)_2]_2\}_2$. It has the dimensions:

$$a_0 = 33.005(3) \text{ A.}; \quad b_0 = 7.766(2) \text{ A.}; \quad c_0 = 15.961(2) \text{ A.}$$
$$\beta = 109.31(2)°$$

The space group is C_{2h}^6 ($C2/c$). The phosphorus atom is in:

(*4e*) $\pm(0 \; u \; 1/4; \; 1/2, u+1/2, 1/4)$ with $u = 0.1284(1)$

All the other atoms are in:

(*8f*) $\pm(xyz; \; x,\bar{y},z+1/2; \; x+1/2,y+1/2,z; \; x+1/2,1/2-y,z+1/2)$

The parameters, including those for hydrogen, have been given the values of Table XVE,5.

TABLE XVE,5

Parameters of Atoms in $P(C_6H_5)_4\{C_6H_4[C(CN)_2]_2\}_2$

Atom	x	y	z
C(1)	0.03324(4)	−0.0112(2)	0.33426(9)
C(2)	0.05449(5)	−0.1453(2)	0.30811(11)
C(3)	0.07920(6)	−0.2598(3)	0.36978(13)
C(4)	0.08291(6)	−0.2426(3)	0.45758(13)
C(5)	0.06206(6)	−0.1126(3)	0.48404(12)
C(6)	0.03687(5)	0.0059(2)	0.42292(10)

(continued)

TABLE XVE,5 (*continued*)

Parameters of Atoms in $P(C_6H_5)_4\{C_6H_4[C(CN)_2]_2\}_2$

Atom	x	y	z
C(7)	0.03101(4)	0.2690(2)	0.20598(9)
C(8)	0.00956(6)	0.4019(2)	0.14970(12)
C(9)	0.03255(8)	0.5194(3)	0.11865(14)
C(10)	0.07665(8)	0.5066(3)	0.14402(15)
C(11)	0.09793(7)	0.3764(3)	0.19969(14)
C(12)	0.07526(5)	0.2554(2)	0.23060(11)
C(13)	0.21056(4)	0.2388(2)	0.40398(10)
C(14)	0.24403(5)	0.1545(2)	0.38118(10)
C(15)	0.27466(5)	0.2450(2)	0.36110(10)
C(16)	0.27460(5)	0.4286(2)	0.36206(10)
C(17)	0.24080(5)	0.5130(2)	0.38360(11)
C(18)	0.21018(5)	0.4229(2)	0.40346(11)
C(19)	0.17946(5)	0.1458(2)	0.42622(10)
C(20)	0.14584(5)	0.2293(2)	0.44837(11)
C(21)	0.17998(5)	−0.0375(2)	0.42832(12)
C(22)	0.30634(5)	0.5244(2)	0.34330(11)
C(23)	0.30692(5)	0.7070(2)	0.34622(12)
C(24)	0.34081(5)	0.4476(2)	0.32163(12)
N(1)	0.11874(5)	0.2973(2)	0.46498(12)
N(2)	0.18111(5)	−0.1841(2)	0.43049(13)
N(3)	0.30770(5)	0.8538(2)	0.34750(13)
N(4)	0.36910(5)	0.3915(2)	0.30464(13)
H(1)	0.0521(6)	−0.150(2)	0.2411(14)
H(2)	0.0944(7)	−0.358(3)	0.3480(14)
H(3)	0.1006(6)	−0.326(3)	0.5007(13)
H(4)	0.0648(7)	−0.098(3)	0.5431(15)
H(5)	0.0232(5)	0.105(2)	0.4440(11)
H(6)	−0.0218(7)	0.415(3)	0.1310(13)
H(7)	0.0191(7)	0.621(3)	0.0824(15)
H(8)	0.0931(7)	0.594(3)	0.1232(14)
H(9)	0.1290(7)	0.363(3)	0.2173(14)
H(10)	0.0912(6)	0.169(3)	0.2690(13)
H(11)	0.2441(5)	0.036(2)	0.3791(11)
H(12)	0.2975(5)	0.187(2)	0.3474(11)
H(13)	0.2402(5)	0.639(2)	0.3830(10)
H(14)	0.1870(5)	0.479(2)	0.4159(11)

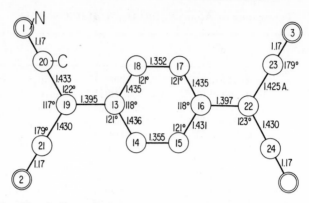

Fig. XVE,3.   Bond dimensions in the anion of tetraphenylphosphonium bis
(tetracyanoquinodimethanide).

The resulting structure is ionic. Its cation, $P(C_6H_5)_4^+$, has four planar C_6H_5 rings tetrahedrally distributed about the phosphorus atom; in it P–C = 1.796 A. In the rings C–C lies between 1.370 and 1.406 A. The anion can be considered as the molecular ion C_6H_4-$[C(CN)_2]_2^{1/2-}$ or as its centrosymmetric dimer. Bond dimensions in a ring appreciably lacking in regularity are those of Figure XVE,3. In the structure as a whole rows of cations and of anions lie parallel to one another along the a_0 axis.

XV,e6.   Crystals of *tris(cis-1,2-diphenylethylene-1,2-dithiolato)vanadium*, $V[S_2C_2(C_6H_5)_2]_3$, are monoclinic with a tetramolecular unit of the dimensions:

$$a_0 = 19.25(3) \text{ A.}; \quad b_0 = 11.31(2) \text{ A.}; \quad c_0 = 18.01(3) \text{ A.}$$
$$\beta = 106.3(2)°$$

The space group is C_{2h}^6 (C2/c). Vanadium atoms are in the positions:

V : (4e) $\pm(0\ u\ 1/4;\ 1/2, u+1/2, 1/4)$ with $u = 0.2196(2)$

All other atoms are in

(8f) $\pm(xyz;\ x, \bar{y}, z+1/2;\ x+1/2, y+1/2, z;\ x+1/2, 1/2-y, z+1/2)$

with the parameters of Table XVE,6.

TABLE XVE,6

Parameters of the Atoms in $V[S_2C_2(C_6H_5)_2]_3$

Atom	x	y	z
S(1)	−0.0445(1)	0.3762(2)	0.1647(2)
S(2)	−0.1145(1)	0.1520(2)	0.1814(1)
S(3)	−0.0336(1)	0.1345(2)	0.3531(1)
C(1)	−0.0220(5)	0.5069(9)	0.2098(5)
C(2)	−0.1557(5)	0.0910(8)	0.2433(5)
C(3)	−0.1194(5)	0.0815(8)	0.3210(5)
		Ring carbons	
1C,1	−0.0483(6)	0.6148(9)	0.1660(6)
1C,2	−0.0435(5)	0.6290(9)	0.0921(6)
1C,3	−0.0670(7)	0.733(1)	0.0510(7)
1C,4	−0.0973(7)	0.821(1)	0.0851(8)
1C,5	−0.1060(8)	0.809(1)	0.1573(8)
1C,6	−0.0806(6)	0.703(1)	0.2019(7)
2C,1	−0.2310(5)	0.0503(8)	0.2071(5)
2C,2	−0.2803(5)	0.1232(8)	0.1578(6)
2C,3	−0.3493(6)	0.0825(9)	0.1209(6)
2C,4	−0.3697(6)	−0.033(1)	0.1343(7)
2C,5	−0.3229(6)	−0.102(1)	0.1843(7)
2C,6	−0.2531(6)	−0.0624(9)	0.2226(6)
3C,1	−0.1510(5)	0.0268(8)	0.3791(5)
3C,2	−0.2173(6)	0.069(1)	0.3850(7)
3C,3	−0.2499(7)	0.014(1)	0.4392(8)
3C,4	−0.2148(7)	−0.081(1)	0.4790(7)
3C,5	−0.1516(7)	−0.119(1)	0.4763(7)
3C,6	−0.1181(6)	−0.0639(9)	0.4241(6)

A portion of the structure is shown in Figure XVE,4. The most significant bond dimensions of the molecules are those of Figure XVE,5. The central vanadium atom has six equidistant sulfur neighbors (V–S = 2.34 A.); their distribution is trigonal prismatic with angles S–V–S = ca. 81.7°. The S–S distances lie between 2.927 and 3.178 A. The existence of a twofold axis of symmetry in the molecule makes one $(C_6H_5CS)_2$ group crystallographically unlike the other two but the differences are minimal. Each VS_2C_2 chelating group is nearly planar and these planes make dihedral angles of 46.2, 46.7, and 57.0° with their attached planar phenyl rings.

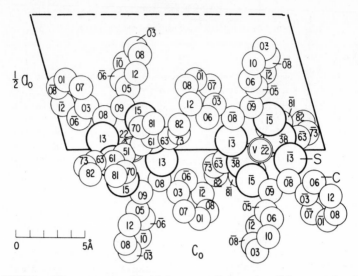

Fig. XVE,4. The contents of half the monoclinic unit cell in crystals of tris(*cis*-1,2-diphenylethylene-1,2-dithiolato)vanadium projected along the b_0 axis.

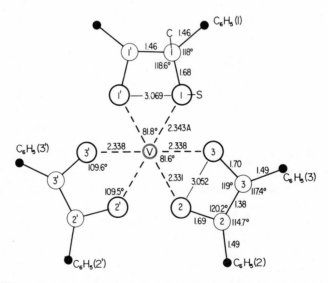

Fig. XVE,5. Some bond dimensions in the molecule of tris(*cis*-1,2-diphenylethylene-1,2-dithiolato)vanadium.

XV,e7. The triclinic crystals of *tris(cis-1,2-diphenylethylene-1,2-dithiolato)rhenium,* $Re[S_2C_2(C_6H_5)_2]_3$, have a bimolecular cell of the dimensions:

$$a_0 = 19.73(4) \text{ A.;} \quad b_0 = 11.94(3) \text{ A.;} \quad c_0 = 9.87(3) \text{ A.}$$
$$\alpha = 120°6(6)'; \quad \beta = 73°36(6)'; \quad \gamma = 102°30(6)'$$

All atoms are in the positions $(2i) \pm (xyz)$ of C_i^1 $(P\bar{1})$. Their parameters are given in Table XVE,7.

TABLE XVE,7

Parameters of the Atoms in $Re[S_2C_2(C_6H_5)_2]_3$

Atom	x	y	z
Re	0.2486(1)	0.0169(1)	0.0239(1)
S(1)	0.2477(6)	0.1561(6)	0.2939(8)
S(2)	0.1545(6)	−0.0831(6)	0.1363(8)
S(3)	0.3197(6)	−0.0852(6)	0.0740(8)
S(4)	0.3354(6)	−0.0406(6)	−0.2090(8)
S(5)	0.1765(7)	−0.0526(7)	−0.1599(9)
S(6)	0.2577(6)	0.2133(6)	0.0199(8)
C(1)	0.256(2)	0.317(2)	0.329(3)
C(2)	0.090(2)	−0.137(2)	0.026(3)
C(3)	0.385(2)	−0.145(2)	−0.077(3)
C(4)	0.391(2)	−0.130(2)	−0.206(3)
C(5)	0.105(2)	−0.131(2)	−0.110(3)
C(6)	0.256(2)	0.339(2)	0.208(3)
	Derived parameters of phenyl carbons		
C(1,1)	0.255	0.413	0.502
C(2,1)	0.208	0.509	0.583
C(3,1)	0.206	0.597	0.744
C(4,1)	0.251	0.588	0.825
C(5,1)	0.298	0.492	0.743
C(6,1)	0.300	0.405	0.582
C(1,2)	0.022	−0.193	0.101
C(2,2)	−0.017	−0.113	0.250
C(3,2)	−0.081	−0.162	0.314
C(4,2)	−0.106	−0.291	0.230
C(5,2)	−0.067	−0.372	0.080
C(6,2)	−0.003	−0.323	0.016

(continued)

TABLE XVE,7 (*continued*)

Parameters of the Atoms in Re[$S_2C_2(C_6H_5)_2$]$_3$

Atom	x	y	z
C(1,3)	0.442	−0.203	−0.060
C(2,3)	0.515	−0.182	−0.111
C(3,3)	0.563	−0.240	−0.089
C(4,3)	0.537	−0.320	−0.017
C(5,3)	0.464	−0.341	0.034
C(6,3)	0.417	−0.283	0.013
C(1,4)	0.449	−0.179	−0.358
C(2,4)	0.451	−0.313	−0.449
C(3,4)	0.505	−0.365	−0.585
C(4,4)	0.556	−0.282	−0.629
C(5,4)	0.554	−0.148	−0.538
C(6,4)	0.501	−0.096	−0.402
C(1,5)	0.050	−0.188	−0.218
C(2,5)	0.077	−0.254	−0.382
C(3,5)	0.032	−0.304	−0.489
C(4,5)	−0.040	−0.289	−0.433
C(5,5)	−0.067	−0.223	−0.269
C(6,5)	−0.022	−0.172	−0.161
C(1,6)	0.261	0.477	0.233
C(2,6)	0.208	0.511	0.193
C(3,6)	0.206	0.640	0.234
C(4,6)	0.255	0.736	0.315
C(5,6)	0.308	0.702	0.355
C(6,6)	0.310	0.573	0.313

Note: In this table C(4,6) for instance refers to the fourth C atom in the phenyl group attached to C(6).

The structure is built up of molecules that have the bond dimensions of Figure XVE,6. The central rhenium atom has a trigonal prismatic coordination with its six sulfur neighbors. Each of the chelating C_2S_2 groups is coplanar with the rhenium atom; the phenyl group attached to each carbon is rotated about its C–C bond so that the whole has a paddle-wheel configuration. Bond dimensions within the benzene rings and between molecules are normal.

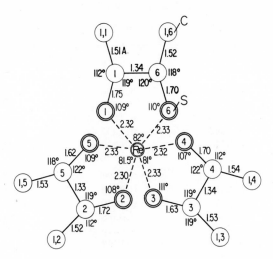

Fig. XVE,6. Some bond dimensions in the molecule of tris(*cis*-1,2-diphenyl-ethylene-1,2-dithiolato)rhenium.

XV,e8. Crystals of *nitridodichloro bis(triphenylphosphine)rhenium-(V)*, $ReNCl_2[P(C_6H_5)_3]_2$, are monoclinic with a tetramolecular unit of the edge lengths:

$$a_0 = 15.712(5) \text{ A.}; \quad b_0 = 9.531(4) \text{ A.}; \quad c_0 = 22.168(9) \text{ A.}$$
$$\beta = 103°26(1)' \qquad (28°\text{C.})$$

The space group is C_{2h}^6 in the orientation $I2/a$. Atoms are in the positions:

Re : (4e) $\pm (1/4\ u\ 0;\ 3/4, u + 1/2, 1/2)$ with $u = 0.26567(5)$
N : (4e)   with $u = 0.0976(10)$

All other atoms are in

$$(8f) \quad \pm (xyz;\ x + 1/2, \bar{y}, z); \quad \text{B.C.}$$

their parameters being those of Table XVE,8. In this table values designated Ph refer to the centers of benzene rings.

The structure is illustrated by Figure XVE,7. It is composed of independent molecules. Each molecule centered about a rhenium atom possesses a twofold axis of symmetry. The coordination about this atom is fivefold (in contrast to that described in **XV,e7**). Bond

TABLE XVE,8

Parameters of Atoms in $ReNCl_2[P(C_6H_5)_3]_2$

Atom	x	y	z
Cl	0.14683(15)	0.34962(23)	0.05463(10)
P	0.36453(14)	0.30334(22)	0.09443(10)
Ph(1)	0.3413(3)	0.1361(4)	0.2156(2)
Ph(2)	0.5546(3)	0.1955(4)	0.0845(1)
Ph(3)	0.3745(3)	0.6318(4)	0.1278(2)
		Phenyl carbons	
C(1)	0.3510	0.2103	0.1636
C(2)	0.2752	0.1330	0.1615
C(3)	0.2655	0.0588	0.2136
C(4)	0.3316	0.0619	0.2677
C(5)	0.4073	0.1392	0.2697
C(6)	0.4170	0.2134	0.2177
C(7)	0.4735	0.2472	0.0891
C(8)	0.5449	0.3371	0.0965
C(9)	0.6260	0.2854	0.0920
C(10)	0.6357	0.1438	0.0800
C(11)	0.5643	0.0539	0.0726
C(12)	0.4831	0.1056	0.0771
C(13)	0.3722	0.4891	0.1144
C(14)	0.3564	0.5354	0.1703
C(15)	0.3586	0.6781	0.1837
C(16)	0.3767	0.7745	0.1413
C(17)	0.3926	0.7282	0.0854
C(18)	0.3903	0.5855	0.0719

dimensions are those of Figure XVE,8. The angle P–Re–P = 163.1° and the distribution about the phosphorus atoms is essentially tetrahedral.

XV,e9. Crystals of *bis(triphenylphosphine) ethylene nickel(0)*, $Ni[(C_6-H_5)_3P]_2 \cdot C_2H_4$, have triclinic symmetry. Their bimolecular units possess the dimensions:

$$a_0 = 10.40(2) \text{ A.}; \quad b_0 = 17.58(10) \text{ A.}; \quad c_0 = 10.04(2) \text{ A.}$$
$$\alpha = 102.6(3)°; \quad \beta = 116.9(7)°; \quad \gamma = 91.8(3)°$$

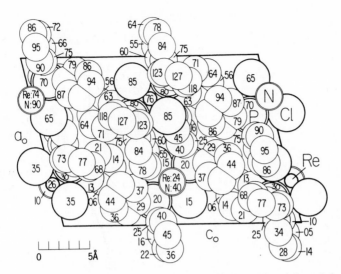

Fig. XVE,7.   The monoclinic structure of nitridodichloro bis(triphenylphosphine) rhenium projected along its b_0 axis.

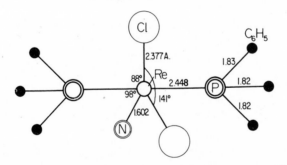

Fig. XVE,8.   Some bond dimensions in the molecule of nitridodichloro bis (triphenylphosphine)rhenium.

TABLE XVE,9

Parameters of the Atoms in Bis(triphenylphosphine) Ethylene Nickel

Atom	x	y	z
Ni	0.3990	0.2279	0.0119
P(1)	0.7630	0.8634	0.0308
P(2)	0.4161	0.3291	0.1891
C(1)	0.9509	0.8426	0.1120
C(2)	0.0678	0.8956	0.1493

(continued)

TABLE XVE,9 *(continued)*

Parameters of the Atoms in Bis(triphenylphosphine) Ethylene Nickel

Atom	x	y	z
C(3)	0.2093	0.8786	0.2100
C(4)	0.2373	0.8037	0.2379
C(5)	0.1207	0.7504	0.2125
C(6)	0.9789	0.7695	0.1484
C(7)	0.7776	0.9615	0.1607
C(8)	0.8074	0.9637	0.3064
C(9)	0.8279	0.0346	0.4118
C(10)	0.8283	0.1047	0.3711
C(11)	0.8036	0.1039	0.2229
C(12)	0.7873	0.0326	0.1211
C(13)	0.2785	0.1127	0.1460
C(14)	0.1917	0.1162	0.2174
C(15)	0.2317	0.1024	0.3606
C(16)	0.3719	0.0865	0.4461
C(17)	0.4658	0.0830	0.3828
C(18)	0.4231	0.0966	0.2380
C(19)	0.3410	0.3241	0.3147
C(20)	0.1871	0.3141	0.2551
C(21)	0.1274	0.3042	0.3516
C(22)	0.2200	0.3063	0.5019
C(23)	0.6215	0.6761	0.4297
C(24)	0.4245	0.3273	0.4703
C(25)	0.3378	0.4144	0.1136
C(26)	0.2805	0.4712	0.1853
C(27)	0.2222	0.5335	0.1231
C(28)	0.2244	0.5428	0.9934
C(29)	0.7063	0.5086	0.0659
C(30)	0.6593	0.5817	0.0217
C(31)	0.6037	0.3673	0.3267
C(32)	0.7056	0.3128	0.3780
C(33)	0.8441	0.3377	0.4853
C(34)	0.1039	0.5794	0.4560
C(35)	0.8018	0.4735	0.4900
C(36)	0.6585	0.4492	0.3857
C(37)	0.4363	0.7448	0.0191
C(38)	0.5196	0.8165	0.1240

The space group is C_i^1 ($P\bar{1}$) with atoms in the positions $(2i)$ $\pm(xyz)$. Determined parameters are listed in Table XVE,9; values for the hydrogen atoms of the benzene rings are given in the original article.

The distribution of the atoms in the central portion of the molecules of this structure and bond dimensions involving these atoms are shown in Figure XVE,9.

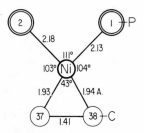

Fig. XVE,9. Bond dimensions in the central portion of the molecule of bis (triphenylphosphine) ethylene nickel.

XV,e10. The monoclinic crystals of *palladium bis(triphenylphosphine) carbon disulfide*, $Pd[(C_6H_5)_3P]_2CS_2$, have a tetramolecular unit of the dimensions:

$$a_0 = 11.55(3) \text{ A.}; \quad b_0 = 19.52(2) \text{ A.}; \quad c_0 = 18.38(3) \text{ A.}$$
$$\beta = 128°53(10)'$$

Atoms are in the positions

$$(4e) \quad \pm(xyz; x,{}^1\!/_2-y,z+{}^1\!/_2)$$

of the space group C_{2h}^5 ($P2_1/c$). The parameters are listed in Table XVE,10.

TABLE XVE,10

Parameters of the Atoms in $Pd[(C_6H_5)_3P]_2CS_2$

Atom	x	y	z
Pd	0.0866	0.1042	0.2757
S(1)	−0.2125	0.1108	0.2743
S(2)	0.0363	0.2051	0.3163
P(1)	−0.0114	−0.0025	0.2101
P(2)	0.3202	0.1172	0.3046

(continued)

TABLE XVE,10 (*continued*)

Parameters of the Atoms in Pd[(C_6H_5)$_3$P]$_2$CS$_2$

Atom	x	y	z
C(1)	−0.068	0.137	0.285
C(1,1)	0.094	−0.060	0.190
C(1,2)	0.135	−0.126	0.231
C(1,3)	0.215	−0.165	0.206
C(1,4)	0.245	−0.141	0.150
C(1,5)	0.201	−0.074	0.112
C(1,6)	0.127	−0.033	0.132
C(2,1)	−0.196	0.000	0.091
C(2,2)	−0.247	−0.060	0.031
C(2,3)	−0.384	−0.058	−0.053
C(2,4)	−0.470	0.004	−0.089
C(2,5)	−0.410	0.062	−0.032
C(2,6)	−0.271	0.061	0.063
C(3,1)	−0.035	−0.052	0.286
C(3,2)	−0.140	−0.106	0.248
C(3,3)	−0.147	−0.143	0.313
C(3,4)	−0.063	−0.124	0.402
C(3,5)	0.043	−0.069	0.439
C(3,6)	0.055	−0.034	0.378
C(4,1)	0.456	0.170	0.407
C(4,2)	0.408	0.230	0.423
C(4,3)	0.507	0.270	0.505
C(4,4)	0.659	0.253	0.570
C(4,5)	0.708	0.197	0.546
C(4,6)	0.606	0.155	0.468
C(5,1)	0.431	0.043	0.331
C(5,2)	0.429	−0.009	0.383
C(5,3)	0.531	−0.064	0.415
C(5,4)	0.619	−0.071	0.386
C(5,5)	0.617	−0.020	0.332
C(5,6)	0.522	0.038	0.306
C(6,1)	0.298	0.166	0.212
C(6,2)	0.424	0.202	0.230
C(6,3)	0.399	0.237	0.157
C(6,4)	0.267	0.237	0.067
C(6,5)	0.145	0.199	0.051
C(6,6)	0.166	0.162	0.125

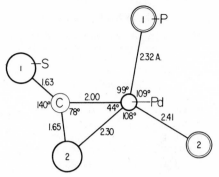

Fig. XVE,10.   Bond dimensions in the central portion of the molecule of palladium bis(triphenylphosphine) carbon disulfide.

In this structure the atomic distribution about the palladium atom (Fig. XVE,10) is unusual. The two phosphorus, the S(2) and the C(1) atoms about the palladium atom are coplanar but this coordination is not square; instead the C(1)–Pd–S(2) angle is the very small 44°. The bond S(1)–C(1) makes an angle of 7° with the coordination plane and there is an angle of 140° between S(1)–C(1) and C(1)–S(2). The phenyl groups have their usual pinwheel-like distribution about the phosphorus atoms, with P–C lying between 1.80 and 1.86 A.

XV,e11.   Crystals of *dibromo tris(diphenylphosphine) cobalt(II)*, $CoBr_2[P(C_6H_5)_2]_3$, have a triclinic, bimolecular unit of the dimensions:

$$a_0 = 11.05 \text{ A.}; \quad b_0 = 11.47 \text{ A.}; \quad c_0 = 15.41 \text{ A.}$$
$$\alpha = 98°0'; \qquad \beta = 82°54'; \qquad \gamma = 118°30'$$

The space group is C_i^1 ($P\bar{1}$) with all atoms in the positions (2i) $\pm(xyz)$. The determined parameters are listed in Table XVE,11, those for carbon being considered approximate only.

TABLE XVE,11

Parameters of the Atoms in $CoBr_2[P(C_6H_5)_2]_3$

Atom	x	y	z
Co	0.3606(4)	0.0541(3)	0.2248(3)
Br(1)	0.3151(4)	0.1774(3)	0.1184(2)
Br(2)	0.1943(3)	0.9032(3)	0.3181(2)
P(1)	0.3185(8)	0.8921(6)	0.1163(5)
P(2)	0.3974(8)	0.2210(6)	0.3259(5)
P(3)	0.5811(8)	0.1306(6)	0.1929(5)

(continued)

TABLE XVE,11 (*continued*)

Parameters of the Atoms in $CoBr_2[P(C_6H_5)_2]_3$

Atom	x	y	z
	Ring carbons		
C(1,1)	0.152(4)	0.811(3)	0.064(3)
C(2,1)	0.057(4)	0.859(3)	0.081(3)
C(3,1)	0.936(5)	0.802(4)	0.030(3)
C(4,1)	0.906(4)	0.692(4)	0.967(3)
C(5,1)	0.994(4)	0.642(4)	0.954(3)
C(6,1)	0.111(4)	0.693(3)	0.006(2)
C(1,2)	0.351(3)	0.754(3)	0.133(2)
C(2,2)	0.282(4)	0.668(3)	0.192(3)
C(3,2)	0.320(5)	0.566(4)	0.205(3)
C(4,2)	0.423(6)	0.567(5)	0.154(4)
C(5,2)	0.501(5)	0.653(4)	0.096(3)
C(6,2)	0.469(4)	0.760(4)	0.091(3)
C(1,3)	0.512(3)	0.258(3)	0.420(2)
C(2,3)	0.535(5)	0.171(5)	0.447(4)
C(3,3)	0.630(5)	0.193(4)	0.508(3)
C(4,3)	0.699(5)	0.326(4)	0.550(3)
C(5,3)	0.667(5)	0.416(5)	0.531(3)
C(6,3)	0.575(5)	0.392(4)	0.460(3)
C(1,4)	0.247(4)	0.220(3)	0.378(3)
C(2,4)	0.201(5)	0.187(4)	0.463(3)
C(3,4)	0.078(6)	0.178(5)	0.504(4)
C(4,4)	0.015(5)	0.215(4)	0.453(3)
C(5,4)	0.043(5)	0.256(5)	0.363(3)
C(6,4)	0.180(5)	0.258(4)	0.324(3)
C(1,5)	0.670(3)	0.047(3)	0.220(2)
C(2,5)	0.601(4)	0.947(3)	0.288(3)
C(3,5)	0.678(7)	0.884(6)	0.296(5)
C(4,5)	0.802(6)	0.904(5)	0.283(4)
C(5,5)	0.864(5)	0.992(5)	0.221(3)
C(6,5)	0.802(5)	0.070(4)	0.183(3)
C(1,6)	0.700(4)	0.309(3)	0.219(2)
C(2,6)	0.681(5)	0.392(4)	0.165(3)
C(3,6)	0.782(5)	0.531(4)	0.191(3)
C(4,6)	0.868(5)	0.571(4)	0.254(3)
C(5,6)	0.878(5)	0.480(4)	0.314(3)
C(6,6)	0.792(5)	0.344(4)	0.285(3)

Note: In this table C(3,6) for instance refers to the third carbon atom in benzene ring 6. Rings 1 and 2 are attached to P(1), 3 and 4 to P(2) and 5 and 6 to P(3).

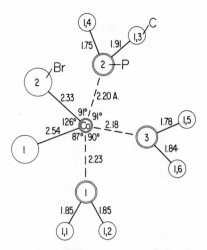

Fig. XVE,11.   Some bond dimensions in the molecule of dibromo tris(diphenylphosphine) cobalt.

In this structure the apparent coordination of the cobalt atom is fivefold. It has the form of a pyramid which is described as something between a trigonal bipyramid and a square pyramid. Bond dimensions are shown in Figure XVE,11. As in certain other phenyl substituted phosphine compounds, the benzene rings are twisted in a rather irregular fashion.

Identical cell dimensions have been assigned $NiBr_2[P(C_6H_5)_2]_3$.

A partial structure has been determined for the isostructural *diiodo tris(diphenylphosphine) nickel(II)*, $NiI_2[P(C_6H_5)_2]_3$. Its unit has the dimensions:

$$a_0 = 11.16 \text{ A.}; \quad b_0 = 11.93 \text{ A.}; \quad c_0 = 15.53 \text{ A.}$$
$$\alpha = 100°48'; \quad \beta = 81°42'; \quad \gamma = 118°18'$$

Parameters for the heavy atoms have been found to be as follows:

Atom	x	y	z
Ni	0.359(2)	0.043(1)	0.224(1)
I(1)	0.289(1)	0.152(1)	0.106(1)
I(2)	0.187(1)	0.896(1)	0.324(1)
P(1)	0.311(4)	0.879(2)	0.122(2)
P(2)	0.394(4)	0.218(2)	0.319(2)
P(3)	0.573(4)	0.111(3)	0.198(2)

The analogous $CoI_2[P(C_6H_5)_2]_3$ has the same cell dimensions.

TABLE XVE,12
Parameters of the Atoms in NiBr$_2$[P(C$_6$H$_5$)$_3$]$_2$

Atom	x	y	z
Ni	0.1730(2)	0.1214(1)	0.2625(2)
Br(1)	−0.0037(3)	0.1324(1)	0.0211(2)
Br(2)	0.4264(2)	0.1087(1)	0.3363(3)
P(1)	0.0911(5)	0.0707(1)	0.3418(4)
P(2)	0.1766(5)	0.1710(1)	0.4075(5)
C(1)	−0.1100(16)	0.0612(4)	−0.7422(16)
C(2)	−0.1595(18)	0.0257(4)	−0.7413(17)
C(3)	−0.3155(19)	0.0182(5)	−0.7985(19)
C(4)	−0.4127(19)	0.0462(5)	−0.8535(19)
C(5)	−0.3686(18)	0.0809(4)	−0.8574(18)
C(6)	−0.2119(18)	0.0891(4)	−0.8044(17)
C(7)	−0.8221(16)	0.0309(4)	−0.6913(16)
C(8)	−0.7424(17)	0.0062(4)	−0.5794(17)
C(9)	−0.6783(22)	−0.0245(5)	−0.6149(22)
C(10)	−0.7002(21)	−0.0308(5)	−0.7579(21)
C(11)	−0.7760(21)	−0.0061(5)	−0.8704(21)
C(12)	−0.8425(20)	0.0243(5)	−0.8395(20)
C(13)	0.1397(16)	0.0721(4)	0.5388(15)
C(14)	0.0342(16)	0.0687(4)	0.5935(16)
C(15)	0.0765(18)	0.0723(4)	0.7478(18)
C(16)	0.2277(19)	0.0806(4)	0.8375(18)
C(17)	0.3341(19)	0.0830(5)	0.7868(19)
C(18)	0.2893(17)	0.0787(4)	0.6331(16)
C(19)	0.2966(18)	0.1698(4)	0.6059(17)
C(20)	0.4456(22)	0.1646(5)	0.6451(21)
C(21)	0.5490(27)	0.1634(7)	0.8031(26)
C(22)	0.4936(24)	0.1708(6)	0.8945(24)
C(23)	0.3444(26)	0.1756(6)	0.8675(25)
C(24)	0.2358(20)	0.1762(5)	0.7097(20)
C(25)	0.2471(17)	0.2100(4)	0.3492(17)
C(26)	0.3049(23)	0.2402(6)	0.4424(23)
C(27)	0.3579(24)	0.2712(6)	0.3955(23)
C(28)	0.3487(23)	0.2719(6)	0.2588(22)
C(29)	0.2389(21)	0.2117(5)	0.2043(21)
C(30)	0.2948(24)	0.2446(6)	0.1616(24)
C(31)	−0.0109(17)	0.1824(4)	0.3853(16)
C(32)	−0.0796(20)	0.2142(5)	0.3139(19)
C(33)	−0.2351(23)	0.2202(6)	0.2957(22)
C(34)	−0.3037(23)	0.1947(6)	0.3461(22)
C(35)	−0.2339(23)	0.1648(6)	0.4206(22)
C(36)	−0.0815(20)	0.1582(5)	0.4422(20)

XV,e12. The crystals of *dibromo bis(triphenylphosphine) nickel*, $NiBr_2[P(C_6H_5)_3]_2$, have monoclinic symmetry. Their unit is tetra-molecular with the dimensions:

$$a_0 = 9.828(2) \text{ A.}; \quad b_0 = 37.178(9) \text{ A.}; \quad c_0 = 10.024(3) \text{ A.}$$
$$\beta = 114.65(2)°$$

Atoms are in the positions

$$(4e) \quad \pm(xyz; x+{}^1\!/_2, {}^1\!/_2-y, z+{}^1\!/_2)$$

of C_{2h}^5 in the orientation $P2_1/n$. Parameters are stated in Table XVE,12.

In the molecules of this structure coordination about the nickel atom is approximately tetrahedral, with P–Ni–P = 110° and Br–Ni–Br = 126°. The bond lengths are Ni–Br = 2.346 and 2.329 A., Ni–P = 2.323 and 2.343 A. Coordination is also tetrahedral about the atoms of phosphorus, with P–C lying between 1.806 and 1.840 A.

XV,e13. The *oxygen adduct* of *iodocarbonyl bis(triphenylphosphine)-iridium-dichloromethane*, $IrIO_2(CO)[P(C_6H_5)_3]_2 \cdot CH_2Cl_2$, forms monoclinic crystals whose tetramolecular unit has the dimensions:

$$a_0 = 10.693(8) \text{ A.}; \quad b_0 = 22.864(15) \text{ A.}; \quad c_0 = 14.997(11) \text{ A.}$$
$$\beta = 93.05(2)°$$

The space group is C_{2h}^5 ($P2_1/n$) with atoms in the positions:

$$(4e) \quad \pm(xyz; x+{}^1\!/_2, {}^1\!/_2-y, z+{}^1\!/_2)$$

Some parameters are listed in Table XVE,13. Values for the centers of the phenyl groups and derived parameters for their carbon atoms are given in the original paper.

The resulting structure is made up of molecules which have bond dimensions stated in Figure XVE,12; they resemble those in IrO_2-$Cl(CO)[P(C_6H_5)_3]_2$ (**XV,e15**); in the present compound there is a disorder in the distribution of the iodine and carbonyl about iridium; as a result, two kinds of molecule are present.

TABLE XVE,13

Parameters of Atoms in $IrIO_2(CO)[P(C_6H_5)_3]_2 \cdot CH_2Cl_2$

Atom	x	y	z
Ir	0.18567(13)	0.15625(6)	0.25456(9)
I′	−0.0683(4)	0.1482(2)	0.2673(3)
I″	0.2442(5)	0.1450(2)	0.0778(3)
P(1)	0.1928(8)	0.2605(3)	0.2426(5)
P(2)	0.2138(7)	0.0532(3)	0.2707(5)
Cl(1)	−0.3522(17)	0.1813(8)	0.3755(11)
Cl(2)	−0.2437(14)	0.2895(6)	0.3344(10)
O(1)	0.2260(23)	0.1656(9)	0.3879(14)
O(2)	0.3454(23)	0.1664(10)	0.3387(16)
C	−0.3178(43)	0.2477(21)	0.4089(30)
C′(CO)	0.020	0.147	0.22
O′(CO)	−0.075	0.147	0.20
C″(CO)	0.165	0.147	0.13
O″(CO)	0.155	0.147	0.06

Note: Primes and double-primes designate atoms in the two kinds of molecule.

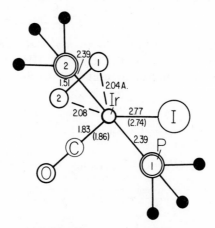

Fig. XVE,12.   Some bond dimensions in the molecule of iodocarbonyl bis(triphenylphosphine)iridium-dichloromethane.

XV,e14. *Chlorocarbonyl (sulfur dioxide) bis(triphenylphosphine) iridium,* $IrCl(CO)(SO_2)[P(C_6H_5)_3]_2$, forms monoclinic crystals whose tetramolecular unit has the dimensions:

$$a_0 = 12.11(2) \text{ A.}; \quad b_0 = 16.79(2) \text{ A.}; \quad c_0 = 17.09(2) \text{ A.}$$
$$\beta = 103°0(30)′$$

Atoms are in the positions

$$(4e) \quad \pm(xyz; \ x, \ 1/2-y, z+1/2)$$

of the space group C_{2h}^5 ($P2_1/c$). The parameters are those of Table XVE,14.

TABLE XVE,14

Parameters of the Atoms in $IrCl(CO)(SO_2)[P(C_6H_5)_3]_2$

Atom	x	y	z
Ir	0.21335(13)	0.27084(9)	0.26122(11)
S	0.2683(7)	0.3352(6)	0.3955(5)
Cl	0.2689(8)	0.1393(6)	0.3029(5)
P(1)	0.3959(7)	0.2879(6)	0.2366(6)
P(2)	0.0276(6)	0.2388(6)	0.2648(5)
O(1)	0.3232(18)	0.4060(14)	0.3811(13)
O(2)	0.1654(20)	0.3434(14)	0.4266(13)
O(3)	0.1321(20)	0.4267(17)	0.1835(15)
C	0.1604(30)	0.3741(25)	0.2132(21)
Derived parameters of phenyl carbons			
C(1,P1,1)	0.511	0.300	0.329
C(2,P1,1)	0.509	0.256	0.397
C(3,P1,1)	0.593	0.267	0.467
C(4,P1,1)	0.681	0.321	0.467
C(5,P1,1)	0.683	0.364	0.389
C(6,P1,1)	0.598	0.354	0.329
C(1,P1,2)	0.437	0.202	0.182
C(2,P1,2)	0.552	0.184	0.195
C(3,P1,2)	0.588	0.119	0.157
C(4,P1,2)	0.509	0.072	0.105
C(5,P1,2)	0.394	0.091	0.091
C(6,P1,2)	0.358	0.156	0.130
C(1,P1,3)	0.399	0.368	0.167
C(2,P1,3)	0.394	0.357	0.086
C(3,P1,3)	0.393	0.422	0.035
C(4,P1,3)	0.395	0.499	0.066
C(5,P1,3)	0.399	0.511	0.147
C(6,P1,3)	0.401	0.445	0.198

(continued)

TABLE XVE,14 (*continued*)

Parameters of the Atoms in $IrCl(CO)(SO_2)[P(C_6H_5)_3]_2$

Atom	x	y	z
C(1,P2,1)	−0.015	0.163	0.186
C(2,P2,1)	0.017	0.171	0.113
C(3,P2,1)	−0.019	0.116	0.052
C(4,P2,1)	−0.088	0.053	0.064
C(5,P2,1)	−0.120	0.045	0.136
C(6,P2,1)	−0.083	0.100	0.197
C(1,P2,2)	−0.083	0.311	0.236
C(2,P2,2)	−0.066	0.382	0.279
C(3,P2,2)	−0.141	0.445	0.257
C(4,P2,2)	−0.233	0.437	0.191
C(5,P2,2)	−0.249	0.366	0.149
C(6,P2,2)	−0.174	0.303	0.171
C(1,P2,3)	−0.004	0.195	0.361
C(2,P2,3)	0.079	0.155	0.416
C(3,P2,3)	0.052	0.118	0.482
C(4,P2,3)	−0.059	0.121	0.493
C(5,P2,3)	−0.142	0.161	0.437
C(6,P2,3)	−0.114	0.198	0.371

Note: In the above table (3,P2,3) for instance is the third carbon atom in the third ring about P2.

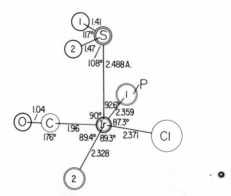

Fig. XVE,13.   Some bond dimensions in the molecule of chlorocarbonyl (sulfur dioxide) bis(triphenylphosphine) iridium.

In this structure the iridium atom has the coordination of a fivefold tetragonal pyramid, with the bond dimensions of Figure XVE,13. The base of this pyramid, involving P(1), P(2), Cl, and O, is planar to within ca. 0.01 A.; the iridium atom is 0.21 A. above this plane in the direction of the sulfur atom. Intermolecular separations are normal.

XV,e15. The *oxygen adduct* of *chlorocarbonyl bis(triphenylphosphine) iridium*, $IrO_2Cl(CO)[P(C_6H_5)_3]_2$, is triclinic with a bimolecular unit of the dimensions:

$$a_0 = 19.02(3) \text{ A.}; \quad b_0 = 9.83(2) \text{ A.}; \quad c_0 = 9.93(2) \text{ A.}$$
$$\alpha = 94.0(1)°; \quad \beta = 64.9(1)°; \quad \gamma = 93.2(1)°$$

Atoms are in the positions $(2i) \pm (xyz)$ of C_i^1 $(P\bar{1})$. The determined parameters are listed in Table XVE,15; as indicated there, the chlorine atoms and (CO) groups [designated as X(1) and X(2) in the table] are distributed in a disordered fashion.

TABLE XVE,15

Parameters of the Atoms in $IrO_2Cl(CO)[P(C_6H_5)_3]_2$

Atom	x	y	z
Ir	0.2342(1)	0.2100(2)	0.0068(2)
P(1)	0.1335(6)	0.3207(13)	0.2136(11)
P(2)	0.3430(7)	0.1266(15)	−0.2024(13)
X(1)	0.2873(9)	0.1153(18)	0.1618(16)
X(2)	0.1467(10)	0.0293(21)	0.0019(19)
O(1)	0.224(1)	0.366(3)	−0.117(2)
O(2)	0.279(2)	0.397(3)	−0.073(3)
	Derived parameters of phenyl carbons		
C(1,P1,1)	0.039	0.317	0.203
C(2,P1,1)	0.036	0.325	0.066
C(3,P1,1)	−0.036	0.321	0.058
C(4,P1,1)	−0.104	0.309	0.188
C(5,P1,1)	−0.101	0.302	0.324
C(6,P1,1)	−0.030	0.306	0.332
C(1,P1,2)	0.111	0.257	0.394
C(2,P1,2)	0.098	0.118	0.410
C(3,P1,2)	0.085	0.063	0.545
C(4,P1,2)	0.085	0.148	0.663
C(5,P1,2)	0.098	0.288	0.646
C(6,P1,2)	0.111	0.342	0.512

(*continued*)

TABLE XVE,15 (continued)

Parameters of the Atoms in $IrO_2Cl(CO)[P(C_6H_5)_3]_2$

Atom	x	y	z
C(1,P1,3)	0.160	0.499	0.238
C(2,P1,3)	0.111	0.601	0.243
C(3,P1,3)	0.132	0.737	0.256
C(4,P1,3)	0.202	0.772	0.263
C(5,P1,3)	0.251	0.670	0.258
C(6,P1,3)	0.230	0.534	0.245
C(1,P2,1)	0.361	0.217	−0.363
C(2,P2,1)	0.435	0.271	−0.448
C(3,P2,1)	0.449	0.334	−0.579
C(4,P2,1)	0.390	0.342	−0.625
C(5,P2,1)	0.316	0.288	−0.540
C(6,P2,1)	0.302	0.225	−0.409
C(1,P2,2)	0.431	0.138	−0.178
C(2,P2,2)	0.444	0.257	−0.107
C(3,P2,2)	0.509	0.271	−0.075
C(4,P2,2)	0.561	0.166	−0.114
C(5,P2,2)	0.548	0.046	−0.185
C(6,P2,2)	0.483	0.033	−0.216
C(1,P2,3)	0.333	−0.052	−0.250
C(2,P2,3)	0.352	−0.102	−0.396
C(3,P2,3)	0.349	−0.241	−0.427
C(4,P2,3)	0.326	−0.331	−0.313
C(5,P2,3)	0.307	−0.282	−0.167
C(6,P2,3)	0.310	−0.142	−0.136

Note: In the above table (4,P1,2) for instance, is the fourth C atom in the second ring about P1. The values for X(1) and X(2) apply to the disordered distribution of the Cl atoms and CO radicals.

The central part of the molecule found in this structure has the bond dimensions of Figure XVE,14. In it the two oxygen atoms (separated from one another by 1.30 A.) lie −0.04 and 0.06 A. on either side of the plane through Ir–X(1)–X(2).

XV,e16. Crystals of *oxopentachloropropionato bis(triphenylphosphine) dirhenium(IV)*, $Re_2OCl_5(C_2H_5CO_2)[P(C_6H_5)_3]_2$, are tetragonal with a tetramolecular unit of the edge lengths:

$$a_0 = 10.669(3) \text{ A.}, \quad c_0 = 34.898(7) \text{ A.}$$

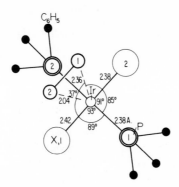

Fig. XVE,14. Some bond dimensions in the molecule of chlorocarbonyl bis(triphenylphosphine) iridium-oxygen.

All atoms are in the positions

$$(4a) \quad xyz;\ \bar{x},\bar{y},z+{}^1/_2;\ \bar{y},x,z+{}^3/_4;\ y,\bar{x},z+{}^1/_4$$

of the space group C_4^4 $(P4_3)$. The parameters are those of Table XVE,16.

TABLE XVE,16

Parameters of the Atoms in $Re_2OCl_5(C_2H_5CO_2)[P(C_6H_5)_3]_2$

Atom	x	y	z
Re(1)	0.14245(7)	0.31929(7)	0.00000
Re(2)	0.16479(7)	0.37300(7)	−0.07005(3)
Cl(1)	0.0251(4)	0.1371(4)	−0.0044(2)
Cl(2)	0.0540(4)	0.2092(4)	−0.0964(1)
Cl(3)	0.2492(4)	0.2454(5)	0.0540(1)
Cl(4)	0.3042(4)	0.3803(5)	−0.1229(1)
Cl(B)	0.3072(4)	0.2310(4)	−0.0393(1)
P(1)	−0.0173(4)	0.4027(4)	0.0446(1)
P(2)	0.0403(4)	0.5327(4)	−0.1064(1)
C(1,1)	−0.0025(18)	0.3401(17)	0.0936(6)
C(2,1)	−0.0283(21)	0.2185(21)	0.0990(7)
C(3,1)	−0.0128(25)	0.1606(25)	0.1358(8)
C(4,1)	0.0264(30)	0.2437(32)	0.1646(10)
C(5,1)	0.0528(27)	0.3592(28)	0.1595(9)
C(6,1)	0.0408(20)	0.4214(21)	0.1228(6)

(*continued*)

TABLE XVE,16 (*continued*)

Parameters of the Atoms in $Re_2OCl_5(C_2H_5CO_2)[P(C_6H_5)_3]_2$

Atom	x	y	z
C(1,2)	0.0356(17)	0.5256(17)	−0.1584(5)
C(2,2)	0.0047(19)	0.6340(19)	−0.1787(6)
C(3,2)	−0.0067(25)	0.6206(25)	−0.2206(8)
C(4,2)	0.0082(25)	0.5065(25)	−0.2370(7)
C(5,2)	0.0389(26)	0.3967(26)	−0.2176(8)
C(6,2)	0.0486(21)	0.4063(20)	−0.1778(7)
C(1,3)	−0.0111(17)	0.5711(17)	0.0465(5)
C(2,3)	−0.1122(21)	0.6419(20)	0.0317(6)
C(3,3)	−0.1072(26)	0.7761(26)	0.0327(8)
C(4,3)	−0.0009(26)	0.8348(26)	0.0482(9)
C(5,3)	0.1006(26)	0.7608(26)	0.0645(8)
C(6,3)	0.0955(21)	0.6306(21)	0.0610(7)
C(1,4)	0.1110(17)	0.6793(17)	−0.0956(6)
C(2,4)	0.2007(21)	0.7307(21)	−0.1205(7)
C(3,4)	0.2675(25)	0.8424(25)	−0.1105(8)
C(4,4)	0.2302(28)	0.9049(27)	−0.0759(9)
C(5,4)	0.1425(28)	0.8561(29)	−0.0511(9)
C(6,4)	0.0828(19)	0.7425(20)	−0.0612(6)
C(1,5)	−0.1777(17)	0.3646(17)	0.0347(5)
C(2,5)	−0.2202(16)	0.3359(15)	−0.0032(5)
C(3,5)	−0.3441(18)	0.3112(19)	−0.0088(6)
C(4,5)	−0.4315(19)	0.3161(18)	0.0211(6)
C(5,5)	−0.3899(20)	0.3424(20)	0.0563(6)
C(6,5)	−0.2661(19)	0.3681(19)	0.0642(6)
C(1,6)	−0.1300(18)	0.5383(19)	−0.0968(6)
C(2,6)	−0.1962(20)	0.4278(20)	−0.1025(6)
C(3,6)	−0.3337(21)	0.4337(22)	−0.1007(7)
C(4,6)	−0.3869(23)	0.5432(23)	−0.0907(7)
C(5,6)	−0.3199(25)	0.6477(25)	−0.0833(8)
C(6,6)	−0.1882(20)	0.6528(20)	−0.0871(6)
C(1)	0.3044(18)	0.5400(18)	−0.0159(6)
C(2)	0.3836(20)	0.6578(20)	−0.0069(7)
C(3)	0.4746(23)	0.6333(23)	0.0237(7)
O(1)	0.2580(11)	0.4746(11)	0.0104(3)
O(2)	0.2811(11)	0.5187(11)	−0.0514(4)
O(B)	0.0473(10)	0.4238(11)	−0.0329(3)

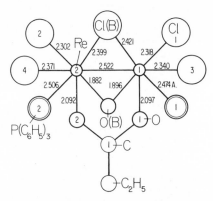

Fig. XVE,15.   Some bond dimensions in the molecule of oxopentachloropropionato bis(triphenylphosphine) dirhenium.

The basic features of the coordination and some of the bond lengths
in the molecules of this structure are shown in Figure XVE,15. As it
indicates, each metal atom has seven neighbors including the other
metal, and these two rhenium atoms are bridged by an oxygen and a
chlorine atom.

TABLE XVE,17

Parameters of the Atoms in $(C_6H_5)_3PAu \cdot Mn(CO)_4 \cdot P(OC_6H_5)_3$

Atom	x	y	z
Au	0.4837	0.2003	0.1656
Mn	0.3034	0.2818	0.2555
P(1)	0.651	0.177	0.086
P(2)	0.338	0.109	0.335
O(1)	0.083	0.401	0.342
O(2)	0.506	0.406	0.334
O(3)	0.168	0.130	0.142
O(4)	0.239	0.483	0.137
O(5)	0.459	0.016	0.333
O(6)	0.237	0.014	0.323
O(7)	0.351	0.127	0.429
C(1)	0.172	0.358	0.311
C(2)	0.426	0.354	0.306
C(3)	0.223	0.170	0.189
C(4)	0.272	0.395	0.177

(*continued*)

TABLE XVC,17 *(continued)*

Parameters of the Atoms in $(C_6H_5)_3PAu \cdot Mn(CO)_4 \cdot P(OC_6H_5)_3$

Atom	x	y	z
C(5)	0.753	0.019	0.082
C(6)	0.881	0.018	0.087
C(7)	0.961	−0.106	0.094
C(8)	0.904	−0.208	0.075
C(9)	0.785	−0.208	0.070
C(10)	0.700	−0.092	0.070
C(11)	0.751	0.285	0.118
C(12)	0.833	0.359	0.073
C(13)	0.894	0.436	0.090
C(14)	0.913	0.464	0.169
C(15)	0.841	0.407	0.227
C(16)	0.772	0.325	0.198
C(17)	0.601	0.220	−0.018
C(18)	0.520	0.324	−0.030
C(19)	0.477	0.367	−0.110
C(20)	0.531	0.283	−0.174
C(21)	0.618	0.167	−0.165
C(22)	0.654	0.134	−0.077
C(23)	0.577	0.020	0.340
C(24)	0.646	0.110	0.368
C(25)	0.762	0.114	0.382
C(26)	0.863	0.036	0.366
C(27)	0.802	−0.054	0.345
C(28)	0.669	−0.066	0.321
C(29)	0.245	−0.101	0.347
C(30)	0.336	−0.161	0.417
C(31)	0.337	−0.279	0.437
C(32)	0.265	−0.345	0.391
C(33)	0.186	−0.291	0.316
C(34)	0.175	−0.165	0.301
C(35)	0.262	0.207	0.474
C(36)	0.151	0.180	0.492
C(37)	0.066	0.262	0.533
C(38)	0.106	0.361	0.563
C(39)	0.230	0.388	0.546
C(40)	0.302	0.306	0.501

XV,e17. The complex *triphenylphosphinegold tetracarbonylman-ganese trioxyphenylphosphine*, $(C_6H_5)_3PAu \cdot Mn(CO)_4 \cdot P(OC_6H_5)_3$, forms triclinic crystals. The bimolecular unit chosen for them has the dimensions:

$$a_0 = 10.76(1) \text{ A.}; \quad b_0 = 10.63(1) \text{ A.}; \quad c_0 = 16.74(1) \text{ A.}$$
$$\alpha = 93.6(1)°; \quad \beta = 97.0(1)°; \quad \gamma = 77.8(1)°$$

The space group is C_i^1 ($P\bar{1}$) with atoms in the positions $(2i) \pm(xyz)$. The chosen parameters are listed in Table XVE,17.

Important bond dimensions in the molecules are given in Figure XVE,16. The manganese atom is octahedrally, the phosphorus atoms tetrahedrally coordinated. The atom of gold has its usual twofold coordination, with Au–Mn–P(1) = 166.5°.

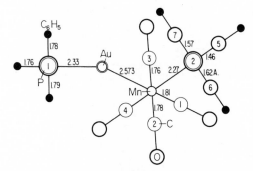

Fig. XVE,16.   Some bond dimensions in the molecule of $(C_6H_5)_3PAu \cdot Mn(CO)_4 \cdot P(OC_6H_5)_3$

XV,e18. *Nitrosyl dicarbonyl bis(triphenylphosphine) manganese*, $Mn(NO)(CO)_2[P(C_6H_5)_3]_2$, forms orthorhombic crystals. Their large unit, containing eight molecules, has the dimensions:

$$a_0 = 18.147(8) \text{ A.}; \quad b_0 = 17.071(8) \text{ A.}; \quad c_0 = 21.860(11) \text{ A.} \quad (22°C.)$$

Atoms are in the positions:

$$(8c) \quad \pm(xyz; \, x+^1/_2,^1/_2-y,\bar{z}; \, \bar{x},y+^1/_2,^1/_2-z; \, ^1/_2-x,\bar{y},z+^1/_2)$$

of the space group V_h^{15} (*Pbca*). Their parameters are stated in Table XVE,18.

The structure is made up of molecules, some of whose bond dimensions are shown in Figure XVE,17. The bonds Mn–P(1) and Mn–P(2) are almost in a straight line and the distribution of the (NO) and (CO)

TABLE XVE,18

Parameters of Atoms in $Mn(NO)(CO)_2[P(C_6H_5)_3]_2$

Atom	x	y	z
Mn	0.4104(1)	0.3176(1)	0.6471(1)
P(1)	0.4385(2)	0.2554(2)	0.5579(2)
P(2)	0.3919(2)	0.3978(2)	0.7289(2)
C(1)	0.4938(9)	0.3711(8)	0.6362(8)
C(2)	0.3282(10)	0.3510(9)	0.6141(7)
N	0.4127(7)	0.2332(6)	0.6907(5)
O(1)	0.5482(7)	0.4052(7)	0.6307(6)
O(2)	0.2734(7)	0.3735(7)	0.5947(6)
O(3)	0.4120(7)	0.1756(6)	0.7205(5)
Ph(1)	0.6090(4)	0.2038(4)	0.5465(3)
Ph(2)	0.4014(3)	0.3577(4)	0.4368(4)
Ph(3)	0.3570(3)	0.0906(4)	0.5317(3)
Ph(4)	0.2606(4)	0.3502(3)	0.8210(3)
Ph(5)	0.5322(4)	0.4123(4)	0.8187(3)
Ph(6)	0.3479(4)	0.5762(4)	0.6939(3)

Ph designates the centers of each of the phenyl groups.

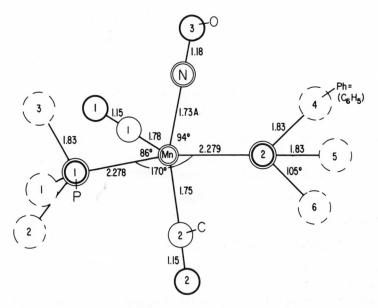

Fig. XVE,17. Some bond dimensions in the molecule of nitrosyl dicarbonyl bis(triphenylphosphine) manganese.

radicals normal to them is triangular. Bonds about the phosphorus atoms are tetrahedral, with the three benzene rings attached to each in the usual pinwheel fashion. The parameters for all the carbon atoms in these rings, as stated in the original article, involve a C–C separation of 1.397 A.

XV,e19. Crystals of *triphenyltin tetracarbonylmanganese triphenylphosphine*, $(C_6H_5)_3Sn \cdot Mn(CO)_4 \cdot P(C_6H_5)_3$, are monoclinic with a tetramolecular cell of the dimensions:

$$a_0 = 10.45(1) \text{ A.}; \quad b_0 = 26.48(4) \text{ A.}; \quad c_0 = 12.65(2) \text{ A.}$$
$$\beta = 99°34(3)'$$

The space group is C_{2h}^5 in the orientation $P2_1/n$. Atoms thus are in the positions:

$$(4e) \quad \pm (xyz; x + 1/2, 1/2 - y, z + 1/2)$$

The final parameters are those of Table XVE,19.

TABLE XVE,19

Parameters of the Atoms in $(C_6H_5)_3Sn \cdot Mn(CO)_4 \cdot P(C_6H_5)_3$

Atom	x	y	z
C(1)	0.4287	0.1294	0.1330
C(2)	0.1044	0.0908	0.0362
C(3)	0.3303	0.0456	0.0753
C(4)	0.2350	0.1751	0.0774
C(5)	0.1988	0.1575	−0.1956
C(6)	0.0856	0.1789	−0.1658
C(7)	0.0061	0.2115	−0.2362
C(8)	0.0309	0.2203	−0.3429
C(9)	0.1367	0.1956	−0.3710
C(10)	0.2203	0.1630	−0.2994
C(11)	0.3063	0.0594	−0.1832
C(12)	0.3963	0.0527	−0.2469
C(13)	0.3952	0.0043	−0.3045
C(14)	0.3070	−0.0296	−0.2914
C(15)	0.2207	−0.0242	−0.2253
C(16)	0.2161	0.0188	−0.1643
C(17)	0.4614	0.1467	−0.1023
C(18)	0.4785	0.1944	−0.1342
C(19)	0.5887	0.2153	−0.1254
C(20)	0.7080	0.1879	−0.0911

(*continued*)

TABLE XVE,19 (*continued*)

Parameters of the Atoms in $(C_6H_5)_3Sn \cdot Mn(CO)_4 \cdot P(C_6H_5)_3$

Atom	x	y	z
C(21)	0.7002	0.1372	−0.0599
C(22)	0.5775	0.1164	−0.0612
C(23)	0.1645	0.0223	0.3048
C(24)	0.1487	−0.0129	0.2272
C(25)	0.1050	−0.0620	0.2526
C(26)	0.0989	−0.0759	0.3663
C(27)	0.1232	−0.0385	0.4457
C(28)	0.1579	0.0127	0.4078
C(29)	0.0359	0.1388	0.2843
C(30)	−0.0026	0.1397	0.3852
C(31)	−0.1196	0.1599	0.3994
C(32)	−0.1917	0.1859	0.3142
C(33)	−0.1586	0.1865	0.2152
C(34)	−0.0387	0.1656	0.1980
C(35)	0.3575	0.1262	0.3915
C(36)	0.4393	0.0913	0.4550
C(37)	0.5420	0.1057	0.5334
C(38)	0.5513	0.1616	0.5556
C(39)	0.4668	0.1923	0.4935
C(40)	0.3744	0.1749	0.4117
O(1)	0.5291	0.1429	0.1835
O(2)	0.0040	0.0749	0.0156
O(3)	0.3773	0.0089	0.0836
O(4)	0.2090	0.2183	0.0924
Sn	0.2102	0.0990	0.2631
Mn	0.2673	0.1109	0.0700
P	0.3055	0.1177	−0.1007

The structure, indicated in Figure XVE,18, is molecular. Its phosphorus, manganese, and tin atoms are in a nearly linear array, with the four carbonyl groups that are in a square distribution about manganese lying in a plane normal to the line P–Mn–Sn. The three benzene rings about phosphorus and about tin have their usual pinwheel distribution (Fig. XVE,19). The separations P–C = 1.80 – 1.86 A., Sn–C = 2.16–2.17 A., Mn–C = 1.74–1.85 A., P–Mn = 2.267 A., Mn–Sn = 2.627 A., and C–O = 1.09–1.20 A.; the C–C distances are normal.

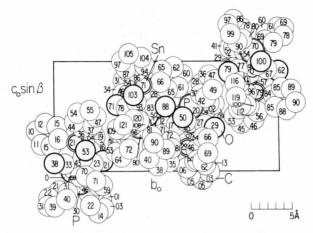

Fig. XVE,18.   The monoclinic structure of triphenyltin tetracarbonylmanganese triphenylphosphine projected along its a_0 axis. The manganese atoms scarcely show.

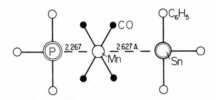

Fig. XVE,19.   The distribution of atoms in the center of the molecule of triphenyltin tetracarbonylmanganese triphenylphosphine.

XV,e20.   *Bis(triphenylmethylarsonium) tetrachloro nickel(II)*, [As(C$_6$-H$_5$)$_3$CH$_3$]$_2$NiCl$_4$, forms cubic crystals whose tetramolecular unit has the edge length:

$$a_0 = 15.557(4) \text{ A.}$$

The space group is T^4 (P2$_1$3) with atoms in the positions:

(4a) uuu; $u + 1/2, 1/2 - u, \bar{u}$; $\bar{u}, u + 1/2, 1/2 - u$; $1/2 - u, \bar{u}, u + 1/2$

(12b) xyz; $x + 1/2, 1/2 - y, \bar{z}$; $z + 1/2, 1/2 - x, \bar{y}$; $y + 1/2, 1/2 - z, \bar{x}$;   tr

Determined parameters are listed in Table XVE,20.

The structure is an association of complex ions. The cations have the dimensions to be expected from those found in similar compounds; in

TABLE XVE,20

Positions and Parameters of the Atoms in [As(C_6H_5)$_3$CH$_3$]$_2$NiCl$_4$

Atom	Position	x	y	z
Ni	(4a)	0.0597(3)	0.0597(3)	0.0597(3)
Cl(1)	(12b)	0.1815(4)	0.0938(4)	−0.0127(4)
Cl(2)	(4a)	−0.0246(4)	−0.0246(4)	−0.0246(4)
As(1)	(4a)	0.7824(2)	0.7824(2)	0.7824(2)
C(1)	(12b)	0.750(1)	0.900(1)	0.770(2)
C(2)	(12b)	0.806(2)	0.962(2)	0.793(2)
C(3)	(12b)	0.783(2)	0.050(2)	0.790(2)
C(4)	(12b)	0.700(2)	0.074(2)	0.761(2)
C(5)	(12b)	0.647(2)	0.010(2)	0.738(2)
C(6)	(12b)	0.666(2)	0.920(2)	0.740(2)
CH$_3$(7)	(4a)	0.710(1)	0.710(1)	0.710(1)
As(2)	(4a)	0.5092(2)	0.5092(2)	0.5092(2)
C(8)	(12b)	0.393(1)	0.555(1)	0.505(1)
C(9)	(12b)	0.329(1)	0.499(2)	0.483(2)
C(10)	(12b)	0.245(2)	0.530(2)	0.478(2)
C(11)	(12b)	0.229(2)	0.621(2)	0.494(2)
C(12)	(12b)	0.300(2)	0.673(2)	0.515(2)
C(13)	(12b)	0.382(2)	0.643(2)	0.522(2)
CH$_3$(14)	(4a)	0.582(2)	0.582(2)	0.582(2)

this case they are regularly tetrahedral with As–C = 1.91–1.95 A. The NiCl$_4^{2-}$ anion is also a regular tetrahedron with Ni–Cl = 2.27 A.

Several compounds involving metals other than nickel are isostructural. Their unit cubes have the edges:

For [As(C_6H_5)$_3$CH$_3$]$_2$MnCl$_4$: a_0 = 15.63 A.
For [As(C_6H_5)$_3$CH$_3$]$_2$FeCl$_4$: a_0 = 15.65 A.
For [As(C_6H_5)$_3$CH$_3$]$_2$CoCl$_4$: a_0 = 15.53 A.
For [As(C_6H_5)$_3$CH$_3$]$_2$ZnCl$_4$: a_0 = 15.55 A.

XV,e21. Crystals of the addition compound *mercuric chloride·triphenylarsenicoxide*, [HgCl$_2$·(C_6H_5)$_3$AsO]$_2$, are orthorhombic with a unit that contains four of these molecules and has the edge lengths:

a_0 = 19.60(3) A.; b_0 = 17.51(3) A.; c_0 = 11.16(2) A.

All atoms are in the positions

(8c) $\pm (xyz; \; ^1/_2 - x, y + ^1/_2, z; \; x, ^1/_2 - y, z + ^1/_2; \; x + ^1/_2, y, ^1/_2 - z)$

of the space group V_h^{15} (*Pbca*). The determined parameters are those of Table XVE,21.

TABLE XVE,21

Parameters of the Atoms in $[HgCl_2 \cdot (C_6H_5)_3AsO]_2$

Atom	x	y	z
Hg(1)	0.42255	0.55410	0.06013
As(2)	0.4094	0.3598	−0.0809
Cl(3)	0.4176	0.5447	0.2675
Cl(4)	0.3644	0.5935	−0.1103
O(5)	0.4562	0.4255	−0.0126
C(6)	0.3325	0.3310	0.0134
C(7)	0.3256	0.3753	0.1225
C(8)	0.2670	0.3515	0.1983
C(9)	0.2271	0.2918	0.1609
C(10)	0.2435	0.2509	0.0616
C(11)	0.2921	0.2727	−0.0280
C(12)	0.4689	0.2734	−0.1023
C(13)	0.4531	0.2233	−0.1978
C(14)	0.4947	0.1557	−0.1996
C(15)	0.5400	0.1328	−0.1229
C(16)	0.5538	0.1875	−0.0373
C(17)	0.5118	0.2560	−0.0157
C(18)	0.3710	0.3945	−0.2287
C(19)	0.3037	0.4126	−0.2308
C(20)	0.2794	0.4416	−0.3390
C(21)	0.3202	0.4495	−0.4295
C(22)	0.3864	0.4284	−0.4331
C(23)	0.4116	0.3939	−0.3278

A portion of the resulting structure is shown in Figure XVE,20. Its centrosymmetric, dimeric molecules have the bond dimensions of Figure XVE,21. The coordination about both arsenic and mercury is tetrahedral but that about mercury is very distorted, with angles ranging from 79 to 145°.

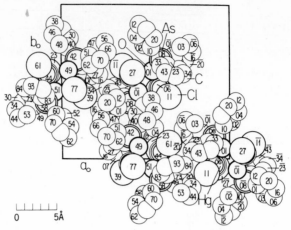

Fig. XVE,20.   Part of the contents of the orthorhombic unit cell of crystals of
[HgCl₂·(C₆H₅)₃AsO]₂ projected along the c₀ axis.

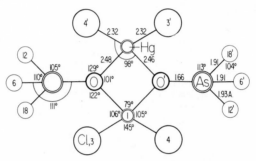

Fig. XVE,21.   Bond dimensions in the molecule of [HgCl₂·(C₆H₅)₃AsO]₂. Only
one carbon atom of each phenyl group is shown.

XV,e22.   The addition compound *mercuric chloride·bis(triphenyl-arsenicoxide)*, $HgCl_2 \cdot 2[(C_6H_5)_3AsO]$, is monoclinic with a tetramolecular unit of the dimensions.

$$a_0 = 18.40(2) \text{ A.}; \quad b_0 = 10.58(1) \text{ A.}; \quad c_0 = 18.23(2) \text{ A.}$$
$$\beta = 106.0(2)°$$

The space group is C_{2h}^5 $(P2_1/c)$ with all atoms in the positions:

$$(4e) \quad \pm (xyz; \ x, \tfrac{1}{2} - y, z + \tfrac{1}{2})$$

The parameters are stated in Table XVE,22.

TABLE XVE,22

Parameters of the Atoms in $HgCl_2 \cdot 2[(C_6H_5)_3AsO]$

Atom	x	y	z
Hg(1)	0.1985	0.2425	0.1179
Cl(2)	0.0711	0.2039	0.0579
Cl(3)	0.2964	0.3882	0.1428
O(4)	0.210	0.149	0.236
O(5)	0.260	0.072	0.076
As(6)	0.2059	0.1966	0.3232
As(7)	0.2889	0.9269	0.1109
C(8)	0.300	0.255	0.387
C(9)	0.348	0.325	0.351
C(10)	0.421	0.352	0.392
C(11)	0.446	0.327	0.469
C(12)	0.395	0.263	0.502
C(13)	0.329	0.220	0.465
C(14)	0.134	0.333	0.321
C(15)	0.148	0.409	0.384
C(16)	0.091	0.508	0.386
C(17)	0.028	0.507	0.312
C(18)	0.021	0.442	0.257
C(19)	0.077	0.346	0.255
C(20)	0.171	0.058	0.370
C(21)	0.179	−0.062	0.344
C(22)	0.151	−0.172	0.382
C(23)	0.118	−0.151	0.434
C(24)	0.106	−0.026	0.458
C(25)	0.133	0.082	0.427
C(26)	0.318	−0.169	0.035
C(27)	0.320	−0.098	−0.030
C(28)	0.344	−0.163	−0.088
C(29)	0.359	−0.282	−0.081
C(30)	0.355	−0.360	−0.021
C(31)	0.334	−0.299	0.039
C(32)	0.379	−0.059	0.199
C(33)	0.434	−0.160	0.212
C(34)	0.499	−0.146	0.276
C(35)	0.503	−0.040	0.324

(continued)

TABLE XVE,22 (continued)
Parameters of the Atoms in $HgCl_2 \cdot 2[(C_6H_5)_3AsO]$

Atom	x	y	z
C(36)	0.445	0.051	0.310
C(37)	0.377	0.040	0.248
C(38)	0.212	−0.165	0.138
C(39)	0.140	−0.141	0.105
C(40)	0.084	−0.205	0.122
C(41)	0.097	−0.318	0.167
C(42)	0.173	−0.349	0.196
C(43)	0.236	−0.275	0.187

A portion of the resulting structure is shown in Figure XVE,22. Bond dimensions in the central non-phenyl part of the molecule are

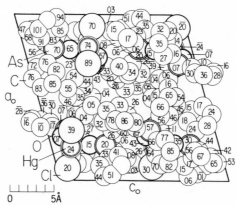

Fig. XVE,22. A portion of the monoclinic structure of $HgCl_2 \cdot 2[(C_6H_5)_3AsO]$ projected along its b_0 axis.

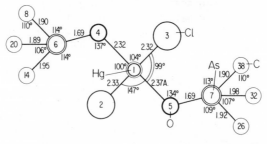

Fig. XVE,23. Some bond dimensions in the molecule of $HgCl_2 \cdot 2[(C_6H_5)_3AsO]$.

given in Figure XVE,23. The structure is to be compared with that of **XV,e21**; as in the previously described crystal, the coordination about the arsenic and mercury atoms is tetrahedral, that about mercury being highly distorted. All the benzene rings in this molecule are essentially planar, the maximum departure from the best plane through any of them being 0.05 A.

XV,e23. A structure has been described for *dioxodinitrato bis-(triphenylarsenicoxide) uranium(VI)*, $UO_2(NO_3)_2 \cdot 2[(C_6H_5)_3AsO]$. The symmetry of its crystals is monoclinic with a bimolecular unit of the dimensions:

$$a_0 = 11.09(4) \text{ A.}; \quad b_0 = 19.28(6) \text{ A.}; \quad c_0 = 10.88(4) \text{ A.}$$
$$\beta = 128°10(10)'$$

The space group is C_{2h}^5 $(P2_1/c)$ with uranium atoms in the positions:

$$(2a) \quad 000; \, 0 \, 1/2 \, 1/2$$

All other atoms are in the general positions

$$(4e) \quad \pm(xyz; \, x, 1/2-y, z+1/2)$$

with the parameters stated in Table XVE,23.

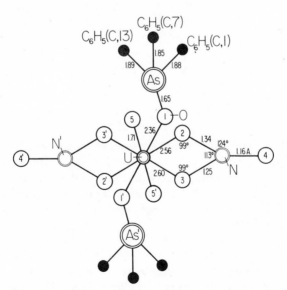

Fig. XVE,24. Some bond dimensions in the molecule of dioxodinitrato bis(triphenylarsenicoxide) uranium.

TABLE XVE,23

Parameters of Atoms in $UO_2(NO_3)_2 \cdot 2[(C_6H_5)_3AsO]$

Atom	x	y	z
As	0.3132(4)	0.1385(3)	0.1221(5)
O(5)	0.102(3)	−0.068(2)	0.006(3)
O(4)	0.142(5)	−0.055(3)	0.448(5)
O(3)	−0.008(3)	−0.068(2)	0.202(3)
O(2)	0.173(4)	0.008(3)	0.298(4)
O(1)	0.218(3)	0.072(2)	0.120(4)
N	0.104(4)	−0.040(3)	0.325(6)
C(1)	0.466(4)	0.156(3)	0.336(5)
C(2)	0.563(6)	0.216(4)	0.382(6)
C(3)	0.676(6)	0.232(4)	0.545(7)
C(4)	0.688(7)	0.189(4)	0.641(8)
C(5)	0.602(7)	0.132(4)	0.606(7)
C(6)	0.492(6)	0.107(4)	0.448(7)
C(7)	0.405(4)	0.118(3)	0.032(5)
C(8)	0.566(6)	0.098(3)	0.137(6)
C(9)	0.630(4)	0.071(3)	0.068(5)
C(10)	0.555(5)	0.071(3)	−0.092(5)
C(11)	0.407(6)	0.089(3)	−0.188(6)
C(12)	0.324(5)	0.133(3)	−0.125(5)
C(13)	0.197(5)	0.221(3)	0.028(5)
C(14)	0.078(4)	0.235(2)	0.033(4)
C(15)	−0.010(5)	0.300(2)	−0.020(5)
C(16)	0.023(9)	0.355(5)	−0.078(9)
C(17)	0.133(8)	0.338(4)	−0.095(8)
C(18)	0.220(5)	0.279(5)	−0.038(9)

The resulting arrangement is composed of molecules having the bond dimensions of Figure XVE,24. Coordination about the uranium atoms is eightfold, with the uranyl oxygens much closer than the others to uranium. The angles O–U–O range from 50 to 92°. Coordination about the arsenic atoms is, as expected, tetrahedral, with O–As–C and C–As–C between 103 and 115°.

XV,e24. *Hexaphenyl benzene*, $C_6(C_6H_5)_6$, is dimorphous, both forms having orthorhombic symmetry. A structure has been determined for the B-form. Its tetramolecular unit has the dimensions:

$$a_0 = 11.080(8) \text{ A.}; \quad b_0 = 21.777(10) \text{ A.}; \quad c_0 = 12.553(8) \text{ A.}$$

The space group has been found to be C_{2v}^9 ($Pna2_1$) with all atoms in the positions:

(4a) xyz; $\bar{x},\bar{y},z+{}^1/_2$; ${}^1/_2-x,y+{}^1/_2,z+{}^1/_2$; $x+{}^1/_2,{}^1/_2-y,z$

Parameters are those of Table XVE,24.

TABLE XVE,24

Parameters of the Atoms in Hexaphenyl Benzene

Atom	x	y	z
C(1)	0.2286	0.3881	0.1054
C(2)	0.2349	0.4309	0.0221
C(3)	0.2252	0.4106	−0.0834
C(4)	0.2081	0.3492	−0.1055
C(5)	0.1964	0.3056	−0.0244
C(6)	0.2047	0.3259	0.0838
C(A,1)	0.2475	0.4082	0.2191
C(A,2)	0.1659	0.4448	0.2718
C(A,3)	0.1897	0.4621	0.3784
C(A,4)	0.2950	0.4416	0.4287
C(A,5)	0.3755	0.4057	0.3747
C(A,6)	0.3523	0.3881	0.2699
C(B,1)	0.2548	0.4968	0.0468
C(B,2)	0.1572	0.5383	0.0254
C(B,3)	0.1791	0.6025	0.0539
C(B,4)	0.2882	0.6196	0.1005
C(B,5)	0.3834	0.5773	0.1177
C(B,6)	0.3634	0.5159	0.0918
C(C,1)	0.2332	0.4562	−0.1740
C(C,2)	0.1330	0.4658	−0.2402
C(C,3)	0.1392	0.5076	−0.3225
C(C,4)	0.2441	0.5427	−0.3386
C(C,5)	0.3437	0.5339	−0.2714
C(C,6)	0.3383	0.4906	−0.1892
C(D,1)	0.2121	0.3278	−0.2217
C(D,2)	0.1071	0.3084	−0.2718
C(D,3)	0.1090	0.2921	−0.3801
C(D,4)	0.2189	0.2968	−0.4346
C(D,5)	0.3281	0.3147	−0.3827
C(D,6)	0.3220	0.3306	−0.2760

(continued)

TABLE XVE,24 *(continued)*

Parameters of the Atoms in Hexaphenyl Benzene

Atom	x	y	z
C(E,1)	0.1742	0.2402	−0.0471
C(E,2)	0.0644	0.2130	−0.0179
C(E,3)	0.0439	0.1513	−0.0389
C(E,4)	0.1297	0.1161	−0.0896
C(E,5)	0.2411	0.1418	−0.1179
C(E,6)	0.2632	0.2046	−0.0968
C(F,1)	0.1846	0.2814	0.1714
C(F,2)	0.0850	0.2881	0.2382
C(F,3)	0.0624	0.2465	0.3201
C(F,4)	0.1458	0.1978	0.3380
C(F,5)	0.2485	0.1924	0.2730
C(F,6)	0.2663	0.2333	0.1890

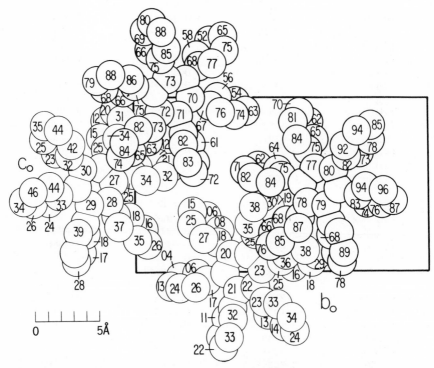

Fig. XVE,25. The structure of the B, orthorhombic, form of hexaphenyl benzene projected along its a_0 axis.

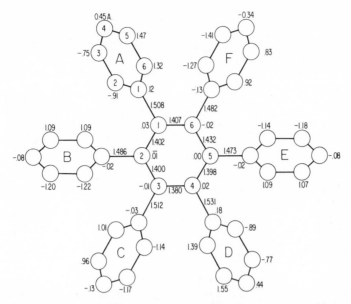

Fig. XVE,26. Bond lengths in the molecule of hexaphenyl benzene. The numbers beside the circles are the distances in angstrom units of these atoms from the best plane through the entire molecule.

The pseudotrigonal layer structure that results is shown in Figure XVE,25. The central benzene ring of the molecule is nearly normal to the a_0 axis of the crystal, whereas the attached rings are turned, propellor-fashion, through 62–71° about the bonds joining them to their central carbon atoms (Fig. XVE,26). Average bond lengths within the rings lie between 1.391 and 1.416 A.; the bonds joining the rings and the central ring range between 1.473 and 1.531 A.

XV,e25. The form of *dimeric 3-bromo-2,4,6-triphenyl phenoxyl*, $[C_6HBr(C_6H_5)_3O]_2$, which crystallizes with 1.5 molecules of benzene, has triclinic symmetry. A structure for it has been described in terms of a bimolecular cell of the dimensions:

$$a_0 = 14.09(2) \text{ A.}; \quad b_0 = 12.85(2) \text{ A.}; \quad c_0 = 12.96(2) \text{ A.}$$
$$\alpha = 85.7(2)°; \quad \beta = 99.5(2)°; \quad \gamma = 103.0(2)°$$

Atoms are in the general positions $(2i) \pm (xyz)$ of C_i^1 $(P\bar{1})$ with the assigned parameters of Table XVE,25.

TABLE XVE,25

Parameters of the Atoms in $[C_6HBr(C_6H_5)_3O]_2$

Atom	x	y	z
O(1)	0.0693	0.5810	0.8297
C(I,1)	0.1066	0.6888	0.8035
C(I,2)	0.0456	0.7362	0.7273
C(I,3)	0.0805	0.8423	0.6958
C(I,4)	0.1741	0.9006	0.7368
C(I,5)	0.2303	0.8504	0.8151
C(I,6)	0.1973	0.7467	0.8523
C(II,1)	−0.0553	0.6774	0.6615
C(II,2)	−0.0684	0.6256	0.5895
C(II,3)	−0.1646	0.5707	0.5477
C(II,4)	−0.2416	0.5666	0.5983
C(II,5)	−0.2301	0.6211	0.6889
C(II,6)	−0.1353	0.6753	0.7303
C(III,1)	0.2593	0.7067	0.9439
C(III,2)	0.3598	0.7203	0.9460
C(III,3)	0.4227	0.6935	0.0334
C(III,4)	0.3800	0.6489	0.1233
C(III,5)	0.2788	0.6282	0.1218
C(III,6)	0.2171	0.6589	0.0325
C(IV,1)	0.2180	0.0123	0.7011
C(IV,2)	0.2223	0.0403	0.5961
C(IV,3)	0.2687	0.1462	0.5660
C(IV,4)	0.3035	0.2177	0.6400
C(IV,5)	0.3006	0.1945	0.7459
C(IV,6)	0.2581	0.0884	0.7762
Br(1)	−0.0073	0.9109	0.6030
Br(2)	0.0831	0.4033	0.9973
C(B,1)	0.0096	0.0072	−0.1046
C(B,2)	0.0205	−0.0916	0.0603
C(B,3)	−0.0315	0.0790	0.0400
O(2)	0.4041	0.4581	0.8321
C(V,1)	0.3204	0.4765	0.8260
C(V,2)	0.2573	0.4283	0.9056
C(V,3)	0.1663	0.4506	0.8963
C(V,4)	0.1170	0.5027	0.7998
C(V,5)	0.1881	0.5517	0.7260
C(V,6)	0.2828	0.5444	0.7366

(*continued*)

TABLE XVE,25 (*continued*)

Parameters of the Atoms in [C₆HBr(C₆H₅)₃O]₂

Atom	x	y	z
C(VI,1)	0.2951	0.3566	0.9867
C(VI,2)	0.3801	0.3897	0.0584
C(VI,3)	0.4094	0.3188	0.1363
C(VI,4)	0.3600	0.2134	0.1408
C(VI,5)	0.2733	0.1796	0.0696
C(VI,6)	0.2414	0.2516	0.9927
C(VII,1)	0.3473	0.6012	0.6609
C(VII,2)	0.3270	0.7020	0.6175
C(VII,3)	0.3811	0.7535	0.5371
C(VII,4)	0.4558	0.7156	0.5096
C(VII,5)	0.4773	0.6210	0.5539
C(VII,6)	0.4221	0.5633	0.6292
C(VIII,1)	0.0398	0.4115	0.7437
C(VIII,2)	0.0731	0.3326	0.7018
C(VIII,3)	0.0062	0.2491	0.6521
C(VIII,4)	−0.0907	0.2429	0.6437
C(VIII,5)	−0.1303	0.3212	0.6868
C(VIII,6)	−0.0592	0.4096	0.7369
C(B,4)	0.4765	0.9854	0.3231
C(B,5)	0.4686	0.9450	0.2315
C(B,6)	0.3621	0.9115	0.1795
C(B,7)	0.2965	0.9335	0.2274
C(B,8)	0.3177	0.9858	0.3219
C(B,9)	0.4171	0.0115	0.3672

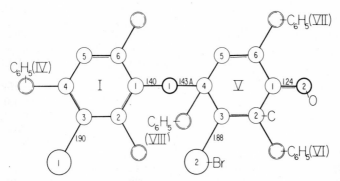

Fig. XVE,27. Some bond lengths in the molecule of dimeric 3-bromo-2,4,6-triphenyl phenoxyl.

The resulting structure is made up of molecules, the two halves of which are dissimilar. They have the composition indicated in Figure XVE,27.

TABLE XVE,26

Parameters of the Atoms in $[As(C_6H_5)_4]_2[Co(NO_3)_4]$

Atom	x	y	z
O(1)	0.5579(6)	0.954(1)	0.2299(8)
O(2)	0.6419(6)	0.941(1)	0.2060(8)
O(3)	0.5920(8)	0.792(1)	0.2276(9)
O(4)	0.5350(6)	0.703(1)	0.3310(7)
O(5)	0.5596(12)	0.863(2)	0.3868(12)
O(6)	0.5808(8)	0.693(2)	0.4457(8)
N(1)	0.6006(6)	0.895(2)	0.2210(8)
N(2)	0.5611(7)	0.756(2)	0.3898(9)
C(1,1)	0.8832(6)	0.766(1)	0.0881(8)
C(1,2)	0.8745(6)	0.810(2)	0.0151(8)
C(1,3)	0.9051(7)	0.767(1)	−0.0347(9)
C(1,4)	0.9451(7)	0.674(2)	−0.0140(10)
C(1,5)	0.9543(8)	0.626(2)	0.0607(11)
C(1,6)	0.9246(7)	0.666(2)	0.1107(9)
C(2,1)	0.8903(6)	0.837(1)	0.2521(8)
C(2,2)	0.9260(7)	0.935(2)	0.2673(10)
C(2,3)	0.9564(8)	0.958(2)	0.3443(10)
C(2,4)	0.9543(7)	0.881(2)	0.4013(9)
C(2,5)	0.9204(8)	0.776(2)	0.3836(9)
C(2,6)	0.8900(7)	0.750(1)	0.3056(9)
C(3,1)	0.7884(6)	0.680(1)	0.1529(8)
C(3,2)	0.7461(6)	0.695(2)	0.2001(8)
C(3,3)	0.7064(8)	0.606(2)	0.2029(10)
C(3,4)	0.7083(9)	0.550(2)	0.1587(11)
C(3,5)	0.7489(8)	0.488(1)	0.1144(10)
C(3,6)	0.7875(7)	0.578(1)	0.1123(9)
C(4,1)	0.7936(7)	0.946(1)	0.1126(8)
C(4,2)	0.8211(7)	0.053(1)	0.1122(10)
C(4,3)	0.7882(7)	0.155(1)	0.0899(10)
C(4,4)	0.7262(8)	0.149(1)	0.0735(9)
C(4,5)	0.7003(6)	0.044(2)	0.0721(9)
C(4,6)	0.7304(6)	0.939(1)	0.0947(9)
As	0.8401(1)	0.806(1)	0.1509(1)

XV,e26. *Tetraphenylarsonium tetranitratocobaltate(II)*, [As(C$_6$H$_5$)$_4$]$_2$-[Co(NO$_3$)$_4$], forms monoclinic crystals whose tetramolecular unit has the dimensions:

$$a_0 = 23.47(3) \text{ A.}; \quad b_0 = 11.34(1) \text{ A.}; \quad c_0 = 18.57(2) \text{ A.}$$
$$\beta = 107°0(6)'$$

The space group is C$_{2h}^6$ (*C2/c*) with atoms in the positions:

Co : (4e)　±(0 *u* $^1/_4$; $^1/_2$,*u*+$^1/_2$,$^1/_4$)　　with *u* = 0.834(1)

All other atoms are in:

(8f)　±(*xyz*; *x*,*ȳ*,*z*+$^1/_2$; *x*+$^1/_2$,*y*+$^1/_2$,*z*; *x*+$^1/_2$,$^1/_2$−*y*,*z*+$^1/_2$)

The determined parameters are those of Table XVE,26.

In this structure the cobalt atom is dodecahedrally surrounded by eight oxygen atoms, two from each of the four NO$_3$ groups of the Co(NO$_3$)$_4^{2-}$ anion. Half of them are nearer the cobalt atoms than the others, the nearer being 2.03 or 2.11 A., the farther 2.36 and 2.54 A. away. The NO$_3$ groups have their accustomed triangular planar shape, with N–O = 1.20–1.25 A. In the As(C$_6$H$_5$)$_4^+$ cation the planar benzene rings are, as usual, tetrahedrally distributed about the central arsenic, with As–C = 1.81–1.94 A.

XV,e27. The triclinic crystals of *bis(tetraphenylarsonium) tri-μ-chlorooctachlorotrirhenate(III)*, [As(C$_6$H$_5$)$_4$]$_2$Re$_3$Cl$_{11}$, have been assigned a bimolecular unit of the dimensions:

$$a_0 = 12.01 \text{ A.}; \quad b_0 = 26.22 \text{ A.}; \quad c_0 = 9.87 \text{ A.}$$
$$\alpha = 88°30'; \quad \beta = 66°12'; \quad \gamma = 105°24'$$

Atoms are in the positions (2i) ±(*xyz*) of C$_i^1$ (*P$\bar{1}$*). The parameters are given in Table XVE,27.

TABLE XVE,27

Parameters of the Atoms in [As(C$_6$H$_5$)$_4$]$_2$Re$_3$Cl$_{11}$

Atom	x	y	z
Re(1)	0.6007	0.2036	0.4976
Re(2)	0.7327	0.2979	0.4605
Re(3)	0.6125	0.2463	0.7100
Cl(1)	0.759	0.1609	0.457
Cl(2)	0.418	0.2132	0.502

(*continued*)

TABLE XVE,27 (*continued*)

Parameters of the Atoms in $[As(C_6H_5)_4]_2Re_3Cl_{11}$

Atom	x	y	z
Cl(3)	0.926	0.2836	0.404
Cl(4)	0.585	0.3370	0.453
Cl(5)	0.765	0.2192	0.744
Cl(6)	0.435	0.2764	0.793
Cl(7)	0.712	0.2612	0.252
Cl(8)	0.732	0.3318	0.689
Cl(9)	0.489	0.1614	0.740
Cl(10)	0.524	0.1232	0.378
Cl(11)	0.880	0.3863	0.295
Cations (cation 2 in parentheses)			
As(1,2)	0.3237 (0.0064)	0.4252 (0.0796)	−0.0903 (−0.2913)
C(1,25)	0.35 (−0.17)	0.385 (0.038)	0.05 (−0.27)
C(2,26)	0.50 (−0.21)	0.399 (0.007)	0.01 (−0.12)
C(3,27)	0.49 (−0.35)	0.362 (−0.035)	0.13 (−0.06)
C(4,28)	0.41 (−0.41)	0.326 (−0.014)	0.22 (−0.20)
C(5,29)	0.29 (−0.33)	0.309 (0.014)	0.26 (−0.34)
C(6,30)	0.25 (−0.20)	0.346 (0.049)	0.17 (−0.39)
C(7,31)	0.24 (0.10)	0.478 (0.032)	0.03 (−0.27)
C(8,32)	0.21 (0.25)	0.468 (0.072)	0.20 (−0.33)
C(9,33)	0.15 (0.30)	0.520 (0.039)	0.24 (−0.33)
C(10,34)	0.13 (0.26)	0.552 (−0.036)	0.13 (−0.26)
C(11,35)	0.15 (0.11)	0.538 (−0.076)	0.00 (−0.20)
C(12,36)	0.20 (0.04)	0.497 (−0.024)	−0.05 (−0.27)
C(13,37)	0.22 (−0.01)	0.372 (0.134)	−0.18 (−0.17)
C(14,38)	0.07 (−0.13)	0.367 (0.127)	−0.09 (−0.06)
C(15,39)	0.02 (−0.12)	0.332 (0.166)	−0.16 (0.02)
C(16,40)	0.06 (−0.01)	0.305 (0.196)	−0.30 (0.04)
C(17,41)	0.18 (0.13)	0.332 (0.204)	−0.37 (−0.12)
C(18,42)	0.26 (0.09)	0.354 (0.164)	−0.29 (−0.15)
C(19,43)	0.48 (0.08)	0.472 (0.122)	−0.26 (−0.49)
C(20,44)	0.55 (0.18)	0.436 (0.095)	−0.34 (−0.58)
C(21,45)	0.67 (0.24)	0.467 (0.137)	−0.46 (−0.75)
C(22,46)	0.72 (0.18)	0.530 (0.165)	−0.50 (−0.80)
C(23,47)	0.61 (0.11)	0.560 (0.179)	−0.42 (−0.68)
C(24,48)	0.51 (0.03)	0.529 (0.154)	−0.28 (−0.52)
"H$_2$O"	0.54	0.239	0.98

The structure is built up of $As(C_6H_5)_4^+$ and $Re_3Cl_{11}^{2-}$ ions. The cations are of low symmetry, with $As-C = 1.84-2.05$ A. The complex anion has the bond dimensions of Figure XVE,28. There is some evidence for the presence of a molecule of water in these crystals.

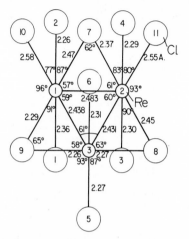

Fig. XVE,28.   Bond dimensions in the Re_3Cl_{11} anion in bis(tetraphenylarsonium) tri-μ-chlorooctachlorotrirhenate.

XV,e28.   *Tetraphenylarsonium bis(N-cyanodithiocarbimato) nickelate,* $[(C_6H_5)_4As]_2[Ni(S_2CNCN)_2]$, forms triclinic crystals whose non-primitive, bimolecular cell has the dimensions:

$$a_0 = 10.50(3) \text{ A.;} \quad b_0 = 16.90(3) \text{ A.;} \quad c_0 = 13.63(4) \text{ A.}$$
$$\alpha = 90.02(5)°; \quad \beta = 101.74(7)°; \quad \gamma = 89.51(7)°$$

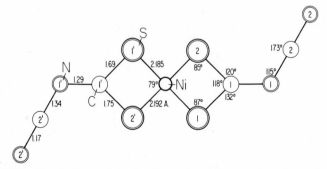

Fig. XVE,29.   Bond dimensions in the anion of tetraphenylarsonium bis(N-cyanodithiocarbimato) nickelate.

The space group is C_i^1 but with the unusual orientation $A\bar{1}$, which places all but the nickel atoms in the general positions:

$$(4i) \quad \pm(xyz;\ x,y+\tfrac{1}{2},z+\tfrac{1}{2})$$

The nickel atoms are in the special positions $000;\ 0\ \tfrac{1}{2}\ \tfrac{1}{2}$. Parameters for the other atoms are those of Table XVE,28.

TABLE XVE,28

Parameters of Atoms in $[(C_6H_5)_4As]_2[Ni(S_2CNCN)_2]$

Atom	x	y	z
S(1)	0.0495	0.9856	0.8529
S(2)	0.0154	0.8709	0.9921
C(1)	0.0587	0.8871	0.8767
N(1)	0.0893	0.3282	0.3257
C(2)	0.1276	0.3477	0.2414
N(2)	0.1628	0.3566	0.1662
As	0.4773	0.4318	0.7709
C(1,1)	0.3399	0.3735	0.8058
C(1,2)	0.3100	0.3893	0.8966
C(1,3)	0.2073	0.3446	0.9288
C(1,4)	0.1507	0.2858	0.8639
C(1,5)	0.1819	0.2711	0.7729
C(1,6)	0.2778	0.3149	0.7421
C(2,1)	0.4901	0.4087	0.6381
C(2,2)	0.3826	0.4260	0.5587
C(2,3)	0.3915	0.4090	0.4612
C(2,4)	0.5053	0.3765	0.4432
C(2,5)	0.6174	0.3619	0.5188
C(2,6)	0.6044	0.3788	0.6185
C(3,1)	0.6361	0.4042	0.8577
C(3,2)	0.6428	0.3438	0.9244
C(3,3)	0.7677	0.3212	0.9860
C(3,4)	0.8743	0.3635	0.9791
C(3,5)	0.8710	0.4228	0.9118
C(3,6)	0.7516	0.4446	0.8476
C(4,1)	0.4368	0.5376	0.7870
C(4,2)	0.5277	0.5908	0.8344
C(4,3)	0.4936	0.6696	0.8420
C(4,4)	0.3710	0.6983	0.8045
C(4,5)	0.2765	0.6446	0.7569
C(4,6)	0.3063	0.5654	0.7476

In this structure the $(C_6H_5)_4As$ cation has its usual pinwheel-like arrangement, with As–C = 1.86–1.89 A. The anion has the bond dimensions of Figure XVE,29; in it coordination about the nickel atom is square.

XV,e29. The monoclinic crystals of *dichloro tris(triphenylphosphine) ruthenium(II)*, $RuCl_2[P(C_6H_5)_3]_3$, have a tetramolecular unit of the dimensions:

$$a_0 = 18.01(4) \text{ A.}; \quad b_0 = 20.22(4) \text{ A.}; \quad c_0 = 12.36(2) \text{ A.}$$
$$\beta = 90.5(3)°$$

All atoms are in the positions

$$(4e) \quad \pm(xyz; \; x, 1/2-y, z+1/2)$$

of the space group C_{2h}^5 ($P2_1/c$). The parameters for the heavy atoms and for the centers of the phenyl groups are listed in Table XVE,29; calculated positions for the individual carbon atoms of these groups are stated in the original article.

TABLE XVE,29

Parameters of the Atoms in $RuCl_2[P(C_6H_5)_3]_3$

Atom	x	y	z
Ru	0.23919(7)	0.48707(6)	0.30760(10)
P(1)	0.2410(2)	0.5878(2)	0.4062(3)
P(2)	0.1891(2)	0.3833(2)	0.2444(3)
P(3)	0.3279(2)	0.5057(2)	0.1866(3)
Cl(1)	0.1230(2)	0.5276(2)	0.2408(3)
Cl(2)	0.3234(2)	0.4353(2)	0.4315(3)
Ph(1)[a]	0.138	0.542	0.607
Ph(2)	0.162	0.715	0.300
Ph(3)	0.380	0.654	0.537
Ph(4)	0.084	0.333	0.438
Ph(5)	0.266	0.238	0.245
Ph(6)	0.096	0.394	0.022
Ph(7)	0.468	0.589	0.284
Ph(8)	0.290	0.589	−0.030
Ph(9)	0.411	0.378	0.085

[a] Ph(1), etc., refer to the centers of phenyl groups.

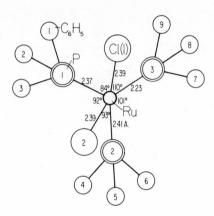

Fig. XVE,30.   Some bond dimensions in the molecule of dichloro tris(triphenylphosphine) ruthenium.

The molecules in this structure have the bond dimensions of Figure
XVE,30. The coordination of the ruthenium atom is fivefold, square
pyramidal: the two chlorine and two of the phosphorus atoms form an
approximately planar base, with P(3) at its apex. The ruthenium atom
is 0.46 A. above this base.

XV,e30. *Tris(triphenylphosphine) rhodium carbonyl hydride*, [P(C$_6$-
H$_5$)$_3$]$_3$RhH(CO), is monoclinic with a large tetramolecular unit of the
dimensions:

$$a_0 = 10.11(5) \text{ A.}; \quad b_0 = 33.31(15) \text{ A.}; \quad c_0 = 13.33(7) \text{ A.}$$
$$\beta = 90.0(1)°$$

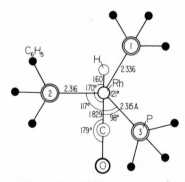

Fig. XVE,31.   Some bond dimensions in the molecule of tris(triphenylphosphine)
rhodium carbonyl hydride.

The space group is C_{2h}^5 in the orientation $P2_1/n$. Atoms accordingly are in the positions:

$$(4e) \quad \pm (xyz; \; x+\tfrac{1}{2}, \tfrac{1}{2}-y, z+\tfrac{1}{2})$$

The parameters are given in Table XVE,30.

TABLE XVE,30

Parameters of the Atoms in $[P(C_6H_5)_3]_3RhH(CO)$

Atom	x	y	z
Rh	0.22696(17)	0.11330(6)	0.22315(14)
P(1)	0.2816(6)	0.1114(2)	0.0528(5)
P(2)	0.1809(6)	0.1762(2)	0.2882(5)
P(3)	0.3134(6)	0.0664(2)	0.3333(5)
C(1)	0.064(3)	0.0904(8)	0.2064(19)
O	−0.0415(18)	0.0763(5)	0.1966(13)
H	0.377(12)	0.129(4)	0.217(9)
	C atoms about P(1)		
C(2)	−0.3225(15)	0.3580(5)	0.4706(13)
C(3)	−0.3593(17)	0.3202(6)	0.5053(10)
C(4)	−0.4427(17)	0.2962(4)	0.4479(14)
C(5)	−0.4893(15)	0.3099(5)	0.3559(13)
C(6)	−0.4526(17)	0.3477(6)	0.3212(10)
C(7)	−0.3692(17)	0.3718(4)	0.3785(14)
C(8)	−0.2244(18)	0.4365(4)	0.4878(12)
C(9)	−0.3470(14)	0.4551(5)	0.4777(13)
C(10)	−0.3574(13)	0.4912(5)	0.4259(13)
C(11)	−0.2452(18)	0.5088(4)	0.3841(12)
C(12)	−0.1226(14)	0.4902(5)	0.3941(13)
C(13)	−0.1122(13)	0.4540(5)	0.4460(13)
C(14)	0.4483(13)	0.1280(5)	0.0236(13)
C(15)	0.5532(18)	0.1133(5)	0.0805(11)
C(16)	0.6826(15)	0.1235(5)	0.0556(13)
C(17)	0.7071(13)	0.1484(5)	−0.0263(14)
C(18)	0.6022(18)	0.1631(5)	−0.0832(11)
C(19)	0.4728(15)	0.1529(5)	−0.0582(12)
	C atoms about P(2)		
C(2′)	0.0100(12)	0.1933(5)	0.2967(13)
C(3′)	−0.0272(15)	0.2215(5)	0.3686(11)
C(4′)	−0.1550(17)	0.2371(4)	0.3683(11)
C(5′)	−0.2455(12)	0.2245(5)	0.2960(13)
C(6′)	−0.2083(14)	0.1964(5)	0.2241(11)
C(7′)	−0.0805(17)	0.1808(5)	0.2244(11)

(*continued*)

TABLE XVE,30 (*continued*)

Parameters of the Atoms in [P(C$_6$H$_5$)$_3$]$_3$RhH(CO)

Atom	x	y	z
C(8′)	0.2626(15)	0.2191(4)	0.2223(12)
C(9′)	0.1939(11)	0.2546(5)	0.2039(12)
C(10′)	0.2554(15)	0.2860(4)	0.1532(13)
C(11′)	0.3856(15)	0.2819(4)	0.1208(12)
C(12′)	0.4544(11)	0.2464(5)	0.1392(12)
C(13′)	0.3929(15)	0.2150(4)	0.1899(13)
C(14′)	0.2312(20)	0.1833(6)	0.4159(11)
C(15′)	0.3184(19)	0.2135(6)	0.4448(14)
C(16′)	0.3541(16)	0.2176(5)	0.5451(17)
C(17′)	0.3027(20)	0.1914(6)	0.6166(11)
C(18′)	0.2156(19)	0.1612(6)	0.5878(14)
C(19′)	0.1798(17)	0.1571(5)	0.4874(17)
	C atoms about P(3)		
C(2″)	−0.5139(12)	0.0757(5)	0.3678(13)
C(3″)	−0.4142(17)	0.0478(4)	0.3501(13)
C(4″)	−0.2826(15)	0.0581(5)	0.3669(13)
C(5″)	−0.2508(12)	0.0963(5)	0.4015(13)
C(6″)	−0.3505(17)	0.1243(4)	0.4192(12)
C(7″)	−0.4821(14)	0.1140(5)	0.4024(13)
C(8″)	0.1890(16)	0.5133(4)	0.2021(13)
C(9″)	0.2306(16)	0.5037(5)	0.2988(12)
C(10″)	0.2398(16)	0.4637(5)	0.3282(10)
C(11″)	0.2072(15)	0.4333(4)	0.2610(13)
C(12″)	0.1655(16)	0.4429(5)	0.1643(12)
C(13″)	0.1564(16)	0.4829(5)	0.1349(10)
C(14″)	−0.2612(19)	0.4399(6)	−0.0411(11)
C(15″)	−0.3974(18)	0.4459(6)	−0.0377(12)
C(16″)	−0.4605(13)	0.4513(5)	0.0542(15)
C(17″)	−0.3873(18)	0.4506(6)	0.1427(11)
C(18″)	−0.2511(18)	0.4445(6)	0.1393(12)
C(19″)	−0.1881(13)	0.4392(6)	0.0474(15)

The distribution of the non-ring atoms is indicated in Figure XVE,31. The fivefold coordination of rhodium is that of a trigonal bipyramid with the three phosphorus atoms forming the base. In the triphenylphosphine groups P–C lies between 1.79 and 1.87 A.

XV,e31. Crystals of *tris(o-diphenylarsinophenyl) arsine ruthenium dibromide*, $As_4(C_6H_5)_6(C_6H_4)_3RuBr_2$, are orthorhombic with a tetra-molecular unit of the edge lengths:

$$a_0 = 31.53(8) \text{ A.}; \quad b_0 = 11.40(10) \text{ A.}; \quad c_0 = 13.03(2) \text{ A.}$$

The space group is C_{2v}^9 (*Pna2₁*) with atoms in the positions:

(4a) xyz; $\bar{x},\bar{y},z+^1/_2$; $^1/_2-x,y+^1/_2,z+^1/_2$; $x+^1/_2,^1/_2-y,z$

The parameters are listed in Table XVE,31.

TABLE XVE,31

Parameters of the Atoms in $As_4(C_6H_5)_6(C_6H_4)_3RuBr_2$

Atom	x	y	z
Ru	0.1202	0.1718	0.0008
Br(1)	0.0599	0.0213	0.0393
As(1)	0.0721	0.3196	0.0337
As(2)	0.1711	0.3238	−0.0305
Br(2)	0.1747	0.0055	−0.0405
As(3)	0.1269	0.1712	0.1893
As(4)	0.0863	0.1753	−0.1704
C(1)	0.098	0.472	0.013
C(2)	0.141	0.469	−0.015
C(3)	0.163	0.580	−0.016
C(4)	0.139	0.692	−0.006
C(5)	0.097	0.686	0.016
C(6)	0.076	0.579	0.037
C(7)	0.023	0.321	−0.059
C(8)	0.030	0.238	−0.140
C(9)	−0.003	0.211	−0.211
C(10)	−0.043	0.287	−0.202
C(11)	−0.047	0.362	−0.121
C(12)	−0.015	0.384	−0.051
C(13)	0.050	0.310	0.173
C(14)	0.071	0.231	0.236
C(15)	0.057	0.205	0.337
C(16)	0.020	0.264	0.373
C(17)	−0.002	0.346	0.308
C(18)	0.012	0.368	0.210

(continued)

TABLE XVE,31 (*continued*)

Parameters of the Atoms in $As_4(C_6H_5)_6(C_6H_4)_3RuBr_2$

Atom	x	y	z
C(19)	0.226	0.331	0.054
C(20)	0.247	0.227	0.052
C(21)	0.287	0.231	0.112
C(22)	0.299	0.336	0.158
C(23)	0.275	0.431	0.162
C(24)	0.236	0.438	0.105
C(25)	0.200	0.335	−0.163
C(26)	0.214	0.225	−0.206
C(27)	0.243	0.236	−0.288
C(28)	0.254	0.334	−0.345
C(29)	0.238	0.431	−0.302
C(30)	0.209	0.442	−0.217
C(31)	0.162	0.281	0.266
C(32)	0.147	0.396	0.269
C(33)	0.174	0.482	0.326
C(34)	0.213	0.444	0.369
C(35)	0.227	0.328	0.362
C(36)	0.201	0.246	0.303
C(37)	0.137	0.034	0.270
C(38)	0.143	0.043	0.377
C(39)	0.153	−0.069	0.432
C(40)	0.156	−0.169	0.381
C(41)	0.147	−0.180	0.272
C(42)	0.139	−0.074	0.212
C(43)	0.079	0.038	−0.256
C(44)	0.090	−0.066	−0.215
C(45)	0.082	−0.165	−0.272
C(46)	0.068	−0.170	−0.371
C(47)	0.055	−0.060	−0.415
C(48)	0.062	0.043	−0.356
C(49)	0.102	0.293	−0.277
C(50)	0.086	0.401	−0.264
C(51)	0.101	0.485	−0.342
C(52)	0.124	0.461	−0.424
C(53)	0.139	0.341	−0.434
C(54)	0.129	0.259	−0.360

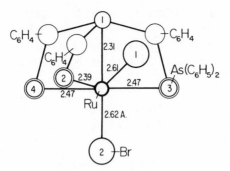

Fig. XVE,32.   Some bond dimensions in the molecule of tris(*o*-diphenylarsino-
phenyl) arsine ruthenium dibromide.

The molecules in the resulting structure have the bond dimensions
indicated in Figure XVE,32. Coordination about the ruthenium atom
is octahedral and around the arsenic atoms, tetrahedral. The atoms
Ru, As(1), As(2), Br(1), and Br(2) are coplanar, and this plane forms
an approximate mirror plane for the molecule. The distances between
arsenic and the attached benzene ring carbons lie between 1.90 and
2.05 A.

XV,e32.   A structure has been described for crystals of *trihydrido
bis(triphenylphosphine)ethylene   bis(diphenylphosphine)   rhenium(III)*,
$[P(C_6H_5)_3]_2[P(C_6H_5)_2CH_2—]_2ReH_3$. The symmetry is monoclinic, with
a tetramolecular unit having the dimensions:

$$a_0 = 18.75(5) \text{ A.}; \quad b_0 = 14.71(3) \text{ A.}; \quad c_0 = 20.03(5) \text{ A.}$$
$$\beta = 91°15(15)'$$

Atoms are in the positions

$$(4e) \quad \pm (xyz; x, 1/2 - y, z + 1/2)$$

of the space group C_{2h}^5 ($P2_1/c$). Parameters for the rhenium and phosphorus atoms are listed below; those for the numerous carbon atoms
are given in the original article.

Atom	x	y	z
Re	0.2338(1)	−0.0323(1)	0.2422(1)
P(1)	0.2558(3)	−0.1076(4)	0.3453(3)
P(2)	0.3419(2)	−0.0602(4)	0.1841(3)
P(3)	0.1716(3)	0.0179(4)	0.1455(3)
P(4)	0.1933(2)	0.1087(4)	0.2830(3)

In this structure coordination about the rhenium atom is considered to be sevenfold, the distribution of the four phosphorus atoms being distortedly tetrahedral, with Re–P = 2.36–2.39 A. Possible positions for the hydrogen atoms are discussed. For the phenyl groups attached to phosphorus P-C = 1.86–1.91 A.

XV,e33. Crystals of μ_4-*oxo-hexa-μ-chloro-tetrakis*[(*triphenylphosphine oxide) copper(II)*], $Cu_4OCl_6[(C_6H_5)_3PO]_4$, are cubic with a unimolecular cell of the edge length:

$$a_0 = 12.22(2) \text{ A.}$$

The space group is T_d^1 ($P\bar{4}3m$) with atoms in the positions:

O(1) : (1a)	000		
Cu : (4e)	uuu; $u\bar{u}\bar{u}$; $\bar{u}u\bar{u}$; $\bar{u}\bar{u}u$	with $u = 0.0900(3)$	
O(2) : (4e)	with $u = 0.179(2)$		
P : (4e)	with $u = 0.2508(7)$		
Cl : (6f)	$\pm(u00; 0u0; 00u)$	with $u = 0.237(1)$	
C(2) : (12i)	uuv; $\bar{u}u\bar{v}$; $u\bar{u}\bar{v}$; $\bar{u}\bar{u}v$; tr		
	with $u = 0.165(3)$, $v = 0.437(4)$		
C(3) : (12i)	with $u = 0.146(3)$, $v = 0.548(4)$		

The remaining carbon atoms are thought to be in some of the general positions

$$(24j) \quad xyz; \; xzy; \; x\bar{y}\bar{z}; \; x\bar{z}\bar{y}; \; \bar{x}\bar{y}z; \; \bar{x}z\bar{y}; \; \bar{x}\bar{y}z; \; \bar{x}\bar{z}y; \quad \text{tr}$$

with the following parameters:

Atom	x	y	z
C(1)	0.248(3)	0.207(3)	0.394(3)
C(4)	0.236(4)	0.140(4)	0.610(4)
C(5)	0.320(7)	0.182(6)	0.566(7)
C(6)	0.345(5)	0.211(5)	0.459(5)

It is a molecular structure resembling that of $Be_4O(CH_3COO)_6$ [XIV,b77]. The central oxygen atom is tetrahedrally surrounded by four metal atoms, and beyond is an octahedron of chlorine atoms also bound to the metal. In the structure being discussed here Cu–O(1) = 1.905 A. and Cu–Cl = 2.38 A. The presence of the $OP(C_6H_5)_3$ groups makes the coordination of copper fourfold, with Cu–O(2) = 1.89 A. In the attached groups O(2)–P = 1.51 A. and P–C(1) = 1.83 A. The disorder attributed to the carbon atoms in general positions permits two orientations for the phenyl radicals turned 27° in either direction about the P–C bond.

F. DERIVATIVES OF NAPHTHALENE

XV,f1. *Naphthalene*, $C_{10}H_8$, has been repeatedly studied with x-rays. Its crystals are monoclinic with a bimolecular unit of the dimensions:

$$a_0 = 8.235 \text{ A.;} \quad b_0 = 6.003 \text{ A.;} \quad c_0 = 8.658 \text{ A.;} \quad \beta = 122°55'$$

The space group is C_{2h}^5 ($P2_1/a$) and all atoms are in the positions:

$$(4e) \quad \pm(xyz; x+{}^1/_2, {}^1/_2-y, z)$$

The most recently redetermined parameters have values much like the original ones but more precise; they are as follows:

Atom	x	y	z
C(1)	0.0856	0.0186	0.3251
C(2)	0.1148	0.1588	0.2200
C(3)	0.0472	0.1025	0.0351
C(4)	0.0749	0.2471	−0.0784
C(5)	0.0116	0.1869	−0.2541

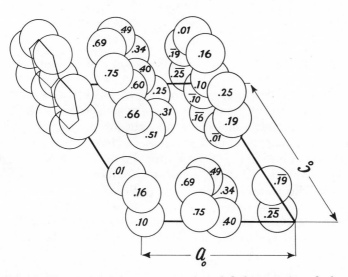

Fig. XVF,1a. The monoclinic structure of naphthalene projected along its b_0 axis. Left-hand axes.

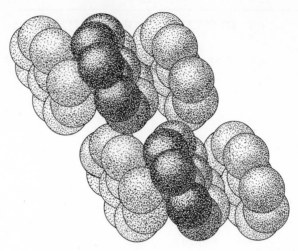

Fig. XVF,1b.   A packing drawing of the monoclinic naphthalene arrangement
viewed along its b_0 axis. Left-hand axes.

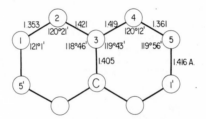

Fig. XVF,2.   Bond dimensions in the molecule of naphthalene.

The resulting structure, shown in Figure XVF,1, has planar molecules arranged nearly parallel to the c_0 axis. The bond dimensions in
these centrosymmetric molecules are indicated in Figure XVF,2.

Naphthalene is typical of a group of condensed ring hydrocarbons,
all with the same type of structure. Some others are anthracene
(**XV,g1**), coronene (**XV,h22**), ovalene (**XV,h26**), biphenyl (**XV,bI,1**),
terphenyl (**XV,c1**), and quaterphenyl (**XV,h23**).

XV,f2.   Crystals of the complex *naphthalene-tetracyanoethylene*,
$C_{10}H_8 \cdot C_2(CN)_4$, are monoclinic with a tetramolecular unit of the
dimensions:

$$a_0 = 7.26(1) \text{ A.;} \quad b_0 = 12.69(1) \text{ A.;} \quad c_0 = 7.21(1) \text{ A.}$$
$$\beta = 94.4(2)°$$

The space group is C_{2h}^3 ($C2/m$) with atoms in the positions:

C(1) : (4h) $\pm (0\ u\ {}^1/_2;\ {}^1/_2,u+{}^1/_2,{}^1/_2)$ with $u = 0.0532(5)$
C(3) : (4g) $\pm (0u0;\ {}^1/_2,u+{}^1/_2,0)$ with $u = 0.0552(6)$

The other atoms are in

(8j) $\pm (xyz;\ x\bar{y}z;\ x+{}^1/_2,y+{}^1/_2,z;\ x+{}^1/_2,{}^1/_2-y,z)$

with the following parameters:

Atom	x	y	z
C(2)	0.1547(8)	0.1109(4)	0.5806(8)
C(4)	0.1591(14)	0.1079(7)	0.0840(10)
C(5)	0.2993(10)	0.0474(15)	0.1594(10)
N	0.2791(7)	0.1548(4)	0.6452(8)

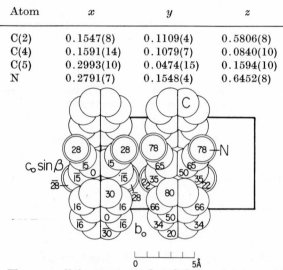

Fig. XVF,3.   The monoclinic structure of naphthalene-tetracyanoethylene projected along its a_0 axis.

The structure (Fig. XVF,3) is built up of alternate naphthalene and tetracyanoethylene molecules stacked along the c_0 axis. These molecules, both of which are planar, have the bond dimensions of Figure XVF,4. Between molecules the shortest interatomic distances are C(1)–C(4) = 3.37 A., C(3)–C(2) = 3.38 A., C(5)–C(2) = 3.39 A., and C(5)–C(1) = C(4)–N = 3.40 A., all within a stack. Between stacks the shortest atomic separation is C(5)–C(5) = 3.54 A.

XV,f3.   The 1 : 1 complex *naphthalene-1,2,4,5-tetracyanobenzene*, $C_{10}H_8 \cdot C_6H_2(CN)_4$, forms monoclinic crystals whose bimolecular unit has the dimensions:

$$a_0 = 9.39(1)\ \text{A.};\quad b_0 = 12.66(2)\ \text{A.};\quad c_0 = 6.87(2)\ \text{A.}$$
$$\beta = 107.2(3)°$$

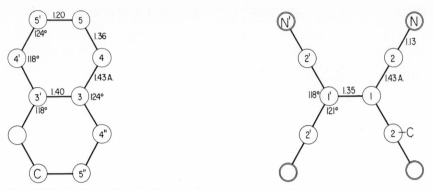

Fig. XVF,4a *(left)*.   Bond dimensions of the naphthalene molecule in naphthalene-tetracyanoethylene.

Fig. XVF,4b *(right)*.   Bond dimensions of the tetracyanoethylene molecule in
naphthalene-tetracyanoethylene.

The space group has been chosen as C_{2h}^3 $(C2/m)$ with atoms in the
positions:

$$C(1) : (4g) \quad \pm (0u0; {}^1/_2, u + {}^1/_2, 0) \qquad \text{with } u = 0.1121(5)$$

All other atoms have the coordinates:

$$(8j) \quad \pm (xyz; x\bar{y}z; x + {}^1/_2, y + {}^1/_2, z; x + {}^1/_2, {}^1/_2 - y, z)$$

The atoms C(2), C(3), and N(4) which, together with C(1) constitute
the $C_6H_2(CN)_4$ molecule, have the parameters:

Atom	x	y	z
C(2)	0.1321(5)	0.0563(4)	0.0564(7)
C(3)	0.2727(6)	0.1125(4)	0.1163(8)
N	0.3810(5)	0.1574(4)	0.1584(9)

The naphthalene molecules are considered to be disordered, their atoms
occupying half the positions of $(8j)$ with the parameters:

Atom	x	y	z
C(5)	0.0214(13)	0.0543(5)	0.5134(21)
C(6)	0.1783(14)	0.0801(10)	0.5822(17)
C(7)	0.0898(14)	0.1363(10)	0.5323(15)
C(8)	0.2382(13)	0.1102(15)	0.5977(20)
C(9)	0.2796(11)	0.0058(27)	0.6240(15)

In the arrangement thus described the two planar molecules are stacked alternately one above the other along c_0. The naphthalene takes one or the other of the two orientations corresponding to the disorder. Expected bond lengths prevail.

XV,f4. The triclinic crystals of *naphthalene-chromium tricarbonyl,* $C_{10}H_8 \cdot Cr(CO)_3$, have a bimolecular unit of the dimensions:

$$a_0 = 12.37(3) \text{ A.;} \quad b_0 = 6.58(2) \text{ A.;} \quad c_0 = 7.36(2) \text{ A.}$$
$$\alpha = 107°22(8)'; \quad \beta = 96°44(5)'; \quad \gamma = 92°50(4)'$$

Atoms are in the positions $(2i)$ $\pm(xyz)$ of the space group C_i^1 $(P\bar{1})$. Determined parameters are those of Table XVF,1; values for the hydrogen atoms are stated in the original article.

TABLE XVF,1

Parameters of the Atoms in $C_{10}H_8 \cdot Cr(CO)_3$

Atom	x	y	z
Cr	0.3724(2)	0.1895(2)	0.6254(4)
O(1)	0.1819(10)	0.2007(15)	0.8153(22)
O(2)	0.4326(13)	0.0164(14)	0.9197(24)
O(3)	0.4480(11)	0.3645(13)	0.9049(20)
C(1)	0.3602(15)	0.4270(16)	0.3360(29)
C(2)	0.2700(18)	0.4668(19)	0.2926(29)
C(3)	0.1953(15)	0.3894(25)	0.2681(31)
C(4)	0.2042(14)	0.2693(19)	0.2822(30)
C(5)	0.2956(14)	0.2280(16)	0.3317(27)
C(6)	0.3131(14)	0.1110(16)	0.3595(37)
C(7)	0.4057(14)	0.0706(16)	0.3854(38)
C(8)	0.4830(15)	0.1423(19)	0.4066(29)
C(9)	0.4680(12)	0.2588(16)	0.3940(31)
C(10)	0.3752(13)	0.3056(15)	0.3450(17)
C(11)	0.2571(13)	0.1961(19)	0.7481(21)
C(12)	0.4137(15)	0.0844(18)	0.8063(29)
C(13)	0.4197(13)	0.2987(17)	0.7894(24)

The structure is shown in Figure XVF,5. The chromium atom is centered over one of the benzene rings, with Cr–C(5–10) = 2.19–2.34 A. (Fig. XVF,6). Between molecules the shortest interatomic distances are C–O = 3.34 and 3.48 A. and C–C = 3.53 A.

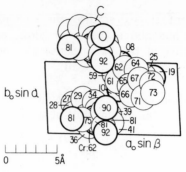

Fig. XVF,5. The triclinic structure of naphthalene-chromium tricarbonyl projected along its c_0 axis. The large doubly ringed chromium barely shows.

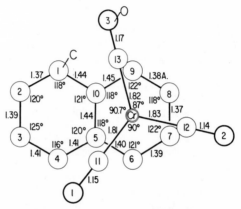

Fig. XVF, 6.   Bond dimensions in naphthalene-chromium tricarbonyl.

XV,f5.   A recent redetermination of the structure of α-*naphthol*, $C_{10}H_7OH$, has led to an arrangement showing important differences from that initially proposed. The crystals are monoclinic with a tetra-molecular unit of the dimensions:

$$a_0 = 13.20(15) \text{ A.}; \quad b_0 = 4.78(2) \text{ A.}; \quad c_0 = 13.20(15) \text{ A.}$$
$$\beta = 117°3(9)'$$

All atoms are in the positions

$$(4e) \quad \pm (xyz; x + {}^1/_2, {}^1/_2 - y, z)$$

of the space group C_{2h}^5 $(P2_1/a)$. The parameters are those of Table XVF,2.

TABLE XVF,2

Parameters of the Atoms in α-Naphthol

Atom	x	y	z
O	0.2747	0.277	0.4611
C(1)	0.2258	0.296	0.3427
C(2)	0.1307	0.129	0.2750
C(3)	0.0836	0.181	0.1544
C(4)	0.1320	0.359	0.1083
C(5)	0.2833	0.696	0.1365
C(6)	0.3784	0.866	0.2071
C(7)	0.4227	0.833	0.3253
C(8)	0.3755	0.634	0.3725
C(9)	0.2762	0.485	0.3002
C(10)	0.2286	0.503	0.1818
H(2)	0.092	0.975	0.312
H(3)	0.007˙	0.050	0.096
H(4)	0.094	0.395	0.017
H(5)	0.274	0.725	0.045
H(6)	0.418	0.025	0.173
H(7)	0.497	0.957	0.380
H(8)	0.410	0.605	0.465

The resulting structure is indicated in Figure XVF,7. Bond dimensions in the molecules, though not known with great accuracy, appear normal. The molecules themselves are held together in the crystals by hydrogen bonds of length O–H–O = 2.79 A.

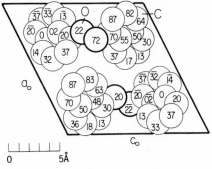

Fig. XVF,7.   The monoclinic structure of α-naphthol projected along its b_0 axis.

XV,f6. Crystals of β-*naphthol*, $C_{10}H_7OH$, are monoclinic with a unit that contains eight molecules and has the dimensions:

$$a_0 = 8.185(15) \text{ A.}; \quad b_0 = 5.950(3) \text{ A.}; \quad c_0 = 36.29(1) \text{ A.}$$
$$\beta = 119°52(7)'$$

The space group is the low symmetry C_s^4 in the uncommon orientation *Ia*. Atoms are thus in the positions:

$$(4a) \quad xyz; x + {}^1/_2, \bar{y}, z; \quad \text{B.C.}$$

The approximate parameters are given in Table XVF,3.

TABLE XVF,3

Parameters of the Atoms in β-Naphthol

Atom	x	y	z
O(1)	0.379	0.212	0.238
O(2)	0.131	0.131	0.269
C(1)	0.169	0.255	0.046
C(2)	0.253	0.054	0.059
C(3)	0.308	0.000	0.102
C(4)	0.279	0.146	0.128
C(5)	0.352	0.975	0.171
C(6)	0.315	0.247	0.195
C(7)	0.231	0.454	0.179
C(8)	0.166	0.494	0.139
C(9)	0.200	0.356	0.113
C(10)	0.131	0.403	0.069
C(11)	0.187	0.212	0.309
C(12)	0.150	0.081	0.336
C(13)	0.217	0.146	0.377
C(14)	0.180	0.001	0.407
C(15)	0.248	0.072	0.447
C(16)	0.342	0.294	0.462
C(17)	0.393	0.430	0.438
C(18)	0.319	0.360	0.393
C(19)	0.357	0.481	0.366
C(20)	0.291	0.418	0.324

The structure (Fig. XVF,8) has its molecules arranged in sheets. Within a sheet, they are tied together by O–H–O bonds of lengths

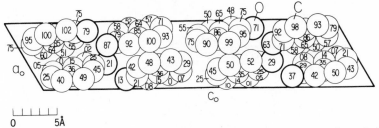

Fig. XVF,8.   The elongated monoclinic structure of β-naphthol projected along its b_0 axis.

2.72 and 2.79 A. The chains thus produced are held together by van der Waals forces only.

XV,f7.   The monoclinic crystals of *1-naphthoic acid*, $C_{10}H_7COOH$, possess a tetramolecular unit of the dimensions:

$$a_0 = 31.12(10) \text{ A.}; \quad b_0 = 3.87(1) \text{ A.}; \quad c_0 = 6.92(2) \text{ A.}$$
$$\beta = 92°12(12)'$$

Atoms are in the positions

$$(4e) \quad \pm(xyz; x+1/2, 1/2-y, z)$$

of C_{2h}^5 ($P2_1/a$). The parameters are listed in Table XVF,4.

TABLE XVF,4

Parameters of the Atoms in 1-Naphthoic Acid

Atom	x	y	z
C(1)	0.0800	−0.176	−0.3081
C(2)	0.0677	−0.335	−0.4902
C(3)	0.0979	−0.448	−0.6335
C(4)	0.1420	−0.433	−0.5893
C(5)	0.1999	−0.286	−0.3749
C(6)	0.2135	−0.134	−0.2095
C(7)	0.1846	−0.045	−0.0607
C(8)	0.1393	−0.050	−0.0962
C(9)	0.1261	−0.167	−0.2731
C(10)	0.1544	−0.306	−0.4095
C(11)	0.0478	−0.101	−0.1783
O(1)	0.0515	0.072	−0.0257
O(2)	0.0082	−0.176	−0.2159

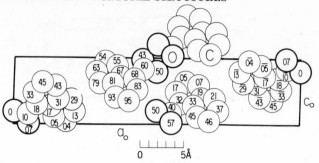

Fig. XVF,9.   The monoclinic structure of 1-naphthoic acid projected along its b_0 axis.

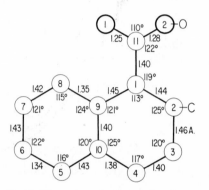

Fig. XVF,10.   Bond dimensions in the molecule of 1-naphthoic acid.

The structure (Fig. XVF,9) is composed of molecules having the bond dimensions of Figure XVF,10. These molecules are planar except for their oxygens. The carboxyl groups are twisted about the C(1)–C(11) bond so that the oxygens are 0.20 A. above and below the plane of the rest. In the structure they are tied together into pairs by O–H–O bonds of length 2.58 A. The shortest atomic distances between such pairs are C–O = 3.54 and 3.68 A. and C–C = 3.81 and 3.87 A.

XV,f8.   Crystals of *2-naphthoic acid*, $C_{10}H_7COOH$, have monoclinic symmetry. Their unit is tetramolecular and has the dimensions:

$$a_0 = 30.59(10) \text{ A.}; \quad b_0 = 5.00(2) \text{ A.}; \quad c_0 = 5.63(2) \text{ A.}$$
$$\beta = 92.6(2)°$$

The space group is C_{2h}^5 ($P2_1/n$). All atoms are in the general positions

$$(4e) \quad \pm(xyz; x+1/2, 1/2-y, z+1/2)$$

with the parameters as listed in Table XVF,5.

TABLE XVF,5

Parameters of the Atoms in 2-Naphthoic Acid

Atom	x	y	z
C(1)	0.1157	−0.5204	0.0266
C(2)	0.0771	−0.4804	−0.1261
C(3)	0.0714	−0.6468	−0.3321
C(4)	0.1023	−0.8488	−0.3819
C(5)	0.1717	−0.0872	−0.2796
C(6)	0.2065	−0.1272	−0.1261
C(7)	0.2139	−0.9612	0.0764
C(8)	0.1845	−0.7564	0.1364
C(9)	0.1461	−0.7196	−0.0266
C(10)	0.1400	−0.8856	−0.2242
C(11)	0.0453	−0.2832	−0.0654
O(1)	0.0084	−0.2424	−0.1961
O(2)	0.0486	−0.1164	0.1265

The structure is that of Figure XVF,11. In it the molecules occur in pairs held together by short hydrogen bonds of length 2.54 A. between carboxyl oxygens. The bond dimensions are those of Figure XVF,12, departures from the mean plane through the molecules not exceeding 0.06 A. Between molecules other short separations are C–O = 3.42 A. and C–C = 3.72 A.

The close similarity in structure between this compound and the 1-naphthoic acid is apparent from a comparison of Figures XVF,11 and XVF,9.

XV,f9. The orthorhombic crystals of *1-aminonaphthalene-chromium tricarbonyl*, $C_{10}H_7NH_2 \cdot Cr(CO)_3$, possess a unit that contains eight

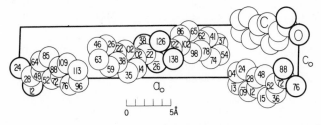

Fig. XVF,11.   The monoclinic structure of 2-naphthoic acid projected along its b_0 axis.

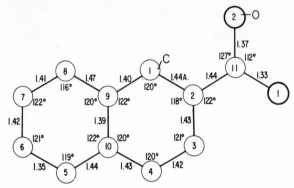

Fig. XVF,12.   Bond dimensions in the molecule of 2-naphthoic acid.

molecules and has the edge lengths:

$$a_0 = 14.88 \text{ A.;} \quad b_0 = 11.43 \text{ A.;} \quad c_0 = 13.96 \text{ A.}$$

The space group is V_h^{15} (*Pbca*) with atoms in the positions:

$$(8c) \quad \pm(xyz; \, 1/2-x, y+1/2, z; \, x, 1/2-y, z+1/2; \, x+1/2, y, 1/2-z)$$

The parameters are stated in Table XVF,6.

TABLE XVF,6

Parameters of the Atoms in $C_{10}H_7NH_2 \cdot Cr(CO)_3$

Atom	x	y	z
C(1)	0.4364	0.6716	0.4273
C(2)	0.5246	0.6967	0.4345
C(3)	0.5927	0.6223	0.4053
C(4)	0.5832	0.5158	0.3711
C(5)	0.4671	0.3739	0.3159
C(6)	0.3778	0.3297	0.3029
C(7)	0.3038	0.4010	0.3274
C(8)	0.3204	0.5157	0.3636
C(9)	0.4102	0.5577	0.3833
C(10)	0.4842	0.4846	0.3580
N(11)	0.3644	0.7295	0.4448
C(12)	0.4889	0.5266	0.1497
O(13)	0.5520	0.5420	0.1087
C(14)	0.3405	0.4405	0.1239
O(15)	0.3001	0.4108	0.0544
C(16)	0.3476	0.6317	0.1966
O(17)	0.3154	0.7299	0.1756
Cr	0.3912	0.5010	0.2269

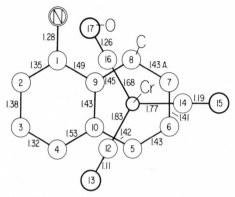

Fig. XVF,13.   Bond lengths in 1-aminonaphthalene-chromium tricarbonyl.

In this structure the $Cr(CO)_3$ trigonal pyramid is centered over the unsubstituted ring of naphthalene. Bond dimensions are those of Figure XVF,13. Distances from chromium to the ring carbon atoms beneath it are 2.19–2.30 A. In $Cr(CO)_3$, C–Cr–C = 86–91° and Cr–C–O = 172–179°.

XV,f10. Crystals of *N-α-naphthyl-1,2,3,6-tetrahydrophthalamic acid*, $C_{10}H_7(NH)(CO)C_6H_4COOH$, are monoclinic. Their tetramolecular unit has the dimensions

$$a_0 = 17.43(3) \text{ A.}; \quad b_0 = 4.97(2) \text{ A.}; \quad c_0 = 22.90(4) \text{ A.}$$
$$\beta = 132.0(3)°$$

All atoms are in the positions

$$(4e) \quad \pm(xyz; x, 1/2 - y, z + 1/2)$$

of the space group C_{2h}^5 $(P2_1/c)$. The chosen parameters are given in Table XVF,7.

TABLE XVF,7

Parameters of the Atoms in N-α-Naphthyl-1,2,3,6-tetrahydrophthalamic Acid

Atom	x	y	z
C(1)	0.2325	−0.2327	0.2773
C(2)	0.1631	−0.3525	0.2809
C(3)	0.0527	−0.2994	0.2181
C(4)	0.0193	−0.1228	0.1597
C(5)	0.0892	−0.0007	0.1564

(continued)

TABLE XVF,7 (*continued*)

Parameters of the Atoms in *N*-α-Naphthyl-1,2,3,6-tetrahydrophthalamic Acid

Atom	x	y	z
C(6)	0.0545	0.1838	0.0952
C(7)	0.1208	0.3083	0.0907
C(8)	0.2305	0.2514	0.1500
C(9)	0.2652	0.0738	0.2105
C(10)	0.1985	−0.0529	0.2162
C(11)	0.4482	0.2090	0.3063
C(12)	0.5601	0.0968	0.3697
C(13)	0.6359	0.3158	0.3864
C(14)	0.7473	0.2594	0.4617
C(15)	0.7718	0.0916	0.5172
C(16)	0.6961	−0.0720	0.5134
C(17)	0.5799	0.0087	0.4455
C(18)	0.5478	0.2275	0.4712
O(1)	0.4322	0.4502	0.2953
O(2)	0.4472	0.2258	0.4325
O(3)	0.6075	0.3919	0.5235
N	0.3743	0.0143	0.2686

The structure, shown in Figure XVF,14, is composed of molecules with the bond dimensions of Figure XVF,15. These molecules are tied into pairs by hydrogen bonds of length 2.61 A. between carboxyl oxygens; there are also short N–O = 2.76 A. involving the N and O(1) atoms.

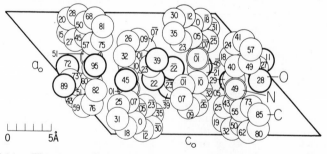

Fig. XVF,14.　The monoclinic structure of *N*-α-naphthyl-1,2,3,6-tetrahydro-phthalamic acid projected along its b_0 axis.

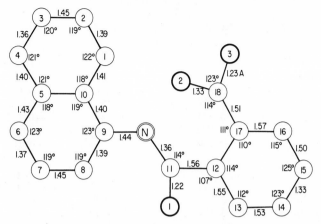

Fig. XVF,15.   Bond dimensions in the molecule of N-α-naphthyl-1,2,3,6-tetra-
hydrophthalamic acid.

A formally similar structure has been described for the analogous
compound α-*naphthyl-4-chlorophthalamic acid*, $C_{10}H_7(NH)(CO)C_6H_3$-
ClCOOH. Its tetramolecular monoclinic unit has the dimensions:

$$a_0 = 13.60(3)\ A.; \quad b_0 = 4.80(2)\ A.; \quad c_0 = 25.20(5)\ A.$$
$$\beta = 115.0(3)°$$

The parameters of the atoms, in the general positions of C_{2h}^5 ($P2_1/c$),
are those of Table XVF,8.

The resulting structure is shown in Figure XVF,16.

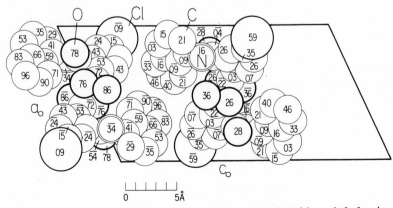

Fig. XVF,16.   The monoclinic structure of α-naphthyl-4-chlorophthalamic acid
projected along its b_0 axis.

TABLE XVF,8

Parameters of the Atoms in α-Naphthyl-4-chlorophthalamic Acid

Atom	x	y	z
C(1)	0.321	−0.530	−0.086
C(2)	0.326	−0.433	−0.137
C(3)	0.251	−0.243	−0.173
C(4)	0.172	−0.146	−0.157
C(5)	0.166	−0.236	−0.107
C(6)	0.239	−0.427	−0.071
C(7)	0.221	−0.539	−0.021
C(8)	0.402	−0.723	−0.048
C(9)	0.197	−0.413	0.066
C(10)	0.106	−0.288	0.065
C(11)	0.066	−0.348	0.108
C(12)	0.125	−0.526	0.153
C(13)	0.221	−0.656	0.156
C(14)	0.281	−0.834	0.202
C(15)	0.374	−0.965	0.203
C(16)	0.414	−0.895	0.161
C(17)	0.358	−0.715	0.118
C(18)	0.261	−0.592	0.113
O(1)	0.198	−0.785	−0.018
O(2)	0.424	−0.757	0.003
O(3)	0.455	−0.859	−0.074
N	0.230	−0.345	0.020
Cl	0.080	0.094	−0.202

XV,f11. Crystals of *1,5-dimethylnaphthalene*, $C_{10}H_6(CH_3)_2$, are monoclinic with a tetramolecular unit of the dimensions:

$$a_0 = 6.18(1) \text{ A.}; \quad b_0 = 8.91(1) \text{ A.}; \quad c_0 = 16.77(2) \text{ A.}$$
$$\beta = 101°24(3)'$$

The space group is C_{2h}^5 ($P2_1/c$) with atoms in the positions:

$$(4e) \quad \pm(xyz; x,1/2-y,z+1/2)$$

The determined parameters are those of Table XVF,9.

TABLE XVF,9

Parameters of the Atoms in $C_{10}H_6(CH_3)_2$

Atom	x	y	z
C(1)	0.3758	0.4054	0.3264
C(2)	0.2398	0.3254	0.3665
C(3)	0.0523	0.2449	0.3239
C(4)	0.0024	0.2432	0.2399
C(5)	0.0898	0.3275	0.1090
C(6)	0.2216	0.4070	0.0688
C(7)	0.4082	0.4878	0.1106
C(8)	0.4593	0.4884	0.1951
C(9)	0.3209	0.4056	0.2385
C(10)	0.1434	0.3256	0.1963
$CH_3(1')$	0.5676	0.4926	0.3728
$CH_3(5')$	−0.1049	0.2381	0.0614
H(2)	0.275	0.323	0.433
H(3)	−0.052	0.183	0.358
H(4)	−0.140	0.181	0.207
H(6)	0.184	0.409	0.003
H(7)	0.514	0.549	0.077
H(8)	0.602	0.550	0.228
$H[CH_3(1),1]$	0.576	0.479	0.438
$H[CH_3(1),2]$	0.546	0.611	0.357
$H[CH_3(1),3]$	0.720	0.452	0.357
$H[CH_3(5),1]$	−0.112	0.252	−0.004
$H[CH_3(5),2]$	−0.258	0.279	0.077
$H[CH_3(5),3]$	−0.084	0.119	0.077

Note: For the atoms of carbon, $\sigma(x) = 0.0014$, $\sigma(y) = 0.0006$, $\sigma(z) = 0.0003$ if the molecule is taken to be centrosymmetric.

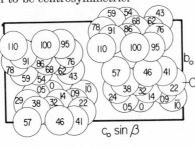

Fig. XVF,17.   The monoclinic structure of 1,5-dimethylnaphthalene projected along its a_0 axis.

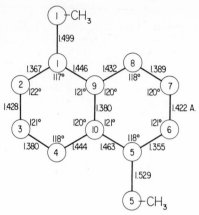

Fig. XVF,18. Bond dimensions in the molecule of 1,5-dimethylnaphthalene.

The structure is shown in Figure XVF,17. Its molecule has the bond dimensions of Figure XVF,18 and is planar, no carbon atom departing by as much as 0.02 A. from its best plane. Between molecules the shortest C–H = 2.86 A. and the shortest H–H = 2.49 A.

XV,f12. The bimolecular, monoclinic unit of *2,6-dichloronaphthalene*, $C_{10}H_6Cl_2$, has the dimensions:

$$a_0 = 5.98(2) \text{ A.}; \quad b_0 = 4.94(1) \text{ A.}; \quad c_0 = 14.95(3) \text{ A.}$$
$$\beta = 98°36(10)'$$

All atoms are in the positions

$$(4e) \quad \pm (xyz; x, 1/2 - y, z + 1/2)$$

of C_{2h}^5 ($P2_1/c$). The stated parameters are those of Table XVF,10.

TABLE XVF,10

Parameters of the Atoms in 2,6-Dichloronaphthalene

Atom	x	y	z
Cl	0.3768	0.5902	0.1706
C(1)	0.247	0.247	0.031
C(2)	0.190	0.351	0.108
C(3)	0.000	0.260	0.143
C(4)	−0.146	0.088	0.098
C(9)	0.096	0.043	−0.019

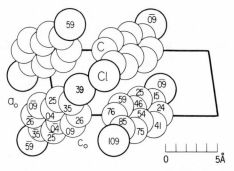

Fig. XVF,19. The monoclinic structure of 2,6-dichloronaphthalene projected along its b_0 axis.

The structure, illustrated in Figure XVF,19, is built up of molecules that are planar within the limit of accuracy of the determination, with C–Cl = 1.79 A.

XV,f13. Crystals of *1,4-naphthohydroquinone*, $C_{10}H_6(OH)_2$, are orthorhombic with a tetramolecular unit of the edge lengths:

$$a_0 = 12.67(2) \text{ A.}; \quad b_0 = 12.95(2) \text{ A.}; \quad c_0 = 4.80(1) \text{ A.}$$

Atoms are in the positions

$$(8d) \quad \pm (xyz; \, 1/2-x,y+1/2,z+1/2; \, x,1/2-y,z; \, x+1/2,y,1/2-z)$$

of the space group V_h^{16} (*Pnma*). The parameters, including those for hydrogen, are given in Table XVF,11.

TABLE XVF,11

Parameters of the Atoms in $C_{10}H_6(OH)_2$

Atom	x	y	z
C(1)	0.2082(5)	0.1433(4)	−0.1785(12)
C(2)	0.2755(4)	0.1952(4)	−0.3469(11)
C(7)	−0.0012(5)	0.1965(5)	0.3436(14)
C(8)	0.0662(5)	0.1431(6)	0.1765(14)
C(9)	0.1363(4)	0.1958(4)	0.0017(10)
O(1)	0.2072(3)	0.0370(3)	−0.1689(10)
H(2)	0.327(4)	0.161(3)	−0.465(9)
H(7)	−0.055(4)	0.162(3)	0.452(10)
H(8)	0.059(4)	0.070(4)	0.146(11)
H(1)	0.240(4)	0.012(4)	−0.333(10)

The resulting structure, shown in Figure XVF,20, consists of molecules that are planar and possess a plane of symmetry. Their bond dimensions are those of Figure XVF,21. In the crystals these molecules are in stacks tied together along b_0 by two hydrogen bonds involving each hydroxyl (O–H–O = 2.74 A.); the hydrogen atom does not lie along the O–O line but at 11° to it.

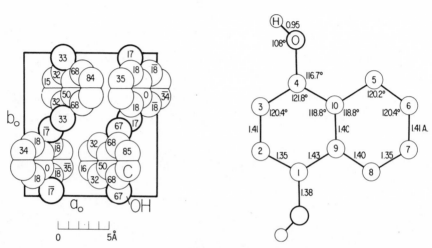

Fig. XVF,20 (*left*).   The orthorhombic structure of 1,4-naphthohydroquinone projected along its c_0 axis.

Fig. XVF,21 (*right*).   Average bond dimensions in the molecules of 1,4-naphthohydroquinone.

TABLE XVF,12

Parameters of the Atoms in 1,5-Dinitronaphthalene

Atom	x	y	z
C(1)	0.117	0.099	−0.017
C(2)	−0.009	0.152	−0.209
C(3)	−0.182	0.121	−0.319
C(4)	−0.213	0.041	−0.256
C(5)	0.087	0.016	0.048
N	0.295	0.136	0.092
O(1)	0.418	0.097	0.048
O(2)	0.307	0.200	0.271

XV,f14. The monoclinic crystals of *1,5-dinitronaphthalene*, $C_{10}H_6$-$(NO_2)_2$, possess a bimolecular unit of the dimensions:

$$a_0 = 7.76(2) \text{ A.}; \quad b_0 = 16.32(4) \text{ A.}; \quad c_0 = 3.70(1) \text{ A.}$$
$$\beta = 101°48(10)'$$

The space group is C_{2h}^5 ($P2_1/a$) with atoms in the positions:

$$(4e) \quad \pm (xyz; x+{}^1\!/_2, {}^1\!/_2-y, z)$$

The parameters determined in the latest study are those of Table XVF, 12.

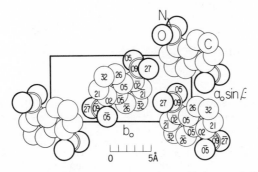

Fig. XVF,22. The monoclinic structure of 1,5-dinitronaphthalene projected along its c_0 axis.

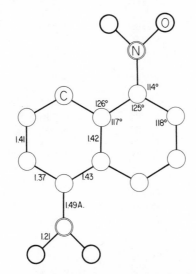

Fig. XVF,23. Some bond dimensions in the molecule of 1,5-dinitronaphthalene.

The structure (Fig. XVF,22) consists of molecules with the bond dimensions of Figure XVF,23. They are centrosymmetric and planar except for the oxygen atoms; planes through the NO_2 radicals make 48.7° with the rest of the molecule. Between molecules the shortest atomic separations are $C-O = 3.23$ and 3.28 A.

XV,f15. The orthorhombic *1,8-dinitronaphthalene*, $C_{10}H_6(NO_2)_2$, has a tetramolecular unit of the edge lengths:

$$a_0 = 11.352(2) \text{ A.}; \quad b_0 = 14.934(2) \text{ A.}; \quad c_0 = 5.376(1) \text{ A.}$$

Atoms are in the positions

$$(4a) \quad xyz; \; 1/2-x,\bar{y},z+1/2; \; x+1/2,1/2-y,\bar{z}; \; \bar{x},y+1/2,1/2-z$$

of the space group V^4 $(P2_12_12_1)$. The parameters are listed in Table XVF,13.

TABLE XVF,13

Parameters of the Atoms in 1,8-Dinitronaphthalene

Atom	x	y	z
C(1)	0.600	0.648	0.119
C(2)	0.660	0.609	0.307
C(3)	0.639	0.517	0.360
C(4)	0.565	0.468	0.223
C(5)	0.425	0.452	−0.126
C(6)	0.365	0.484	−0.319
C(7)	0.370	0.578	−0.373
C(8)	0.439	0.632	−0.227
C(9)	0.518	0.597	−0.037
C(10)	0.504	0.505	0.017
N(1)	0.638	0.737	0.033
N(2)	0.424	0.730	−0.263
O(1)	0.656	0.750	−0.189
O(2)	0.658	0.794	0.193
O(3)	0.408	0.754	−0.480
O(4)	0.420	0.777	−0.074
H(2)	0.721	0.647	0.416
H(3)	0.684	0.486	0.516
H(4)	0.551	0.399	0.270
H(5)	0.413	0.383	−0.075
H(6)	0.313	0.440	−0.434
H(7)	0.321	0.394	0.474

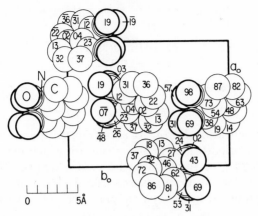

Fig. XVF,24.   The orthorhombic structure of 1,8-dinitronaphthalene projected along its c_0 axis.

The structure is illustrated in Figure XVF,24. Bond dimensions found for its molecules are shown in Figure XVF,25. In these molecules the nitro groups, including their nitrogen atoms, are considerably displaced from the naphthalene plane: thus N(1) is 0.36 A. and N(2) 0.41 A. on either side with the O(1) and O(4) atoms more than an angstrom unit from it. The intermolecular separations, excluding the hydrogen atoms, range upwards from an O–O = 3.21 A.

XV,f16.   The red form (unstable below 117°C.) of the complex *picric acid-1-bromo-2-aminonaphthalene*, $C_6H_2(NO_2)_3OH \cdot C_{10}H_6Br(NH_2)$,

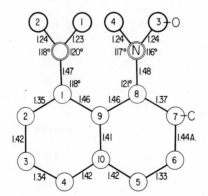

Fig. XVF,25.   Bond dimensions in the molecule of 1,8-dinitronaphthalene.

is monoclinic at room temperature with a tetramolecular unit of the dimensions:

$$a_0 = 14.24(3) \text{ A.}; \quad b_0 = 16.99(3) \text{ A.}; \quad c_0 = 7.01(1) \text{ A.}$$
$$\beta = 96.6(1)°$$

Atoms are in the positions

$$(4e) \quad \pm(xyz; x+{}^1/_2,{}^1/_2-y,z+{}^1/_2)$$

of C_{2h}^5 in the orientation $P2_1/n$. It is considered that there is disorder in the molecular orientations in this structure which can be expressed by saying that 83% of them have the parameters of Table XVF,14. In the alternative orientation, possessed by the remaining 17%, atomic parameters (unrefined) have been assigned the values of the B section of this table.

TABLE XVF,14

Parameters of the Atoms in the Red Form of Picric Acid-1-bromo-2-aminonaphthalene

Atom	x	y	z
A. Principal position (83%)			
Br	0.1361	0.1993	0.0106
C(1)	0.2419	0.1310	0.0401
C(2)	0.3277	0.1557	0.0194
C(3)	0.4050	0.1090	0.0740
C(4)	0.3954	0.0308	0.1233
C(5)	0.2120	0.4212	0.3116
C(6)	0.2956	0.3935	0.2913
C(7)	0.3770	0.4420	0.3240
C(8)	0.1287	0.0181	0.1240
C(9)	0.2224	0.0497	0.1011
C(10)	0.1962	0.4976	0.3593
C(11)	0.1689	0.0599	0.5918
C(12)	0.1745	0.1362	0.5398
C(13)	0.2571	0.1734	0.5218
C(14)	0.3406	0.1280	0.5525
C(15)	0.3411	0.0499	0.6010
C(16)	0.2539	0.0164	0.6207
N(1)	0.3413	0.2334	0.9604
N(2)	0.0892	0.1838	0.5052
N(3)	0.4282	0.1681	0.5369

(continued)

TABLE XVF,14 (*continued*)

Parameters of the Atoms in the Red Form of Picric Acid-1-bromo-2-aminonaphthalene

Atom	x	y	z
N(4)	0.2454	0.4307	0.8238
O(1)	0.0882	0.0244	0.6174
O(2)	0.0175	0.1522	0.4784
O(3)	0.0988	0.2515	0.5033
O(4)	0.4266	0.2314	0.4622
O(5)	0.0000	0.3690	0.0865
O(6)	0.1730	0.3973	0.8045
O(7)	0.3194	0.4032	0.7981
B. Alternate position (17%)			
Br	0.4478	0.1027	0.0744
O(1)	0.421	0.014	0.635
C(1)	0.307	0.108	0.062
C(2)	0.269	0.178	0.002
C(3)	0.165	0.182	0.027
C(4)	0.114	0.117	0.068
C(5)	0.390	0.474	0.350
C(6)	0.336	0.404	0.310
C(7)	0.231	0.407	0.303
C(8)	0.184	0.467	0.333
C(9)	0.268	0.034	0.123
C(10)	0.162	0.039	0.114

XV,f17. Crystals of *2-methyl-4-amino-1-naphthol hydrochloride* (vitamin K_5), $C_{10}H_5(CH_3)(NH_2)OHCl$, are orthorhombic. Their tetra-molecular unit possesses the edge lengths:

$$a_0 = 20.55(2) \text{ A.}; \quad b_0 = 10.60(2) \text{ A.}; \quad c_0 = 4.84(2) \text{ A.}$$

The space group is V^4 ($P2_12_12_1$) and accordingly all atoms are in the positions:

$$(4a) \quad xyz; \; {}^1/_2 - x, \bar{y}, z + {}^1/_2; \; x + {}^1/_2, {}^1/_2 - y, \bar{z}; \; \bar{x}, y + {}^1/_2, {}^1/_2 - z$$

The parameters as given in a preliminary note are those of Table XVF,15.

TABLE XVF,15

Parameters of the Atoms in $C_{10}H_5(CH_3)(NH_2)OHCl$

Atom	x	y	z
C(1)	0.474	0.344	0.064
C(2)	0.486	0.255	−0.128
C(3)	0.434	0.170	−0.196
C(4)	0.377	0.180	−0.074
C(5)	0.301	0.284	0.262
C(6)	0.291	0.375	0.457
C(7)	0.342	0.462	0.527
C(8)	0.401	0.451	0.394
C(9)	0.413	0.358	0.198
C(10)	0.362	0.271	0.130
C(11)	0.551	0.244	−0.286
O	0.520	0.431	0.144
N	0.325	0.086	−0.139
Cl	0.672	0.398	0.155

The resulting structure is shown in figure XVF,26. Each oxygen atom has one chlorine neighbor, whereas every nitrogen is about equidistant (ca. 3.14 A.) from three chlorines.

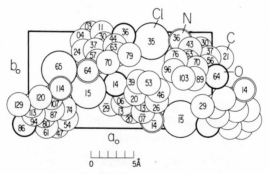

Fig. XVF,26. The orthorhombic structure of 2-methyl-4-amino-1-naphthol hydrochloride projected along its c_0 axis.

XV,f18. The monoclinic crystals of *1,4,5,8-tetrabromonaphthalene*, $C_{10}H_4Br_4$, have a tetramolecular unit of the dimensions:

$$a_0 = 9.55(3) \text{ A.}; \quad b_0 = 15.75(5) \text{ A.}; \quad c_0 = 7.50(3) \text{ A.}; \quad \beta = 94(1)°$$

All atoms are in the positions

$$(4e) \quad \pm (xyz; x, {}^1/_2 - y, z + {}^1/_2)$$

of the space group C_{2h}^5 ($P2_1/c$). The chosen parameters are listed in Table XVF,16.

TABLE XVF,16

Parameters of the Atoms in $C_{10}H_4Br_4$

Atom	x	y	z
Br(1)	−0.065	0.128	0.232
Br(2)	0.186	0.016	0.397
Br(3)	0.332	0.462	0.394
Br(4)	0.600	0.346	0.331
C(1)	0.071	0.217	0.282
C(2)	0.009	0.291	0.274
C(3)	0.099	0.363	0.303
C(4)	0.242	0.356	0.347
C(5)	0.454	0.263	0.339
C(6)	0.510	0.184	0.353
C(7)	0.432	0.106	0.363
C(8)	0.290	0.116	0.355
C(9)	0.221	0.200	0.329
C(10)	0.309	0.275	0.331

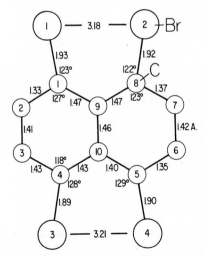

Fig. XVF,27.   Bond dimensions in the molecule of 1,4,5,8-tetrabromonaphthalene.

Bond dimensions in the molecule thus defined are given in Figure XVF,27. Bromine atoms are so close together that the naphthalene ring is no longer strictly planar; its carbon atoms are up to 0.20 A. on either side of the best plane through the ring.

The unit cell of this monoclinic crystal has almost the same dimensions as those of the orthorhombic chlorine derivative (**XV,f20**).

XV,f19. Crystals of *1,5-dibromo-4,8-dichloronaphthalene*, $C_{10}H_4$-Br_2Cl_2, are monoclinic with a tetramolecular pseudo-orthorhombic unit of the dimensions:

$$a_0 = 9.57(2) \text{ A.}; \quad b_0 = 15.42(1) \text{ A.}; \quad c_0 = 7.17(1) \text{ A.}; \quad \beta = 90.0(5)°$$

The space group is C_{2h}^5 ($P2_1/c$). Atoms in the positions

$$(4e) \quad \pm(xyz; x, 1/2 - y, z + 1/2)$$

have the parameters listed in Table XVF,17.

TABLE XVF,17

Parameters of the Atoms in 1,5-Dibromo-4,8-dichloronaphthalene

Atom	x	y	z
Br(1)	0.095	0.346	0.839
Br(2)	0.421	0.139	0.778
Cl(1)	0.680	0.020	0.873
Cl(2)	0.835	0.466	0.876
C(1)	0.950	0.262	0.838
C(2)	0.005	0.178	0.850
C(3)	0.912	0.103	0.848
C(4)	0.769	0.118	0.847
C(5)	0.557	0.223	0.807
C(6)	0.517	0.306	0.808
C(7)	0.600	0.378	0.837
C(8)	0.740	0.363	0.843
C(9)	0.802	0.281	0.836
C(10)	0.707	0.203	0.822
H(2)	0.106	0.165	0.850
H(3)	0.967	0.042	0.848
H(6)	0.407	0.319	0.808
H(7)	0.559	0.442	0.837

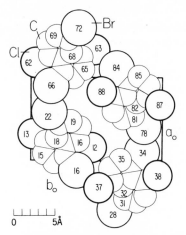

Fig. XVF,28. The monoclinic structure of 1,5-dibromo-4,8-dichloronaphthalene projected along its c_0 axis.

The atomic arrangement is indicated in Figure XVF,28. According to the original description, the molecules show a considerable departure from planarity, but a recent recalculation indicates that the departure, though real, is small. The structure is layered like that of the tetrachloro compound (**XV,f20**), with interatomic distances between molecules ranging upward from ca. 3.50 A.

Though the cell dimensions are close to those of the tetrabromo derivative (**XV,f18**) and the chosen space group is the same $C_{2h}{}^5$, different parameters have been assigned.

XV,f20. The orthorhombic *1,4,5,8-tetrachloronaphthalene*, $C_{10}H_4Cl_4$, has a tetramolecular unit of the edge lengths:

$$a_0 = 9.359(1) \text{ A.}; \quad b_0 = 15.100(1) \text{ A.}; \quad c_0 = 7.070(9) \text{ A.}$$

The space group, selected as $V_h{}^{10}$ (*Pccn*), places all atoms in the positions:

$$(8e) \quad \pm (xyz; \, x + {}^1/_2, y + {}^1/_2, \bar{z}; \, x + {}^1/_2, \bar{y}, {}^1/_2 - z; \, x, {}^1/_2 - y, z + {}^1/_2)$$

There have been two agreeing determinations; the parameters from one (1962: G & H) are listed in Table XVF,18.

The molecules are stacked along c_0 in piles that are distributed in an approximately hexagonal array (Fig. XVF,29). The bond dimensions are those of Figure XVF,30. Crowding of the chlorine atoms in the molecules leads to some distortion of the naphthalene ring and a

TABLE XVF,18

Parameters of the Atoms in 1,4,5,8-$C_{10}H_4Cl_4$

Atom	x	y	z
Cl(1)	$-0.0783(3)$	$0.1478(2)$	$0.1771(5)$
Cl(2)	$0.1711(3)$	$0.0276(2)$	$0.1226(4)$
C(1)	$0.0011(11)$	$0.3121(7)$	$0.1566(15)$
C(2)	$0.0514(8)$	$0.2288(5)$	$0.1603(13)$
C(3)	$0.2028(8)$	$0.2129(5)$	$0.1535(12)$
C(4)	$0.2686(10)$	$0.1257(5)$	$0.1448(13)$
C(5)	$0.4100(11)$	$0.1139(6)$	$0.1509(16)$

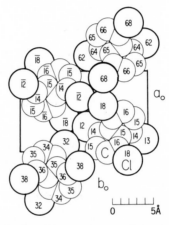

Fig. XVF,29. The orthorhombic structure of 1,4,5,8-tetrachloronaphthalene projected along its c_0 axis.

bending of the C–Cl bonds. Departures from the mean plane of the naphthalene ring thus are as much as 0.2 A.

XV,f21. According to a preliminary note, crystals of *octachloro-naphthalene*, $C_{10}Cl_8$, are monoclinic with a tetramolecular unit of the dimensions:

$$a_0 = 19.48(3) \text{ A.}; \quad b_0 = 7.30(4) \text{ A.}; \quad c_0 = 9.76(1) \text{ A.}$$
$$\beta = 111°33(5)'$$

Atoms are in the positions

$$(4e) \quad \pm(xyz; x+{}^1\!/_2, {}^1\!/_2-y, z)$$

of the space group C_{2h}^5 ($P2_1/a$). The parameters are those of Table XVF,19.

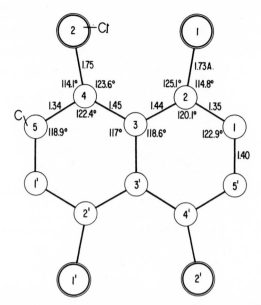

Fig. XVF,30. Bond dimensions in the molecule of 1,4,5,8-tetrachloronaphthalene.

TABLE XVF,19

Parameters of the Atoms in $C_{10}Cl_8$

Atom	x	y	z
Cl(1)	0.1262(2)	0.3128(6)	−0.0198(4)
Cl(2)	0.0594(3)	0.3764(7)	−0.3481(5)
Cl(3)	0.1557(3)	0.4851(7)	−0.5232(5)
Cl(4)	0.3231(3)	0.4925(7)	−0.3750(5)
Cl(5)	0.4325(3)	0.2454(7)	−0.1576(5)
Cl(6)	0.5087(3)	0.2636(7)	0.1762(6)
Cl(7)	0.4228(3)	0.3777(7)	0.3716(5)
Cl(8)	0.2585(2)	0.4270(6)	0.2423(5)
C(1)	0.1864(9)	0.363(3)	−0.106(2)
C(2)	0.1534(9)	0.390(2)	−0.260(2)
C(3)	0.1983(8)	0.424(2)	−0.338(2)
C(4)	0.2727(8)	0.419(2)	−0.273(2)
C(5)	0.3819(8)	0.320(2)	−0.059(2)
C(6)	0.4166(9)	0.318(2)	0.096(2)
C(7)	0.3773(9)	0.356(2)	0.182(2)
C(8)	0.3007(8)	0.382(2)	0.123(2)
C(9)	0.2636(7)	0.370(2)	−0.035(2)
C(10)	0.3076(8)	0.370(2)	−0.122(2)

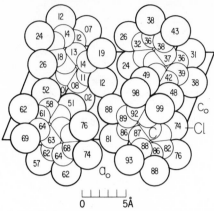

Fig. XVF,31.   The monoclinic structure of octachloronaphthalene projected
along its b_0 axis.

The structure is shown in Figure XVF,31. Overcrowding due to the
many chlorine atoms causes the molecule to depart seriously from planarity. This is indicated in Figure XVF,32, where the circled numbers
are the departures (in angstrom units) from the mean molecular plane.
The bond lengths, given in Figure XVF,33, are normal, however.
Within molecules the Cl–Cl distances lie between 3.00 and 3.08 A.

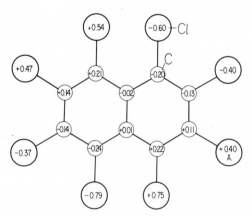

Fig. XVF,32.   Numbers in the circles of this molecular drawing for octachloronaphthalene indicate the atomic distances from the best plane through the
molecule.

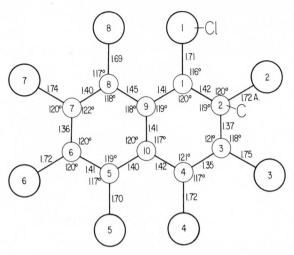

Fig. XVF,33.   Bond dimensions in the molecule of octachloronaphthalene.

XV,f22.   Crystals of *octamethylnaphthalene*, $C_{10}(CH_3)_8$, are ortho-rhombic with a tetramolecular cell of the edge lengths:

$$a_0 = 16.66 \text{ A.}; \quad b_0 = 11.31 \text{ A.}; \quad c_0 = 7.64 \text{ A.}$$

The space group is V_h^{22} (*Ccca*) and atoms are in the following positions:

$$(8e) \quad \pm(u00; \; u \, ^1/_2 \, ^1/_2; \; u+^1/_2,^1/_2,0; \; u+^1/_2,0,^1/_2)$$
$$(16i) \quad xyz; \; \bar{x}\bar{y}z; \; \bar{x},^1/_2-y,^1/_2-z; \; x,y+^1/_2,^1/_2-z;$$
$$x\bar{y}\bar{z}; \; \bar{x}y\bar{z}; \; \bar{x},y+^1/_2,z+^1/_2; \; x,^1/_2-y,z+^1/_2;$$

plus eight similar positions around $^1/_2 \, ^1/_2 \, 0$. The parameters are:

Atom	Position	x	y	z
C(1)	8(e)	0.041	0	0
C(2)	16(i)	0.079	0.110	0.034
C(3)	16(i)	0.042	0.212	−0.012
C(4)	16(i)	0.168	0.115	0.095
C(5)	16(i)	0.086	0.331	−0.034

The molecular packing that results is illustrated in Figure XVF,34. The molecule itself, as shown in Figure XVF,35, is not strictly planar, the methyl carbons being displaced as much as 0.73 A. from the plane of the naphthalene nucleus. The data are insufficient to show if there is also a minor warping of this naphthalene core.

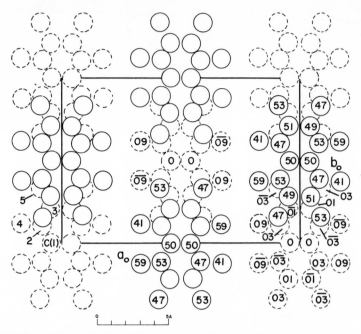

Fig. XVF,34. The orthorhombic structure of octamethylnaphthalene projected along its c_0 axis. Left-hand axes.

TABLE XVF,20

Parameters of the Atoms in 1,4-Naphthoquinone

Atom	x	y	z
C(1)	0.3436	−0.0934	0.1536
C(2)	0.2800	−0.2715	0.1554
C(3)	0.1759	−0.3325	0.0680
C(4)	0.1316	−0.2230	−0.0327
C(5)	0.1478	0.0555	−0.1337
C(6)	0.2165	0.2229	−0.1417
C(7)	0.3253	0.2845	−0.0496
C(8)	0.3591	0.1842	0.0520
C(9)	0.2984	0.0173	0.0561
C(10)	0.1935	−0.0471	−0.0395
O(1)	0.4312	−0.0387	0.2387
O(4)	0.0381	−0.2808	−0.1150

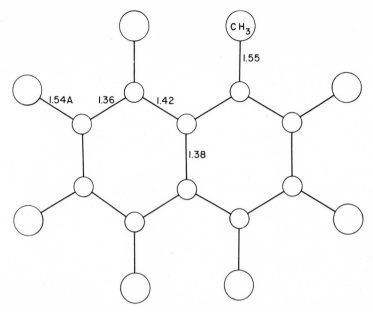

Fig. XVF,35.   Some bond lengths in the molecule of octamethylnaphthalene.

XV,f23.   Crystals of *1,4(α)-naphthoquinone*, $C_{10}H_6O_2$, are mono-clinic with a tetramolecular cell of the dimensions:

$$a_0 = 8.27(2) \text{ A.}; \quad b_0 = 7.76(2) \text{ A.}; \quad c_0 = 11.71(2) \text{ A.}$$
$$\beta = 99°30(20)'$$

The space group C_{2h}^5 ($P2_1/c$) puts all atoms in the positions:

$$(4e) \quad \pm (xyz; x, {}^1\!/_2 - y, z + {}^1\!/_2)$$

Their parameters are stated in Table XVF,20.

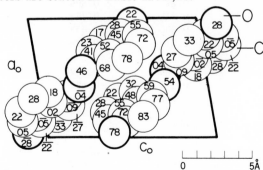

Fig. XVF,36.   The monoclinic structure of 1,4(α)-naphthoquinone projected along its b_0 axis.

The resulting structure, illustrated in Figure XVF,36, is built of essentially planar molecules with the bond dimensions of Figure XVF,37. The closest approaches of atoms in neighboring molecules are C–O = 3.27, 3.31, and 3.49 A. but there are also C–C separations almost as short: 3.50 and 3.55 A.

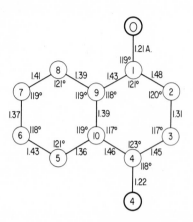

Fig. XVF,37.   Bond dimensions in the molecule of 1,4(α)-naphthoquinone.

TABLE XVF,21

Parameters of the Atoms in $C_{10}H_5(CH_3)(OH)_2$

Atom	x	y	z
C(1)	0.143	0.680	0.234
C(2)	0.212	0.806	0.309
C(3)	0.318	0.738	0.322
C(4)	0.350	0.553	0.261
C(5)	0.310	0.230	0.118
C(6)	0.239	0.109	0.043
C(7)	0.131	0.177	0.029
C(8)	0.099	0.360	0.092
C(9)	0.172	0.486	0.170
C(10)	0.279	0.423	0.183
C(11)	0.395	0.879	0.408
O(1)	0.038	0.736	0.219
O(4)	0.45	0.499	0.274

XV,f24. According to a preliminary note, crystals of *2-methyl-α-naphthohydroquinone*, $C_{10}H_5(CH_3)(OH)_2$, are monoclinic with a tetramolecular unit of the dimensions:

$$a_0 = 13.50(1) \text{ A.}; \quad b_0 = 4.49(2) \text{ A.}; \quad c_0 = 14.89(1) \text{ A.}; \quad \beta = 104°$$

The space group is C_{2h}^5 ($P2_1/c$) with atoms in the positions:

$$(4e) \quad \pm(xyz; x, 1/2 - y, z + 1/2)$$

The parameters are given in Table XVF,21.

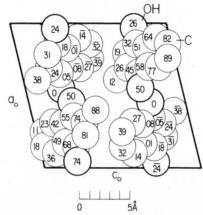

Fig. XVF,38.   The monoclinic structure of 2-methyl-α-naphthohydroquinone projected along its b_0 axis.

The structure, shown in Figure XVF,38, is reported to be undergoing further refinement.

XV,f25. The monoclinic *2-bromo-1,4-naphthoquinone*, $C_{10}H_5BrO_2$, has a tetramolecular unit of the dimensions:

$$a_0 = 13.88(2) \text{ A.}; \quad b_0 = 3.98(2) \text{ A.}; \quad c_0 = 15.74(2) \text{ A.}$$
$$\beta = 104°0(30)'$$

All atoms are in the positions

$$(4e) \quad \pm(xyz; x, 1/2 - y, z + 1/2)$$

of C_{2h}^5 ($P2_1/c$). Determined parameters are listed in Table XVF,22.

TABLE XVF,22

Parameters of the Atoms in $C_{10}H_5BrO_2$

Atom	x	y	z
C(1)	0.7207	0.0005	0.5301
C(2)	0.8265	0.1020	0.5491
C(3)	0.8879	0.0953	0.6280
C(4)	0.8486	−0.0351	0.7040
C(5)	0.7140	−0.3085	0.7542
C(6)	0.6201	−0.4309	0.7414
C(7)	0.5543	−0.4225	0.6566
C(8)	0.5872	−0.2708	0.5868
C(9)	0.6859	−0.1570	0.6037
C(10)	0.7482	−0.1555	0.6880
O(1)	0.6680	0.0209	0.4587
O(4)	0.9073	−0.0161	0.7780
Br	0.8690	0.2881	0.4553

The structure (Fig. XVF,39) is composed of molecules possessing the bond dimensions of Figure XVF,40. Only oxygen departs by more than 0.05 A. from the molecular plane; one of these atoms is 0.07 A., the other 0.13 A. below the plane. Oxygens are nearest the atoms of other molecules, with Br–O(1) = 3.10 A. and C(3)–O(1) = 3.26 A.

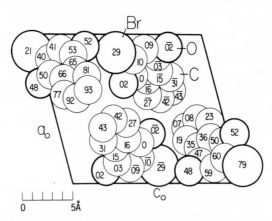

Fig. XVF,39. The monoclinic structure of 2-bromo-1,4-naphthoquinone projected along its b_0 axis.

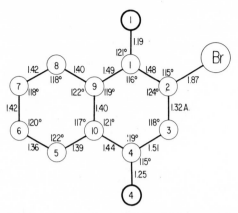

Fig. XVF,40.   Bond dimensions in the molecule of 2-bromo-1,4-naphthoquinone.

XV,f26. According to a preliminary announcement, the crystals of *3-amino-1,4-naphthoquinone*, $C_{10}H_5(NH_2)O_2$, are monoclinic with a tetramolecular unit of the dimensions:

$$a_0 = 17.03(2) \text{ A.}; \quad b_0 = 3.96(1) \text{ A.}; \quad c_0 = 14.72(2) \text{ A.}$$
$$\beta = 125°30'$$

The space group is C_{2h}^5 ($P2_1/c$) with atoms in the positions:

$$(4e) \quad \pm (xyz; x, 1/2 - y, z + 1/2)$$

The proposed parameters are listed in Table XVF,23.

TABLE XVF,23
Parameters of the Atoms in $C_{10}H_5(NH_2)O_2$

Atom	x	y	z
C(1)	0.152	0.433	0.422
C(2)	0.105	0.262	0.315
C(3)	0.147	0.239	0.261
C(4)	0.247	0.369	0.314
C(5)	0.391	0.667	0.477
C(6)	0.438	0.817	0.581
C(7)	0.393	0.847	0.636
C(8)	0.299	0.723	0.583
C(9)	0.252	0.565	0.478
C(10)	0.296	0.538	0.423
O(1)	0.112	0.463	0.469
O(4)	0.284	0.334	0.264
N(3)	0.104	0.083	0.159

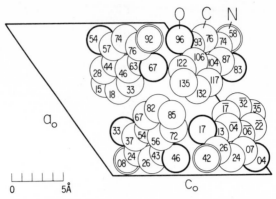

Fig. XVF,41.   The monoclinic structure of 3-amino-1,4-naphthoquinone projected along its b_0 axis.

The structure to which they lead is illustrated in Figure XVF,41. As it indicates, O(1) and N atoms of adjacent molecules are less than 3 A. apart.

XV,f27.   A preliminary announcement reports that crystals of *4-amino-1,2-naphthoquinone hemihydrate*, $C_{10}H_5(NH_2)O_2 \cdot 1/2H_2O$, are monoclinic with a unit containing eight molecules. The cell dimensions are:

$$a_0 = 26.38(2) \text{ A.;} \quad b_0 = 4.67(1) \text{ A.;} \quad c_0 = 14.08(2) \text{ A.}$$
$$\beta = 105°10'$$

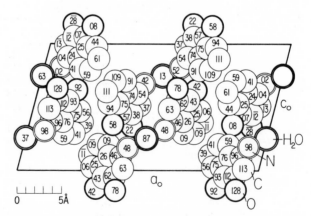

Fig. XVF,42.   The monoclinic structure of 4-amino-1,2-naphthoquinone hemihydrate projected along its b_0 axis.

The space group is C_{2h}^6 ($C2/c$) with the water oxygens in:

$$(4e) \quad \pm(0 \; u \; {}^1/_4; {}^1/_2, u+{}^1/_2, {}^1/_4) \qquad \text{with } u = 0.3705$$

The other atoms are in

$$(8f) \quad \pm(xyz; x, \bar{y}, z+{}^1/_2; x+{}^1/_2, y+{}^1/_2, z; x+{}^1/_2, {}^1/_2-y, z+{}^1/_2)$$

with the parameters of Table XVF,24.

TABLE XVF,24

Parameters of Atoms in 4-Amino-1,2-naphthoquinone Hemihydrate

Atom	x	y	z
C(1)	0.3647	0.4259	0.5548
C(2)	0.4102	0.6231	0.5596
C(3)	0.4332	0.6254	0.4815
C(4)	0.4135	0.4573	0.3970
C(5)	0.3496	0.0913	0.3042
C(6)	0.3072	−0.0935	0.2994
C(7)	0.2837	−0.1071	0.3765
C(8)	0.3027	0.0626	0.4606
C(9)	0.3448	0.2454	0.4655
C(10)	0.3688	0.2597	0.3880
O(1)	0.3444	0.4193	0.6220
O(2)	0.4248	0.7799	0.6350
N	0.4348	0.4752	0.3220

Molecular distribution in the structure is shown in Figure XVF,42.

XV,f28. The A form of *5,8-dihydroxy-1,4-naphthoquinone* (naphthazarin), $C_{10}H_4O_2(OH)_2$, is monoclinic with a bimolecular unit of the dimensions:

$$a_0 = 3.75(2) \text{ A.}; \quad b_0 = 7.66(2) \text{ A.}; \quad c_0 = 14.47(5) \text{ A.}$$
$$\beta = 97°0(30)'$$

Its space group C_{2h}^5 ($P2_1/c$) places all atoms in the positions:

$$(4e) \quad \pm(xyz; x, {}^1/_2-y, z+{}^1/_2)$$

The parameters, listed in Table XVF,25, include values for hydrogen.

TABLE XVF,25

Parameters of the Atoms in A-Naphthazarin

Atom	x	y	z
C(1)	−0.303	0.156	−0.151
C(2)	−0.149	−0.013	−0.129
C(3)	0.014	−0.067	−0.037
C(4)	0.173	−0.230	−0.017
C(5)	0.317	−0.274	0.079
O(1)	−0.133	−0.128	−0.192
O(2)	0.193	−0.345	−0.084
H(1)	−0.43	0.27	−0.25
H(2)	0.47	−0.45	0.08
H(3)	0.07	−0.29	−0.17

The resulting structure, shown in Figure XVF,43, is composed of essentially planar centrosymmetric molecules with the expected bond dimensions. Between them the shortest atomic separations are C–C = 3.34 and 3.45 A.

XV,f29. The monoclinic B modification of *5,8-dihydroxy-1,4-naphthoquinone* (naphthazarin), $C_{10}H_4O_2(OH)_2$, possesses a bimolecular unit of the dimensions:

$$a_0 = 5.41(1) \text{ A.}; \quad b_0 = 6.40(2) \text{ A.}; \quad c_0 = 11.85(2) \text{ A.} \quad \beta = 91°23(5)'$$

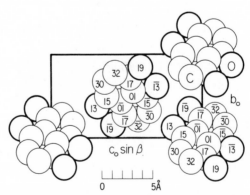

Fig. XVF,43.   The monoclinic structure of the A form of naphthazarin projected along its a_0 axis.

Atoms are in the positions

$$(4e) \quad \pm (xyz; x, {}^1\!/_2 - y, z + {}^1\!/_2)$$

of C_{2h}^5 ($P2_1/c$). The determined parameters, including values for the atoms of hydrogen, are listed in Table XVF,26.

TABLE XVF,26

Parameters of the B-Form of Naphthazarin

Atom	x	y	z
C(1)	-0.354	-0.048	-0.148
C(2)	-0.165	0.100	-0.131
C(3)	0.006	0.087	-0.036
C(4)	0.199	0.247	-0.017
C(5)	0.368	0.218	0.078
O(1)	-0.159	0.267	-0.197
O(2)	0.207	0.408	-0.083
H(1)	-0.46	-0.02	-0.24
H(2)	0.50	0.37	0.09
H(3)	0.12	0.40	-0.18

The structure, as shown in Figure XVF,44, closely resembles that of the A form. Its planar molecules have the usual bond dimensions. Between molecules there are shortest atomic separations of O–O = 3.20 A. and C–O = 3.32 and 3.42 A.

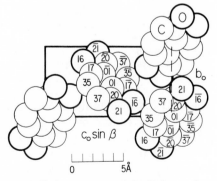

Fig. XVF,44.   The monoclinic structure of the B form of naphthazarin projected along its a_0 axis.

XV,f30. The C modification of *5,8-dihydroxy-1,4-naphthoquinone* (naphthazarin), $C_{10}H_4O_2(OH)_2$, is also monoclinic with a tetramolecular unit of the dimensions·

$$a_0 = 7.906(5) \text{ A.}; \quad b_0 = 7.324(5) \text{ A.}; \quad c_0 = 14.05(1) \text{ A.}$$
$$\beta = 96°35(6)'$$

Its atoms are in the positions

$$(4e) \quad \pm (xyz; x + {}^1/_2, {}^1/_2 - y, z + {}^1/_2)$$

of C_{2h}^5 in the orientation $P2_1/n$. The parameters are listed in Table XVF,27.

TABLE XVF,27

Parameters of the Atoms in C-Naphthazarin

Atom	x	y	z
Molecule I			
C(1)	−0.0540	0.1688	−0.1604
C(2)	0.0406	0.2130	−0.0728
C(3)	0.0464	0.0859	0.0081
C(4)	0.1354	0.1245	0.0958
C(5)	0.1400	−0.0068	0.1754
O(1)	0.1236	0.3667	−0.0644
O(2)	0.2243	0.2760	0.1164
Molecule II			
C(1')	0.6845	0.2380	0.0967
C(2')	0.6102	0.2304	0.0022
C(3')	0.5061	0.0737	−0.0340
C(4')	0.4327	0.0633	−0.1274
C(5')	0.3292	−0.0953	−0.1611
O(1')	0.6279	0.3691	−0.0564
O(2')	0.4440	0.1960	−0.1910

The structure (Fig. XVF,45) is built of molecules of two crystallographic sorts, each with a center of symmetry. Both are planar within 0.02 A. The average of their closely agreeing bond dimensions is given in Figure XVF,46. Of particular interest is the fact that all the C–O bonds are the same; thus there is no distinction between the phenolic and quinone oxygens. Between molecules the shortest atomic separations are O–O = 3.05 and 3.11 A. and C–O = 3.21 and 3.32 A.

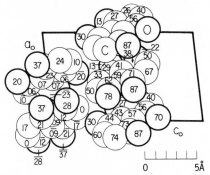

Fig. XVF,45.   The monoclinic structure of the C form of naphthazarin projected
along its b_0 axis.

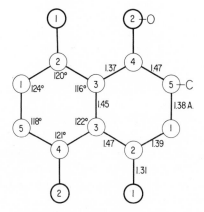

Fig. XVF,46.   Bond dimensions of the molecule of naphthazarin in its C-type
crystals.

XV,f31.   Crystals of phthiocol, *methyl-2-hydroxy-3-naphthoquinone*,
$C_{10}H_4O_2(CH_3)(OH)$, are monoclinic. Their unit is bimolecular with the
dimensions:

$$a_0 = 11.85(2) \text{ A.}; \quad b_0 = 4.85(2) \text{ A.}; \quad c_0 = 7.71(2) \text{ A.}; \quad \beta = 90°30'$$

All atoms are in the positions

$$(2a) \quad xyz; \; \bar{x}, y + 1/2, \bar{z}$$

of the low symmetry space group C_2^2 $(P2_1)$. The determined parameters
are those of Table XVF,28.

TABLE XVF,28

Parameters of the Atoms in Phthiocol

Atom	x	y	z
C(1)	0.2807	0.1873	0.4564
C(2)	0.1846	0.3781	0.4438
C(3)	0.1231	0.3920	0.2982
C(4)	0.1461	0.2155	0.1434
C(5)	0.2678	−0.1368	0.0127
C(6)	0.3605	−0.3109	0.0238
C(7)	0.4277	−0.3231	0.1678
C(8)	0.4006	−0.1614	0.3091
C(9)	0.3082	0.0131	0.3022
C(10)	0.2426	0.0278	0.1527
O(1)	0.3375	0.1720	0.5896
O(4)	0.0834	0.2333	0.0170
OH	0.0368	0.5732	0.2794
C(CH₃)	0.1599	0.5546	0.5993

In this structure (Fig. XVF,47) the molecules have the bond dimensions of Figure XVF,48. The naphthalene ring is planar within 0.02 A., with the quinone oxygens 0.03 and 0.04 A. outside this plane. The methyl radical is in the plane of the rest, but the hydroxyl is ca. 0.10 A. outside. Molecules are held together in the crystal by hydrogen bonds involving the hydroxyl group and one of the quinone oxygens (4), with

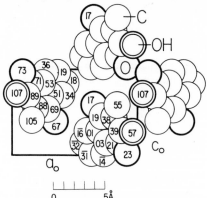

Fig. XVF,47.   The monoclinic structure of phthiocol projected along its b_0 axis.

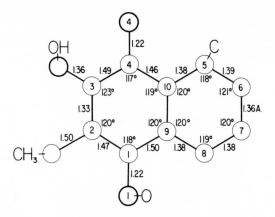

Fig. XVF,48.   Bond dimensions in the molecule of phthiocol.

O–H···O = 2.79 A. Within the molecule there is also a short O–H···
O = 2.67 A., which, it is proposed, is a second hydrogen bond. Between
molecules there are an OH–OH = 3.14 A. and O–C separations ranging
upward from 3.29 A.

XV,f32.   The monoclinic crystals of *2-chloro-3-hydroxy-1,4-naph-
thoquinone*, $C_{10}H_4O_2ClOH$, have a bimolecular unit of the dimensions:

$$a_0 = 8.25(2) \text{ A.}; \quad b_0 = 3.92(2) \text{ A.}; \quad c_0 = 14.39(3) \text{ A.}; \quad \beta = 113°20'$$

Atoms are in the positions

$$(2a) \quad xyz; x, \bar{y}, z + 1/2$$

of the low symmetry space group C_s^2 (*Pc*). The parameters have been
determined to be those of Table XVF,29.

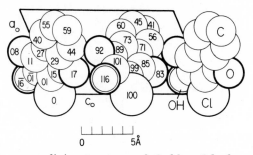

Fig. XVF,49.   The monoclinic structure of 2-chloro-3-hydroxy-1,4-naphtho-
quinone projected along its b_0 axis.

TABLE XVF,29

Parameters of the Atoms in $C_{10}H_4O_2Cl(OH)$ and in Parentheses of
$C_{10}H_4O_2Cl(NH_2)$

Atom	x	y	z
C(1)	0.3050 (0.3047)	0.1483 (0.1689)	0.1219 (0.1249)
C(2)	0.2090 (0.2127)	0.0110 (0.0231)	0.0254 (0.0291)
C(3)	0.2703 (0.2885)	−0.0100 (0.0003)	−0.0478 (−0.0372)
C(4)	0.4528 (0.4744)	0.1108 (0.1321)	−0.0213 (−0.0121)
C(5)	0.7242 (0.7449)	0.3980 (0.4183)	0.0938 (0.1082)
C(6)	0.8200 (0.8368)	0.5515 (0.5732)	0.1877 (0.1997)
C(7)	0.7461 (0.7499)	0.5900 (0.6046)	0.2589 (0.2650)
C(8)	0.5759 (0.5765)	0.4394 (0.4571)	0.2381 (0.2400)
C(9)	0.4774 (0.4837)	0.2918 (0.3091)	0.1459 (0.1495)
C(10)	0.5561 (0.5697)	0.2721 (0.2854)	0.0725 (0.0839)
O(1)	0.2394 (0.2330)	0.1672 (0.1815)	0.1885 (0. 1854)
O(4)	0.5063 (0.5408)	0.0814 (0.1028)	−0.0920 (−0.0737)
OH (or NH₂)	0.1858 (0.2083)	−0.1609 (−0.1424)	−0.1347 (−0.1286)
Cl	0.0000 (0.0000)	−0.1428 (−0.1264)	0.0000 (0.0000)

The resulting atomic arrangement (Fig. XVF,49) is built up of
molecules that are planar within 0.05 A. and have the bond dimensions
of Figure XVF,50. Between molecules there is a short O–O = 2.75 A.,
suggesting hydrogen bonding; other intermolecular separations range

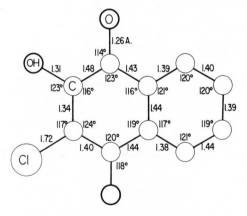

Fig. XVF,50.   Bond dimensions in the molecule of 2-chloro-3-hydroxy-1,4-
naphthoquinone.

Fig. XVF,51. Bond dimensions in the molecule of 2-chloro-3-amino-1,4-naphthoquinone.

upward from Cl–C = 3.56 A., O–O = 3.21 A., O–C = 3.30 A., and C–C = 3.50 A.

The corresponding amino compound, *2-chloro-3-amino-1,4-naphthoquinone*, $C_{10}H_4O_2ClNH_2$, is isostructural. Its atomic parameters are given in parentheses in Table XVF,29. Bond dimensions are those of Figure XVF,51. Between molecules there is a short N–N = 2.85 A.

XV,f33. Orthorhombic *2,3-dibromo-1,4-naphthoquinone*, $C_{10}H_4Br_2$-O_2, possesses a tetramolecular unit of the edge lengths:

$$a_0 = 15.17(3) \text{ A.}; \quad b_0 = 3.97(2) \text{ A.}; \quad c_0 = 15.83(3) \text{ A.}$$

The space group is the low symmetry C_{2v}^5 (*Pca2_1*) with atoms in the positions:

$$(4a) \quad xyz; \bar{x},\bar{y},z+1/2; 1/2-x,y,z+1/2; x+1/2,\bar{y},z$$

The determined parameters are those of Table XVF,30.

The structure, shown in Figure XVF,52, has its molecules (Fig. XVF,53) stacked along the b_0 axis; and, as is the case with other derivatives of 1,4-naphthoquinone, these stacks are close-packed. Within a stack the shortest intermolecular separations are C–C = 3.55 and 3.56 A. and C–Br = 3.64 A. Between stacks there is a short Br–O = 3.15 A.; other close approaches are Br–O = 3.22 A. and C–O = 3.31 and 3.42 A.

TABLE XVF,30

Parameters of the Atoms in $C_{10}H_4Br_2O_2$

Atom	x	y	z
C(1)	−0.1652(10)	−0.358(5)	0.0814(11)
C(2)	−0.0813(10)	−0.187(6)	0.0523(12)
C(3)	−0.0142(12)	−0.102(6)	0.1033(12)
C(4)	−0.0172(12)	−0.158(6)	0.1952(13)
C(5)	−0.0993(11)	−0.330(7)	0.2277(11)
C(6)	−0.1062(16)	−0.399(7)	0.3158(12)
C(7)	−0.1808(18)	−0.535(9)	0.3499(18)
C(8)	−0.2535(18)	−0.615(7)	0.2957(18)
C(9)	−0.2445(13)	−0.553(6)	0.2107(14)
C(10)	−0.1689(11)	−0.416(6)	0.1721(12)
O(1)	−0.2232(9)	−0.425(6)	0.0356(11)
O(4)	0.0439(8)	−0.084(4)	0.2390(12)
Br(2)	−0.0783(1)	−0.1259(7)	−0.0675(2)
Br(3)[a]	0.090	0.091	0.061

[a] An error in 1967 : BL gives Br(3) the same parameters as Br(2). The values stated here are from the preliminary note.

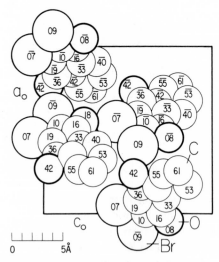

Fig. XVF,52.   The orthorhombic structure of 2,3-dibromo-1,4-naphthoquinone projected along its b_0 axis.

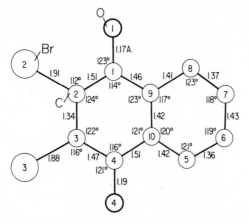

Fig. XVF,53.   Bond dimensions in the molecule of 2,3-dibromo-1,4-naphtho-quinone.

XV,f34.   Crystals of *2,3-dichloro-1,4-naphthoquinone*, $C_{10}H_4O_2Cl_2$, are triclinic. They have been assigned a large tetramolecular unit of the dimensions:

$$a_0 = 18.18 \text{ A.}; \quad b_0 = 8.29 \text{ A.}; \quad c_0 = 7.35 \text{ A.}$$
$$\alpha = 112°30'; \quad \beta = 73°; \quad \gamma = 117°30'$$

The space group is C_i^1 $(P\bar{1})$ with atoms in the positions $(2i) \pm (xyz)$. The parameters are those of Table XVF,31.

The dimensions of the two crystallographically different molecules in this structure (Fig. XVF,54) are probably the same within the limit

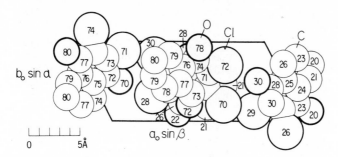

Fig. XVF,54.   The triclinic structure of 2,3-dichloro-1,4-naphthoquinone pro-jected along its c_0 axis.

of error of the determination; their averaged bond lengths are those of Figure XVF,55.

TABLE XVF,31

Parameters of the Atoms in 2,3-Dichloro-1,4-naphthoquinone

Atom	x	y	z
Molecule I			
Cl(2)	0.121	0.108	0.745
Cl(3)	0.235	0.898	0.712
C(1)	0.989	0.774	0.770
C(2)	0.083	0.869	0.740
C(3)	0.129	0.780	0.730
C(4)	0.091	0.576	0.720
C(5)	0.961	0.297	0.740
C(6)	0.876	0.199	0.765
C(7)	0.824	0.286	0.795
C(8)	0.866	0.473	0.795
C(9)	0.953	0.577	0.760
C(10)	0.002	0.488	0.745
O(1)	0.943	0.856	0.800
O(4)	0.140	0.506	0.700
Molecule II			
Cl(2′)	0.615	0.264	0.702
Cl(3′)	0.731	0.691	0.722
C(1′)	0.485	0.350	0.735
C(2′)	0.578	0.431	0.707
C(3′)	0.627	0.613	0.711
C(4′)	0.588	0.743	0.740
C(5′)	0.463	0.805	0.788
C(6′)	0.376	0.742	0.805
C(7′)	0.325	0.551	0.791
C(8′)	0.354	0.414	0.780
C(9′)	0.443	0.477	0.765
C(10′)	0.495	0.668	0.760
O(1′)	0.440	0.190	0.720
O(4′)	0.638	0.905	0.780

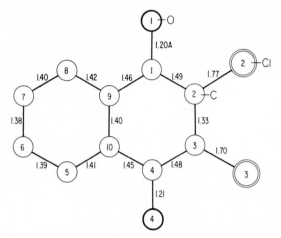

Fig. XVF,55.   Average bond lengths for the two crystallographically different
molecules of 2,3-dichloro-1,4-naphthoquinone.

XV,f35.   The tetramolecular, monoclinic unit of *2-bromo-3-amino-
1,4-naphthoquinone*, $C_{10}H_4O_2BrNH_2$, has the dimensions:

$$a_0 = 15.18(2) \text{ A.}; \quad b_0 = 3.93(2) \text{ A.}; \quad c_0 = 15.92(2) \text{ A.}$$
$$\beta = 110°0(20)'$$

All atoms are in the positions

$$(4e) \quad \pm (xyz; x, 1/2 - y, z + 1/2)$$

of C_{2h}^5 ($P2_1/c$). The parameters are listed in Table XVF,32.

The structure, shown in Figure XVF,56, consists of molecules with
the bond dimensions of Figure XVF,57. Except for nitrogen and bromine, none of the atoms departs by more than ca. 0.03 A. from the

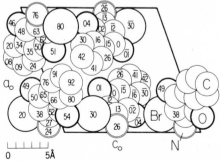

Fig. XVF,56.   The monoclinic structure of 2-bromo-3-amino-1,4-naphthoquinone
projected along its b_0 axis.

TABLE XVF,32
Parameters of the Atoms in $C_{10}H_4O_2BrNH_2$

Atom	x	y	z
C(1)	0.2872	0.0036	0.4088
C(2)	0.1950	0.1256	0.3768
C(3)	0.1359	0.1273	0.4233
C(4)	0.1726	−0.0258	0.5167
C(5)	0.2994	−0.2977	0.6400
C(6)	0.3922	−0.4182	0.6734
C(7)	0.4511	−0.4095	0.6229
C(8)	0.4160	−0.2642	0.5364
C(9)	0.3250	−0.1485	0.5006
C(10)	0.2667	−0.1648	0.5531
O(1)	0.3401	0.0061	0.3661
O(4)	0.1179	−0.0351	0.5567
N(2)	0.0466	0.2592	0.3954
Br(3)	0.1487	0.2991	0.2573

plane of the rest; the bromine is 0.06 A. and the nitrogen 0.08 A. on either side of the plane. It is possible that the short N–O(4) = 2.69 A. corresponds to an internal hydrogen bond; between molecules the N–O(4) = 2.98 A. probably represents another hydrogen bond. Other intermolecular separations range upwards from a Br–O = 3.20 A.

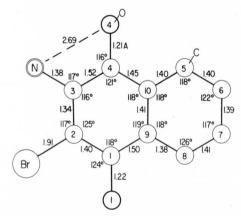

Fig. XVF,57. Bond dimensions in the molecule of 2-bromo-3-amino-1,4-naphthoquinone.

XV,f36. Crystals of *2-bromo-3-methyl-1,4-naphthoquinone*, $C_{10}H_4O_2$-$BrCH_3$, are monoclinic with a tetramolecular unit of the dimensions:

$$a_0 = 7.46(2) \text{ A.}; \quad b_0 = 8.42(2) \text{ A.}; \quad c_0 = 14.87(3) \text{ A.}; \quad \beta = 88°17'$$

The space group is the low symmetry C_2^2 ($P2_1$) with atoms in the positions:

$$(2a) \quad xyz; \bar{x},y+{}^1\!/_2,\bar{z}$$

TABLE XVF,33

Parameters of the Atoms in $C_{10}H_4O_2BrCH_3$

Atom	x	y	z
C(1)	0.039	0.300	0.700
C(2)	0.965	0.285	0.800
C(3)	0.930	0.420	0.845
C(4)	0.955	0.575	0.805
C(5)	0.030	0.595	0.715
C(6)	0.055	0.745	0.680
C(7)	0.125	0.765	0.595
C(8)	0.165	0.635	0.545
C(9)	0.144	0.470	0.570
C(10)	0.068	0.460	0.665
X(1)[a]	0.850	0.385	0.955
X(2)	0.930	0.090	0.835
O(1)	0.915	0.685	0.850
O(2)	0.080	0.175	0.655
C(1′)	−0.540	−0.295	0.795
C(2′)	−0.460	−0.275	0.700
C(3′)	−0.435	−0.420	0.665
C(4′)	−0.455	−0.585	0.690
C(5′)	−0.530	−0.595	0.790
C(6′)	−0.555	−0.760	0.820
C(7′)	−0.625	−0.770	0.900
C(8′)	−0.670	−0.630	0.960
C(9′)	−0.645	−0.480	0.925
C(10′)	−0.570	−0.455	0.835
X(1′)[a]	−0.350	−0.420	0.535
X(2′)	−0.435	−0.080	0.665
O(1′)	−0.410	−0.700	0.645
O(2′)	−0.575	−0.190	0.840

[a] Atomic positions X(1) and X(2′) are 60% occupied by C, 40% by Br; the reverse is true for X(2) and X(1′).

It is said that there is disorder in the way the bromine and CH_3 substituents are distributed in the positions assigned them. In accordance with this, parameters have been given the values of Table XVF,33. A refinement of this structure is promised.

The corresponding *chloro* compound, $C_{10}H_4O_2ClCH_3$, has a unit of nearly the same dimensions:

$$a_0 = 7.43(2) \text{ A.}; \quad b_0 = 8.30(2) \text{ A.}; \quad c_0 = 14.79(3) \text{ A.}; \quad \beta = 87°30'$$

TABLE XVF,34

Parameters of the Atoms in $C_{10}H_4O_2ClCH_3$

Atom	x	y	z
C(1)	0.281	0.293	0.044
C(2)	0.210	0.284	0.047
C(3)	0.181	0.413	0.097
C(4)	0.204	0.575	−0.059
C(5)	0.302	0.744	−0.069
C(6)	0.367	0.765	−0.160
C(7)	0.405	0.622	−0.210
C(8)	0.384	0.466	−0.175
C(9)	0.312	0.455	−0.086
C(10)	0.275	0.591	−0.033
C(11)	0.105	0.405	0.205
Cl(1)	0.185	0.096	0.087
O(1)	0.159	0.696	0.104
O(2)	0.323	0.171	−0.091
C(1′)	0.781	0.207	0.544
C(2′)	0.710	0.216	0.547
C(3′)	0.681	0.087	0.597
C(4′)	0.704	−0.075	0.441
C(5′)	0.802	−0.244	0.431
C(6′)	0.867	−0.265	0.340
C(7′)	0.905	−0.122	0.290
C(8′)	0.884	0.034	0.325
C(9′)	0.812	0.045	0.414
C(10′)	0.775	−0.091	0.467
C(11′)	0.685	0.404	0.587
Cl(1′)	0.605	0.095	0.705
O(1′)	0.659	−0.196	0.604
O(2′)	0.823	0.329	0.591

The space group is also C_2^2 $(P2_1)$ but, as Table XVF,34 indicates, different parameters, leading to somewhat different molecular orientations, have been assigned. There would not appear to be the same molecular disorder in these crystals as has been described for the bromine compound.

XV,f37. The monoclinic crystals of *2,3-dimethyl-1,4-naphthoquinone*, $C_{10}H_4O_2(CH_3)_2$, possess a tetramolecular unit of the dimensions:

$$a_0 = 7.52(2) \text{ A.}; \quad b_0 = 8.35(2) \text{ A.}; \quad c_0 = 14.97(3) \text{ A.}$$
$$\beta = 91°48(30)'$$

TABLE XVF,35

Parameters of the Atoms in $C_{10}H_4O_2(CH_3)_2$

Atom	x	y	z
C(1)	0.2826(15)	0.2993(15)	0.0438(8)
C(2)	0.2151(14)	0.2847(13)	−0.0494(7)
C(3)	0.1814(14)	0.4174(13)	−0.0980(7)
C(4)	0.2070(15)	0.5755(14)	−0.0587(7)
C(5)	0.3039(17)	0.7462(16)	0.0713(8)
C(6)	0.3676(16)	0.7587(16)	0.1594(8)
C(7)	0.4039(17)	0.6220(16)	0.2079(8)
C(8)	0.3799(15)	0.4675(14)	0.1737(7)
C(9)	0.3159(14)	0.4544(13)	0.0848 —
C(10)	0.2770(13)	0.5943(13)	0.0355(6)
O(1)	0.3146(12)	0.1755(12)	0.0876(6)
O(2)	0.1651(12)	0.6967(11)	−0.1024(6)
C(13)	0.1794(24)	0.1196(20)	−0.0852(11)
C(14)	0.1129(21)	0.4151(19)	−0.1931(9)
H(5)	0.277	0.846	0.035
H(6)	0.386	0.867	0.186
H(7)	0.449	0.632	0.269
H(8)	0.407	0.368	0.210
H(13,1)	0.036	0.102	−0.076
H(13,2)	0.287	0.063	−0.068
H(13,3)	0.223	0.119	−0.144
H(14,1)	0.143	0.305	−0.216
H(14,2)	−0.035	0.441	−0.225
H(14,3)	0.170	0.479	−0.230

Atoms are in the positions

$$(4e) \quad \pm (xyz; \; x + 1/2, 1/2 - y, z + 1/2)$$

of the space group C_{2h}^5 in the axial orientation $P2_1/n$. The parameters are those of Table XVF,35.

The structure, as shown in Figure XVF,58, is a stacking of molecules along a_0, each stack having about it six similar stacks. Between piles the shortest distances are C–O = 3.54 A. and C–C = 3.61 A.; within one

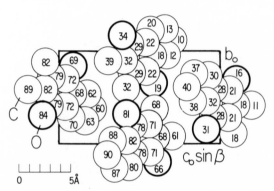

Fig. XVF,58.   The monoclinic structure of 2,3-dimethyl-1,4-naphthoquinone projected along its a_0 axis.

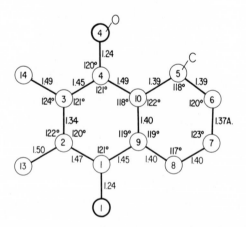

Fig. XVF,59.   Bond dimensions in the molecule of 2,3-dimethyl-1,4-naphthoquinone.

they are C–O = 3.51 A. and C–C = 3.58 A. The molecules, which have the bond dimensions of Figure XVF,59, are planar except for O(4) and one methyl carbon, which are 0.075 A. outside this plane.

XV,f38. Crystals of *2-methyl-3-amino-1,4-naphthoquinone*, $C_{10}H_4O_2$-$(CH_3)NH_2 \cdot 1/4 H_2O$, are monoclinic with a large unit containing eight molecules. Its dimensions are:

$$a_0 = 17.06(2) \text{ A.}; \quad b_0 = 3.91(2) \text{ A.}; \quad c_0 = 30.95(3) \text{ A.}$$
$$\beta = 116°0(20)'$$

The space group is C_{2h}^5 $(P2_1/c)$ with all atoms in the positions:

$$(4e) \quad \pm (xyz; x, 1/2 - y, z + 1/2)$$

Parameters for the two kinds of molecule are listed in Table XVF,36. Positions have been assigned to some but not all the hydrogen atoms.

TABLE XVF,36

Parameters of the Atoms in 2-Methyl-3-amino-1,4-naphthoquinone

Atom	x	y	z
	Molecule A		
C(1)	0.2665	0.5225	0.0844
C(2)	0.1989	0.5210	0.1017
C(3)	0.2185	0.6366	0.1465
C(4)	0.3073	0.7726	0.1785
C(5)	0.4560	0.9343	0.1900
C(6)	0.5180	0.9520	0.1710
C(7)	0.4971	0.8342	0.1245
C(8)	0.4153	0.6855	0.0954
C(9)	0.3524	0.6672	0.1136
C(10)	0.3734	0.7918	0.1600
O(1)	0.2490	0.3996	0.0437
C(CH$_3$)	0.1094	0.3795	0.0690
N(NH$_2$)	0.1600	0.6533	0.1659
O(4)	0.3211	0.8673	0.2192
H(1)	0.1082	0.5389	0.1475
H(2)	0.1845	0.7654	0.1952

(continued)

TABLE XVF,36 (*continued*)

Parameters of the Atoms in 2-Methyl-3-amino-1,4-naphthoquinone

Atom	x	y	z
Molecule B			
C(1')	0.0986	0.8497	0.3697
C(2')	0.1513	0.6793	0.4155
C(3')	0.2383	0.6226	0.4308
C(4')	0.2826	0.7441	0.4003
C(5')	0.2704	0.0067	0.3242
C(6')	0.2200	0.1603	0.2799
C(7')	0.1307	0.2188	0.2651
C(8')	0.0911	0.1236	0.2945
C(9')	0.1419	0.9632	0.3398
C(10')	0.2306	0.9093	0.3536
O(1')	0.0205	0.9076	0.3570
C(CH₃')	0.1059	0.5544	0.4465
N(NH₂')	0.2916	0.4509	0.4713
O(4')	0.3616	0.7033	0.4157
H(1')	0.3511	0.4163	0.4703
H(2')	0.2680	0.3161	0.4894

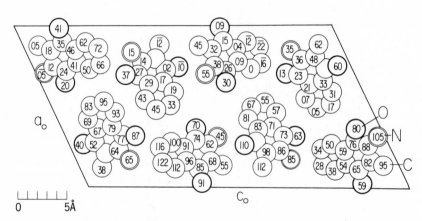

Fig. XVF,60a.   The monoclinic structure of 2-methyl-3-amino-1,4-naphthoquinone projected along its b_0 axis.

The molecules in this structure (Fig. XVF,60) are not exactly planar, atoms attached to the naphthalene rings departing by as much as 0.2 A. from their planes. Between nitrogen atoms and the oxygens of neighboring molecules there are separations of 2.98 and 2.99 A. which presumably represent hydrogen bonds. Possible positions were discussed for the water molecules which may be statistically distributed in succeeding unit cells.

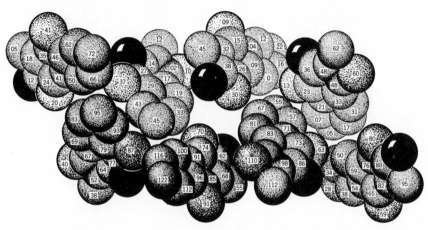

Fig. XVF,60b. A packing drawing of the monoclinic 2-methyl-3-amino-1,4-naphthoquinone arrangement seen along its b_0 axis. The nitrogen atoms are black, the oxygens are heavily outlined and more coarsely dotted than the atoms of carbon.

XV,f39. *Sodium naphthionate tetrahydrate,* $Na(C_{10}H_6NH_2SO_3) \cdot 4H_2O$, is monoclinic with a tetramolecular cell of the dimensions:

$$a_0 = 11.613 \text{ A.}; \quad b_0 = 12.053 \text{ A.}; \quad c_0 = 10.045 \text{ A.}; \quad \beta = 98°48'$$

All atoms are in the positions

$$(4e) \quad \pm (xyz; x, 1/2 - y, z + 1/2)$$

of C_{2h}^5 ($P2_1/c$). The parameters are listed in Table XVF,37.

The resulting structure (Fig. XVF,61) contains organic anions with the bond dimensions of Figure XVF,62. Their naphthalene rings are planar, with the sulfur atom 0.194 A. and the nitrogen atom 0.116 A. to one side of the plane. Six octahedrally coordinated oxygens surround each Na+ ion, four of them belonging to water molecules; these Na–O separations range between 2.300 and 2.583 A. Hydrogen bond lengths are 2.731 A. and greater.

TABLE XVF,37

Parameters of the Atoms in Sodium Naphthionate Tetrahydrate

Atom	x	y	z
C(1)	0.5606	0.0320	0.2996
C(2)	0.4951	−0.0385	0.2110
C(3)	0.3735	−0.0444	0.2073
C(4)	0.3182	0.0213	0.2885
C(5)	0.3816	0.1015	0.3759
C(6)	0.3298	0.1740	0.4612
C(7)	0.3953	0.2472	0.5442
C(8)	0.5171	0.2533	0.5460
C(9)	0.5695	0.1841	0.4662
C(10)	0.5044	0.1062	0.3797
N	0.6831	0.0324	0.3076
O(1)	0.1559	−0.0286	0.4287
O(2)	0.1077	0.1006	0.2470
O(3)	0.1356	−0.0929	0.1975
O(4)	0.1220	0.4406	0.4459
O(5)	0.0092	0.6854	0.0176
O(6)	0.1816	0.6923	0.3388
O(7)	0.1570	0.3295	0.2185
Na	0.0467	0.6267	0.4648
S	0.1682	−0.0019	0.2897

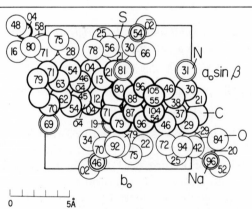

Fig. XVF,61. The monoclinic structure of sodium naphthionate tetrahydrate projected along its c_0 axis. No distinction is made between the sulfonate and water oxygens.

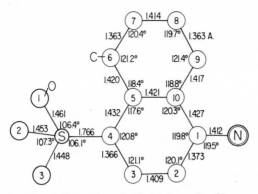

Fig. XVF,62.   Bond dimensions in the anion of sodium naphthionate.

XV,f40.   *Calcium 1-naphthylphosphate trihydrate*, $Ca(C_{10}H_7HPO_4)_2 \cdot 3H_2O$, is triclinic with a large bimolecular unit of the dimensions:

$$a_0 = 7.244(2) \text{ A.}; \quad b_0 = 8.994(3) \text{ A.}; \quad c_0 = 18.725(4) \text{ A.}$$
$$\alpha = 95°41(4)'; \quad \beta = 101°28(3)'; \quad \gamma = 88°52(5)'$$

All atoms are in the general positions $(2i) \pm (xyz)$ of C_i^1 $(P\bar{1})$, with the parameters of Table XVF,38.

The structure is shown in Figure XVF,63. Bond dimensions in the two crystallographically different anions are given in Figure XVF,64, those for one of the molecules being in parentheses. These anions are tied together in the crystal by hydrogen bonds between phosphate

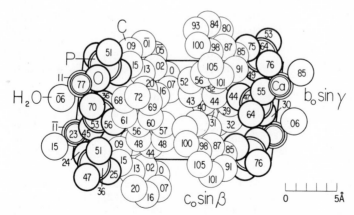

Fig. XVF,63.   The triclinic structure of calcium 1-naphthylphosphate trihydrate projected along its a_0 axis.

TABLE XVF,38

Parameters of the Atoms in $Ca(C_{10}H_7HPO_4)_2 \cdot 3H_2O$

Atom	x	y	z
Ca(1)	0.2309(2)	0.2500(3)	−0.0001(1)
P(2)	0.5257(3)	0.4336(4)	0.8905(1)
P(3)	0.3651(3)	−0.0661(4)	0.8905(1)
O(4)	0.3618(9)	0.5410(9)	0.8697(4)
O(5)	0.4548(8)	0.3056(9)	0.9236(4)
O(6)	0.6969(9)	0.5133(9)	0.9339(4)
O(7)	0.5642(10)	0.3592(9)	0.8141(4)
O(8)	0.2364(8)	0.0152(9)	−0.0675(4)
O(9)	0.5068(8)	0.0431(8)	0.8685(4)
O(10)	0.4678(9)	−0.1934(9)	0.9233(4)
O(11)	0.2481(9)	−0.1414(10)	0.8149(4)
O(12)	0.1538(10)	0.1140(12)	0.0960(5)
O(13)	0.0600(10)	0.3880(12)	−0.0970(5)
O(14)	−0.1061(11)	0.2490(14)	−0.0012(6)
C(15)	0.6826(21)	0.5747(27)	0.7661(8)
C(16)	0.7187(23)	0.6339(24)	0.7069(12)
C(17)	0.6870(29)	0.5580(33)	0.6399(12)
C(18)	0.6044(23)	0.4157(27)	0.6288(9)
C(19)	0.5664(32)	0.3370(35)	0.5588(12)
C(20)	0.4840(32)	0.2050(38)	0.5487(9)
C(21)	0.4427(24)	0.1336(29)	0.6061(9)
C(22)	0.4782(17)	0.2116(23)	0.6749(7)
C(23)	0.5625(18)	0.3537(22)	0.6872(7)
C(24)	0.6056(16)	0.4376(21)	0.7574(7)
C(25)	0.0874(19)	0.0746(22)	0.7672(8)
C(26)	−0.0126(21)	0.1366(25)	0.7059(11)
C(27)	−0.0478(24)	0.0639(32)	0.6396(12)
C(28)	0.0217(21)	−0.0837(29)	0.6292(8)
C(29)	−0.0053(28)	−0.1739(35)	0.5576(9)
C(30)	0.0657(37)	−0.3027(34)	0.5477(9)
C(31)	0.1627(23)	−0.3661(25)	0.6071(8)
C(32)	0.1955(20)	−0.2923(22)	0.6751(8)
C(33)	0.1264(17)	−0.1530(21)	0.6882(7)
C(34)	0.1517(14)	−0.0633(17)	0.7574(7)

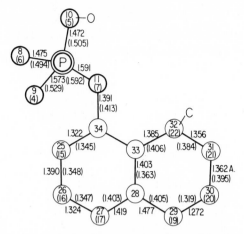

Fig. XVF,64. Bond lengths in the two crystallographically different anions in calcium 1-naphthylphosphate trihydrate.

oxygens, with O–O = 2.533 and 2.568 A. and two more remote. The Ca^{2+} cations have a sevenfold coordination, being distant from three water oxygens (2.422–2.441 A.) and four phosphate oxygens (2.355–2.454 A.). Bond angles around phosphorus range between 103 and 117°, those within the naphthyl rings, between 115 and 127°.

XV,f41. The monoclinic crystals of *bis(benzene-azo-β-naphtholato) copper(II)*, $Cu(C_{16}H_{11}N_2O)_2$, possess a bimolecular unit of the dimensions:

$$a_0 = 17.34 \text{ A.}; \quad b_0 = 3.90 \text{ A.}; \quad c_0 = 17.46 \text{ A.}; \quad \beta = 96°54'$$

The space group is C_{2h}^5 ($P2_1/n$) with the copper atoms in:

$$(2a) \quad 000; {}^1/_2 \, {}^1/_2 \, {}^1/_2$$

The other atoms are in

$$(4e) \quad \pm(xyz; x+{}^1/_2, {}^1/_2-y, z+{}^1/_2)$$

with the parameters of Table XVF,39.

The structure, shown in Figure XVF,65, consists of centrosymmetric molecules with the bond dimensions of Figure XVF,66. The planar naphthalene and benzene rings make an angle of 19° with one another. The oxygen and two nitrogen atoms are within 0.11 A. of the naphthalene plane, whereas N(17) lies in the benzene plane. The coordination of the copper atom is square, and there are two oxygens of adjacent molecules distant 3.00 A. from this coordination plane.

TABLE XVF,39

Parameters of the Atoms in $Cu(C_{16}H_{11}N_2O)_2$

Atom	x	y	z
C(1)	0.1785	0.0820	−0.0179
C(2)	0.1228	0.2523	−0.0719
C(3)	0.1455	0.3672	−0.1439
C(4)	0.2203	0.3059	−0.1609
C(5)	0.3528	0.0923	−0.1303
C(6)	0.4064	−0.0720	−0.0779
C(7)	0.3871	−0.1895	−0.0084
C(8)	0.3120	−0.1336	0.0121
C(9)	0.2558	0.0331	−0.0381
C(10)	0.2767	0.1413	−0.1096
C(11)	0.0975	−0.0823	0.1549
C(12)	0.0364	0.0536	0.1880
C(13)	0.0345	−0.0010	0.2683
C(14)	0.0956	−0.1713	0.3102
C(15)	0.1564	−0.3041	0.2771
C(16)	0.1585	−0.2554	0.1972
N(17)	0.0973	−0.0213	0.0750
N(18)	0.1675	−0.0092	0.0536
O(19)	0.0527	0.3269	−0.0580

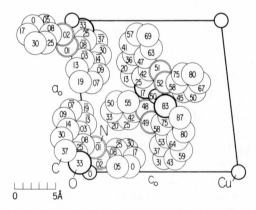

Fig. XVF,65. The monoclinic structure of bis(benzene-azo-β-naphtholato)copper projected along its b_0 axis.

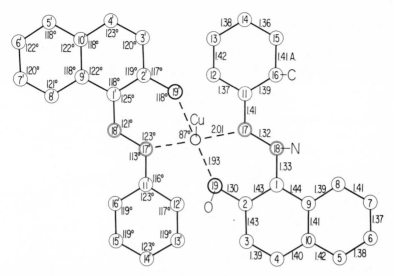

Fig. XVF,66.   Bond dimensions in the molecule of bis(benzene-azo-β-naphtho-lato)copper.

XV,f42.   Crystals of *bis(1-m-tolylazo-2-naphtholato)nickel*, [$C_{10}H_6$-(O)NNC$_6$H$_4$CH$_3$]$_2$Ni, are monoclinic. Their unit is bimolecular and has the dimensions:

$$a_0 = 16.532(6) \text{ A.}; \quad b_0 = 5.082(1) \text{ A.}; \quad c_0 = 19.382(8) \text{ A.}$$
$$\beta = 125.04(3)°$$

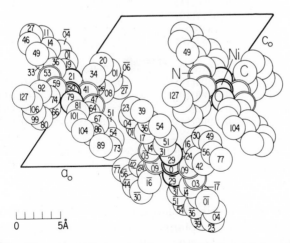

Fig. XVF,67.   The monoclinic structure of bis(1-*m*-tolylazo-2-naphtholato) nickel projected along its b_0 axis.

Nickel atoms are in the positions

$$(2a) \quad 000; \; 0 \; \tfrac{1}{2} \; \tfrac{1}{2}$$

of the space group C_{2h}^5 ($P2_1/c$). All other atoms are in the positions

$$(4e) \quad \pm(xyz; \; x,\tfrac{1}{2}-y,z+\tfrac{1}{2})$$

with the parameters of Table XVF,40.

TABLE XVF,40

Parameters of the Atoms in $[C_{10}H_6(O)NNC_6H_4CH_3]_2Ni$

Atom	x	y	z
O	0.0532(7)	0.287(2)	0.0716(6)
N(1)	0.1244(8)	−0.086(2)	0.0245(7)
N(2)	0.2104(7)	−0.030(3)	0.0915(7)
C(1)	a	0.140(3)	0.1513(8)
C(2)	0.1409(11)	0.308(3)	0.1353(9)
C(3)	0.1644(11)	0.511(5)	0.1943(9)
C(4)	0.2549(12)	0.543(4)	0.2651(10)
C(5)	0.3353(11)	0.361(4)	0.2854(9)
C(6)	0.4318(12)	0.392(4)	0.3609(10)
C(7)	0.5077(13)	0.231(4)	0.3782(11)
C(8)	0.4876(12)	0.045(4)	0.3193(11)
C(9)	0.3966(11)	0.010(5)	0.2447(9)
C(10)	0.3177(10)	0.170(3)	0.2261(9)
C(11)	0.1340(9)	−0.236(3)	−0.0356(8)
C(12)	0.0753(10)	−0.165(3)	−0.1167(9)
C(13)	0.0878(10)	−0.299(3)	−0.1734(9)
C(14)	0.1604(10)	−0.491(4)	−0.1435(9)
C(15)	0.2209(10)	−0.564(3)	−0.0591(9)
C(16)	0.2090(9)	−0.419(3)	−0.0033(8)
C(17)	0.2996(11)	−0.769(4)	−0.0275(9)

a The x parameter for C(1) [0.3198] as stated in the original is clearly in error. The drawing of Figure XVF,67 has been made on the assumption that it should be 0.2198.

The structure that results is shown in Figure XVF,67. Coordination about the nickel atom is planar and almost exactly square, with N–Ni–O = 90.8°. Bond lengths within one-half the molecule, which has a center in nickel, are given in Figure XVF,68.

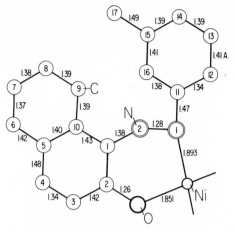

Fig. XVF,68.   Bond lengths in half the centrosymmetric molecule of bis(1-*m*-tolylazo-2-naptholato)nickel.

XV,f43.   Orthorhombic α-*naphthylphenylmethylfluorosilane*, $(C_{10}H_7)(C_6H_5)(CH_3)SiF$, has a tetramolecular unit of the edge lengths:

$$a_0 = 8.774 \text{ A.}; \quad b_0 = 19.890 \text{ A.}; \quad c_0 = 8.105 \text{ A.}$$

All atoms are in the positions

$$(4a) \quad xyz; \tfrac{1}{2}-x, \bar{y}, z+\tfrac{1}{2}; x+\tfrac{1}{2}, \tfrac{1}{2}-y, \bar{z}; \bar{x}, y+\tfrac{1}{2}, \tfrac{1}{2}-z$$

of the space group V^4 ($P2_12_12_1$). The parameters, including values for hydrogen, are listed in Table XVF,41.

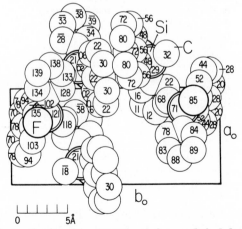

Fig. XVF,69.   The orthorhombic structure of α-naphthylphenylmethylfluorosilane projected along its c_0 axis.

TABLE XVF,41

Parameters of the Atoms in α-Naphthyl Phenyl Methyl Silanes

Atom	x	y	z
Si	0.7749 (0.7685)	0.1707 (0.1714)	0.7112 (0.7207)
F(H)	0.8649 (—)	0.1282 (—)	0.8501 (—)
αC(1)	0.7766 (0.7674)	0.1192 (0.1199)	0.5156 (0.5232)
αC(2)	0.6410 (0.6329)	0.1115 (0.1100)	0.4361 (0.4372)
αC(3)	0.6340 (0.6305)	0.0784 (0.0761)	0.2762 (0.2779)
αC(4)	0.7637 (0.7599)	0.0564 (0.0541)	0.2034 (0.2082)
αC(5)	0.9072 (0.9011)	0.0647 (0.0630)	0.2832 (0.2877)
αC(6)	0.0462 (0.0388)	0.0422 (0.0397)	0.2044 (0.2168)
αC(7)	0.1800 (0.1717)	0.0484 (0.0472)	0.2793 (0.2951)
αC(8)	0.1860 (0.1799)	0.0767 (0.0768)	0.4453 (0.4542)
αC(9)	0.0565 (0.0495)	0.0989 (0.1003)	0.5209 (0.5292)
αC(10)	0.9115 (0.9063)	0.0944 (0.0948)	0.4407 (0.4498)
φC(1)	0.5770 (0.5662)	0.1818 (0.1833)	0.7836 (0.7899)
φC(2)	0.4938 (0.4858)	0.1271 (0.1296)	0.8400 (0.8537)
φC(3)	0.3419 (0.3339)	0.1337 (0.1356)	0.8887 (0.8975)
φC(4)	0.2726 (0.2630)	0.1935 (0.1965)	0.8810 (0.8809)
φC(5)	0.3506 (0.3408)	0.2486 (0.2508)	0.8272 (0.8230)
φC(6)	0.5029 (0.4924)	0.2437 (0.2448)	0.7761 (0.7770)
CH₃	0.8770 (0.8685)	0.2499 (0.2517)	0.6823 (0.6802)
αH(2)	0.520 (0.527)	0.135 (0.129)	0.481 (0.480)
αH(3)	0.550 (0.536)	0.067 (0.077)	0.242 (0.231)
αH(4)	0.774 (0.769)	0.029 (0.029)	0.102 (0.096)
αH(6)	0.006 (0.028)	0.029 (0.020)	0.070 (0.087)
αH(7)	0.286 (0.262)	0.035 (0.029)	0.241 (0.244)
αH(8)	0.285 (0.275)	0.090 (0.086)	0.526 (0.502)
αH(9)	0.064 (0.050)	0.126 (0.119)	0.634 (0.652)
φH(2)	0.523 (0.542)	0.082 (0.079)	0.853 (0.871)
φH(3)	0.322 (0.259)	0.095 (0.096)	0.908 (0.952)
φH(4)	0.167 (0.163)	0.200 (0.199)	0.925 (0.909)
φH(5)	0.335 (0.292)	0.287 (0.290)	0.823 (0.805)
φH(6)	0.572 (0.563)	0.288 (0.280)	0.744 (0.751)

Note: Values in parentheses apply to the hydrogen silane.

The structure, indicated in Figure XVF,69, contains molecules that have the bond dimensions of Figure XVF,70. Silicon exhibits its expected tetrahedral coordination and the benzene ring is planar,

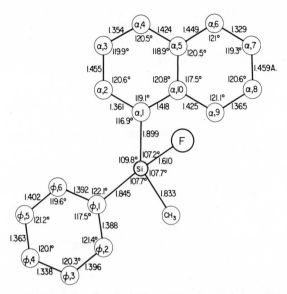

Fig. XVF,70.   Bond dimensions in the molecule of α-naphthylphenylmethylfluorosilane.

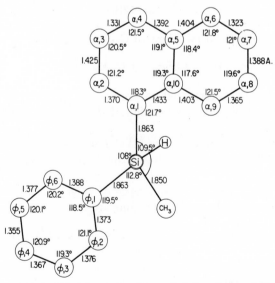

Fig. XVF,71.   Bond dimensions in the molecule of α-naphthylphenylmethyl-
silane.

but there is a definite though small distortion of the naphthyl group. The silicon atom is ca. 0.25 A. outside the best plane through this group and the three atoms C(α1,2 and 9) are 0.03–0.06 A. from it.

The corresponding *silane* $(C_{10}H_7)(C_6H_5)(CH_3)SiH$ is isostructural. Its unit has the edges:

$$a_0 = 8.727 \text{ A.}; \quad b_0 = 19.896 \text{ A.}; \quad c_0 = 7.884 \text{ A.}$$

The parameters are given in parentheses in the table. Bond dimensions in its molecules are shown in Figure XVF,71.

G.   DERIVATIVES OF ANTHRACENE

XV,g1.   *Anthracene,* $C_{14}H_{10}$, with the same type of structure as naphthalene (**XV,f1**), has almost the same a_0 and b_0 axes and β angle. The dimensions of its bimolecular monoclinic unit are:

$$a_0 = 8.562(6) \text{ A.}; \quad b_0 = 6.038(8) \text{ A.}; \quad c_0 = 11.184(8) \text{ A.}$$
$$\beta = 124°42(6)' \qquad (290°K.)$$

Atoms are in the general positions of C_{2h}^5 $(P2_1/a)$:

$$(4e) \quad \pm(xyz; x+{}^1\!/_2, {}^1\!/_2-y, z)$$

The most recently redetermined parameters, as listed in Table XVG,1, are close to those found in the original determination but are more precise.

The structure is shown in Figure XVG,1. Its centrosymmetric planar (to within 0.012 A.) molecules have the bond dimensions of Figure XVG,2.

A detailed study has also been made of this structure at 95°K. At this low temperature the cell dimensions are:

$$a_0 = 8.443(6) \text{ A.}; \quad b_0 = 6.002(7) \text{ A.}; \quad c_0 = 11.124(8) \text{ A.}$$
$$\beta = 125°36(8)'$$

Parameters, very little changed from those at room temperature, are stated in square brackets in the table.

TABLE XVG,1

Parameters of the Atoms in Anthracene

Atom	x	y	z
C(1)	0.0873(6) [0.0862]	0.0271(10) [0.0261]	0.3656(5) [0.3681]
C(2)	0.1187(6) [0.1179]	0.1578(8) [0.1585]	0.2807(4) [0.2835]
C(3)	0.0586(5) [0.0589]	0.0803(7) [0.0790]	0.1382(4) [0.1403]
C(4)	0.0879(5) [0.0878]	0.2094(8) [0.2092]	0.0474(4) [0.0508]
C(5)	0.0304(5) [0.0301]	0.1307(7) [0.1349]	−0.0899(4) [−0.0897]
C(6)	0.0606(6) [0.0612]	0.2594(8) [0.2663]	−0.1835(4) [−0.1821]
C(7)	0.0034(6) [0.0039]	0.1806(9) [0.1876]	−0.3166(5) [−0.3183]
H(1)	0.131(6) [0.141]	0.108(8) [0.097]	0.472(4) [0.475]
H(2)	0.196(6) [0.181]	0.325(8) [0.319]	0.317(4) [0.315]
H(4)	0.155(6) [0.146]	0.376(8) [0.385]	0.085(4) [0.068]
H(6)	0.121(6) [0.128]	0.430(8) [0.448]	−0.143(4) [−0.141]
H(7)	0.024(6) [0.022]	0.324(9) [0.345]	−0.379(4) [−0.382]

Note: Values in square brackets were obtained at 95°K.

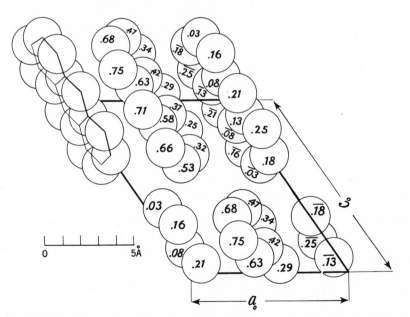

Fig. XVG,1a.   The monoclinic structure of anthracene projected along its b_0 axis. Left-hand axes.

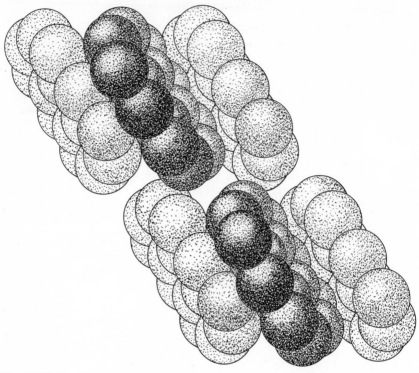

Fig. XVG,1b.   A packing drawing of the monoclinic anthracene structure viewed along its b_0 axis. Left-hand axes.

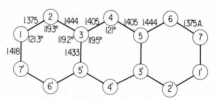

Fig. XVG,2.   Averaged bond dimensions in the molecule of anthracene corrected for rotational effects.

XV,g2.   The molecular complex *7,7,8,8-tetracyanoquinodimethane-anthracene*, $C_6H_4[C(CN)_2]_2 \cdot C_{14}H_{10}$, forms monoclinic crystals whose bimolecular unit has the dimensions:

$$a_0 = 11.476(11) \text{ A.}; \quad b_0 = 12.947(13) \text{ A.}; \quad c_0 = 7.004(7) \text{ A.}$$
$$\beta = 105.4(2)°$$

The space group is C_{2h}^3 ($C2/m$) with atoms in the positions:

C(1) : (4g) $\pm(0u0;\ {}^1/_2,u+{}^1/_2,0)$ with $u = 0.1088(9)$
C(5) : (4i) $\pm(u0v;\ u+{}^1/_2,{}^1/_2,v)$
          with $u = 0.1271(9)$, $v = 0.5509(14)$
C(7) : (4i)   with $u = 0.2506(9)$, $v = 0.6026(15)$

The rest of the atoms are in

(8j) $\pm(xyz;\ x\bar{y}z;\ x+{}^1/_2,y+{}^1/_2,z;\ x+{}^1/_2,{}^1/_2-y,z)$

with the parameters of Table XVG,2.

TABLE XVG,2

Parameters of the Atoms in $C_6H_4[C(CN)_2]_2 \cdot C_{14}H_{10}$

Atom	x	y	z
C(2)	0.1077(7)	0.0551(6)	0.0475(12)
C(3)	0.2214(10)	0.1127(12)	0.0938(16)
C(4)	0.3228(9)	0.0499(13)	0.1382(15)
C(6)	0.0617(6)	0.0975(5)	0.5241(11)
C(8)	0.3198(6)	0.0929(6)	0.6325(12)
N	0.3764(6)	0.1657(6)	0.6574(12)

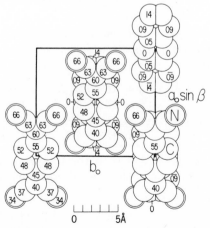

Fig. XVG,3.   The monoclinic structure of 7,7,8,8-tetracyanoquinodimethane-anthracene projected along its c_0 axis.

The structure is that of Figure XVG,3. Its planar molecules, stacked along the c_0 axis, are 3.50 A. apart. Bond dimensions within the molecules are those of Figure XVG,4.

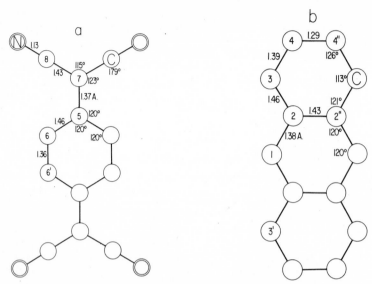

Fig. XVG,4a (*left*).   Bond dimensions of the 7,7,8,8-tetracyanoquinodimethane molecule in its complex with anthracene.

Fig. XVG,4b (*right*).   Bond dimensions of the anthracene molecule in its complex with 7,7,8,8-tetracyanoquinodimethane.

XV,g3.   The complex *anthracene-s-trinitrobenzene*, $C_{14}H_{10} \cdot C_6H_3$-$(NO_2)_3$, forms monoclinic crystals which have a tetramolecular unit of the dimensions:

$$a_0 = 11.70(2) \text{ A.}; \quad b_0 = 16.20(2) \text{ A.}; \quad c_0 = 13.22(2) \text{ A.}$$
$$\beta = 132.8(5)° \qquad \text{(room temperature)}$$
$$a_0 = 11.35(2) \text{ A.}; \quad b_0 = 16.27(2) \text{ A.}; \quad c_0 = 13.02(2) \text{ A.}$$
$$\beta = 133.0(5)° \qquad (-100° \text{ C.})$$

The space group is C_{2h}^6 ($C2/c$) with the following three atoms of the trinitrobenzene component in the special positions:

$$C(10) : (4e) \quad \pm (0 \; u \; 1/4; \; 1/2, u+1/2, 1/4)$$
$$\text{with } u = -0.009 \; (-0.004)$$
$$C(11) : (4e) \quad \text{with } u = 0.168 \; (0.1660)$$
$$N(2) : (4e) \quad \text{with } u = 0.247 \; (0.2556)$$

All other atoms are in the general positions:

(8f) $\pm (xyz;\ x,\bar{y},z+{}^{1}/_{2};\ x+{}^{1}/_{2},y+{}^{1}/_{2},z;\ x+{}^{1}/_{2},{}^{1}/_{2}-y,z+{}^{1}/_{2})$

Parameters were established at both room and low temperatures, those at $-100°C$. being refined from the room temperature values. They are listed in Table XVG,3, values for $-100°C$. being in parentheses.

TABLE XVG,3

Parameters of Atoms in $C_{14}H_{10} \cdot C_6H_3(NO_2)_3$

Atom	x	y	z
C(1)	−0.011 (−0.0091)	0.087 (0.0862)	−0.018 (−0.0175)
C(2)	0.143 (0.1431)	0.044 (0.0464)	0.097 (0.0915)
C(3)	0.283 (0.2842)	0.098 (0.0975)	0.182 (0.1833)
C(4)	0.265 (0.2762)	0.181 (0.1810)	0.156 (0.1655)
C(5)	0.116 (0.1258)	0.222 (0.2196)	0.053 (0.0547)
C(6)	−0.020 (−0.0143)	0.172 (0.1733)	−0.035 (−0.0355)
C(7)	−0.149 (−0.1501)	0.035 (0.0383)	−0.107 (−0.1083)
C(8)	0.145 (0.1466)	0.123 (0.1260)	0.349 (0.3517)
C(9)	0.136 (0.1400)	0.041 (0.0427)	0.344 (0.3471)
N(1)	0.317 (0.2978)	−0.007 (−0.0035)	0.467 (0.4524)
O(1)	0.120 (0.1288)	0.292 (0.2910)	0.330 (0.3388)
O(2)	0.275 (0.2876)	−0.075 (−0.0778)	0.432 (0.4360)
O(3)	0.410 (0.4129)	0.041 (0.0368)	0.541 (0.5419)

Part of the structure is shown in Figure XVG,5. It consists of the separate molecules alternating in stacks along the c_0 axis. These molecules have the bond dimensions (at the lower temperature) of

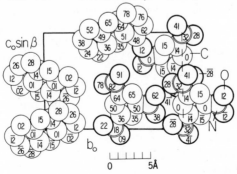

Fig. XVG,5.   A portion of the monoclinic structure of anthracene-*s*-trinitrobenzene projected along its a_0 axis.

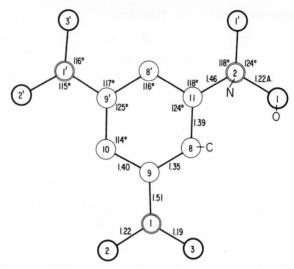

Fig. XVG,6a. Bond dimensions in the *s*-trinitrobenzene molecule in its complex with anthracene.

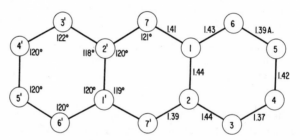

Fig. XVG,6b. Bond dimensions in the anthracene molecule in its complex with *s*-trinitrobenzene.

Figure XVG,6. The anthracene portion is planar. The benzene ring of the nitro component and the $N(1),O(1),O(1')$ group are planar, but the NO_2 groups $[N(1),O(2),O(3)]$ are turned through 18° and the $C(9)$–$N(1)$ bond is inclined 2.5° to the ring. Between molecules in the stacks the shortest atomic separations are C–O = 3.28 A. and C–C = 3.30 A.

XV,g4. Crystals of *anthracene-chromium tricarbonyl*, $C_{14}H_{10} \cdot Cr(CO)_3$, are orthorhombic with a large unit that contains eight molecules and has the edge lengths:

$$a_0 = 13.36(4) \text{ A.;} \quad b_0 = 15.95(6) \text{ A.;} \quad c_0 = 12.94(3) \text{ A.}$$

All atoms are in the general positions

$$(8c) \quad \pm (xyz; \; x+{}^1/_2,{}^1/_2-y,\bar{z}; \; \bar{x},y+{}^1/_2,{}^1/_2-z; \; {}^1/_2-x,\bar{y},z+{}^1/_2)$$

of the space group $V_h{}^{15}$ (*Pbca*). The determined parameters are those of Table XVG,4.

TABLE XVG,4

Parameters of the Atoms in $C_{14}H_{10} \cdot Cr(CO)_3$

Atom	x	y	z
Cr	0.0673(1)	0.1787(1)	0.2018(1)
O(1)	−0.0497(6)	0.1429(6)	0.3917(6)
O(2)	−0.0381(7)	0.3444(5)	0.1941(7)
O(3)	0.2259(6)	0.2636(6)	0.3275(6)
C(1)	0.1789(7)	0.0842(6)	0.1521(7)
C(2)	0.1805(7)	0.1515(6)	0.0807(7)
C(3)	0.0899(7)	0.1764(6)	0.0323(7)
C(4)	0.0006(6)	0.1272(5)	0.0471(6)
C(5)	−0.0894(7)	0.1460(5)	−0.0055(7)
C(6)	−0.1738(6)	0.0976(5)	0.0100(6)
C(7)	−0.2648(8)	0.1124(7)	−0.0468(8)
C(8)	−0.3490(7)	0.0641(8)	−0.0306(8)
C(9)	−0.3470(7)	−0.0012(7)	0.0431(8)
C(10)	−0.2636(7)	−0.0173(6)	0.0990(7)
C(11)	−0.1741(6)	0.0310(5)	0.0846(6)
C(12)	−0.0865(6)	0.0134(5)	0.1379(6)
C(13)	0.0030(6)	0.0598(5)	0.1221(6)
C(14)	0.0921(6)	0.0427(5)	0.1765(6)
C(15)	−0.0034(7)	0.1554(6)	0.3191(7)
C(16)	0.1636(8)	0.2302(6)	0.2772(7)
C(17)	0.0030(8)	0.2801(6)	0.1942(8)

The molecules in this structure have the bond dimensions of Figure XVG,7. As in the analogous phenanthrene compound (**XV,h3, h4**), the $Cr(CO)_3$ portion is approximately centered over one of the end phenyl groups, with Cr–C distances lying between 2.215 and 2.340 A. The two larger of these separations are between Cr and C(4) and C(13). The anthracene part is not strictly planar; instead its two planar parts are bent by 3.2° along a line through C(5) and C(12).

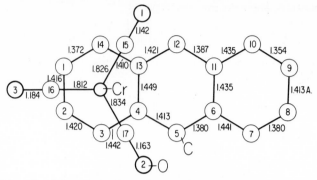

Fig. XVG,7. Bond lengths in the molecule of anthracene-chromium tricarbonyl.

XV,g5. The symmetry of *1,9-5,10-diperinaphthylene-anthracene,* $C_{34}H_{18}$, is monoclinic. Its bimolecular unit has the dimensions:

$$a_0 = 11.95(3) \text{ A.}; \quad b_0 = 7.83(2) \text{ A.}; \quad c_0 = 11.17(3) \text{ A.}$$
$$\beta = 92°18(20)'$$

The space group is the low symmetry C_2^2 ($P2_1$) with atoms in the positions:

$$(2a) \quad xyz; \ \bar{x}, y + \frac{1}{2}, \bar{z}$$

Parameters of the carbon atoms have been given as those of Table XVG,5.

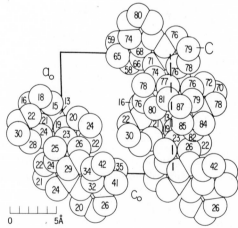

Fig. XVG,8. The monoclinic structure of 1,9-5,10-diperinaphthylene projected along its b_0 axis.

The structure is shown in Figure XVG,8. There is a very wide variation in the bond dimensions of its molecules and it seems likely that the parameters are no more than approximate.

TABLE XVG,5

Parameters of the Atoms in $C_{34}H_{18}$

Atom	x	y	z
C(1)	0.095	0.286	0.147
C(2)	0.040	0.286	0.037
C(3)	0.200	0.261	0.142
C(4)	0.263	0.225	0.252
C(5)	0.388	0.242	0.255
C(6)	0.448	0.197	0.145
C(7)	−0.070	0.242	0.038
C(8)	−0.138	0.242	0.155
C(9)	−0.267	0.203	0.163
C(10)	−0.322	0.197	0.282
C(11)	−0.265	0.261	0.370
C(12)	−0.141	0.258	0.367
C(13)	−0.093	0.411	0.472
C(14)	0.015	0.350	0.483
C(15)	0.073	0.422	0.370
C(16)	0.030	0.342	0.263
C(17)	−0.088	0.325	0.262
C(18)	0.215	0.250	−0.080
C(19)	0.260	0.233	0.033
C(20)	0.108	0.242	−0.070
C(21)	0.045	0.217	−0.170
C(22)	−0.067	0.208	−0.178
C(23)	−0.138	0.244	−0.065
C(24)	0.387	0.186	0.028
C(25)	0.440	0.128	−0.075
C(26)	0.570	0.147	−0.093
C(27)	0.623	0.181	−0.202
C(28)	0.575	0.161	−0.298
C(29)	0.459	0.219	−0.298
C(30)	0.412	0.219	−0.402
C(31)	0.307	0.303	−0.410
C(32)	0.242	0.283	−0.297
C(33)	0.287	0.239	−0.197
C(34)	0.382	0.214	−0.192

XV,g6. A partially disordered structure has been established for *anthrone*, $C_{14}H_{10}O$. Its monoclinic crystals possess a bimolecular unit of the dimensions:

$$a_0 = 15.80(3) \text{ A.}; \quad b_0 = 3.998(5) \text{ A.}; \quad c_0 = 7.86(2) \text{ A.}$$
$$\beta = 101°40(10)'$$

The requirements for the space group C_{2h}^5 ($P2_1/a$) were satisfied and atoms therefore were placed in the positions

$$(4e) \quad \pm (xyz; \ x+\tfrac{1}{2}, \tfrac{1}{2}-y, z)$$

with oxygens distributed statistically over half these positions. The parameters are those of Table XVG,6.

TABLE XVG,6

Parameters of the Atoms in Anthrone

Atom	x	y	z
C(1)	0.1358	0.3934	0.3904
C(2)	0.0550	0.2641	0.3404
C(3)	0.0278	0.1333	0.1714
C(4)	0.0605	0.0107	−0.1265
C(5)	0.0847	0.1463	0.0576
C(6)	0.1666	0.2816	0.1133
C(7)	0.1915	0.4077	0.2787
$\tfrac{1}{2}$O	0.1052	0.0238	−0.2118

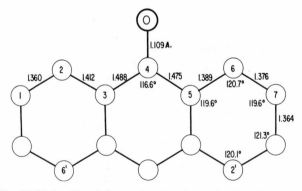

Fig. XVG,9. Bond dimensions in the molecule of anthrone.

The bond dimensions are those of Figure XVG,9. The diffuseness of reflections indicative of disorder is compatible with the structure, described above, which places the molecules in one or the other of two orientations, depending on whether the oxygen atom is attached to C(4) or C(4′).

XV,g7. Crystals of *9-anthraldehyde*, $C_{14}H_9CHO$, are orthorhombic with a tetramolecular unit that has been given the edge lengths:

$$a_0 = 4.29(1) \text{ A.}; \quad b_0 = 14.11(4) \text{ A.}; \quad c_0 = 17.18(4) \text{ A.}$$

Atoms are in the positions

$$(4a) \quad xyz; \; 1/2-x,\bar{y},z+1/2; \; x+1/2,1/2-y,\bar{z}; \; \bar{x},y+1/2,1/2-z$$

of the space group V^4 ($P2_12_12_1$). The determined parameters are listed in Table XVG,7.

TABLE XVG,7

Parameters of the Atoms in 9-Anthraldehyde

Atom	x	y	z
C(1)	0.7427	−0.0117	0.4223
C(2)	0.6500	0.0015	0.3437
C(3)	0.4783	−0.0650	0.3064
C(4)	0.3747	−0.0564	0.2295
C(5)	0.2011	−0.1247	0.1923
C(6)	0.0996	−0.1240	0.1098
C(7)	−0.0951	−0.1919	0.0733
C(8)	−0.2012	−0.2699	0.1173
C(9)	−0.1161	−0.2751	0.1967
C(10)	0.0856	−0.2072	0.2329
C(11)	0.1895	−0.2125	0.3137
C(12)	0.3642	−0.1475	0.3473
C(13)	0.4337	−0.1533	0.4294
C(14)	0.6234	−0.0892	0.4667
C(15)	0.5244	0.0236	0.1906
O	0.4314	0.0655	0.1334

The structure (Fig. XVG,10) is composed of molecules that have the bond dimensions of Figure XVG,11. They are planar except for the aldehyde group for which the C(15) is 0.10 A. on one side and the oxygen 0.26 A. on the other side of the plane. Between molecules the shortest atomic separations are C–O = 3.46 A. and C–C = 3.69 and 3.72 A.

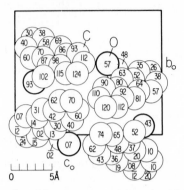

Fig. XVG,10.   The orthorhombic structure of 9-anthraldehyde projected along its a_0 axis.

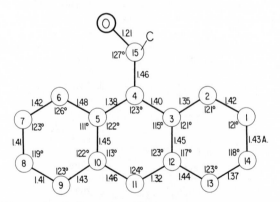

Fig. XVG,11.   Bond dimensions in the molecule of 9-anthraldehyde.

XV,g8.   The tetramolecular monoclinic unit of *9-nitroanthracene*, $C_{14}H_9(NO_2)$, has the dimensions:

$$a_0 = 7.49(2) \text{ A.}; \quad b_0 = 13.77(3) \text{ A.}; \quad c_0 = 11.44(3) \text{ A.}$$
$$\beta = 115°11(10)'$$

Atoms are in the positions

$$(4e) \quad \pm (xyz; \, x+\tfrac{1}{2}, \tfrac{1}{2}-y, z)$$

of C_{2h}^5 ($P2_1/a$) with the parameters of Table XVG,8.

TABLE XVG,8

Parameters of the Atoms in $C_{14}H_9(NO_2)$

Atom	x	y	z
C(1)	0.3165	0.1432	−0.0977
C(2)	0.3331	0.1037	0.0176
C(3)	0.3454	0.1715	0.1153
C(4)	0.3599	0.1396	0.2370
C(5)	0.3727	0.1994	0.3373
C(6)	0.3860	0.1634	0.4586
C(7)	0.3980	0.2283	0.5563
C(1')	0.3942	0.3314	0.5358
C(2')	0.3802	0.3710	0.4168
C(3')	0.3703	0.3018	0.3245
C(4')	0.3556	0.3358	0.2017
C(5')	0.3425	0.2769	0.1031
C(6')	0.3265	0.3114	−0.0222
C(7')	0.3143	0.2432	−0.1136
N	0.3631	0.0335	0.2592
O(1)	0.2129	−0.0103	0.2417
O(2)	0.5242	−0.0046	0.2958

The structure, shown in Figure XVG,12, is composed of molecules with the average bond dimensions of Figure XVG,13. They are planar except for the oxygen atoms which lie ca. 1.0 A. on either side of the plane of the ring, corresponding to a tilt of 85° about the C–N bond. Between molecules the shortest atomic separation is a C–O = 3.05 A.;

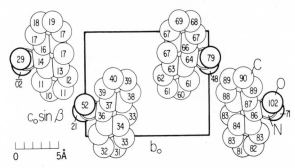

Fig. XVG,12. The monoclinic structure of 9-nitroanthracene projected along its a_0 axis.

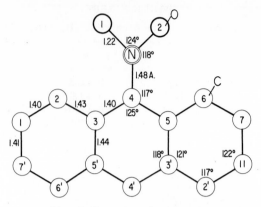

Fig. XVG,13. Averaged bond dimensions in the molecule of 9-nitroanthracene.

other values range upward from a C–O = 3.48 A. The surprisingly open character of this arrangement is noteworthy.

XV,g9. The triclinic crystals of *9,10-dinitroanthracene*, $C_{14}H_8(NO_2)_2$, have a unimolecular cell of the dimensions:

$$a_0 = 3.95(1) \text{ A.}; \quad b_0 = 8.68(2) \text{ A.}; \quad c_0 = 8.76(2) \text{ A.}$$
$$\alpha = 106°46(10)'; \quad \beta = 98°59(10)'; \quad \gamma = 98°1(10)'$$

All atoms are in the positions $(2i) \pm (xyz)$ of C_i^1 $(P\bar{1})$. Their parameters are stated to be those of Table XVG,9.

TABLE XVG,9

Parameters of the Atoms in $C_{14}H_8(NO_2)_2$

Atom	x	y	z
C(1)	0.6210	0.3702	0.2983
C(2)	0.4213	0.3188	0.1455
C(3)	0.2158	0.1603	0.0654
C(4)	0.0086	0.1006	−0.0906
C(5)	−0.1931	−0.0579	−0.1700
C(6)	−0.4006	−0.1166	−0.3291
C(7)	−0.6058	−0.2662	−0.3941
N	0.0281	0.2101	−0.1878
O(1)	−0.0746	0.3416	−0.1443
O(2)	0.1571	0.1741	−0.3108

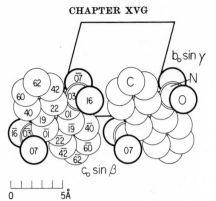

Fig. XVG,14.   The triclinic structure of 9,10-dinitroanthracene projected along
its a_0 axis.

The structure is shown in Figure XVG,14. Its centrosymmetric
molecules have the bond dimensions of Figure XVG,15. Except for the
oxygen atoms, they are planar; the departures of the oxygens from this
plane are such that the NO₂ groups appear twisted by 64° about the
C–N bond. Between molecules the shortest atomic separations are
O–O = 3.06 A. and C–O = 3.36 and 3.42 A.

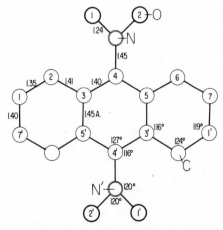

Fig. XVG,15.   Bond dimensions in the molecule of 9,10-dinitroanthracene.

XV,g10. Crystals of *9,10-dibromoanthracene*, $C_{14}H_8Br_2$, have a
triclinic, bimolecular unit of the dimensions:

$$a_0 = 4.06(1) \text{ A.}; \quad b_0 = 8.88(2) \text{ A.}; \quad c_0 = 16.15(4) \text{ A.}$$
$$\alpha = 98°50(10)'; \quad \beta = 97°5(10)'; \quad \gamma = 100°21(10)'$$

All atoms are in the positions $(2i)$ $\pm (xyz)$ of C_i^1 $(P\bar{1})$. The parameters are those of Table XVG,10, there being two crystallographically different molecules, each with a center of symmetry.

TABLE XVG,10

Parameters of the Atoms in $C_{14}H_8Br_2$

Atom	x	y	z
Molecule I			
Br	−0.0671	0.3517	0.1722
C(1)	0.6305	0.8757	0.0896
C(2)	0.4508	0.7765	0.0089
C(3)	0.2242	0.6414	0.0007
C(4)	−0.0270	0.4420	0.0730
C(5)	0.1791	0.5805	0.0786
C(6)	0.3863	0.6570	0.1614
C(7)	0.6050	0.7951	0.1621
Molecule II			
Br′	0.0268	0.6700	0.3811
C(1′)	0.3850	0.1365	0.3130
C(2′)	0.2567	0.0009	0.3457
C(3′)	0.1283	0.0005	0.4227
C(4′)	0.0155	0.8676	0.4511
C(5′)	0.1283	0.1532	0.4743
C(6′)	0.2567	0.2922	0.4448
C(7′)	0.3850	0.2796	0.3711

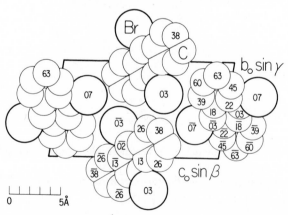

Fig. XVG,16.   The triclinic structure of 9,10-dibromoanthracene projected along its a_0 axis.

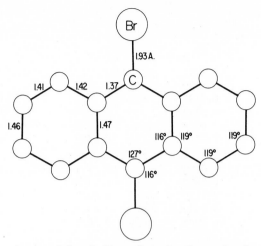

Fig. XVG,17. Averaged bond dimensions in the molecule of 9,10-dibromo-anthracene.

The structure is that of Figure XVG,16. Its molecules, planar and of substantially the same dimensions, have, on averaging, the bonds of Figure XVG,17. Between them the shortest atomic separations are C–O = 3.70 and 3.71 A. and C–C = 3.73 A.

The isostructural 9-chloro-10-bromoanthracene, $C_{14}H_8ClBr$, has a cell of the dimensions:

$$a_0 = 4.06(1) \text{ A.}; \quad b_0 = 8.87(2) \text{ A.}; \quad c_0 = 16.14(3) \text{ A.}$$
$$\alpha = 98°27(10)'; \quad \beta = 97°10(10)'; \quad \gamma = 100°39(10)'$$

Atomic positions have not been established.

XV,g11. The tetramolecular, monoclinic unit of *9,10-dichloro-anthracene*, $C_{14}H_8Cl_2$, has the dimensions:

$$a_0 = 7.04(2) \text{ A.}; \quad b_0 = 17.93(4) \text{ A.}; \quad c_0 = 8.63(2) \text{ A.}$$
$$\beta = 102°56(10)'$$

The space group is C_{2h}^5 ($P2_1/a$) with atoms in the positions:

$$(4e) \quad \pm (xyz; x+1/2, 1/2-y, z)$$

The parameters are those of Table XVG,11.

TABLE XVG,11

Parameters of the Atoms in $C_{14}H_8Cl_2$

Atom	x	y	z
Cl(1)	0.2620	0.2976	0.5505
Cl(1′)	0.1137	0.1223	−0.0939
C(1)	0.2540	0.0540	0.5170
C(2)	0.2523	0.1292	0.5177
C(3)	0.2180	0.1708	0.3732
C(4)	0.2180	0.2471	0.3749
C(5)	0.1850	0.2880	0.2268
C(6)	0.1830	0.3644	0.2353
C(7)	0.1460	0.4060	0.0901
C(1′)	0.1093	0.3691	−0.0556
C(2′)	0.1130	0.2924	−0.0580
C(3′)	0.1489	0.2494	0.0761
C(4′)	0.1530	0.1735	0.0822
C(5′)	0.1830	0.1298	0.2213
C(6′)	0.1830	0.0477	0.2403
C(7′)	0.2180	0.0086	0.3803

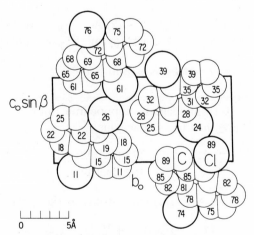

Fig. XVG,18.   The monoclinic structure of 9,10-dichloroanthracene projected along its a_0 axis.

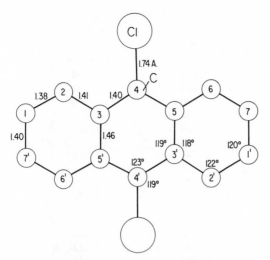

Fig. XVG,19. Averaged bond dimensions in the molecule of 9,10-dichloroan-thracene.

The structure, shown in Figure XVG,18, is built up of molecules with the averaged bond dimensions of Figure XVG,19. These molecules are planar within the limit of accuracy of the determination. Between them the shortest C–C = 3.52 A., Cl–C = 3.70 A., and Cl–Cl = 3.73 A.

XV,g12. Crystals of *9-ethyl-10-bromoanthracene*, $C_{14}H_8Br(C_2H_5)$, have orthorhombic symmetry. The edge lengths of their tetramolecular unit are:

$$a_0 = 16.78(7) \text{ A.}; \quad b_0 = 14.47(3) \text{ A.}; \quad c_0 = 5.13(2) \text{ A.}$$

Atoms are in the positions:

(4a) $xyz; \ 1/2-x, \bar{y}, z+1/2; \ x+1/2, 1/2-y, \bar{z}; \ \bar{x}, y+1/2, 1/2-z$

of the space group V^4 ($P2_12_12_1$). Assigned parameters are listed in Table XVG,12.

The structure is that of Figure XVG,20; it makes C(16) and Br of neighboring molecules only 2.97 A. apart. Its molecules, having the bond dimensions of Figure XVG,21, are reported to show appreciable departures from planarity.

The compound *9-cyanoanthracene*, $C_{14}H_9(CN)$, has the same symmetry and space group and a cell of much the same size as the foregoing.

TABLE XVG,12

Parameters of the Atoms in $C_{14}H_8Br(C_2H_5)$

Atom	x	y	z
C(1)	0.154	0.360	0.914
C(2)	0.168	0.437	0.070
C(3)	0.108	0.464	0.259
C(4)	0.033	0.420	0.259
C(5)	0.860	0.157	0.000
C(6)	0.857	0.079	0.849
C(7)	0.919	0.047	0.677
C(8)	0.990	0.100	0.660
C(9)	0.066	0.239	0.755
C(10)	0.946	0.293	0.121
C(11)	0.998	0.181	0.814
C(12)	0.930	0.214	0.962
C(13)	0.080	0.312	0.940
C(14)	0.014	0.346	0.090
C(15)	0.125	0.209	0.570
C(16)	0.193	0.141	0.635
Br	0.8675	0.3230	0.3710

Its unit has the edges:

$$a_0 = 17.15(2) \text{ A.}; \quad b_0 = 15.11 \text{ A.}; \quad c_0 = 3.93(2) \text{ A.}$$

The determined parameters are those of Table XVG,13.

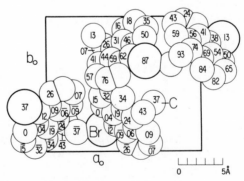

Fig. XVG,20. The orthorhombic structure of 9-ethyl-10-bromoanthracene projected along its c_0 axis.

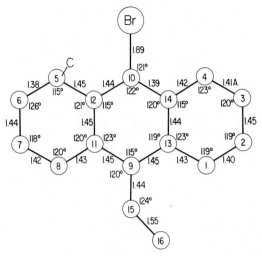

Fig. XVG,21. The bond dimensions in the molecule of 9-ethyl-10-bromoan-thracene.

TABLE XVG,13

Parameters of the Atoms in 9-Cyanoanthracene

Atom	x	y	z
C(1)	0.630	0.382	0.64
C(2)	0.670	0.450	0.80
C(3)	0.625	0.526	0.88
C(4)	0.547	0.532	0.82
C(5)	0.309	0.403	0.40
C(6)	0.270	0.334	0.23
C(7)	0.311	0.259	0.11
C(8)	0.390	0.248	0.145
C(9)	0.510	0.316	0.40
C(10)	0.428	0.463	0.60
C(11)	0.550	0.385	0.56
C(12)	0.507	0.462	0.67
C(13)	0.392	0.392	0.44
C(14)	0.430	0.316	0.315
C(15)	0.552	0.240	0.285
N	0.584	0.178	0.19

Note: $\sigma(x) = \sigma(y) = 0.002$, $\sigma(z) = 0.01$ for all atoms.

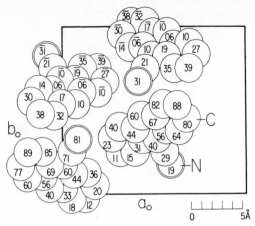

Fig. XVG,22.   The orthorhombic structure of 9-cyanoanthracene projected along
its c_0 axis.

As can be seen from Figure XVG,22, though the distribution of the
molecules is the same as in the preceding compound, their orientations
are entirely different and so are the intermolecular contacts. Molecules
are said to be almost planar, with the bond dimensions of Figure
XVG,23.

Crystals of 9-bromo-10-methyl-anthracene, $C_{14}H_8Br(CH_3)$, appear to
have a structure resembling the foregoing. The unit has the dimensions:

$$a_0 = 17.6 \text{ A.}; \quad b_0 = 16.2 \text{ A.}; \quad c_0 = 3.98 \text{ A.}$$

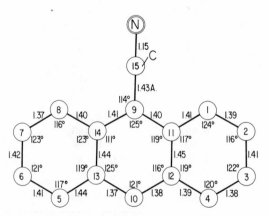

Fig. XVG,23.   Bond dimensions in the molecule of 9-cyanoanthracene.

Only approximate refinement of structure was possible, and this has been attributed to disorder expressed as a statistical orientation of the molecules, the bromine atoms being sometimes in place of the methyl groups and vice versa. It has been inferred that this gives rise to a certain indefiniteness in the positions of the carbon atoms. Approximate values for x and y but not for z (except for Br and CH_3) are given in the original.

XV,g13. The triclinic *9,10-anthrahydroquinone dibenzoate*, $C_{28}H_{18}O_4$, has a unimolecular cell of the dimensions:

$$a_0 = 8.91 \text{ A.}; \quad b_0 = 12.60 \text{ A.}; \quad c_0 = 5.83 \text{ A.}$$
$$\alpha = 105°12'; \quad \beta = 106°30'; \quad \gamma = 58°12'$$

Its atoms are in the positions $(2i) \pm (xyz)$ of C_i^1 $(P\bar{1})$ with the parameters listed in Table XVG,14. Calculated values for hydrogen are given in the original paper.

TABLE XVG,14

Parameters of the Atoms in 9,10-Anthrahydroquinone Dibenzoate

Atom	x	y	z
O(1)	0.1643	0.1404	0.0268
O(2)	−0.0020	0.2973	0.2892
C(1)	0.0782	0.0728	0.0149
C(2)	0.1447	−0.0064	0.1948
C(3)	0.2896	−0.0138	0.3919
C(4)	0.3501	−0.0925	0.5554
C(5)	0.2733	−0.1703	0.5398
C(6)	0.1292	−0.1623	0.3585
C(7)	0.0676	−0.0825	0.1780
C(8)	0.1024	0.2598	0.1593
C(9)	0.1979	0.3241	0.1478
C(10)	0.2937	0.2834	−0.0427
C(11)	0.3781	0.3516	−0.0408
C(12)	0.3624	0.4529	0.1317
C(13)	0.2712	0.4911	0.3015
C(14)	0.1865	0.4263	0.3178

The structure (Fig. XVG,24) has a molecule that possesses a center of symmetry and the bond dimensions of Figure XVG,25. The benzoate

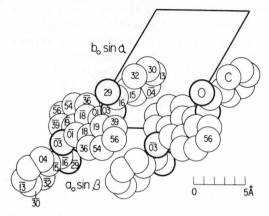

Fig. XVG,24. The triclinic structure of 9,10-anthrahydroquinone dibenzoate projected along its c_0 axis.

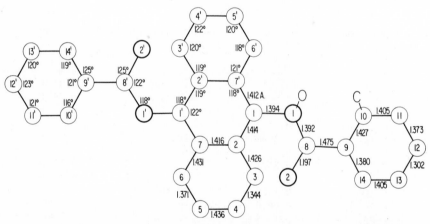

Fig. XVG,25. Bond dimensions in the molecule of 9,10-anthrahydroquinone dibenzoate.

ring is planar except for its oxygen and so is the anthraquinone part of the molecule. Between molecules the shortest C–C distance is 3.24 A.

XV,g14. Crystals of *10-dicyanomethylene anthrone,* $C_{17}H_8N_2O$, possess an orthorhombic, tetramolecular unit of the edge lengths:

$$a_0 = 7.520(14) \text{ A.}; \quad b_0 = 8.843(6) \text{ A.}; \quad c_0 = 18.015(8) \text{ A.}$$

The space group was chosen as V_h^{16} (*Pnam*) with atoms in the positions:

(4c) $\pm\,(u\ v\ {}^1\!/_4;\ u+{}^1\!/_2,{}^1\!/_2-v,{}^1\!/_4)$

(8d) $\pm\,(xyz;\ x+{}^1\!/_2,{}^1\!/_2-y,{}^1\!/_2-z;\ x,y,{}^1\!/_2-z;\ x+{}^1\!/_2,{}^1\!/_2-y,z)$

Parameters are listed in Table XVG,15.

TABLE XVG,15

Positions and Parameters of the Atoms in 10-Dicyanomethylene Anthrone

Atom	Position	x	y	z
C(1)	(8d)	0.5827(14)	0.0601(11)	0.3828(6)
C(2)	(8d)	0.6184(15)	0.1429(12)	0.4511(7)
C(3)	(8d)	0.5940(15)	0.3056(12)	0.4519(7)
C(4)	(8d)	0.5473(15)	0.3794(11)	0.3865(6)
C(5)	(8d)	0.5209(13)	0.2972(12)	0.3206(6)
C(6)	(8d)	0.5367(13)	0.1362(10)	0.3193(6)
C(7)	(4c)	0.4966(19)	0.0534(14)	$^1/_4$
C(8)	(4c)	0.4915(20)	0.3782(15)	$^1/_4$
C(9)	(4c)	0.4070(18)	−0.0807(13)	$^1/_4$
C(10)	(8d)	0.3512(15)	−0.1534(10)	0.3151(6)
N	(8d)	0.2965(13)	−0.2173(11)	0.3659(5)
O	(4c)	0.4487(16)	0.5133(11)	$^1/_4$

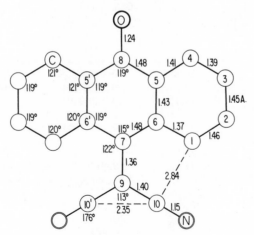

Fig. XVG,26. Bond dimensions in the molecule of 10-dicyanomethylene anthrone.

The resulting structure is built of molecules with the bond dimensions of Figure XVG,26. The central anthracene part is planar within 0.09 A. but the dihedral angle between its strictly planar terminal benzene rings is 152°. The oxygen atom lies in the mean anthracene plane but atom C(9) is 0.79 A. outside it. A plane through the dicyanomethylene group is tilted through 36.5° with respect to the anthracene plane. This warping of the molecule is apparently necessary in order to obtain a sufficient separation between C(1) and C(10). Between molecules the minimum atomic distances are C–O = 3.26 A. and C–C = 3.43 A.

XV,g15. The structure of *anthraquinone*, $C_{14}H_8O_2$, has been repeatedly investigated, most recently at several temperatures between -170 and 20°C. Its crystals are monoclinic with a bimolecular unit of the dimensions:

$$a_0 = 15.810(15) \text{ A.}; \quad b_0 = 3.942(5) \text{ A.}; \quad c_0 = 7.865(10) \text{ A.}$$
$$\beta = 102°43(2)' \qquad \text{(room temperature)}$$

The space group C_{2h}^5 ($P2_1/a$) places its atoms in the positions:

$$(4e) \quad \pm (xyz; \; x + \tfrac{1}{2}, \tfrac{1}{2} - y, z)$$

A recent reworking of earlier data has led to the parameters of Table XVG,16. The values resulting from the work at reduced temperatures differ slightly among themselves, as would be expected; their departures from the parameters of the table are also minor.

TABLE XVG,16

Parameters of the Atoms in Anthraquinone

Atom	x	y	z
C(1)	0.1314(5)	0.4166(25)	0.3985(10)
C(2)	0.0507(5)	0.2707(25)	0.3406(9)
C(3)	0.0277(4)	0.1390(19)	0.1742(9)
C(4)	0.0850(4)	0.1547(19)	0.0616(9)
C(5)	0.1662(5)	0.2960(23)	0.1215(10)
C(6)	0.1894(5)	0.4274(24)	0.2894(11)
C(7)	0.0605(4)	0.0137(23)	−0.1157(9)
O	0.1109(4)	0.0229(21)	−0.2119(8)
H(1)	0.145	0.514	0.519
H(2)	0.007	0.245	0.416
H(3)	0.207	0.323	0.036
H(4)	0.246	0.565	0.337

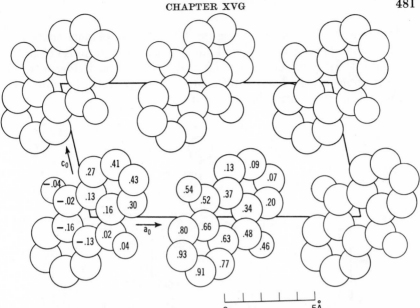

Fig. XVG,27a.   The monoclinic structure of anthraquinone projected along its b_0 axis. The smaller circles are the oxygen. The y parameters shown here are those of an early determination. Left-hand axes.

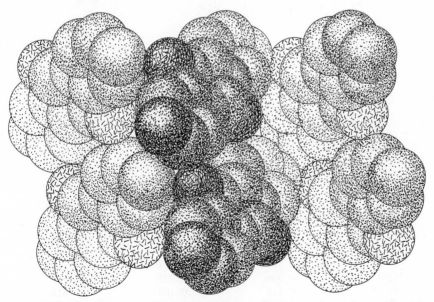

Fig. XVG,27b.   A packing drawing of the monoclinic anthraquinone structure seen along its b_0 axis. The oxygen atoms are line shaded, the carbons are dotted. Left-hand axes.

The structure is shown in Figure XVG,27. Its planar, centrosymmetric molecules have the bond dimensions of Figure XVG,28, the distances C–H lying between 1.00 and 1.04 A. The influence of the oxygen atoms on the bond lengths of the carbon atoms to which they are attached (the central ring) is clear-cut.

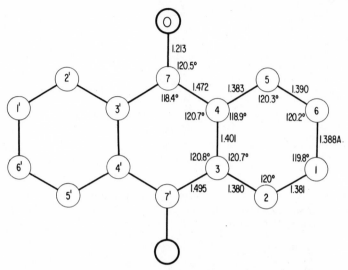

Fig. XVG,28.   Bond dimensions in the molecule of anthraquinone.

TABLE XVG,17

Parameters of the Atoms in 1,5-Dibromoanthraquinone

Atom	x	y	z
C(1)	0.207	−0.108	0.253
C(2)	0.210	−0.214	0.235
C(3)	0.114	−0.264	0.092
C(4)	0.017	−0.214	−0.027
C(5)	0.011	−0.108	−0.014
C(6)	0.108	−0.051	0.144
C(7)	0.098	0.058	0.160
O	0.172	0.109	0.316
Br	0.3473	−0.0489	0.4518

XV,g16. The bimolecular monoclinic unit of *1,5-dibromoanthra-quinone*, $C_{14}H_6Br_2O_2$, has the dimensions:

$$a_0 = 11.24(2) \text{ A.}; \quad b_0 = 13.43(3) \text{ A.}; \quad c_0 = 3.96(1) \text{ A.}$$
$$\beta = 91°26(12)'$$

Atoms are in the positions

$$(4e) \quad \pm (xyz; \; x+{}^1/_2, {}^1/_2-y, z)$$

of the space group C_{2h}^5 ($P2_1/a$). Their parameters have been determined to have the values of Table XVG,17.

The structure is that of Figure XVG,29. The bond dimensions of its molecules (Fig. XVG,30) are recent refinements of the original values.

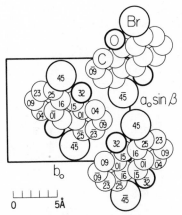

Fig. XVG,29.   The monoclinic structure of 1,5-dibromoanthraquinone projected along its c_0 axis.

XV,g17. According to a preliminary note, crystals of *2,3-dibromo-1,4-anthraquinone*, $C_{14}H_6Br_2O_2$, are monoclinic with a tetramolecular unit of the dimensions:

$$a_0 = 8.86(2) \text{ A.}; \quad b_0 = 5.76(2) \text{ A.}; \quad c_0 = 22.57(2) \text{ A.}; \quad \beta = 92°$$

The space group is C_{2h}^5 ($P2_1/c$) with atoms in the positions:

$$(4e) \quad \pm (xyz; \; x, {}^1/_2-y, z+{}^1/_2)$$

Assigned parameters are listed in Table XVG,18.

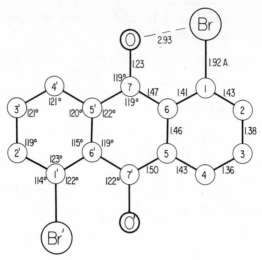

Fig. XVG,30.   Bond dimensions in the molecule of 1,5-dibromoanthraquinone.

TABLE XVG,18

Parameters of the Atoms in $C_{14}H_6Br_2O_2$

Atom	x	y	z
C(1)	0.706	0.217	−0.070
C(2)	0.789	0.307	−0.124
C(3)	0.875	0.496	−0.124
C(4)	0.895	0.646	−0.072
C(5)	0.741	0.826	0.127
C(6)	0.669	0.778	0.178
C(7)	0.577	0.554	0.180
C(8)	0.572	0.415	0.130
C(9)	0.645	0.330	0.029
C(10)	0.817	0.732	0.027
C(11)	0.723	0.384	−0.020
C(12)	0.812	0.584	−0.021
C(13)	0.655	0.472	0.080
C(14)	0.740	0.674	0.081
O(1)	0.626	0.052	−0.073
O(4)	0.976	0.812	−0.073
Br(2)	0.7605	0.1118	−0.1913
Br(3)	0.9828	0.5781	−0.1910

The resulting structure is that of Figure XVG,31.

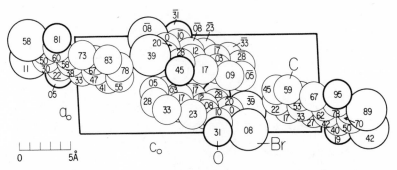

Fig. XVG,31. The monoclinic structure of 2,3-dibromo-1,4-anthraquinone projected along its b_0 axis.

XV,g18. Crystals of *1,5-diiodoanthraquinone*, $C_{14}H_6I_2O_2$, are monoclinic with a bimolecular unit of the dimensions:

$$a_0 = 6.91(2) \text{ A.}; \quad b_0 = 4.27(1) \text{ A.}; \quad c_0 = 22.18(7) \text{ A.}$$
$$\beta = 107°54'$$

Their atoms are in the positions

$$(4e) \quad \pm (xyz; x, 1/2 - y, z + 1/2)$$

of C_{2h}^5 ($P2_1/c$). The recently refined parameters are those of Table XVG,19.

TABLE XVG,19

Parameters of the Atoms in 1,5-Diiodoanthraquinone

Atom	x	y	z
C(1)	0.205	−0.191	0.128
C(2)	0.412	−0.089	0.160
C(3)	0.511	0.081	0.124
C(4)	0.422	0.156	0.063
C(5)	0.207	0.084	0.032
C(6)	0.103	−0.117	0.067
C(7)	−0.113	−0.200	0.031
O	−0.193	−0.403	0.055
I	0.0779	−0.4122	0.1903

The resulting structure is illustrated in Figure XVG,32. Except for oxygen and iodine, the molecule is nearly planar, but these two atoms are displaced by −0.34 and 0.17 A. on either side of the plane. The shortest intermolecular separation is a C–O = 3.55 A.

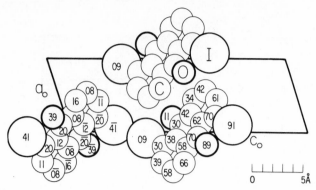

Fig. XVG,32.   The monoclinic structure of 1,5-diiodoanthraquinone projected
along its b_0 axis.

XV,g19. The monoclinic crystals of *1,5-dichloroanthraquinone,*
$C_{14}H_6Cl_2O_2$, possess a bimolecular unit of the dimensions:

$$a_0 = 11.01 \text{ A.}; \quad b_0 = 13.06 \text{ A.}; \quad c_0 = 3.84 \text{ A.}; \quad \beta = 92°7'$$

The space group is C_{2h}^5 $(P2_1/a)$ with atoms in the positions:

$$(4e) \quad \pm (xyz;\ x+{}^1/_2, {}^1/_2-y,z)$$

Determined parameters are listed in Table XVG,20.

TABLE XVG,20

Parameters of the Atoms in $C_{14}H_6Cl_2O_2$

Atom	x	y	z
C(1)	0.2096	0.8903	0.238
C(2)	0.2123	0.7864	0.214
C(3)	0.1117	0.7348	0.089
C(4)	0.0143	0.7864	0.959
C(5)	0.0108	0.8916	0.986
C(6)	0.1082	0.9483	0.128
C(7)	0.0952	0.0565	0.117
O(8)	0.1745	0.1086	0.333
Cl(9)	0.3381	0.9471	0.425

This is a structure (Fig. XVG,33) built of centrosymmetric molecules
with the bond dimensions of Figure XVG,34. This molecule shows a
considerable departure from planarity, the C(7) atom being 0.14 A.

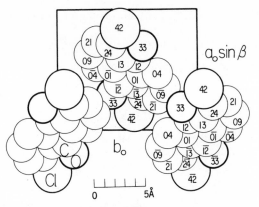

Fig. XVG,33. The monoclinic structure of 1,5-dichloroanthraquinone projected along its c_0 axis.

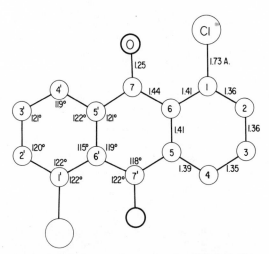

Fig. XVG,34. Bond dimensions in the molecule of 1,5-dichloroanthraquinone.

and the oxygen atom 0.29 A. out of the plane of the rest. The shortest intermolecular distance is 3.1 A.

XV,g20. Crystals of *1,5-difluoroanthraquinone*, $C_{14}H_6F_2O_2$, have a monoclinic, tetramolecular unit of the dimensions:

$$a_0 = 18.13(17) \text{ A.}; \quad b_0 = 3.83(3) \text{ A.}; \quad c_0 = 16.48(15) \text{ A.}$$
$$\beta = 114°46(28)'$$

The space group C_{2h}^5 ($P2_1/c$) puts all atoms in the positions

$$(4e) \quad \pm (xyz;\ x,{}^1/_2 - y, z + {}^1/_2)$$

with the parameters of Table XVG,21.

TABLE XVG,21

Parameters of the Atoms in $C_{14}H_6F_2O_2$

Atom	x	y	z
	Molecule I		
CI(1)	0.1630	0.133	0.1255
CI(2)	0.1820	0.273	0.2109
CI(3)	0.1169	0.361	0.2341
CI(4)	0.0361	0.298	0.1731
CI(5)	0.0210	0.151	0.0878
CI(6)	0.0840	0.061	0.0624
CI(7)	0.0650	−0.086	−0.0266
OI	0.1194	−0.165	−0.0508
FI	0.2264	0.048	0.1073
	Molecule II		
CII(1)	0.3408	−0.269	−0.0385
CII(2)	0.3200	−0.413	0.0252
CII(3)	0.3793	−0.432	0.1141
CII(4)	0.4576	−0.300	0.1355
CII(5)	0.4772	−0.156	0.0670
CII(6)	0.4188	−0.136	−0.0229
CII(7)	0.4384	0.025	−0.0939
OII	0.3887	0.048	−0.1702
FII	0.2818	−0.251	−0.1234

The structure, as shown in Figure XVG,35, is built up of two crystallographically different types of molecule. Their bond dimensions are indicated in Figure XVG,36, those for the II molecule being given in parentheses. These molecules, each with a center of symmetry, are planar. Between molecules the shortest atomic separations are F–C = 3.12 and 3.15 A.; the shortest O–F = 3.63 A.

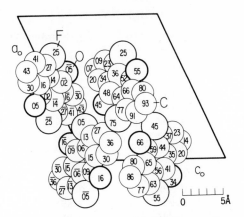

Fig. XVG,35.   The monoclinic structure of 1,5-difluoroanthraquinone projected along its b_0 axis.

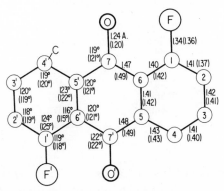

Fig. XVG,36.   Bond dimensions in the molecule of 1,5-difluoroanthraquinone.

XV,g21.   According to a preliminary note, crystals of *1,4-dihydroxy-anthraquinone*, $C_{14}H_6O_2(OH)_2$, have a monoclinic, tetramolecular unit of the dimensions:

$$a_0 = 20.56 \text{ A.}; \quad b_0 = 6.06 \text{ A.}; \quad c_0 = 10.53 \text{ A.}; \quad \beta = 125°10'$$

All atoms are in the positions

$$(4e) \quad \pm (xyz; \ x+1/2, 1/2 - y, z)$$

of C_{2h}^5 ($P2_1/a$). The stated parameters are those of Table XVG,22, values for y having been chosen from assumptions of bond lengths rather than by experiment.

TABLE XVG,22

Parameters of the Atoms in 1,4-Dihydroxyanthraquinone

Atom	x	y^{a}	z
C(1)	0.042	0.351	0.217
C(2)	0.041	0.189	0.133
C(3)	0.102	0.155	0.109
C(4)	0.165	0.336	0.175
C(5)	0.163	0.495	0.260
C(6)	0.104	0.536	0.289
C(7)	0.229	0.299	0.149
C(8)	0.295	0.493	0.223
C(9)	0.293	0.636	0.311
C(10)	0.233	0.679	0.331
C(11)	0.354	0.457	0.192
C(12)	0.422	0.603	0.281
C(13)	0.420	0.799	0.356
C(14)	0.357	0.826	0.382
O(1)	0.229	0.835	0.409
O(2)	0.231	0.124	0.081
O(3)	0.361	0.287	0.118
O(4)	0.359	0.973	0.459

^{a} Calculated

XV,g22. The monoclinic crystals of *1,5-dihydroxyanthraquinone*, $C_{14}H_6O_2(OH)_2$, have a bimolecular unit of the dimensions:

$$a_0 = 15.755(10) \text{ A.}; \quad b_0 = 5.308(5) \text{ A.}; \quad c_0 = 6.003(5) \text{ A.}$$

$$\beta = 93°37(5)'$$

The space group C_{2h}^5 ($P2_1/a$) places atoms in the positions:

$$(4e) \quad \pm (xyz; x + 1/2, 1/2 - y, z)$$

The determined parameters are those of Table XVG,23.

The resulting structure (Fig. XVG,37) is composed of molecules with the bond dimensions of Figure XVG,38. The largest departures of atoms from the mean plane of these molecules is ca. 0.02 A. (for the atoms of oxygen). These oxygens, separated by the rather large 2.62 A., do not appear to be symmetrically hydrogen-bonded to one another.

TABLE XVG,23

Parameters of the Atoms in 1,5-Dihydroxyanthraquinone

Atom	x	y	z
C(1)	0.1120(3)	−0.0267(9)	−0.3363(7)
C(2)	0.1880(3)	0.1430(11)	−0.3315(8)
C(3)	0.1966(3)	0.3233(9)	−0.1621(8)
C(4)	0.1396(3)	0.3397(9)	−0.0005(8)
C(5)	0.0725(3)	0.1728(8)	−0.0033(7)
C(6)	0.0616(3)	−0.0175(9)	−0.1712(7)
C(7)	−0.0103(3)	−0.1935(8)	−0.1741(7)
O(1)	0.1160(3)	−0.1937(8)	−0.5047(6)
O(2)	−0.0195(2)	−0.3558(7)	−0.3161(6)
H(2)	0.2344	0.1309	−0.4596
H(3)	0.2501	0.4528	−0.1578
H(4)	0.1474	0.4789	0.1330

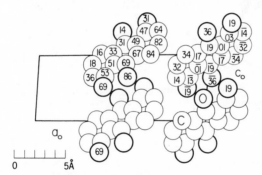

Fig. XVG,37a.   The monoclinic structure of 1,5-dihydroxyanthraquinone projected along its b_0 axis.

The shortest intermolecular distances are O(2)–O(2) = 2.79 A. and O(2)–O(1) = 2.99 A.

In a recent note (1967 : G) the results of an independent determination are compared with the foregoing. The general agreement is good, but it is pointed out that for several atoms the parameter differences considerably exceed the estimated error. In this second study parameters are also given for the H(1) atom; they are $x = 0.0580$, $y = 0.2910$, and $z = 0.4640$.

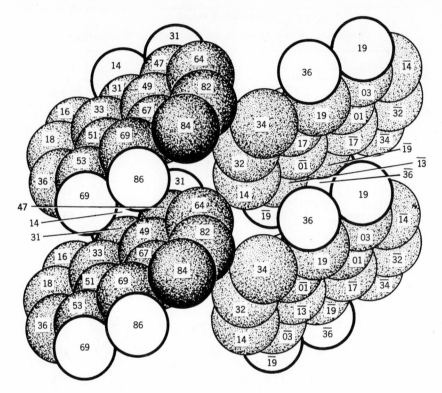

Fig. XVG,37b. A packing drawing of the monoclinic 1,5-dihydroxyanthra-quinone arrangement seen along its b_0 axis. The carbon atoms are dotted, the oxygens are unshaded.

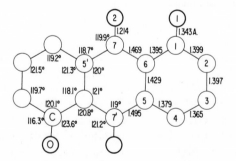

Fig. XVG,38. Bond dimensions in the molecule of 1,5-dihydroxyanthraquinone.

XV,g23. According to a preliminary note, crystals of *1,8-dihydroxy-anthraquinone*, $C_{14}H_6O_2(OH)_2$, are tetragonal with a tetramolecular unit of the edge lengths:

$$a_0 = 5.86 \text{ A.}, \quad c_0 = 31.33 \text{ A.}$$

The space group is C_4^2 ($P4_1$) [or its enantiomorph C_4^4] with atoms in the positions:

$$(4a) \quad xyz; \; \bar{x},\bar{y},z+\tfrac{1}{2}; \; \bar{y},x,z+\tfrac{1}{4}; \; y,\bar{x},z+\tfrac{3}{4}$$

The published parameters are those of Table XVG,24.

TABLE XVG,24

Parameters of the Atoms in $C_{14}H_6O_2(OH)_2$

Atom	x	y	z
C(1)	0.441	0.545	0.095
C(2)	0.248	0.556	0.065
C(3)	0.260	0.392	0.034
C(4)	0.055	0.360	0.004
C(5)	0.067	0.200	−0.034
C(6)	−0.098	0.167	−0.063
C(7)	−0.061	0.033	−0.093
C(8)	0.097	−0.130	−0.096
C(9)	0.259	−0.128	−0.066
C(10)	0.233	0.031	0.032
C(11)	0.404	0.044	−0.003
C(12)	0.431	0.207	0.033
C(13)	0.584	0.253	0.066
C(14)	0.583	0.404	0.096
O(1)	−0.129	0.529	0.003
O(2)	0.578	−0.101	−0.003
O(3)	0.112	0.710	0.063
O(4)	−0.269	0.308	−0.062

XV,g24. *Anthrarufine*, $C_{14}H_6O_2(OH)_2$, is monoclinic with a bi-molecular unit of the dimensions:

$$a_0 = 6.02(1) \text{ A.}; \quad b_0 = 5.31(1) \text{ A.}; \quad c_0 = 15.74(1) \text{ A.}$$
$$\beta = 94.0(1)°$$

The space group C_{2h}^5 ($P2_1/c$) places atoms in the positions

$$(4e) \quad \pm (xyz; \; x,{}^1\!/_2-y,z+{}^1\!/_2)$$

with parameters as listed in Table XVG,25.

TABLE XVG,25

Parameters of the Atoms in Anthrarufine

Atom	x	y	z
C(1)	0.3348(4)	−0.0258(5)	−0.1199(2)
C(2)	0.3302(5)	0.1452(6)	−0.1875(2)
C(3)	0.1643(5)	0.3237(6)	−0.1968(2)
C(4)	−0.0006(5)	0.3384(6)	−0.1394(2)
C(5)	0.0025(4)	0.1708(5)	−0.0717(2)
C(6)	0.1702(4)	−0.0146(5)	−0.0615(2)
C(7)	0.1724(4)	−0.1915(5)	0.0106(2)
O(1)	0.3168(4)	−0.3547(4)	0.0193(1)
O(2)	0.5032(3)	−0.1931(4)	−0.1151(1)
H(1)	0.461(7)	0.124(9)	−0.225(3)
H(2)	0.160(7)	0.447(10)	−0.238(3)
H(3)	−0.096(8)	0.445(11)	−0.139(3)
H(4)	0.464(7)	−0.291(10)	−0.058(3)

The structure, shown in Figure XVG,39, is built of centrosymmetric molecules with the bond dimensions of Figure XVG,40. Between molecules the shortest interatomic distances are O–O = 2.79 and 3.00 A.

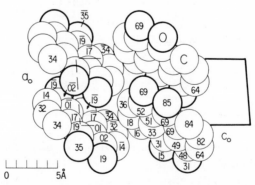

Fig. XVG,39. The monoclinic structure of anthrarufine projected along its b_0 axis.

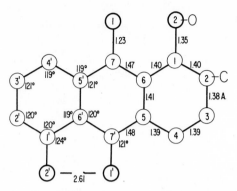

Fig. XVG,40. Bond dimensions in the molecule of anthrarufine.

TABLE XVG,26

Parameters of the Atoms in Alizarin

Atom	x	y	z
	Molecule A		
C(1)	0.119	0.520	0.076
C(2)	0.174	0.655	0.051
C(3)	0.161	0.659	−0.005
C(4)	0.102	0.547	−0.056
C(5)	−0.121	−0.008	−0.061
C(6)	−0.164	−0.134	−0.023
C(7)	−0.148	−0.127	0.043
C(8)	−0.087	−0.027	0.078
C(9)	0.016	0.254	0.078
C(10)	−0.011	0.279	−0.061
C(11)	0.048	0.403	−0.029
C(12)	0.067	0.402	0.045
C(13)	−0.044	0.123	0.045
C(14)	−0.063	0.137	−0.032
O(1)	−0.033	0.302	−0.131
O(2)	0.030	0.256	0.140
O(3)	−0.138	0.013	−0.124
O(4)	−0.221	−0.254	−0.055

(continued)

TABLE XVG,26 (*continued*)

Parameters of the Atoms in Alizarin

Atom	x	y	z
	Molecule B		
C(1,1)	0.433	0.462	0.405
C(1,2)	0.491	0.604	0.376
C(1,3)	0.479	0.628	0.308
C(1,4)	0.418	0.481	0.265
C(1,5)	0.194	−0.011	0.249
C(1,6)	0.162	−0.155	0.291
C(1,7)	0.169	−0.151	0.362
C(1,8)	0.232	−0.030	0.399
C(1,9)	0.341	0.219	0.408
C(1,10)	0.304	0.226	0.251
C(1,11)	0.370	0.365	0.295
C(1,12)	0.383	0.361	0.371
C(1,13)	0.274	0.087	0.364
C(1,14)	0.263	0.107	0.294
O(1,1)	0.293	0.255	0.192
O(1,2)	0.347	0.210	0.406
O(1,3)	0.458	0.460	0.468
O(1,4)	0.551	0.704	0.407
	Molecule C		
C(2,1)	0.287	0.503	0.737
C(2,2)	0.329	0.609	0.697
C(2,3)	0.309	0.612	0.630
C(2,4)	0.246	0.484	0.591
C(2,5)	0.029	−0.007	0.612
C(2,6)	−0.011	−0.119	0.654
C(2,7)	0.014	−0.118	0.722
C(2,8)	0.076	−0.019	0.746
C(2,9)	0.184	0.251	0.748
C(2,10)	0.134	0.238	0.602
C(2,11)	0.205	0.371	0.632
C(2,12)	0.227	0.289	0.705
C(2,13)	0.119	0.111	0.713
C(2,14)	0.096	0.132	0.640
O(2,1)	0.118	0.247	0.533
O(2,2)	0.205	0.271	0.008
O(2,3)	0.229	0.483	0.532
O(2,4)	0.350	0.706	0.596

XV,g25. Crystals of *1,2-dihydroxy-9,10-anthraquinone* (alizarin), $C_{14}H_6O_2(OH)_2$, are monoclinic with an unusual trimolecular unit of the dimensions:

$$a_0 = 21.04(4) \text{ A.}; \quad b_0 = 3.75(1) \text{ A.}; \quad c_0 = 20.12(4) \text{ A.}$$
$$\beta = 104.5(1)°$$

The space group has been chosen as the low symmetry C_s^2 (*Pa*) with atoms in the positions:

$$(2a) \quad xyz; \, x + 1/2, \bar{y}, z$$

Parameters have been assigned the approximate values of Table XVG,26.

Between molecules there are hydrogen bonds O–H–O = 2.6 and 2.8 A.; the next shorter separations are O–O = 3.2 and 3.3 A.

XV,g26. The bimolecular, monoclinic unit of *1,5-dinitro-4,8-di-hydroxyanthraquinone*, $C_{14}H_4O_2(OH)_2(NO_2)_2$, has the dimensions:

$$a_0 = 10.17 \text{ A.}; \quad b_0 = 10.49 \text{ A.}; \quad c_0 = 6.015 \text{ A.}; \quad \beta = 94°34'$$

Its space group C_{2h}^5 ($P2_1/a$) places atoms in the positions

$$(4e) \quad \pm (xyz; \, x + 1/2, 1/2 - y, z)$$

with the parameters listed in Table XVG,27.

TABLE XVG,27

Parameters of the Atoms in $C_{14}H_4O_2(OH)_2(NO_2)_2$

Atom	x	y	z
C(1)	0.1565	0.1297	−0.2910
C(2)	0.1632	0.2612	−0.2974
C(3)	0.0920	0.3320	−0.1629
C(4)	0.0129	0.2710	−0.0163
C(5)	0.0062	0.1392	−0.0092
C(6)	0.0808	0.0649	−0.1507
C(7)	0.0753	−0.0751	−0.1449
N	0.2383	0.0645	−0.4485
O(1)	−0.0550	0.3453	0.1139
O(2)	0.1379	−0.1372	−0.2713
O(3)	0.1908	0.0434	−0.6301
O(4)	0.3470	0.0354	−0.3823
H(1)	−0.1162	0.2980	0.2274
H(2)	0.2255	0.3075	−0.4103
H(3)	0.0961	0.4349	−0.1687

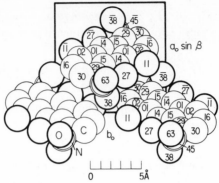

Fig. XVG,41.   The monoclinic structure of 1,5-dinitro-4,8-dihydroxyanthra-
quinone projected along its c_0 axis.

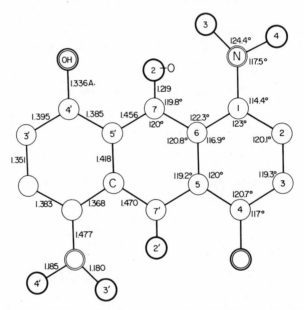

Fig. XVG,42.   Bond dimensions in the molecule of 1,5-dinitro-4,8-dihydroxyan-
thraquinone.

The resulting structure (Fig. XVG,41) is composed of centrosymmetric molecules possessing the bond dimensions of Figure XVG,42.
Except for their NO_2 groups, these molecules are almost planar, the
quinone oxygen being 0.027 A. outside the plane of the rings. The
nitrogen is 0.01 A. to one side and its NO_2 group is twisted by 88° with
respect to this plane.

XV,g27. Crystals of *N,N'-diphenyl-1,5-diaminoanthraquinone*, C_{14}-$H_6O_2(NH)_2(C_6H_5)_2$, are monoclinic with a bimolecular unit of the dimensions:

$$a_0 = 21.02 \text{ A.}; \quad b_0 = 4.855 \text{ A.}; \quad c_0 = 9.34 \text{ A.}; \quad \beta = 93°11'$$

The space group is C_{2h}^5 in the orientation $P2_1/n$ with all atoms in the positions:

$$(4e) \quad \pm (xyz; \ x+{}^1/_2, {}^1/_2-y, z+{}^1/_2)$$

The determined parameters are listed in Table XVG,28.

TABLE XVG,28

Parameters of the Atoms in *N,N'*-Diphenyl-1,5-diaminoanthraquinone

Atom	x	y	z
C(1)	0.0860	0.3548	0.1675
C(2)	0.1242	0.5385	0.0919
C(3)	0.1217	0.5452	−0.0566
C(4)	0.0816	0.3712	−0.1343
C(5)	0.0434	0.1906	−0.0636
C(6)	0.0432	0.1820	0.0871
C(7)	−0.0004	−0.0024	0.1542
C(8)	0.1330	0.4991	0.4046
C(9)	0.1983	0.4526	0.3987
C(10)	0.2408	0.5951	0.4894
C(11)	0.2179	0.7738	0.5875
C(12)	0.1528	0.8206	0.5926
C(13)	0.1100	0.6801	0.5029
N	0.0883	0.3449	0.3140
O	−0.0041	−0.0115	0.2878
H(1)	0.0546	0.2140	0.3647
H(2)	0.1560	0.6768	0.1514
H(3)	0.1514	0.6878	−0.1114
H(4)	0.0799	0.3755	−0.2500
H(9)	0.2156	0.3054	0.3233
H(10)	0.2915	0.5658	0.4829
H(11)	0.2508	0.8795	0.6617
H(12)	0.1356	0.9686	0.6678
H(13)	0.0594	0.7111	0.5094

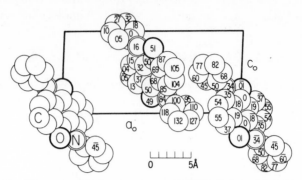

Fig. XVG,43.   The monoclinic structure of N,N'-diphenyl-1,5-diaminoanthra-
quinone projected along its b_0 axis.

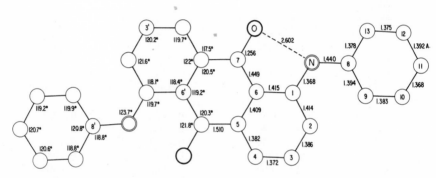

Fig. XVG,44.   Bond dimensions in the molecule of N,N'-diphenyl-1,5-diamino-
anthraquinone.

The structure is shown in Figure XVG,43. Its centrosymmetric
molecules have the bond dimensions of Figure XVG,44. The anthraquinone and two phenyl rings are planar, but the latter are turned so
that there is an angle of 62.4° between their normals and the normal
to the anthraquinone residue. In this structure the molecules are held
together by van der Waals forces only, the shortest intermolecular
distances being C–O = 3.39 A. and C–C = 3.51 A.

XV,g28.   Orthorhombic N,N'-*diphenyl-1,8-diaminoanthraquinone,*
$C_{14}H_6O_2(NH)_2(C_6H_5)_2$, possesses a bimolecular cell of the edge lengths:

$$a_0 = 21.312 \text{ A.}; \quad b_0 = 9.205 \text{ A.}; \quad c_0 = 4.855 \text{ A.}$$

The space group is V^3 $(P2_122_1)$ with atoms in the positions:

(2a) $0u0$; $1/2\ \bar{u}\ 1/2$
(4c) xyz; $\bar{x}y\bar{z}$; $1/2-x,\bar{y},z+1/2$; $x+1/2,\bar{y},1/2-z$

Assigned positions and their parameters are listed in Table XVG,29.

TABLE XVG,29

Positions and Parameters of the Atoms in N,N'-Diphenyl-1,8-diaminoanthraquinone

Atom	Position	x	y	z
C(1)	(4c)	0.0860	0.4478	0.3628
C(2)	(4c)	0.1226	0.3639	0.5419
C(3)	(4c)	0.1199	0.2137	0.5382
C(4)	(4c)	0.0800	0.1441	0.3676
C(5)	(4c)	0.0424	0.2206	0.1882
C(6)	(4c)	0.0434	0.3729	0.1841
C(7)	(2a)	0	0.4496	0
C(8)	(2a)	0	0.1377	0
C(9)	(4c)	0.1345	0.6788	0.5085
C(10)	(4c)	0.1145	0.7815	0.6931
C(11)	(4c)	0.1594	0.8636	0.8325
C(12)	(4c)	0.2219	0.8446	0.7858
C(13)	(4c)	0.2422	0.7399	0.6048
C(14)	(4c)	0.1984	0.6572	0.4594
N	(4c)	0.0892	0.5951	0.3584
O(1)	(2a)	0	0.5854	0
O(2)	(2a)	0	0.0054	0
H(1)	(4c)	0.0554	0.6529	0.2340
H(2)	(4c)	0.1536	0.4174	0.6856
H(3)	(4c)	0.1499	0.1516	0.6732
H(4)	(4c)	0.0775	0.0270	0.3717
H(10)	(4c)	0.0650	0.7985	0.7296
H(11)	(4c)	0.1443	0.9439	0.9806
H(12)	(4c)	0.2557	0.9121	0.8915
H(13)	(4c)	0.2918	0.7217	0.5751
H(14)	(4c)	0.2137	0.5774	0.3109

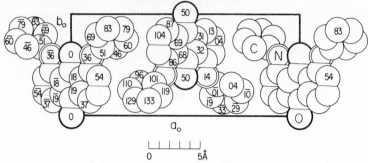

Fig. XVG,45. The orthorhombic structure of *N,N'*-diphenyl-1,8-diaminoanthraquinone projected along its c_0 axis.

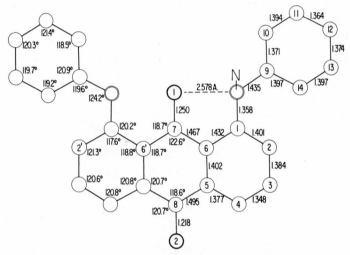

Fig. XVG,46. Bond dimensions in the molecule of *N,N'*-diphenyl-1,8-diaminoanthraquinone.

The structure (Fig. XVG,45) is built of molecules, with a twofold axis of symmetry through oxygen atoms and the bond dimensions of Figure XVG,46. The anthraquinone part is planar within ca. 0.03 A.; the attached benzene rings make an angle of 61.8° with this plane. In this anthraquinone derivative, as in that of the preceding paragraph (**XV,g27**), the molecules are joined by purely van der Waals forces, with closest intermolecular distances of C–O = 3.38 A. and C–C = 3.47 A.

H.   COMPLEX CONDENSED RING COMPOUNDS

XV,h1.   Crystals of *biphenylene*, $C_{12}H_8$, are monoclinic. According to a recent redetermination which confirms but renders more precise the original study, its unit containing six molecules has the dimensions:

$$a_0 = 19.728(10) \text{ A.}; \quad b_0 = 10.578(6) \text{ A.}; \quad c_0 = 5.861(11) \text{ A.}$$
$$\beta = 91°10(3)'$$

TABLE XVH,1

Parameters of the Atoms in Biphenylene

Atom	x	y	z
C(1)	0.0285	0.0040	−0.1469
C(2)	0.0730	−0.0083	−0.3224
C(3)	0.1207	−0.1073	−0.2965
C(4)	0.1233	−0.1853	−0.1074
C(5)	0.0771	−0.1706	0.0754
C(6)	0.0308	−0.0749	0.0491
C(7)	0.3649	0.0000	0.5747
C(8)	0.4090	−0.0140	0.7565
C(9)	0.4603	−0.1065	0.7266
C(10)	0.4651	−0.1781	0.5307
C(11)	0.4187	−0.1638	0.3441
C(12)	0.3693	−0.0742	0.3727
C(13)	0.3068	−0.0099	0.2761
C(14)	0.2611	−0.0006	0.0973
C(15)	0.2096	0.0910	0.1259
C(16)	0.2059	0.1663	0.3189
C(17)	0.2533	0.1540	0.5039
C(18)	0.3026	0.0656	0.4766
H(2)	0.073	0.063	−0.451
H(3)	0.161	−0.119	−0.414
H(4)	0.157	−0.256	−0.098
H(5)	0.078	−0.238	0.216
H(8)	0.406	0.049	0.901
H(9)	0.499	−0.124	0.855
H(10)	0.501	−0.247	0.520
H(11)	0.420	−0.230	0.217
H(14)	0.263	−0.069	−0.047
H(15)	0.172	0.100	−0.001
H(16)	0.167	0.239	0.336
H(17)	0.252	0.222	0.638

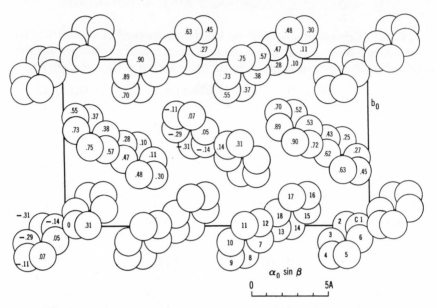

Fig. XVH,1a. The monoclinic structure of biphenylene projected along its c_0 axis. The z parameters shown here are those of the early determination. Left-hand axes.

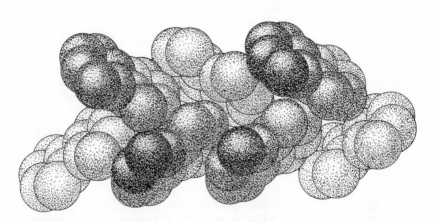

Fig. XVH,1b. A packing drawing of the monoclinic biphenylene structure viewed along its c_0 axis. Only the bottom two rows of molecules of Fig. XVH,1a are shown. Left-hand axes.

Atoms are in the positions

$$(4e) \quad \pm (xyz; x + \tfrac{1}{2}, \tfrac{1}{2} - y, z)$$

of C_{2h}^5 ($P2_1/a$). The recent parameters are those of Table XVH,1.

The structure is shown in Figure XVH,1. One of its molecules has a center of symmetry; the other two are symmetryless. Their bond dimensions are given in Figure XVH,2; the C–H distances lie between 1.00 and 1.12 A.

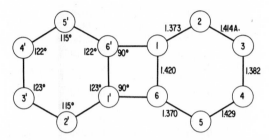

Fig. XVH,2a.   Bond dimensions in the centrosymmetric molecule of biphenylene.

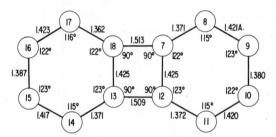

Fig. XVH,2b.   Bond dimensions in the asymmetric molecules of biphenylene.

XV,h2.   Three studies have been made at different times of the structure of *phenanthrene*, $C_{14}H_{10}$. According to the latest, its monoclinic bimolecular unit has the dimensions:

$$a_0 = 8.46 \text{ A.}; \quad b_0 = 6.16 \text{ A.}; \quad c_0 = 9.47 \text{ A.}; \quad \beta = 97.7°$$

The space group is C_2^2 ($P2_1$) with atoms in the positions:

$$(2a) \quad xyz; \bar{x}, y + \tfrac{1}{2}, \bar{z}$$

The parameters (1963:T) are those of Table XVH,2.

TABLE XVH,2

Parameters of the Atoms in Phenanthrene

Atom	x	y	z
C(1)	0.1665	0.3873	−0.3081
C(2)	0.0747	0.2098	−0.3549
C(3)	0.0354	0.0440	−0.2641
C(4)	0.0954	0.0580	−0.1194
C(5)	0.2292	0.0861	0.1872
C(6)	0.2967	0.1046	0.3262
C(7)	0.3858	0.2864	0.3737
C(8)	0.4179	0.4505	0.2823
C(9)	0.3786	0.5849	0.0410
C(10)	0.3228	0.5724	−0.1017
C(11)	0.2278	0.3973	−0.1559
C(12)	0.1919	0.2321	−0.0653
C(13)	0.2544	0.2427	0.0845
C(14)	0.3467	0.4255	0.1355

The structure (Fig. XVH,3) is like that of anthracene (**XV,g1**). Its molecules have the bond dimensions of Figure XVH,4. They are planar within 0.04 A., but there is some evidence of a slight bending toward the C(7) ends.

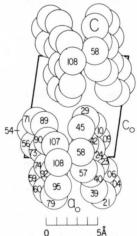

Fig. XVH,3. The monoclinic structure of phenanthrene projected along its b_0 axis.

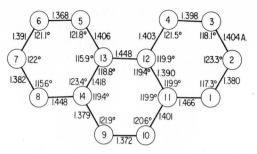

Fig. XVH,4.   Bond dimensions in the molecule of phenanthrene.

XV,h3.   Crystals of the monoclinic form of *phenanthrene-chromium tricarbonyl*, $C_{14}H_{10} \cdot Cr(CO)_3$, possess a tetramolecular unit of the dimensions:

$$a_0 = 14.06(4) \text{ A.}; \quad b_0 = 11.68(4) \text{ A.}; \quad c_0 = 8.63(2) \text{ A.}$$
$$\beta = 102.0(5)°$$

The space group C_{2h}^5 ($P2_1/a$) puts the atoms in the positions:

$$(4e) \quad \pm (xyz; x + \tfrac{1}{2}, \tfrac{1}{2} - y, z)$$

Their parameters are stated in Table XVH,3.

The structure is shown in Figure XVH,5. The phenanthrene group is planar and the $Cr(CO)_3$ part pyramidal, with chromium at the apex of the pyramid. This lies beneath one of the end rings of phenanthrene in such a way that, if one imagines chromium attached to the midpoints

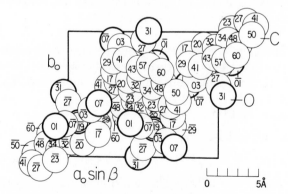

Fig. XVH,5.   The structure of the monoclinic form of phenanthrene-chromium tricarbonyl projected along its c_0 axis. The doubly ringed circles are chromium.

TABLE XVH,3

Parameters of the Atoms in $C_{14}H_{10} \cdot Cr(CO)_3$

Atom	x	y	z
C(1)	0.120	0.274	−0.407
C(2)	0.199	0.258	−0.286
C(3)	0.196	0.174	−0.171
C(4)	0.101	0.104	−0.197
C(5)	−0.087	−0.030	−0.231
C(6)	−0.181	−0.099	−0.269
C(7)	−0.239	−0.064	−0.408
C(8)	−0.222	0.010	−0.501
C(9)	−0.121	0.161	−0.599
C(10)	−0.036	0.226	−0.570
C(11)	0.029	0.195	−0.430
C(12)	0.024	0.111	−0.322
C(13)	−0.062	0.053	−0.342
C(14)	−0.137	0.081	−0.479
C(15)	−0.016	0.257	−0.066
C(16)	0.154	0.372	−0.027
C(17)	0.008	0.421	−0.267
O(1)	−0.075	0.238	0.015
O(2)	0.204	0.423	0.072
O(3)	−0.027	0.501	−0.313
Cr	0.071	0.285	−0.188

of its C–C bonds, its coordination is octahedral. Each Cr–C–O is linear and the angle between these lines is ca. 91°.

XV,h4. There is a more recent and precise determination of the structure of the orthorhombic form of *phenanthrene-chromium tricarbonyl*, $C_{14}H_{10} \cdot Cr(CO)_3$. The unit, containing eight molecules, has the edge lengths:

$$a_0 = 12.14 \text{ A.}; \quad b_0 = 18.08 \text{ A.}; \quad c_0 = 12.34 \text{ A.}$$

All atoms are in the positions

(8c) $\pm (xyz; \; 1/2-x, y+1/2, z; \; x, 1/2-y, z+1/2; \; x+1/2, y, 1/2-z)$

of the space group V_h^{15} (*Pbca*). The determined parameters are those of Table XVH,4; values proposed for the atoms of hydrogen are included in the original article.

TABLE XVH,4

Parameters of the Atoms in Orthorhombic $C_{14}H_{10} \cdot Cr(CO)_3$

Atom	x	y	z
Cr	0.43186(4)	0.67807(3)	0.16117(5)
C(1)	0.4120(4)	0.6803(3)	−0.0170(4)
C(2)	0.3166(4)	0.6512(3)	0.0282(4)
C(3)	0.3228(4)	0.5909(3)	0.0971(4)
C(4)	0.4251(4)	0.5587(2)	0.1236(4)
C(5)	0.6468(5)	0.4921(3)	0.1646(5)
C(6)	0.7519(7)	0.4590(3)	0.1769(6)
C(7)	0.8404(5)	0.4872(3)	0.1169(6)
C(8)	0.8281(4)	0.5455(3)	0.0515(6)
C(9)	0.7125(4)	0.6410(3)	−0.0325(5)
C(10)	0.6157(4)	0.6737(3)	−0.0499(4)
C(11)	0.5172(3)	0.6462(2)	0.0025(3)
C(12)	0.5246(3)	0.5854(2)	0.0737(3)
C(13)	0.6324(4)	0.5520(2)	0.0937(4)
C(14)	0.7242(3)	0.5797(3)	0.0379(4)
C(15)	0.3173(4)	0.7139(2)	0.2417(4)
C(16)	0.4882(4)	0.7728(3)	0.1621(4)
C(17)	0.5102(3)	0.6566(2)	0.2866(4)
O(1)	0.2466(3)	0.7382(3)	0.2948(4)
O(2)	0.5189(5)	0.8327(2)	0.1634(5)
O(3)	0.5572(3)	0.6422(2)	0.3633(3)

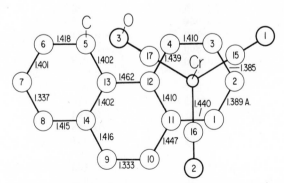

Fig. XVH,6. Bond lengths in the molecule of the orthorhombic form of phenanthrene-chromium tricarbonyl.

Molecules in the structure that results closely resemble those in the monoclinic form, with the $Cr(CO)_3$ portion centered over one of the end rings of the phenanthrene part. Bond dimensions are shown in Figure XVH,6. The phenanthrene carbons are nearly but not exactly co-planar, the maximum departure from the best plane being 0.07 A. The chromium atom is 2.206–2.289 A. from the six carbon atoms of the ring below which it lies; the carbonyl carbons are 2.68–2.90 A. below the plane. In the $Cr(CO)_3$, $Cr–C = 1.83–1.86$ A. and $C–O = 1.14–1.16$ A.

XV,h5. The tetragonal crystals of *chrysazin*, $C_{14}H_8O_4$, possess a tetramolecular unit of the edge lengths:

$$a_0 = 5.76(2) \text{ A.}, \quad c_0 = 31.45(6) \text{ A.}$$

The space group is C_4^2 $(P4_1)$ [or C_4^4 $(P4_3)$] with atoms in the positions:

$$(4a) \quad xyz; \bar{x},\bar{y},z+^1/_2; \bar{y},x,z+^1/_4; y,\bar{x},z+^3/_4$$

The determined parameters are those of Table XVH,5.

TABLE XVH,5

Parameters of the Atoms in Chrysazin

Atom	x	y	z
C(1)	0.446	0.539	0.0951
C(2)	0.256	0.522	0.0661
C(3)	0.280	0.341	0.0327
C(4)	0.089	0.344	0.0008
C(5)	0.113	0.157	−0.0321
C(6)	−0.075	0.153	−0.0623
C(7)	−0.054	−0.017	−0.0944
C(8)	0.109	−0.184	−0.0932
C(9)	0.300	−0.177	−0.0667
C(10)	0.290	−0.013	−0.0333
C(11)	0.477	−0.002	−0.0006
C(12)	0.452	0.176	0.0340
C(13)	0.631	0.194	0.0630
C(14)	0.610	0.375	0.0944
O(1)	−0.061	0.501	0.0000
O(2)	0.631	−0.153	−0.0038
O(3)	0.116	0.692	0.0651
O(4)	−0.242	0.317	−0.0615

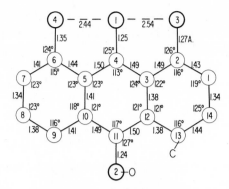

Fig. XVH,7. Bond dimensions in the molecule of chrysazin.

TABLE XVH,6

Parameters of the Atoms in Pyrene

Atom	x	y	z
C(1)	0.2817	−0.0402	0.4119
C(2)	0.2947	0.0246	0.2717
C(3)	0.2296	−0.0077	0.1274
C(4)	0.2389	0.0567	−0.0238
C(5)	0.1783	0.0237	−0.1578
C(6)	0.0990	−0.0738	−0.1606
C(7)	0.0316	−0.1103	−0.3020
C(8)	−0.0449	−0.2090	−0.2966
C(9)	−0.0566	−0.2746	−0.1594
C(10)	0.0071	−0.2396	−0.0131
C(11)	−0.0030	−0.3070	0.1356
C(12)	0.0575	−0.2772	0.2706
C(13)	0.1389	−0.1735	0.2723
C(14)	0.2066	−0.1412	0.4161
C(15)	0.1514	−0.1091	0.1303
C(16)	0.0854	−0.1409	−0.0136
H(1)	0.33	−0.03	0.52
H(2)	0.36	0.10	0.27
H(4)	0.30	0.13	−0.03
H(5)	0.19	0.08	−0.27
H(7)	0.04	−0.06	−0.41
H(8)	−0.10	−0.24	−0.41
H(9)	−0.12	−0.35	−0.16
H(11)	−0.06	−0.38	0.14
H(12)	0.05	−0.33	0.38
H(14)	0.20	−0.19	0.53

The structure is made up of planar molecules with the bond dimensions of Figure XVH,7. They are held together by hydrogen bonds, the shortest of which is O–H–O = 2.68 A.

XV,h6. *Pyrene,* $C_{16}H_{10}$, is monoclinic with a tetramolecular unit of the dimensions:

$$a_0 = 13.649(10) \text{ A.}; \quad b_0 = 9.256(10) \text{ A.}; \quad c_0 = 8.470(10) \text{ A.}$$
$$\beta = 100°28(3)'$$

The space group C_{2h}^5 ($P2_1/a$) places the atoms in the positions:

$$(4e) \quad \pm(xyz; x+^1/_2, ^1/_2-y, z)$$

A recent redetermination has led to the parameters of Table XVH,6.

The structure (Fig. XVH,8) is composed of molecules with the bond dimensions of Figure XVH,9. Between these molecules the shortest distance is 3.55 A. The arrangement resembles that found for benzperylene (**XV,h18**).

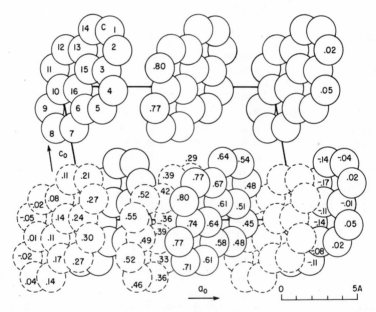

Fig. XVH,8a. The monoclinic structure of pyrene projected along its b_0 axis. The molecules are distributed in pairs (one member of each pair dashed) as indicated in the lower layer; only half of each pair is shown in the upper layer. Left-hand axes.

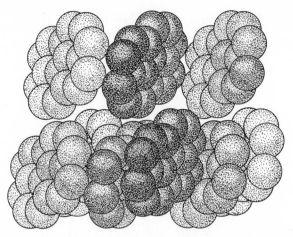

Fig. XVH,8b. A packing drawing of the molecules of the monoclinic pyrene structure (shown in Fig. XVH,8a) viewed along its b_0 axis. Left-hand axes.

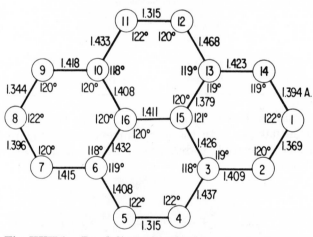

Fig. XVH,9. Bond dimensions in the molecule of pyrene.

XV,h7. The complex *pyrene-tetracyanoethylene*, $C_{16}H_{10} \cdot C_2(CN)_4$, forms monoclinic crystals whose bimolecular unit has the dimensions:

$$a_0 = 14.333(9) \text{ A.}; \quad b_0 = 7.242(8) \text{ A.}; \quad c_0 = 7.978(4) \text{ A.}$$
$$\beta = 92.36(4)°$$

The space group is C_{2h}^5 ($P2_1/a$) with atoms in the positions:

$$(4e) \quad \pm (xyz; x + 1/2, 1/2 - y, z)$$

The determined parameters are listed in Table XVH,7.

TABLE XVH,7

Parameters of the Atoms in $C_{16}H_{10} \cdot C_2(CN)_4$

Atom	x	y	z
Pyrene			
C(1)	0.1433	−0.2742	0.2663
C(2)	0.1750	−0.1041	0.2157
C(3)	0.1193	0.0101	0.1078
C(4)	0.0291	−0.0571	0.0550
C(5)	−0.0027	−0.2304	0.1089
C(6)	0.0558	−0.3402	0.2114
C(7)	0.1502	0.1854	0.0525
C(8)	0.0935	0.2939	−0.0517
Tetracyanoethylene			
C(9)	0.0423	0.0097	0.5328
C(10)	0.0849	−0.1374	0.6360
C(11)	0.0968	0.1788	0.5066
N(1)	0.1213	−0.2454	0.7159
N(2)	0.1406	0.3038	0.4902

The structure is shown in Figure XVH,10. The two components are planar within 0.02 A., with the bond dimensions of Figure XVH,11. Between them there are C–C distances of 3.25 A. and up and a C–N of about 3.43 A.

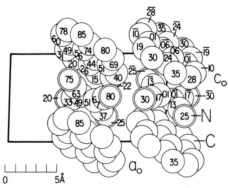

Fig. XVH,10. The monoclinic structure of pyrene-tetracyanoethylene projected along its b_0 axis.

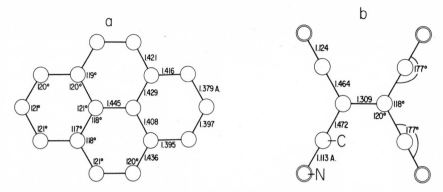

Fig. XVH,11a *(left)*.   Bond dimensions of the pyrene molecule in its complex with tetracyanoethylene.

Fig. XVH,11b *(right)*.   Bond dimensions of the tetracyanoethylene molecule in its complex with pyrene.

XV,h8. The crystal structure of *triphenylene*, $C_{18}H_{12}$, has been repeatedly investigated. Its symmetry is orthorhombic with a tetramolecular unit of the edge lengths:

$$a_0 = 13.17(3) \text{ A.}; \quad b_0 = 16.73(3) \text{ A.}; \quad c_0 = 5.26(1) \text{ A.}$$

Atoms are in the positions

$$(4a) \quad xyz; \; 1/2-x,\bar{y},z+1/2; \; x+1/2,1/2-y,\bar{z}; \; \bar{x},y+1/2,1/2-z$$

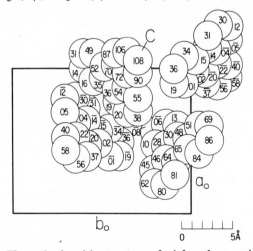

Fig. XVH,12.   The orthorhombic structure of triphenylene projected along its c_0 axis.

of the space group V^4 ($P2_12_12_1$). According to the latest study, the parameters are those of Table XVH,8.

TABLE XVH,8

Parameters of the Atoms in Triphenylene

Atom	x	y	z
C(1)	0.4405	0.6996	0.5767
C(2)	0.5199	0.7012	0.4031
C(3)	0.5327	0.6414	0.2175
C(4)	0.6181	0.6420	0.0451
C(5)	0.6917	0.7052	0.0477
C(6)	0.7703	0.7075	−0.1188
C(7)	0.7817	0.6476	−0.2986
C(8)	0.7132	0.5839	−0.3065
C(9)	0.6309	0.5811	−0.1377
C(10)	0.5558	0.5166	−0.1501
C(11)	0.5642	0.4559	−0.3394
C(12)	0.4918	0.3968	−0.3616
C(13)	0.4088	0.3966	−0.1937
C(14)	0.3997	0.4539	−0.0086
C(15)	0.4718	0.5163	0.0171
C(16)	0.4605	0.5785	0.2017
C(17)	0.3793	0.5793	0.3748
C(18)	0.3692	0.6369	0.5594
H(1)	0.425	0.750	0.740
H(2)	0.578	0.750	0.402
H(5)	0.692	0.751	0.197
H(6)	0.826	0.753	−0.117
H(7)	0.840	0.655	−0.453
H(8)	0.720	0.535	−0.460
H(11)	0.630	0.458	−0.467
H(12)	0.497	0.362	−0.533
H(13)	0.360	0.348	−0.200
H(14)	0.335	0.455	0.130
H(17)	0.328	0.526	0.345
H(18)	0.319	0.635	0.713

The structure is illustrated in Figure XVH,12. Its molecules, though essentially planar, show atomic departures up to 0.055 A. from the best plane. Bond dimensions are given in Figure XVH,13. The C–H bonds

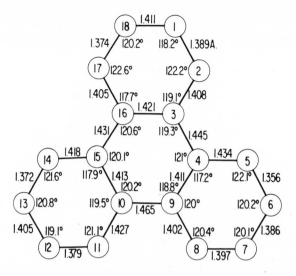

Fig. XVH,13.   Bond dimensions in the molecule of triphenylene.

have lengths of 1.04–1.22 A. Between molecules the atomic separations range upward from C–C = 3.42 A.

XV,h9.   The condensed hydrocarbon *chrysene*, $C_{18}H_{12}$, is monoclinic with a tetramolecular unit of the dimensions:

$$a_0 = 8.386 \text{ A.}; \quad b_0 = 6.196 \text{ A.}; \quad c_0 = 25.203 \text{ A.}; \quad \beta = 116.2°$$

The space group is C_{2h}^6 in the orientation $I2/c$. Atoms are accordingly in the positions:

$$(8f) \quad \pm (xyz; x,\bar{y},z + 1/2); \quad \text{B.C.}$$

The parameters from the more recent refinement are those of Table XVH,9.

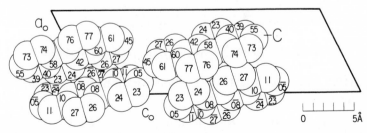

Fig. XVH,14.   The monoclinic structure of chrysene projected along its b_0 axis.

TABLE XVH,9

Parameters of the Atoms in Chrysene

Atom	x	y	z
C(1)	−0.0262	0.0848	0.0126
C(2)	0.0193	0.0799	0.0758
C(3)	−0.0302	0.2437	0.1041
C(4)	0.0133	0.2304	0.1636
C(5)	0.1105	0.0555	0.1969
C(6)	0.1607	−0.1064	0.1710
C(7)	0.1166	−0.0970	0.1095
C(8)	0.1687	−0.2651	0.0823
C(9)	0.1259	−0.2595	0.0234
H(3)	−0.090	0.374	0.085
H(4)	−0.021	0.358	0.183
H(5)	0.138	0.054	0.236
H(6)	0.233	−0.227	0.192
H(8)	0.248	−0.400	0.106
H(9)	0.165	−0.383	0.006

The structure (Fig. XVH,14) is a stacking of molecular layers normal to the $a_0 b_0$ plane. The individual molecules, as would be expected, are planar, with the dimensions of Figure XVH,15.

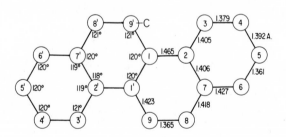

Fig. XVH,15.   Bond dimensions in the molecule of chrysene.

XV,h10.   *Tetracene*, $C_{18}H_{12}$, is triclinic with a bimolecular unit of the dimensions:

$$a_0 = 7.90 \text{ A.}; \quad b_0 = 6.03 \text{ A.}; \quad c_0 = 13.53 \text{ A.}$$
$$\alpha = 100.3°; \quad \beta = 113.2°; \quad \gamma = 86.3°$$

Atoms are in the positions $(2i) \pm (xyz)$ of the space group C_i^1 ($P\bar{1}$). The parameters are given in Table XVH,10.

TABLE XVH,10

Parameters of the Atoms in Tetracene and Pentacene (in parentheses)

Atom	x	y	z
C(1)	0.1194 (0.1484)	0.0331 (0.0183)	0.3890 (0.4071)
C(2)	0.1494 (0.1707)	0.1564 (0.1361)	0.3219 (0.3491)
C(3)	0.0811 (0.1030)	0.0724 (0.0604)	0.2040 (0.2508)
C(4)	0.1094 (0.1274)	0.1899 (0.1650)	0.1360 (0.1894)
C(5)	0.0467 (0.0572)	0.1089 (0.0951)	0.0211 (0.0950)
C(6)	−0.0691 (0.0925)	−0.2228 (0.2095)	0.0516 (0.0335)
C(7)	−0.0035 (−0.0227)	−0.1405 (−0.1227)	0.1637 (0.0567)
C(8)	−0.0378 (−0.0540)	−0.2646 (−0.2325)	0.2355 (0.1225)
C(9)	0.0267 (0.0150)	−0.1832 (−0.1516)	0.3483 (0.2149)
C(10)	— (−0.0152)	— (−0.2634)	— (0.2786)
C(11)	— (0.0519)	— (−0.1800)	— (0.3706)
C(1')	0.5428 (0.5627)	0.5680 (0.5733)	0.3787 (0.4014)
C(2')	0.4799 (0.5035)	0.6861 (0.6805)	0.2930 (0.3286)
C(3')	0.4997 (0.5210)	0.5864 (0.5876)	0.1947 (0.2434)
C(4')	0.4425 (0.4582)	0.7007 (0.7010)	0.1048 (0.1686)
C(5')	0.4638 (0.4810)	0.6140 (0.6062)	0.0052 (0.0846)
C(6')	0.5948 (0.4231)	0.2758 (0.7138)	0.0832 (0.0100)
C(7')	0.5758 (0.5622)	0.3649 (0.3794)	0.1811 (0.0754)
C(8')	0.6306 (0.6103)	0.2568 (0.2743)	0.2685 (0.1483)
C(9')	0.6148 (0.6061)	0.3374 (0.3652)	0.3645 (0.2342)
C(10')	— (0.6549)	— (0.2570)	— (0.3094)
C(11')	— (0.6496)	— (0.3487)	— (0.3925)

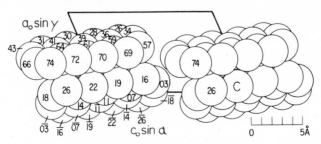

Fig. XVH,16. The triclinic structure of tetracene projected along its b_0 axis.

The structure, shown in Figure XVH,16, is a triclinic distortion of the arrangement that prevails for anthracene and naphthalene. Its molecules, which are crystallographically of two sorts, are planar within ca. 0.04 A. and have a center of symmetry. Average bond lengths for the two molecules are those of Figure XVH,17. Parameters for calculated hydrogen positions are given in the original.

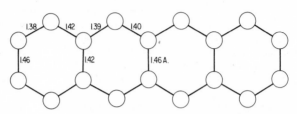

Fig. XVH,17.   Some averaged bond lengths in the molecule of tetracene.

Pentacene, $C_{22}H_{14}$, is isostructural. Its triclinic unit has the dimensions:

$$a_0 = 7.90 \text{ A.}; \quad b_0 = 6.06 \text{ A.}; \quad c_0 = 16.01 \text{ A.}$$
$$\alpha = 101.9°; \quad \beta = 112.6°; \quad \gamma = 85.8°$$

The determined atomic parameters are in parentheses in Table XVH,10; those assumed for hydrogen are stated in the original article. The molecules are almost planar, but one carbon atom, C(5), lies 0.08 A. outside the best plane through the others. Bond lengths, averaged for the two molecules, are shown in Figure XVH,18.

Hexacene, $C_{26}H_{16}$, also has this structure, though its atomic positions have not been established. The dimensions of its unit are:

$$a_0 = 7.9 \text{ A.}; \quad b_0 = 6.1 \text{ A.}; \quad c_0 = 18.4 \text{ A.}$$
$$\alpha = 102.7°; \quad \beta = 112.3°; \quad \gamma = 83.6°$$

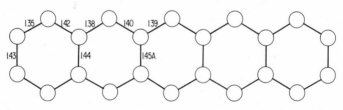

Fig. XVH,18.   Some averaged bond lengths in the molecule of pentacene.

XV,h11. According to a preliminary note, *1,2-benzanthracene*, $C_{18}H_{12}$, forms monoclinic crystals whose bimolecular units have the dimensions:

$$a_0 = 7.95 \text{ A.}; \quad b_0 = 6.50 \text{ A.}; \quad c_0 = 12.12 \text{ A.}; \quad \beta = 100°30'$$

The space group is the low symmetry $C_2{}^2$ ($P2_1$) with carbon atoms in the positions:

$$(2a) \quad xyz; \; \bar{x}, y + \tfrac{1}{2}, \bar{z}$$

The parameters are listed in Table XVH,11.

TABLE XVH,11

Parameters of the Atoms in 1,2-Benzanthracene

Atom	x	y	z
C(1)	0.409	0.432	0.681
C(2)	0.451	0.467	0.796
C(3)	0.415	0.324	0.876
C(4)	0.328	0.131	0.845
C(5)	0.214	−0.070	0.682
C(6)	0.165	−0.134	0.576
C(7)	0.119	−0.053	0.370
C(8)	0.090	0.099	0.176
C(9)	0.116	0.233	0.106
C(10)	0.195	0.414	0.132
C(11)	0.264	0.463	0.241
C(12)	0.303	0.339	0.436
C(13)	0.287	0.112	0.721
C(14)	0.189	0.007	0.489
C(15)	0.154	0.105	0.288
C(16)	0.235	0.293	0.325
C(17)	0.279	0.208	0.513
C(18)	0.329	0.262	0.643

XV,h12. The orthorhombic crystals of the carcinogen *9,10-dimethyl-1,2-benzanthracene*, $C_{18}H_{10}(CH_3)_2$, have a tetramolecular unit of the edge lengths:

$$a_0 = 7.62 \text{ A.}; \quad b_0 = 8.62 \text{ A.}; \quad c_0 = 21.11 \text{ A.}$$

The space group is $C_{2v}{}^9$ ($P2_1nb$) with atoms in the positions:

$$(4a) \quad xyz; \; x + \tfrac{1}{2}, \bar{y}, \bar{z}; \; x + \tfrac{1}{2}, \tfrac{1}{2} - y, z + \tfrac{1}{2}; \; x, y + \tfrac{1}{2}, \tfrac{1}{2} - z$$

A reworking of the original data has led to the parameters of Table XVH,12.

TABLE XVH,12

Parameters of the Atoms in $C_{18}H_{10}(CH_3)_2$

Atom	x	y	z
C(1)	0.2492	0.2514	0.0976
C(2)	0.3255	0.3898	0.1218
C(3)	0.4035	0.4984	0.0798
C(4)	0.3957	0.4799	0.0170
C(5)	0.2436	0.1595	−0.1709
C(6)	0.1902	0.0225	−0.1953
C(7)	0.1600	−0.1057	−0.1504
C(8)	0.1827	−0.0917	−0.0911
C(9)	0.2451	0.0682	0.0041
C(10)	0.3219	0.3233	−0.0773
C(11)	0.3274	0.3403	−0.0118
C(12)	0.2658	0.1817	−0.1035
C(13)	0.2319	0.0522	−0.0631
C(14)	0.2714	0.2157	0.0295
C(15)	0.2425	−0.0783	0.0437
C(16)	0.3759	0.4548	−0.1209
C(1′)	0.1467	0.1613	0.1384
C(2′)	0.1282	0.1980	0.2019
C(3′)	0.2174	0.3270	0.2273
C(4′)	0.3098	0.4209	0.1877

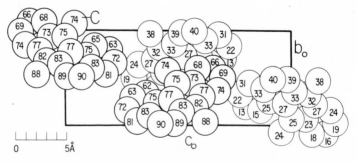

Fig. XVH,19.   The orthorhombic structure of 9,10-dimethyl-1,2-benzanthracene projected along its a_0 axis. The atoms of the near molecules are more heavily outlined than those lower down in the cell.

Fig. XVH,20. Bond dimensions in the molecule of 9,10-dimethyl-1,2-benzanthracene.

The structure (Fig. XVH,19) is built of molecules with the bond dimensions of Figure XVH,20. They are not planar, one planar end ring being tilted through 18.5° with respect to the best plane through the rest. In the anthracene portion ring atoms depart by as much as 0.29 A. and one attached methyl group [C(15)], by 0.57 A. from the mean plane.

XV,h13. The compound *2'-methyl-1,2-benzanthraquinone*, $C_{19}H_{12}O_2$, is dimorphous. Both forms are monoclinic; the one whose structure has been determined (form I) has a bimolecular unit of the dimensions:

$$a_0 = 20.67 \text{ A.}; \quad b_0 = 4.06 \text{ A.}; \quad c_0 = 7.77 \text{ A.}; \quad \beta = 90.8°$$

The space group is C_2^2 ($P2_1$) with atoms in the positions:

$$(2a) \quad xyz; \bar{x}, y + 1/2, \bar{z}$$

The parameters are stated in Table XVH,13; values for hydrogen are to be found in the original article.

The structure (Fig. XVH,21) has molecules with the bond dimensions of Figure XVH,22. They are planar within 0.08 A. except for the attached methyl carbon which is 0.18 A. outside the plane of the other atoms. Departures from strict planarity point to a small twist in the molecule attributable to steric hindrance involving O(2) and the hydrogen atoms attached to C(11).

TABLE XVH,13

Parameters of the Atoms in $C_{19}H_{12}O_2$

Atom	x	y	z
O(1)	0.3935	−0.0576	−0.1593
O(2)	0.2394	0.0106	0.3853
C(1)	0.3591	−0.0463	−0.0344
C(2)	0.3780	−0.1840	0.1315
C(3)	0.4382	−0.3470	0.1507
C(4)	0.4553	−0.4813	0.3068
C(5)	0.4149	−0.4566	0.4452
C(6)	0.3562	−0.2990	0.4293
C(7)	0.3373	−0.1655	0.2732
C(8)	0.2732	0.0111	0.2583
C(9)	0.2531	0.1591	0.0920
C(10)	0.1944	0.3372	0.0686
C(11)	0.1500	0.4064	0.2041
C(12)	0.0943	0.5808	0.1750
C(13)	0.0795	0.6924	0.0107
C(14)	0.1197	0.6362	−0.1244
C(15)	0.1784	0.4512	−0.0979
C(16)	0.2205	0.4101	−0.2367
C(17)	0.2775	0.2475	−0.2113
C(18)	0.2945	0.1273	−0.0493
CH₃	0.0495	0.6618	0.3203

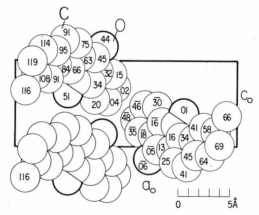

Fig. XVH,21.   The monoclinic structure of form I of 2′-methyl-1,2-benzanthra-quinone projected along its b_0 axis.

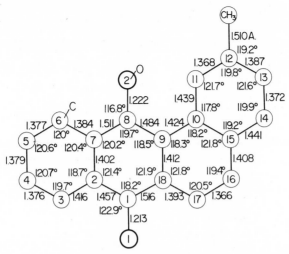

Fig. XVH,22.   Bond dimensions in the molecule of 2′-methyl-1,2-benzanthra-quinone.

XV,h14.   Crystals of *5-methyl-1,2-benzanthraquinone*, $C_{19}H_{12}O_2$, are orthorhombic with a tetramolecular unit of the edge lengths:

$$a_0 = 14.13(1) \text{ A.}; \quad b_0 = 23.27(2) \text{ A.}; \quad c_0 = 3.94(1) \text{ A.}$$

The space group is C_{2v}^9 in the axial orientation $P2_1nb$. Atoms therefore

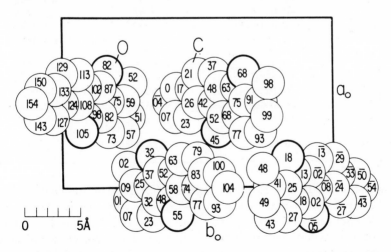

Fig. XVH,23.   The orthorhombic structure of 5-methyl-1,2-benzanthraquinone projected along its c_0 axis.

are in the positions:

$$(4a) \quad xyz; \ x+\tfrac{1}{2},\bar{y},\bar{z}; \ x+\tfrac{1}{2},\tfrac{1}{2}-y,z+\tfrac{1}{2}; \ x,y+\tfrac{1}{2},\tfrac{1}{2}-z$$

The determined parameters are listed in Table XVH,14; assumed positions for the hydrogen atoms are given in the original article.

TABLE XVH,14

Parameters of the Atoms in 5-Methyl-1,2-benzanthraquinone

Atom	x	y	z
O(1)	0.1711	0.1659	0.1771
O(2)	−0.1708	0.0771	−0.0551
C(1)	0.0921	0.1457	0.1320
C(2)	0.0055	0.1778	0.2515
C(3)	0.0085	0.2327	0.4096
C(4)	−0.0766	0.2571	0.4954
C(5)	−0.1639	0.2321	0.4358
C(6)	−0.1659	0.1799	0.2718
C(7)	−0.0818	0.1516	0.1799
C(8)	−0.0927	0.0949	0.0175
C(9)	−0.0049	0.0621	−0.0772
C(10)	−0.0087	0.0052	−0.2396
C(11)	−0.0969	−0.0267	−0.2713
C(12)	−0.0918	−0.0785	−0.4320
C(13)	−0.0066	−0.1059	−0.5439
C(14)	0.0756	−0.0756	−0.5033
C(15)	0.0771	−0.0196	−0.3315
C(16)	0.1630	0.0080	−0.2881
C(17)	0.1660	0.0618	−0.1335
C(18)	0.0826	0.0878	−0.0185
C(19)	0.0996	0.2611	0.4764

The structure is illustrated in Figure XVH,23. The molecule, which has the bond dimensions of Figure XVH,24, is very nearly planar, but the O(2) atom is 0.17 A. from the mean plane and C(11) and C(12) are 0.15 and 0.11 A. on the opposite side of this plane. Between molecules the shortest atomic separations are C–O = 3.34 and 3.44 A.; C–C distances range from 3.51 A. upward.

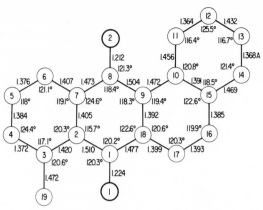

Fig. XVH,24. Bond dimensions in the molecule of 5-methyl-1,2-benzanthra-quinone.

XV,h15. A recent re-examination has confirmed the structure previously found for *perylene*, $C_{20}H_{12}$, and has led to more accurate atomic positions. The symmetry is monoclinic with a tetramolecular unit of the dimensions:

$$a_0 = 11.277(10) \text{ A.}; \quad b_0 = 10.826(10) \text{ A.}; \quad c_0 = 10.263(10) \text{ A.}$$
$$\beta = 100°33(1)'$$

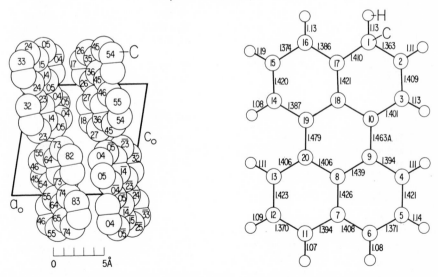

Fig. XVH,25 (*left*). The monoclinic structure of perylene projected along its b_0 axis.

Fig. XVH,26 (*right*). Bond lengths in the molecule of perylene.

All atoms are in the positions

$$(4e) \quad \pm (xyz; x+{}^1/_2, {}^1/_2-y, z)$$

of the space group C_{2h}^5 ($P2_1/a$). The recently established parameters, including those found for hydrogen, are listed in Table XVH,15.

TABLE XVH,15

Parameters of the Atoms in Perylene

Atom	x	y	z
C(1)	0.2624	−0.0474	0.3816
C(2)	0.3005	0.0432	0.3070
C(3)	0.2636	0.0447	0.1683
C(4)	0.1796	0.0428	−0.1185
C(5)	0.1417	0.0386	−0.2582
C(6)	0.0668	−0.0530	−0.3175
C(7)	0.0266	−0.1456	−0.2398
C(8)	0.0651	−0.1452	−0.0996
C(9)	0.1446	−0.0493	−0.0386
C(10)	0.1857	−0.0466	0.1050
C(11)	−0.0517	−0.2387	−0.2959
C(12)	−0.0923	−0.3274	−0.2198
C(13)	−0.0547	−0.3285	−0.0798
C(14)	0.0260	−0.3232	0.2057
C(15)	0.0664	−0.3217	0.3450
C(16)	0.1451	−0.2319	0.4021
C(17)	0.1841	−0.1419	0.3239
C(18)	0.1452	−0.1404	0.1842
C(19)	0.0652	−0.2353	0.1251
C(20)	0.0254	−0.2366	−0.0204
H(1)	0.290	−0.046	0.493
H(2)	0.358	0.114	0.367
H(3)	0.297	0.120	0.108
H(4)	0.243	0.114	−0.070
H(5)	0.173	0.111	−0.325
H(6)	0.038	−0.063	−0.423
H(11)	−0.078	−0.236	−0.402
H(12)	−0.153	−0.400	−0.266
H(13)	−0.086	−0.400	−0.016
H(14)	−0.037	−0.393	0.161
H(15)	0.033	−0.399	0.412
H(16)	0.179	−0.232	0.512

The structure (Fig. XVH,25) is of the same type as that described for pyrene (**XV,h6**). Its molecules are planar to within a few hundredths of an angstrom unit; their bond dimensions are given in Figure XVH,26. The bonds between C(9) and C(10) and between C(19) and C(20), completing the central hexagon, are definitely longer than the others. Internal angles about the carbon atoms lie between 118.3 and 121.7°; angles involving the hydrogen atoms range from 114 to 126°.

XV,h16. The monoclinic crystals of *perylene-tetracyanoethylene,* $C_{20}H_{12} \cdot C_2(CN)_4$, have a bimolecular unit of the dimensions:

$$a_0 = 15.70 \text{ A.}; \quad b_0 = 8.28 \text{ A.}; \quad c_0 = 7.31 \text{ A.}; \quad \beta = 96.1°$$

Atoms are in the positions

$$(4e) \quad \pm (xyz; x + 1/2, 1/2 - y, z)$$

of C_{2h}^5 ($P2_1/a$). The determined parameters are those of Table XVH,16.

TABLE XVH,16

Parameters of the Atoms in $C_{20}H_{12} \cdot C_2(CN)_4$

Atom	x	y	z
	Perylene		
C(1)	−0.182	0.275	0.088
C(2)	−0.130	0.281	0.261
C(3)	−0.052	0.187	0.288
C(4)	−0.026	0.100	0.146
C(5)	−0.078	0.094	−0.026
C(6)	−0.054	−0.005	−0.172
C(7)	−0.108	−0.009	−0.344
C(8)	−0.185	0.091	−0.368
C(9)	−0.207	0.179	−0.225
C(10)	−0.157	0.181	−0.056
	Tetracyanoethylene		
C(11)	−0.037	0.542	−0.009
C(12)	−0.056	0.637	0.145
C(13)	−0.097	0.547	−0.177
N(1)	−0.070	0.708	0.270
N(2)	−0.142	0.550	−0.304

The structure is shown in Figure XVH,27. The bond lengths, though not very accurate, are considered plausible. Between the perylene and

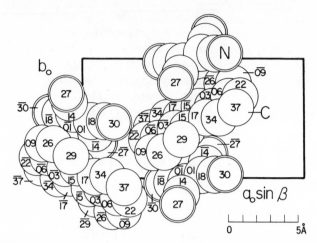

Fig. XVH,27.   The monoclinic structure of perylene-tetracyanoethylene projected along its c_0 axis.

tetracyanoethylene molecules the shortest atomic separations are C–N = 3.30 A. and C–C = 3.31 A.

XV,h17.   The *1 : 1 perylene-fluoranil* complex, $C_{20}H_{12} \cdot C_6F_4O_2$, forms monoclinic crystals. Their bimolecular unit has the dimensions:

$$a_0 = 17.13(1) \text{ A.;} \quad b_0 = 7.49(1) \text{ A.;} \quad c_0 = 6.97(1) \text{ A.}$$
$$\beta = 90.40(5)°$$

The space group is C_{2h}^5 in the orientation $P2_1/n$ with atoms in the positions:

$$(4e) \quad \pm (xyz; x + 1/2, 1/2 - y, z + 1/2)$$

The chosen parameters are stated in Table XVH,17.

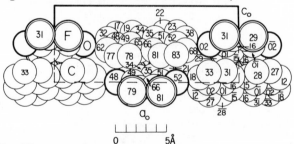

Fig. XVH,28.   The monoclinic structure of perylene-fluoranil projected along its b_0 axis.

TABLE XVH,17

Parameters of the Atoms in $C_{20}H_{12} \cdot C_6F_4O_2$

Atom	x	y	z
C(1)	0.0474	−0.1478	−0.0664
C(2)	0.0834	0.0125	0.0023
C(3)	0.0384	0.1605	0.0677
C(4)	0.0947	−0.2849	−0.1287
C(5)	0.1669	0.0257	0.0052
C(6)	0.0766	0.3114	0.1318
C(7)	0.1773	−0.2708	−0.1239
C(8)	0.2120	−0.1192	−0.0588
C(9)	0.2023	0.1847	0.0725
C(10)	0.1582	0.3252	0.1330
C(11)	0.0414	−0.1522	0.4393
C(12)	0.0858	0.0077	0.4996
C(13)	0.0357	0.1601	0.5629
O(14)	0.1556	0.0150	0.4993
F(15)	0.0850	−0.2896	0.3828
F(16)	0.0744	0.3060	0.6177

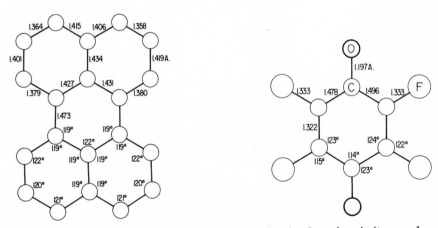

Fig. XVH,29a (*left*). Bond dimensions in the molecule of perylene in its complex with fluoranil.

Fig. XVH,29b (*right*). Bond dimensions in the molecule of fluoranil in its complex with perylene.

The structure is illustrated in Figure XVH,28. The bond dimensions of the two parts of the molecule are given in Figure XVH,29. The fluoranil portion is planar, but two of the carbon atoms of perylene may be slightly outside the plane of the rest.

XV,h18. The structure that was determined years ago for *1,12-benzperylene*, $C_{22}H_{12}$, has since been refined. The symmetry is monoclinic and it has a bimolecular unit of the dimensions:

$$a_0 = 11.72 \text{ A.}; \quad b_0 = 11.88 \text{ A.}; \quad c_0 = 9.89 \text{ A.}; \quad \beta = 98°30'$$

The space group is C_{2h}^5 ($P2_1/a$) with atoms in the positions:

$$(4e) \quad \pm (xyz; x+\tfrac{1}{2}, \tfrac{1}{2}-y, z)$$

The redetermined parameters, which show some differences from the earlier values, are those of Table XVH,18.

TABLE XVH,18

Parameters of the Atoms in 1,12-Benzperylene

Atom	x	y	z
C(1)	0.1597	−0.1755	0.3357
C(2)	0.2318	−0.0885	0.3943
C(3)	0.2620	−0.0014	0.3150
C(4)	0.2263	0.0029	0.1683
C(5)	0.2546	0.0902	0.0862
C(6)	0.2167	0.0928	−0.0525
C(7)	0.1438	0.0081	−0.1173
C(8)	0.1057	0.0090	−0.2632
C(9)	0.0358	−0.0755	−0.3173
C(10)	−0.0011	−0.1651	−0.2402
C(11)	−0.0742	−0.2508	−0.2991
C(12)	−0.1051	−0.3397	−0.2190
C(13)	−0.0693	−0.3423	−0.0754
C(14)	0.0151	−0.3476	0.2139
C(15)	0.0500	−0.3502	0.3575
C(16)	0.1235	−0.2637	0.4130
C(17)	0.1170	−0.1734	0.1920
C(18)	0.1498	−0.0842	0.1085
C(19)	0.1104	−0.0816	−0.0339
C(20)	0.0367	−0.1680	−0.0985
C(21)	0.0032	−0.2563	−0.0180
C(22)	0.0444	−0.2588	0.1306

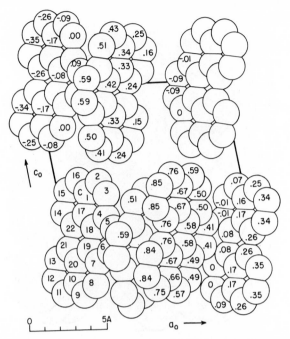

Fig. XVH,30a.   The monoclinic structure of 1,12-benzperylene projected along its b_0 axis. The y parameters of this drawing are those of the earlier determination. Left-hand axes.

The molecular distribution, illustrated in Figure XVH,30, closely resembles that of pyrene (**XV,h6**) and perylene (**XV,h15**). The molecule is planar and has the bond lengths of Figure XVH,31.

XV,h19.   The hydrocarbon *1,2,5,6-dibenznathracene*, $C_{22}H_{14}$, has an orthorhombic modification with a tetramolecular unit of the edge lengths:

$$a_0 = 8.22 \text{ A.}; \quad b_0 = 11.39 \text{ A.}; \quad c_0 = 15.14 \text{ A.}$$

All atoms are in general positions of V_h^{15} in the axial orientation *Pcab*:

(8c) $\pm (xyz; \, {}^1/_2 - x, y + {}^1/_2, \bar{z}; \, x + {}^1/_2, \bar{y}, {}^1/_2 - z; \, \bar{x}, {}^1/_2 - y, z + {}^1/_2)$

The parameters of Table XVH,19 yield a planar molecule with the dimensions of Figure XVH,32. In this figure the ringed carbon atoms (all but two) are those that were assigned positions on the basis of the Fourier projections. The distribution of the elongated molecules in the

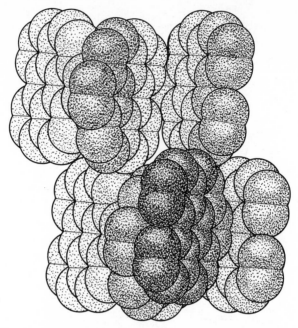

Fig. XVH,30b. A packing drawing of the monoclinic structure of 1,12-benz-perylene seen down its b_0 axis. Left-hand axes.

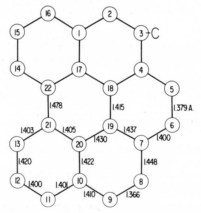

Fig. XVH,31.   Bond lengths in the molecule of 1,12-benzperylene.

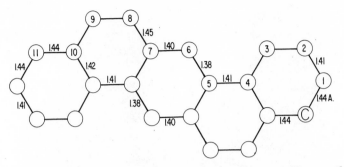

Fig. XVH,32.   Some bond lengths in the molecule of 1,2,5,6-dibenzanthracene
in its orthorhombic modification.

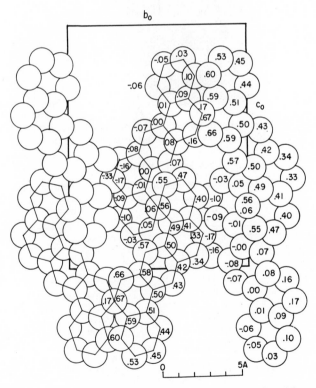

Fig. XVH,33a.   The molecular arrangement in the orthorhombic modification
of 1,2,5,6-dibenzanthracene projected along its a_0 axis.

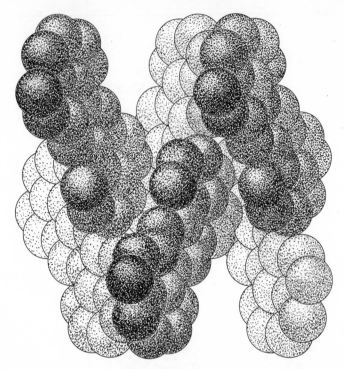

Fig. XVH,33b. A packing drawing of the orthorhombic structure of 1,2,5,6-dibenzanthracene viewed along its a_0 axis.

TABLE XVH,19

Parameters of the Atoms in 1,2,5,6-Dibenzanthracene

Atom	x	y	z
C(1)	−0.026	−0.139	0.362
C(2)	0.054	−0.034	0.337
C(3)	0.061	0.002	0.246
C(4)	−0.011	−0.068	0.181
C(5)	−0.003	−0.031	0.092
C(6)	0.073	0.070	0.065
C(7)	0.079	0.103	−0.025
C(8)	0.161	0.211	0.049
C(9)	0.168	0.245	−0.138
C(10)	0.093	0.174	−0.202
C(11)	0.101	0.210	−0.294

crystal is shown in Figure XVH,33. Intermolecular distances are normal, the closest C–C approach being 3.54 A.

XV,h20. The monoclinic modification of *1,2,5,6-dibenzanthracene*, $C_{22}H_{14}$, has a bimolecular unit of the dimensions:

$$a_0 = 6.59 \text{ A.}; \quad b_0 = 7.84 \text{ A.}; \quad c_0 = 14.17 \text{ A.}; \quad \beta = 103°30'$$

The space group is the low symmetry C_2^2 ($P2_1$) with carbon atoms in the positions:

$$(2a) \quad xyz; \; \bar{x}, y + {}^1/_2, \bar{z}$$

Determined parameters are those of Table XVH,20.

TABLE XVH,20

Parameters of the Atoms in Monoclinic $C_{22}H_{14}$

Atom	x	y	z
C(1)	0.457	−0.017	0.447
C(2)	0.262	−0.073	0.392
C(3)	0.210	−0.072	0.292
C(4)	0.354	−0.012	0.242
C(5)	0.311	−0.007	0.141
C(6)	0.123	−0.061	0.081
C(7)	0.083	−0.056	−0.018
C(8)	−0.114	−0.112	−0.074
C(9)	−0.152	−0.107	−0.174
C(10)	−0.006	−0.047	−0.223
C(11)	−0.055	−0.044	−0.324
C(12)	0.093	0.017	−0.372
C(13)	0.289	0.073	−0.317
C(14)	0.341	0.072	−0.217
C(15)	0.197	0.012	−0.167
C(16)	0.239	0.007	−0.066
C(17)	0.427	0.061	−0.005
C(18)	0.468	0.056	0.094
C(19)	0.664	0.112	0.149
C(20)	0.704	0.107	0.249
C(21)	0.557	0.047	0.298
C(22)	0.606	0.044	0.399

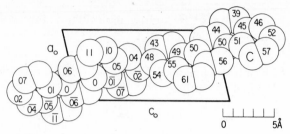

Fig. XVH,34.   The structure of the monoclinic modification of 1,2,5,6-dibenzan-
thracene projected along its b_0 axis.

The resulting structure is shown in Figure XVH,34. Its molecules
have much the same bond dimensions as those in the orthorhombic
form (**XV,h19**).

XV,h21.   Crystals of *3,4-5,6-dibenzophenanthrene*, $C_{22}H_{14}$, have been
assigned an unusually large monoclinic cell containing 12 molecules. Its
dimensions are:

$$a_0 = 26.17 \text{ A.}; \quad b_0 = 8.94 \text{ A.}; \quad c_0 = 19.57 \text{ A.}; \quad \beta = 105°6'$$

The space group is C_{2h}^6 ($A2/a$). All atoms are in the general positions

$$(8f) \quad \pm (xyz; \ x+1/2, \bar{y}, z; \ x, y+1/2, z+1/2; \ x+1/2, 1/2-y, z+1/2)$$

the assigned parameters being those of Table XVH,21.

TABLE XVH,21

Parameters of the Atoms in 3,4-5,6-Dibenzophenanthrene

Atom	x	y	z
C(1)	0.247	0.636	0.031
C(2)	0.233	0.498	0.061
C(3)	0.220	0.498	0.126
C(4)	0.201	0.358	0.158
C(5)	0.198	0.222	0.114
C(6)	0.176	0.089	0.144
C(7)	0.158	0.965	0.094
C(8)	0.163	0.975	0.022
C(9)	0.196	0.095	0.006
C(10)	0.210	0.225	0.044

(continued)

TABLE XVH,21 (*continued*)

Parameters of the Atoms in 3,4-5,6-Dibenzophenanthrene

Atom	x	y	z
C(11)	0.234	0.358	0.022

For C(1'–11'), $x' = \frac{1}{2} - x$, $y' = y$, $z' = \bar{z}$ where x, y and z have the above values.

Atom	x	y	z
C(12)	0.008	0.114	0.162
C(13)	−0.014	0.253	0.132
C(14)	−0.066	0.253	0.095
C(15)	−0.088	0.392	0.067
C(16)	−0.056	0.525	0.069
C(17)	−0.071	0.645	0.022
C(18)	−0.033	0.765	0.014
C(19)	0.019	0.765	0.050
C(20)	0.038	0.637	0.093
C(21)	0.000	0.522	0.107
C(22)	0.012	0.392	0.142
C(12')	0.067	0.117	0.198
C(13')	0.092	0.253	0.234
C(14')	0.148	0.253	0.277
C(15')	0.175	0.392	0.307
C(16')	0.145	0.531	0.298
C(17')	0.166	0.665	0.339
C(18')	0.136	0.793	0.342
C(19')	0.085	0.790	0.301
C(20')	0.061	0.656	0.257
C(21')	0.091	0.533	0.249
C(22')	0.060	0.395	0.211

The resulting structure and the general character of its molecules are indicated in Figure XVH,35. Accurate bond lengths were not established.

XV,h22. In crystals of *coronene*, $C_{24}H_{12}$, the molecules are distributed as in the simpler naphthalene (**XV,f1**) and anthracene (**XV,g1**), but their greater breadth makes them lie flatter in the cell and elongates the a_0 axis. The bimolecular monoclinic unit has the dimensions:

$$a_0 = 16.119(6) \text{ A.}; \quad b_0 = 4.702(4) \text{ A.}; \quad c_0 = 10.102(6) \text{ A.}$$
$$\beta = 110.9(1)°$$

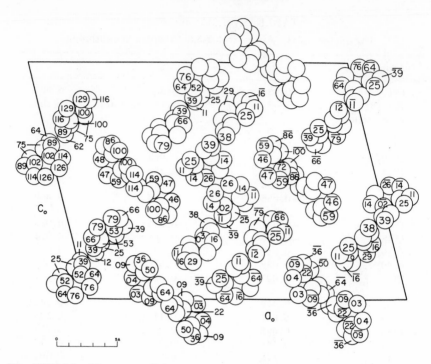

Fig. XVH,35. The monoclinic structure of 3,4-5,6-dibenzophenanthrene projected along its b_0 axis. Left-hand axes.

Atoms are in the general positions of C_{2h}^5 $(P2_1/a)$:

$$(4e) \quad \pm (xyz; x+{}^1\!/_2,{}^1\!/_2-y,z)$$

According to a recent refinement, the parameters are those of Table XVH,22.

The structure is illustrated in Figure XVH,36. Its centrosymmetric molecules have the bond lengths of Figure XVH,37. They are essentially planar, though among the outer atoms there are small departures from this plane.

XV,h23. *Quaterphenyl*, $C_6H_5(C_6H_4)_2C_6H_5$, has a structure like that of terphenyl (**XV,c1**) but with a still more elongated cell. Its bimolecular monoclinic unit has the dimensions:

$$a_0 = 8.05 \text{ A.}; \quad b_0 = 5.55 \text{ A.}; \quad c_0 = 17.81 \text{ A.}; \quad \beta = 95°48'$$

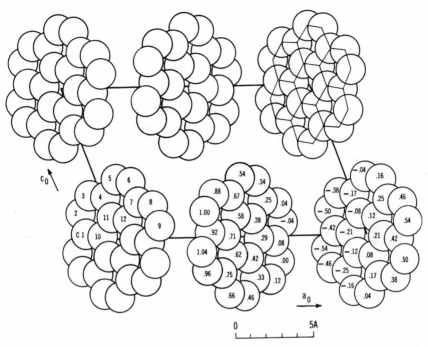

Fig. XVH,36a.   The monoclinic structure of coronene projected along its b_0 axis. Left-hand axes.

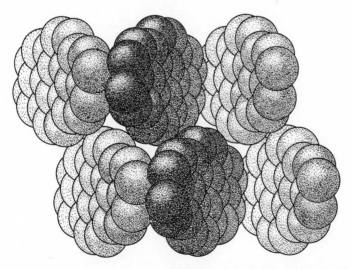

Fig. XVH,36b.   A packing drawing of the monoclinic coronene structure viewed along its b_0 axis. Left-hand axes.

TABLE XVH,22

Parameters of the Atoms in Coronene

Atom	x	y	z
C(1)	−0.1201	−0.4079	0.0381
C(2)	−0.1122	−0.4788	0.1782
C(3)	−0.0497	−0.3600	0.2913
C(4)	0.0121	−0.1607	0.2786
C(5)	0.0799	−0.0339	0.3941
C(6)	0.1364	0.1555	0.3761
C(7)	0.1339	0.2444	0.2410
C(8)	0.1909	0.4490	0.2167
C(9)	0.1843	0.5286	0.0847
C(10)	−0.0606	−0.2029	0.0183
C(11)	0.0057	−0.0823	0.1380
C(12)	0.0666	0.1210	0.1206
H(2)	−0.157	−0.634	0.195
H(3)	−0.044	−0.421	0.394
H(5)	0.083	−0.095	0.495
H(6)	0.188	0.246	0.465
H(8)	0.241	0.544	0.308
H(9)	0.228	0.685	0.070

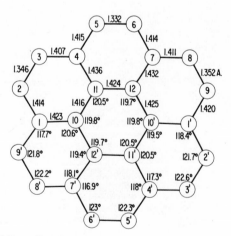

Fig. XVH,37.   Bond dimensions in the molecule of coronene.

The atoms are in the general positions of C_{2h}^5 $(P2_1/a)$:

$$(4e) \quad \pm (xyz; x+\tfrac{1}{2}, \tfrac{1}{2}-y, z)$$

Many years ago they were assigned the parameters of Table XVH,23.

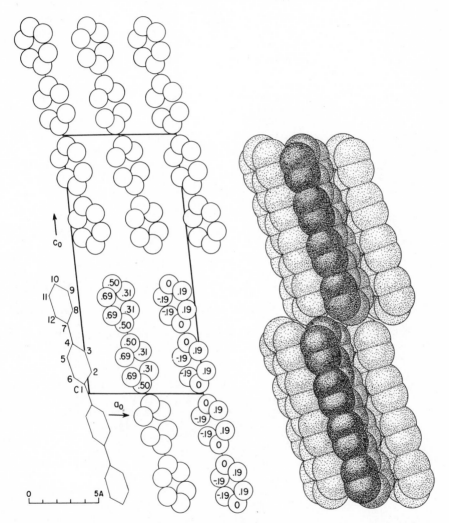

Fig. XVH,38a (*left*). The monoclinic structure of quaterphenyl projected along its b_0 axis. Left-hand axes.

Fig. XVH,38b (*right*). A packing drawing of the monoclinic quarterphenyl structure seen along its b_0 axis. Left-hand axes.

TABLE XVH,23

Parameters of the Atoms in Quaterphenyl

Atom	x	y	z
C(1)	-0.018	0.000	0.040
C(2)	0.043	0.186	0.089
C(3)	0.009	0.186	0.165
C(4)	-0.087	0.000	0.192
C(5)	-0.149	-0.186	0.144
C(6)	-0.115	-0.186	0.067
C(7)	-0.123	0.000	0.272
C(8)	-0.061	0.186	0.321
C(9)	-0.096	0.186	0.398
C(10)	-0.192	0.000	0.425
C(11)	-0.254	-0.186	0.376
C(12)	-0.220	-0.186	0.300

The structure is that of Figure XVH,38.

XV,h24. The monoclinic crystals of *2,3-8,9-dibenzperylene*, $C_{28}H_{16}$, possess a tetramolecular unit of the dimensions:

$$a_0 = 16.51 \text{ A.}; \quad b_0 = 5.23 \text{ A.}; \quad c_0 = 20.52 \text{ A.}; \quad \beta = 107°48(12)'$$

Atoms are in the general positions of C_{2h}^6 in the axial orientation $A2/a$:

$$(8f) \quad \pm(xyz; x+^1/_2, \bar{y}, z; x, y+^1/_2, z+^1/_2; x+^1/_2, ^1/_2-y, z+^1/_2)$$

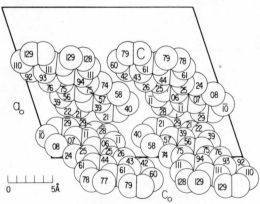

Fig. XVH,39. The monoclinic structure of 2,3-8,9-dibenzperylene projected along its b_0 axis.

The parameters, including those for hydrogen, are given in Table XVH,24.

TABLE XVH,24

Parameters of the Atoms in $C_{28}H_{16}$

Atom	x	y	z
C(1)	−0.05966	0.44047	0.26584
C(2)	−0.01311	0.24868	0.31648
C(3)	0.04540	0.06190	0.30068
C(4)	0.09155	−0.11296	0.34916
C(5)	0.14692	−0.28317	0.33368
C(6)	0.16079	−0.28890	0.27035
C(7)	−0.11545	0.61235	0.28045
C(8)	−0.12793	0.61287	0.34780
C(9)	−0.18169	0.79004	0.36520
C(10)	−0.19180	0.78966	0.43109
C(11)	−0.14999	0.59927	0.47697
C(12)	−0.09760	0.41830	0.46137
C(13)	−0.08665	0.43382	0.39584
C(14)	−0.02978	0.26269	0.37960
H(4)	0.087	−0.179	0.397
H(5)	0.177	−0.338	0.373
H(6)	0.207	−0.323	0.264
H(9)	−0.229	0.912	0.338
H(10)	−0.234	0.879	0.441
H(11)	−0.165	0.599	0.525
H(12)	−0.061	0.314	0.498
H(14)	0.011	0.095	0.423

The structure (shown in Fig. XVH,39) is composed of centrosymmetric molecules with the bond dimensions of Figure XVH,40. They are planar except for atoms C(9)-to-C(13) which lie as much as 0.07 A. from the plane of the rest. Between C(12) atoms of adjacent molecules the separation is the rather short 3.24 A.

XV,h25. Crystals of *5,6-dichloro-11,12-diphenylnaphthacene*, $C_{30}H_{18}Cl_2$, are monoclinic, pseudo-orthorhombic, with a tetramolecular

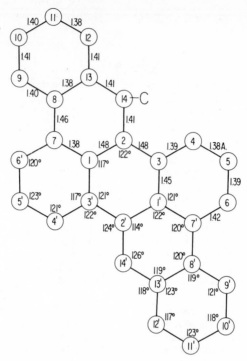

Fig. XVH,40. Bond dimensions in the molecule of 2,3-8,9-dibenzperylene.

unit of the dimensions:

$$a_0 = 16.13(4) \text{ A.}; \quad b_0 = 11.56(4) \text{ A.}; \quad c_0 = 11.42(3) \text{ A.}$$
$$\beta = 90°0'$$

The space group is C_{2h}^6 in the orientation $I2/c$. Two carbon atoms, C(1) and C(10), are in the positions:

$$(4e) \quad \pm(0\ u\ ^1/_4;\ ^1/_2, u + ^1/_2, ^3/_4)$$

with u for C(1) = -0.029 and for C(10), -0.155

All other atoms are in the positions:

$$(8f) \quad \pm(xyz;\ x, \bar{y}, z + ^1/_2); \quad \text{B.C.}$$

with the parameters of Table XVH,25.

The resulting structure is illustrated in Figure XVH,41. The bond dimensions of its molecules, which because of overcrowding show in the

TABLE XVH,25

Parameters of Atoms in 5,6-Dichloro-11,12-diphenylnaphthacene

Atom	x	y	z
Cl	0.0320	0.1722	0.1268
C(2)	0.045	0.020	0.155
C(3)	0.095	−0.032	0.080
C(4)	0.141	0.021	−0.021
C(5)	0.193	−0.040	−0.085
C(6)	0.213	−0.156	−0.053
C(7)	0.172	−0.214	0.033
C(8)	0.112	−0.154	0.105
C(9)	0.063	−0.213	0.188
C(11)	0.082	−0.335	0.214
C(12)	0.065	−0.425	0.125
C(13)	0.086	−0.534	0.161
C(14)	0.126	−0.564	0.262
C(15)	0.148	−0.472	0.338
C(16)	0.125	−0.360	0.314

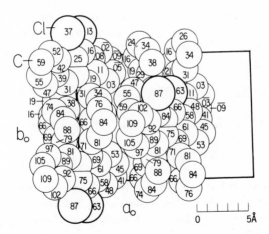

Fig. XVH,41.   The monoclinic, pseudo-orthorhombic, structure of 5,6-dichloro-11,12-diphenylnaphthacene projected along its c_0 axis.

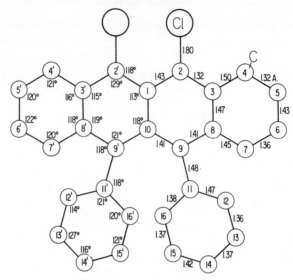

Fig. XVH,42.   Bond dimensions in the molecule of 5,6-dichloro-11,12-diphenyl-
naphthacene.

naphthacene part departures from planarity of more than 0.5 A., are
given in Figure XVH,42. The phenyl rings are planar.

XV,h26. A determination has been made of *ovalene*
(octabenzonaphthalene), $C_{32}H_{14}$. Its structure, of the same type as
coronene (**XV,h22**) and quaterphenyl (**XV,h23**), has a bimolecular unit
of the dimensions:

$$a_0 = 19.47 \text{ A.}; \quad b_0 = 4.70 \text{ A.}; \quad c_0 = 10.12 \text{ A.}; \quad \beta = 105°0'$$

The parameters of the carbon atoms, all in the general positions

$$(4e) \quad \pm (xyz; x + \tfrac{1}{2}, \tfrac{1}{2} - y, z)$$

of $C_{2h}{}^5$ ($P2_1/a$), are listed in Table XVH,26. The y parameters have been
calculated assuming that the molecule is planar; furthermore, all
parameters have been averaged to make the chemically equivalent
bonds of equal length.

These bond lengths are shown in Figure XVH,43.

TABLE XVH,26

Parameters of the Atoms in Ovalene

Atom	x	y	z
C(1)	0.057	−0.002	0.316
C(2)	0.111	0.127	0.424
C(3)	0.157	0.320	0.396
C(4)	0.156	0.409	0.261
C(5)	0.203	0.610	0.232
C(6)	0.199	0.690	0.098
C(7)	0.148	0.581	−0.012
C(8)	0.142	0.645	−0.151
C(9)	0.092	0.531	−0.257
C(10)	0.042	0.324	−0.234
C(11)	−0.009	0.202	−0.341
C(12)	0.002	−0.041	0.069
C(13)	0.055	0.083	0.178
C(14)	0.104	0.289	0.151
C(15)	0.100	0.369	0.015
C(16)	0.047	0.246	−0.095

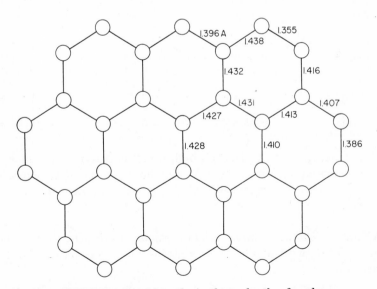

Fig. XVH,43.   Bond lengths in the molecule of ovalene.

XV,h27. Crystals of *1,2-7,8-dibenzocoronene*, $C_{32}H_{16}$, are monoclinic with a tetramolecular unit of the dimensions:

$$a_0 = 22.83(6) \text{ A.}; \quad b_0 = 5.22(1) \text{ A.}; \quad c_0 = 15.77(5) \text{ A.}$$
$$\beta = 103.9(3)°$$

Atoms are in the positions

$$(8f) \quad \pm (xyz; x,\bar{y},z+1/2; x+1/2,y+1/2,z; x+1/2,1/2-y,z+1/2)$$

of the space group C_{2h}^6 ($C2/c$). The determined parameters are those of Table XVH,27.

TABLE XVH,27

Parameters of the Atoms in $C_{32}H_{16}$

Atom	x	y	z
C(1)	0.2604	0.775	0.1641
C(2)	0.3175	0.770	0.1504
C(3)	0.3351	0.607	0.0920
C(4)	0.3929	0.600	0.0785
C(5)	0.4098	0.430	0.0208
C(6)	0.0837	0.769	0.2020
C(7)	0.1405	0.776	0.1884
C(8)	0.1589	0.601	0.1321
C(9)	0.2170	0.599	0.1170
C(10)	0.2341	0.430	0.0595
C(11)	0.2919	0.427	0.0427
C(12)	0.3099	0.248	−0.0135
C(13)	0.3680	0.244	−0.0285
C(14)	0.3842	0.077	−0.0862
C(15)	0.4434	0.077	−0.0978
C(16)	0.4598	−0.091	−0.1574

The structure (Fig. XVH,44) is built up of centrosymmetric molecules that are planar to within 0.04 A. and have the bond dimensions of Figure XVH,45. Intermolecular C–C separations range upwards from 3.59 A.

XV,h28. *Dibenzanthrone* (violanthrone), $C_{34}H_{16}O_2$, forms monoclinic crystals with a unit that contains four molecules and has the dimensions:

$$a_0 = 15.26 \text{ A.}; \quad b_0 = 33.60 \text{ A.}; \quad c_0 = 3.827 \text{ A.}; \quad \beta = 90°50'$$

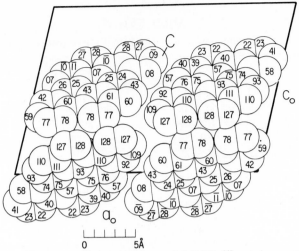

Fig. XVH,44. The monoclinic structure of 1,2-7,8-dibenzocoronene projected along its b_0 axis.

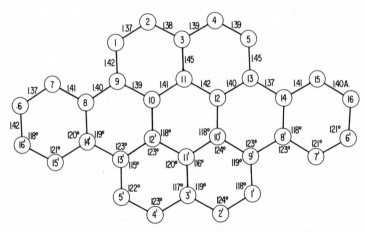

Fig. XVH,45. Bond dimensions in the molecule of 1,2-7,8-dibenzocoronene.

The space group is C_2^2 ($P2_1$) with atoms in the positions:

$$(2a) \quad xyz; \; \bar{x}, y + {}^1/_2, \bar{z}$$

The assigned parameters are those of Table XVH,28; it is clear from the figure referred to below that the decimal point is misplaced for y of O(1). Parameters for the atoms of hydrogen in the original article are also given there.

TABLE XVH,28

Parameters of the Atoms in Violanthrone

Atom	x	y	z
O(1)	0.1235	−0.1026	0.0053
O(2)	0.2008	−0.4237	0.2888
C(1)	0.3767	−0.2217	0.3897
C(2)	0.3863	−0.2628	0.2945
C(3)	0.2300	−0.2367	0.6316
C(4)	0.2415	−0.2791	0.5537
C(5)	0.4397	−0.1941	0.3098
C(6)	0.4303	−0.1534	0.3933
C(7)	0.4118	−0.0719	0.6258
C(8)	0.4629	−0.2781	0.1402
C(9)	0.4723	−0.3187	0.0627
C(10)	0.4967	−0.4045	−0.0977
C(11)	0.2982	−0.2096	0.5560
C(12)	0.3170	−0.2917	0.3868
C(13)	0.2860	−0.1684	0.6465
C(14)	0.3284	−0.3322	0.2898
C(15)	0.2671	−0.0864	0.8227
C(16)	0.3469	−0.4146	0.1099
C(17)	0.2581	−0.0447	0.9128
C(18)	0.3570	−0.4544	0.0181
C(19)	0.3533	−0.1397	0.5678
C(20)	0.4082	−0.3456	0.1358
C(21)	0.3453	−0.0993	0.6582
C(22)	0.4181	−0.3877	0.0434
C(23)	0.2073	−0.1559	0.7990
C(24)	0.2596	−0.3598	0.3747
C(25)	0.1931	−0.1124	0.8790
C(26)	0.2670	−0.4017	0.2523
C(27)	0.3228	−0.0168	0.8619
C(28)	0.4344	−0.4687	−0.1293
C(29)	0.4019	−0.0320	0.7045
C(30)	0.5033	−0.4438	−0.1793
C(31)	0.1440	−0.1813	0.8731
C(32)	0.1875	−0.3459	0.5445
C(33)	0.1563	−0.2227	0.8013

(continued)

TABLE XHV,28 (*continued*)

Parameters of the Atoms in Violanthrone

Atom	x	y	z
C(34)	0.1788	−0.3059	0.6389
O(1′)	−0.3000	−0.4138	−0.5740
O(2′)	−0.3843	−0.0916	0.1358
C(1′)	−0.1226	−0.2291	0.1635
C(2′)	−0.1146	−0.2713	0.0767
C(3′)	−0.2746	−0.2297	−0.1104
C(4′)	−0.2619	−0.2727	−0.1998
C(5′)	−0.0581	−0.2075	0.3375
C(6′)	−0.0671	−0.1675	0.4006
C(7′)	−0.0843	−0.0830	0.5617
C(8′)	−0.0371	−0.2927	0.1701
C(9′)	−0.0269	−0.3326	0.0966
C(10′)	−0.0085	−0.4171	−0.0764
C(11′)	−0.2031	−0.2083	0.0656
C(12′)	−0.1832	−0.2916	−0.1074
C(13′)	−0.2110	−0.1674	0.1502
C(14′)	−0.1735	−0.3320	−0.1807
C(15′)	−0.2332	−0.0855	0.3429
C(16′)	−0.1563	−0.4166	−0.3086
C(17′)	−0.2435	−0.0455	0.4372
C(18′)	−0.1507	−0.4578	−0.3858
C(19′)	−0.1434	−0.1466	0.3186
C(20′)	−0.0936	−0.3534	−0.0798
C(21′)	−0.1536	−0.1044	0.4122
C(22′)	−0.0850	−0.3960	−0.1599
C(23′)	−0.2939	−0.1484	0.0692
C(24′)	−0.2431	−0.3542	−0.3584
C(25′)	−0.3091	−0.1072	0.1887
C(26′)	−0.2376	−0.3959	−0.4191
C(27′)	−0.1747	−0.0239	0.5731
C(28′)	−0.0725	−0.4771	−0.3087
C(29′)	−0.0961	−0.0433	0.6520
C(30′)	−0.0017	−0.4556	−0.1564
C(31′)	−0.3594	−0.1695	−0.1156
C(32′)	−0.3171	−0.3328	−0.4465
C(33′)	−0.3494	−0.2090	−0.2000
C(34′)	−0.3259	−0.2921	−0.3851

The resulting structure, shown in Figure XVH,46, contains two different kinds of molecule. They are stacked along the c_0 axis in a way suggesting the piling of the carbon planes in graphite. Between molecules the shortest C–O = 3.49 A. and C–C = 3.64 A.

XV,h29. Crystals of *isodibenzanthrone* (isoviolanthrone), $C_{34}H_{16}O_2$, are monoclinic with a tetramolecular unit that has the dimensions:

$$a_0 = 15.21 \text{ A.}; \quad b_0 = 3.825 \text{ A.}; \quad c_0 = 33.12 \text{ A.}; \quad \beta = 90°48'$$

The space group is C_{2h}^5 ($P2_1/c$) with atoms in the positions:

$$(4e) \quad \pm(xyz; x, {}^1/_2 - y, z + {}^1/_2)$$

The parameters are recorded in Table XVH,29. As reference to the figure will make clear, y for C(10), as given in the original, is in error. It should be about one-half the 0.4094 stated there. Positions for the hydrogen atoms are also to be found in the original.

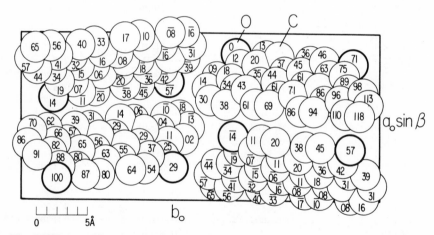

Fig. XVH,46. The monoclinic structure of dibenzanthrone projected along its c_0 axis.

The structure is shown in Figure XVH,47. The molecules are essentially planar, though atoms are as much as 0.14 A. out of these planes. They are stacked along the b_0 axis but their normals make about 27° with it. The shortest intermolecular distances are C–C = 3.46 A. and C–O = 3.60 A. This is a structure which, as a comparison of the figures indicates, closely resembles that of dibenzanthrone (**XV,h28**).

TABLE XVH,29

Parameters of the Atoms in Isoviolanthrone

Atom	x	y	z
O(1)	−0.0980	0.5550	0.1636
C(1)	−0.1642	0.1137	−0.0836
C(2)	−0.0796	0.0988	−0.0212
C(3)	0.0113	0.0795	0.0423
C(4)	0.1020	0.0557	0.1050
C(5)	0.1890	0.0672	0.1688
C(6)	−0.1521	0.1893	−0.0424
C(7)	−0.0690	0.1775	0.0211
C(8)	0.0237	0.1532	0.0838
C(9)	0.1132	0.1514	0.1478
C(10)	0.1995		0.2094
C(11)	−0.1344	0.3550	0.0430
C(12)	−0.0488	0.3205	0.1035
C(13)	0.0448	0.2986	0.1686
C(14)	0.1282	0.3051	0.2286
C(15)	−0.1220	0.4212	0.0833
C(16)	−0.0366	0.3916	0.1472
C(17)	0.0514	0.3775	0.2084
O(1′)	0.5981	0.3396	−0.1747
C(1′)	0.5449	0.3955	−0.1099
C(2′)	0.4912	0.4107	−0.0420
C(3′)	0.4337	0.4214	0.0281
C(4′)	0.3801	0.4337	0.0969
C(5′)	0.3671	0.3354	0.0562
C(6′)	0.4219	0.3226	−0.0137
C(7′)	0.4752	0.3265	−0.0834
C(8′)	0.5344	0.2991	−0.1523
C(9′)	0.4558	0.1346	−0.1662
C(10′)	0.3959	0.1613	−0.0978
C(11′)	0.3445	0.1673	−0.0290
C(12′)	0.4454	0.0437	−0.2068
C(13′)	0.3861	0.0714	−0.1392
C(14′)	0.3340	0.0810	−0.0695
C(15′)	0.3675	−0.0881	−0.2222
C(16′)	0.3072	−0.0714	−0.1555
C(17′)	0.2972	−0.1558	−0.1962

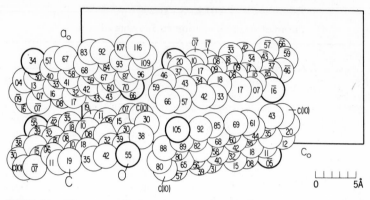

Fig. XVH,47.   The monoclinic structure of isodibenzanthrone projected along
its b_0 axis.

XV,h30.   The α modification of the hydrocarbon *dinaphtho(7′,1′-1,13)*
(1″,7″-6,8)peropyrene, $C_{38}H_{18}$, is monoclinic with a tetramolecular unit
of the dimensions:

$$a_0 = 30.73(5) \text{ A.}; \quad b_0 = 3.855(10) \text{ A.}; \quad c_0 = 19.87(3) \text{ A.}$$
$$\beta = 113°0(30)'$$

The space group has been chosen as the low symmetry C_2^3 ($C2$) with
atoms in the positions:

$$(4c) \quad xyz; \ \bar{x}y\bar{z}; \ x+{}^1/_2,y+{}^1/_2,z; {}^1/_2-x,y+{}^1/_2,\bar{z}$$

Parameters, rendered somewhat uncertain by disorder in the structure,
have been given as those of Table XVH,30.

TABLE XVH,30

Parameters of the Atoms in $C_{38}H_{18}$

Atom	x	y	z
	Molecule I		
C(1)	0.0735	−0.322	−0.1101
C(2)	0.1123	−0.439	−0.1287
C(3)	0.1584	−0.372	−0.0770
C(4)	0.1648	−0.205	−0.0060
C(5)	0.2115	−0.139	0.0472

(continued)

TABLE XVH,30 (*continued*)

Parameters of the Atoms in $C_{38}H_{18}$

Atom	x	y	z
C(6)	0.2191	−0.003	0.1178
C(7)	0.1795	0.070	0.1374
C(8)	0.1865	0.203	0.2066
C(9)	0.1484	0.250	0.2253
C(10)	0.1014	0.192	0.1754
C(11)	0.0625	0.226	0.1972
C(12)	0.0149	0.105	0.1486
C(13)	0.0077	0.005	0.0745
C(14)	0.0470	0.010	0.0534
C(15)	0.0940	0.070	0.1060
C(16)	0.1326	0.013	0.0845
C(17)	0.1261	−0.115	0.0125
C(18)	0.0797	−0.155	−0.0401
C(19)	0.0393	−0.047	−0.0212
	Molecule II		
C(1′)	0.1014	0.141	0.6754
C(2′)	0.1484	0.015	0.7253
C(3′)	0.1865	0.071	0.7066
C(4′)	0.1795	0.207	0.6374
C(5′)	0.2191	0.301	0.6178
C(6′)	0.2115	0.431	0.5472
C(7′)	0.1648	0.509	0.4940
C(8′)	0.5184	0.640	0.4230
C(9′)	0.1123	0.690	0.3713
C(10′)	0.0735	0.623	0.3899
C(11′)	0.0256	0.637	0.3359
C(12′)	−0.0149	0.523	0.3514
C(13′)	−0.0077	0.426	0.4255
C(14′)	0.0393	0.443	0.4788
C(15′)	0.0797	0.510	0.4599
C(16′)	0.1261	0.450	0.5125
C(17′)	0.1326	0.331	0.5845
C(18′)	0.0940	0.297	0.6060
C(19′)	0.0470	0.356	0.5534

XV,h31. Crystals of the stable (α) modification of *quaterrylene*, $C_{40}H_{20}$, are monoclinic with a tetramolecular unit of the dimensions:

$$a_0 = 11.25(4) \text{ A.}; \quad b_0 = 10.66(3) \text{ A.}; \quad c_0 = 19.31(7) \text{ A.}$$
$$\beta = 100.6°$$

Atoms are in the positions

$$(4e) \quad \pm (xyz; x + {}^1/_2, {}^1/_2 - y, z)$$

of C_{2h}^5 ($P2_1/a$). The parameters are listed in Table XVH,31. The y parameters were not established experimentally but were deduced from the determined tilt of the molecule considered as planar.

TABLE XVH,31

Parameters of the Atoms in Quaterrylene

Atom	x	y	z
C(1)	0.2771	−0.0481	0.4420
C(2)	0.3284	0.0453	0.4074
C(3)	0.3109	0.0486	0.3328
C(4)	0.2380	−0.0474	0.2960
C(5)	0.2193	−0.0447	0.2139
C(6)	0.2676	0.0470	0.1785
C(7)	0.2469	0.0451	0.1060
C(8)	0.1758	−0.0465	0.0678
C(9)	0.1529	−0.0486	−0.0124
C(10)	0.2030	0.0432	−0.0500
C(11)	0.1817	0.0428	−0.1247
C(12)	0.1121	−0.0470	−0.1621
C(13)	0.0903	−0.0476	−0.2424
C(14)	0.1386	0.0427	−0.2808
C(15)	0.1199	0.0456	−0.3555
C(16)	0.0492	−0.0473	−0.3913
C(17)	−0.0027	−0.1425	−0.3563
C(18)	−0.0726	−0.2333	−0.3931
C(19)	−0.1231	−0.3243	−0.3588
C(20)	−0.1043	−0.3277	−0.2827

(continued)

TABLE XVH,31 *(continued)*

Parameters of the Atoms in Quaterrylene

Atom	x	y	z
C(21)	−0.0319	−0.2353	−0.2438
C(22)	−0.0090	−0.2319	−0.1634
C(23)	−0.0594	−0.3254	−0.1273
C(24)	−0.0408	−0.3272	−0.0502
C(25)	0.0297	−0.2357	−0.0145
C(26)	0.0516	−0.2349	0.0664
C(27)	0.0018	−0.3257	0.1027
C(28)	0.0228	−0.3259	0.1785
C(29)	0.0908	−0.2368	0.2153
C(30)	0.1143	−0.2347	0.2949
C(31)	0.0614	−0.3297	0.3304
C(32)	0.0807	−0.3327	0.4050
C(33)	0.1514	−0.2378	0.4398
C(34)	0.2046	−0.1416	0.4057
C(35)	0.1840	−0.1430	0.3310
C(36)	0.1433	−0.1415	0.1796
C(37)	0.1243	−0.1400	0.1039
C(38)	0.0802	−0.1426	−0.0513
C(39)	0.0622	−0.1393	−0.1259
C(40)	0.0179	−0.1415	−0.2801

The structure that results is shown in Figure XVH,48. In the molecule the bond lengths range between 1.37 and 1.53 A. depending on the position of the bond; bond angles vary from 115 to 124°.

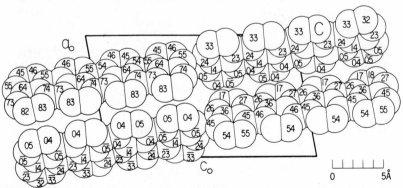

Fig. XVH,48.   The monoclinic structure of quaterrylene projected along its b_0 axis.

XV,h32. The monoclinic crystals of *1,12-2,3-4,5-6,7-8,9-10,11-hexabenzocoronene*, $C_{42}H_{18}$, possess a bimolecular unit of the dimensions:

$$a_0 = 18.42(5) \text{ A.}; \quad b_0 = 5.11(1) \text{ A.}; \quad c_0 = 12.86(3) \text{ A.}$$
$$\beta = 112.5(3)°$$

The space group is C_{2h}^5 ($P2_1/a$) with atoms in the positions:

$$(4e) \quad \pm (xyz; x + 1/2, 1/2 - y, z)$$

The determined parameters are those of Table XVH,32.

TABLE XVH,32

Parameters of the Atoms in $C_{42}H_{18}$

Atom	x	y	z
C(1)	0.0389	−0.178	0.4053
C(2)	0.0965	0.002	0.4773
C(3)	0.1289	0.172	0.4221
C(4)	−0.1379	−0.690	0.2093
C(5)	−0.0794	−0.513	0.2739
C(6)	−0.0483	−0.348	0.2198
C(7)	0.0097	−0.175	0.2874
C(8)	0.0467	0.008	0.2387
C(9)	0.1051	0.173	0.3030
C(10)	0.1397	0.350	0.2507
C(11)	0.1976	0.519	0.3159
C(12)	0.2316	0.698	0.2672
C(13)	−0.1636	−0.695	0.0946
C(14)	−0.1302	−0.520	0.0342
C(15)	−0.0680	−0.344	0.1058
C(16)	−0.0374	−0.175	0.0470
C(17)	0.0232	0.000	0.1166
C(18)	0.0582	0.169	0.0675
C(19)	0.1159	0.343	0.1331
C(20)	0.1519	0.516	0.0832
C(21)	0.2072	0.697	0.1451

The structure that results is that of Figure XVH,49. Its centrosymmetric molecules are planar to ca. 0.06 A.; their bond lengths range from 1.35 to 1.47 A. and bond angles lie between 115 and 126°. Between molecules the shortest C–C separation is 3.49 A.

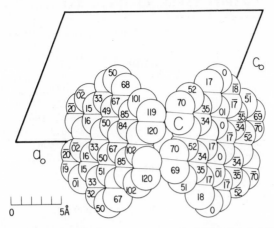

Fig. XVH,49. The monoclinic structure of 1,12-2,3-4,5-6,7-8,9-10,11-hexa-benzocoronene projected along its b_0 axis.

BIBLIOGRAPHY TABLE, CHAPTER XVB-H

Compound	Paragraph	Literature
4-Acetyl-2'-chlorobiphenyl	**bI,4**	1968: S&H
4-Acetyl-2'-fluorobiphenyl	**bI,3**	1968: Y,T&S
Acetyl triphenylgermane	**c8**	1968: H&T
4,4'-di-Amino-3,3'- dimethyl biphenyl	**bI,9**	1968: C,H&S
2,2'-di-Aminodiphenyl disulfide	**bIII,21**	1967: GM
di-Aminodiphenyl sulfone	**bII,30**	1965: A&D
1-Aminonaphthalene-chromium tricarbonyl	**f9**	1968: C,MP&S
3-Amino-1,4-naphthoquinone	**f26**	1966: G&H
4-Amino-1,2-naphthoquinone hemihydrate	**f27**	1968: A,G&H
tetra-p-Anisylethylene dichloroiodide	**d13**	1967: B,B&S

(continued)

BIBLIOGRAPHY TABLE, CHAPTER XVB-H (*continued*)

Compound	Paragraph	Literature
di-*p*-Anisyl-*N*-oxide	**bII,20**	1953: H
Anthracene	**g1**	1921: B; B&J; 1923: B; 1929: R; 1930: B; 1933: R; 1950: M,R&S; R; S,R&M; 1952: A&C; 1953: C&R; K&K; 1956: C; 1957: C; 1964: M
Anthracene-chromium tricarbonyl	**g4**	1968: H&M
Anthracene-*s*-trinitrobenzene	**g3**	1964: B,W&W
9,10-Anthrahydroquinone dibenzoate	**g13**	1962: I&M
9-Anthraldehyde	**g7**	1959: T
Anthraquinone	**g15**	1921: B&J; 1930: H&R; 1938: G; 1945: S; 1948: S; 1955: M; 1960: M; 1966: L,M&ES; 1967: P
Anthrarufine	**g24**	1964: G; 1967: G
Anthrone	**g6**	1959: S; 1961: S; 1962: S; 1964: S
Auramine perchlorate	**bII,12**	1967: F,K&K
di-Azoaminobenzene-copper(I)	**bIV,9**	1961: B&D
cis-Azobenzene	**bIII,13**	1936: R,P&W; 1939: R; 1941: H&R
trans-Azobenzene	**bIII,14**	1921: B&J; 1930: P; 1939: L,R&W; R; 1966: B
p-Azotoluene	**bIII,16**	1958: P,S&K; 1966: B
Benzalazine	**bV,6**	1961: S
1,2-Benzanthracene	**h11**	1938: I; 1956: F&S
1,2,5,6-di-Benzanthracene	**h19,20**	1935: K&B; 1936: I; 1947: R&W; 1956: R&W
di-Benzanthrone	**h28**	1964: B&S
bis(Benzene-azo-β-naphtholato) copper(II)	**f41**	1961: J
Benzil	**bIII,12**	1923: B&R; 1927: A; 1938: B&S; 1965: B&S
Benzil-α-monoxime-*p*- bromobenzoate	**c6**	1967: K,R&S
1,2-7,8-di-Benzocoronene	**h27**	1961: R&T
1,12-2,3-4,5-6,7-8,9-10,11-hexa- benzocoronene	**h32**	1961: R&T

(*continued*)

BIBLIOGRAPHY TABLE, CHAPTER XVB-H (*continued*)

Compound	Paragraph	Literature
3,4-5,6-di-Benzophenanthrene	**h21**	1952: MI,R&V; 1954: MI,R&V
Benzophenone	**bII,8**	1938: B&H; 1952: G,Z,U&G; N,P&S; 1967: V&L
di-Benzoyl methane	**bIV,1**	1966: W
bis(di-Benzoylmethyl)palladium	**bIV,4**	1968: S,S&K
di-Benzoyl peroxide	**bV,4**	1967: S&MM
Benzoyl(triphenylphosphor-anylidene)methyl bromide	**d19**	1965: S
Benzoyl(triphenylphosphor-anylidene)methyl chloride	**d19**	1965: S
Benzoyl(triphenylphosphor-anylidene)methyl iodide	**d20**	1965: S
1,12-Benzperylene	**h18**	1948: W; 1959: T
2,3-8,9-di-Benzperylene	**h24**	1959: L,R&R
di-Benzyl	**bIII,1**	1923: B&R; 1929: H&M; 1934: D; R; 1935: R; 1945: J
di-Benzyl phosphoric acid	**bV,9**	1954: S&C; 1956: D&R
Benzyl sulfide-iodine (1 : 1)	**bIV,6**	1960: R
tri-Benzyl tin acetate	**c9**	1968: A&T
2-Bromo-3-amino-1,4-naphthoquinone	**f35**	1965: G&H; 1966: G&H
9,10-di-Bromoanthracene	**g10**	1958: T
1,5-di-Bromoanthraquinone	**g16**	1963: C,G&Z; 1967: C&G
2,3-di-Bromo-1,4-anthraquinone	**g17**	1967: G,G&H
trans-p,p'-di-Bromoazobenzene	**bIII,17**	1966: A&H
p-Bromobenzeneazo-tribenzoylmethane	**d10**	1966: P,P&C
p-Bromobenzoic anhydride	**bIV,5**	1964: MC&T
3,3'-di-Bromobenzophenone	**bII,9**	1958: R&V
bis(*m*-Bromobenzoyl)methane	**bIV,3**	1962: W,D&R
di-Bromo bis(triphenylphosphine) nickel	**e12**	1968: J,M&O
4,4'-di-Bromocinnamaldazine	**bV,18**	1967: B,E,J,M&R
2,6-di-Bromo-α-cyanostilbene	**bIII,7**	1962: BE&vM
p-di-Bromodiazoaminobenzene	**bIV,12**	1958: K; 1961: K
2,4-di-Bromodiazoaminobenzene	**bIV,11**	1965: O&K
p-Bromodiazoaniline	**bIV,10**	1967: O&K
4,4'di-Bromodibenzoyl peroxide	**bV,5**	1968: CE&A

(*continued*)

BIBLIOGRAPHY TABLE, CHAPTER XVB-H (*continued*)

Compound	Paragraph	Literature
1,5-di-Bromo-4,8-dichloronaphthalene	**f19**	1965: D&S; 1968: D&S
p,p'-di-Bromo-α,α'-difluorostilbene	**bIII,5**	1967: C&G
2-Bromo-4'-dimethylamino-α-cyanostilbene	**bIII,8**	1960: vM&L
N(1-Bromo-3,5-dimethylphenyl) benzenesulfonamide	**bIII,24**	1965: R,D,K&R
Bromodiphenylarsine	**bII,24**	1962: T
2-Bromo-1,1-diphenyl-1-propene	**bII,2**	1966: C,M,M&S
2-Bromo-1,1-di-p-tolylethylene	**bII,3**	1965: M,M&C; 1967: C,M,M&S
3,5-di-Bromo-p-hydroxytriphenyl methane carbinol	**c4**	1949: S&vE; 1966: S; 1968: S
9-Bromo-10-methyl-anthracene	**g12**	1961: P
2-Bromo-3-methyl-1,4-naphthoquinone	**f36**	1965: B
α,4-di-Bromo-α-(4-methyl-2-nitrophenylazo)-acetanilide	**bV,8**	1967: B
N(4-Bromo-2-methylphenyl)-benzenesulfonamide	**bIII,25**	1966: R,D,K&R
1,4,5,8-tetra-Bromonaphthalene	**f18**	1965: D&S; 1968: D&S
2-Bromo-1,4-naphthoquinone	**f25**	1965: G&H
2,3-di-Bromo-1,4-naphthoquinone	**f33**	1966: B; 1967: BL
tri-Bromonitrosobenzene dimer	**bIII,18**	1950: F
N(p-Bromophenyl)ben-zenesulfonamide	**bIII,26**	1967: R; R,D,K&R
N(p-Bromophenyl)-p-chloro-benzenesulfonamide	**bIII,27**	1964: R,D,G,K&R
p-Bromophenyldiphenyl phosphine	**c12**	1966: K&P
3-Bromo-2,4,6-triphenyl phenoxyl dimer	**e25**	1968: A&H
di-Bromo tris(diphenylphosphine) cobalt(II)	**e11**	1966: B&P
di-Bromo tris(diphenylphosphine) nickel(II)	**e11**	1966: B&P
bis(tetra-n-Butylammonium) di[bis(1,2,3,4-tetrachlorobenzene-5,6-dithiolato)cobaltate]	**d35**	1968: BH,D,E&G
$C_6H_5C_2C_6H_5 \cdot Fe_2(CO)_6$	**bIII,11**	1967: D,MP,vM&P
$(C_6H_5)_3P{=}C(CO_2CH_3){-}C(CO_2CH_3){=}N(C_6H_4Br)$	**d22**	1965: M&T

(*continued*)

BIBLIOGRAPHY TABLE, CHAPTER XVB-H *(continued)*

Compound	Paragraph	Literature
Calcium 1-naphthylphosphate trihydrate	**f40**	1965: L&C
bis-(tri-Carbonylchromium) biphenyl	**bI,2**	1960: C&A; 1961: A
tetra-Carbonylcobalt diphenyltin manganesepentacarbonyl	**bII,13**	1968: B,S&S
octa-Carbonyl diphenylvinylidene di-iron	**bII,4**	1968: M&R
tri-Carbonyl(triphenylphosphine)-σ-tetrafluoroethyl cobalt	**c31**	1967: W&P
Cerium tetrakis-dibenzoylmethane	**d7**	1959: S&K; 1960: W&B
Cesium tetracyanoquinodimethanide	**c24**	1964: A; 1966: F&A
tri-Chloro(p-acetylphenylimino) bis(diethylphenylphosphine) rhenium(V)	**c30**	1968: B&I
2-Chloro-3-amino-1,4-naphthoquinone	**f32**	1965: G&H
9,10-di-Chloroanthracene	**g11**	1959: T
1,5-di-Chloroanthraquinone	**g19**	1958: B
2,2'-di-Chlorobenzidine	**bI,8**	1948: S
bis(m-Chlorobenzoyl) methane	**bIV,2**	1964: E&R
Chloro[N,N'-bis(salicylidene) ethylenediamine] iron(III)	**bVI,9**	1967: G&M
9-Chloro-10-bromoanthracene	**g10**	1961: H
Chlorocarbonyl bis(triphenylphosphine) iridium, oxygen adduct	**e15**	1965: LP&I
Chlorocarbonyl (sulfur dioxide) bis(triphenylphosphine) iridium	**e14**	1966: LP&I
tri-μ-Chloro-chloropentakis (diethylphenylphosphine) diruthenium(II)	**e3**	1968: A&R
p-Chlorodiazoaminobenzene	**bIV,12**	1968: K
p-di-Chlorodiazoaminobenzene	**bIV,12**	1958: K; 1961: K
4,4'-di-Chlorodibenzoyl peroxide	**bV,5**	1968: CE&A
4,4'-di-Chloro-α,β-diethyl stilbene	**bIII,9**	1965: B,P&R; 1967: P&R
3,3'-di-Chloro-4,4'-dihydroxy diphenyl methane	**bII,5**	1953: W

(continued)

BIBLIOGRAPHY TABLE, CHAPTER XVB-H (*continued*)

Compound	Paragraph	Literature
p,p'-di-Chlorodiphenoxy-1,2-ethane	**bV,3**	1967: Y,A&K
di-Chlorodiphenoxy titanium(IV)	**bIV,7**	1966: W&C
p,p'-di-Chlorodiphenylamine	**bII,15**	1961: P&R
Chlorodiphenylarsine	**bII,24**	1962: T
p,p'-di-Chlorodiphenyl diselenide	**bIII,23**	1957: K,M&MC
p,p'-di-Chlorodiphenyl ditelluride	**bIII,23**	1957: K,M&MC
5,6-di-Chloro-11,12-diphenyl-naphthacene	**h25**	1963: A,K&S; 1964: A,K&S
4,4'-di-Chloro diphenyl sulfone	**bII,28**	1960: B&C; S&A; 1961: T&A
2-Chloro-3-hydroxy-1,4-naphthoquinone	**f32**	1964: G&H; 1965: G&H
tri-Chloro(*p*-methoxyphenylimino) bis(diethylphenylphosphine) rhenium(V)	**c29**	1968: B&I
2-Chloro-3-methyl-1,4-naphthoquinone	**f36**	1964: B; 1965: B
2,6-di-Chloronaphthalene	**f12**	1964: K&S
1,4,5,8-tetra-Chloronaphthalene	**f20**	1961: D&S; K,S,A&D; 1962: D&S; G&H
octa-Chloronaphthalene	**f21**	1963: G&H
2,3-di-Chloro-1,4-naphthoquinone	**f34**	1961: M
1,6-di-*o*-Chlorophenyl-3,4-dimethyl hexatriene	**bV,14**	1966: S&S
1,6-di-*p*-Chlorophenyl-3,4-dimethyl hexatriene	**bV,13**	1966: S&S
di-*p*-Chlorophenyl hydrogen phosphate	**bIV,13**	1964: C&S
bis(*p*-Chlorophenyl) tellurium diiodide	**bII,35**	1962: C&MC
2-Chloro-*N*-salicylideneaniline	**bIII,20**	1964: B,L&O
N-5-Chlorosalicylideneaniline	**bIII,19**	1964: B,L&S
di-Chlorotetrabenzoato dirhenium(III)-dichloroform	**d6**	1968: B,B,C&R
di-Chloro tris(triphenylphosphine) ruthenium(II)	**e29**	1965: LP&I
Chrysazin	**h5**	1965: P
Chrysene	**h9**	1933: I&R; 1934: I; 1935: B&C; 1956: B&I; 1960: B&I

(*continued*)

BIBLIOGRAPHY TABLE, CHAPTER XVB-H (*continued*)

Compound	Paragraph	Literature
Coronene	**h22**	1944: R&W; 1945: R&W; 1947: R&R; 1966: F&T
9-Cyanoanthracene	**g12**	1959: R&C
1,2,4,5-tetra-Cyanobenzene-hexamethylbenzene	**bVI,1**	1968: N,O&S
10-di-Cyanomethylene anthrone	**g14**	1967: S&Y
7,7,8,8-tetra-Cyanoquinodi-methane-anthracene	**g2**	1968: W&W
trans-α,β-di-Cyanostilbene	**bIII,4**	1961: W
μ-N,N'-Dehydrosemidinatobis (tricarbonyl iron)	**bIII,15**	1967: B&M
Deoxyanisoin	**bIII,28**	1962: N&K
bis(l-Ephedrine) copper(II)-benzene	**bVI,7**	1964: A,O&W
9-Ethyl-10-bromoanthracene	**g12**	1960: H
N,N'-Ethylene bis(l-ephedrine) copper dihydrate	**bV,17**	1968: M,A&U
1,5-di-Fluoroanthraquinone	**g20**	1967: C
Hexacene	**h10**	1962: C,R&T
tri-Hydrido bis (triphenylphosphine) ethylene bis(diphenylphosphine) rhenium(III)	**e32**	1966: A,B&S
Hydridochloro bis (diphenyl-ethylphosphine) platinum	**d23**	1965: E&I
μ-Hydrido-μ-diphenylphosphido-bis(tetracarbonyl manganese)	**bII,22**	1967: D,R&I
1,4-di-Hydroxyanthraquinone	**g21**	1967: S&N
1,5-di-Hydroxyanthraquinone	**g22**	1966: H&N; 1967: G
1,8-di-Hydroxyanthraquinone	**g23**	1962: P
1,2-di-Hydroxy-9,10-anthraquinone	**g25**	1943: N; 1961: G; 1967: G
bis-Hydroxydurylmethane	**bII,6**	1956: C&H
5,8-di-Hydroxy-1,4-naphthoquinone	**f28,29,30**	1934: P&S; 1947: R; 1955: B; 1956: B; 1957: W,O&N; 1958: B; G&Z
4,4'-di-Hydroxythiobenzo-phenone monohydrate	**bII,11**	1965: M&E
1,5-di-Iodoanthraquinone	**g18**	1963: C&G; 1967: C&G

(*continued*)

BIBLIOGRAPHY TABLE, CHAPTER XVB-H (*continued*)

Compound	Paragraph	Literature
Iodocarbonyl bis(triphenyl-phosphine)iridiumdichloro-methane, oxygen adduct	**e13**	1967: MG,D&I
4,4'-di-Iodo diphenyl sulfone	**bII,29**	1951: K&P; 1955: K&P
di-Iodo tris(diphenylphosphine) cobalt(II)	**e11**	1966: B&P
di-Iodo tris(diphenylphosphine) nickel(II)	**e11**	1966: B&P
Isodibenzanthrone	**h29**	1964: B
Mercuric chloride·bis(tri-phenylarsenicoxide)	**e22**	1963: B
Mercuric chloride·triphenyl-arsenicoxide	**e21**	1964: B
d-Methadone hydrobromide	**bII,7**	1958: H&A
p,p'-di-Methoxybenzophenone	**bII,10**	1957: K,H,K&W; 1958: K,H, K&W; 1962: N&K
p-Methoxyindophenol-N-oxide	**bII,19**	1966: R&H
tetra-(p-Methoxyphenyl)tin	**d2**	1949: Z&I; 1950: Z&I
Methoxytetraphenyl antimony	**d34**	1968: S,ME,LP,H&W
di-Methoxytriphenyl antimony	**c23**	1968: S,ME,LP,H&W
4,4'-bis(di-Methylamino)di-phenylamine iodide	**bII,17**	1966: T&O
4,4'-bis(di-Methylamino)di-phenylamine perchlorate	**bII,16**	1967: T,O&H
2-Methyl-4-amino-1-naphthol hydrochloride	**f17**	1967: G,H&S
2-Methyl-3-amino-1,4-naphthoquinone	**f38**	1966: G&H
p,N,N-di-Methylaminophenyl-diazonium chlorozincate	**bV,15**	1966: N,PK,U&K
9,10-di-Methyl-1,2-benzanthracene	**h12**	1960: S&F; 1964: I
2'-Methyl-1,2-benzanthraquinone	**h13**	1963: F&I
5-Methyl-1,2-benzanthraquinone	**h14**	1963: F&I
N,N,N',N'-tetra-Methyl-p-diaminobenzene-chloranil	**bVI,2**	1968: dB&V
p-Methyldiazoaminobenzene	**bIV,12**	1968: K
p-di-Methyldiazoaminobenzene	**bIV,12**	1964: K
4,4'-di-Methyl dibenzyl	**bIII,2**	1954: B

(*continued*)

BIBLIOGRAPHY TABLE, CHAPTER XVB-H (*continued*)

Compound	Paragraph	Literature
p,p'-di-Methyl-α,α'-difluorostilbene	**bIII,6**	1968: C&G
1,2-di-Methyl-1,2-diphenyldi-phosphine disulfide	**bIII,29**	1960: W
Methyl diphenyl thiophosphinite	**bII,23**	1966: SG,L,T,R&C
Methyl-2-hydroxy-3-naphthoquinone	**f31**	1965: G&H
1,5-di-Methylnaphthalene	**f11**	1965: B
octa-Methylnaphthalene	**f22**	1953: D&R
2-Methyl-α-naphthohydroquinone	**f24**	1965: G&H
2,3-di-Methyl-1,4-naphthoquinone	**f37**	1966: B; 1967: BL
N,N,N',N'-tetra-Methyl-p-phenylenediamine-1,2,4,5-tetracyanobenzene	**bVI,3**	1967: O,I&S
di[3-Methyl-1(or 5)-phenyl-5(or 1)-p-tolylformazyl]nickel	**d15**	1967: D
(+)-Methyl-n-propylphenylbenzyl-phosphonium bromide	**bVI,4**	1965: P,H,H&W
tri-Methyl(salicylaldehydato)platinum(IV)	**bVI,8**	1967: T&W
Naphthalene	**f1**	1921: B; B&J; 1923: B; 1928: R; 1929: R; 1930: B; 1933: R; 1947: K; 1949: A,R&W; 1952: A&C; 1953: C&R; K&K; 1955: K&K; 1957: C
Naphthalene-chromium tricarbonyl	**f4**	1967: K&N
Naphthalene-1,2,4,5-tetracyanobenzene(1 : 1)	**f3**	1967: K,I&S
Naphthalene-tetracyanoethylene	**f2**	1967: W&W
1-Naphthoic acid	**f7**	1953: MC; 1960: T
2-Naphthoic acid	**f8**	1954: K&H; 1961: T
1,4-Naphthohydroquinone	**f13**	1965: G&H; 1967: G&H
α-Naphthol	**f5**	1921: B; 1945: K; 1949: K; 1964: R&H
β-Naphthol	**f6**	1921: B; 1939: N; 1945: K; 1947: K; 1957: H&W; 1958: W&H
α-di-Naphtho(7',1'-1,13)(1'',7''-6,8)peropyrene	**h30**	1959: R&T

(*continued*)

BIBLIOGRAPHY TABLE, CHAPTER XVB-H (*continued*)

Compound	Paragraph	Literature
1,4-(α)-naphthoquinone	**f23**	1932: C; 1964: G&H; 1965: G&H
α-Naphthyl-4-chlorophthalamic acid	**f10**	1966: M
α-Naphthylphenylmethylfluorosilane	**f43**	1966: O&A
α-Naphthylphenylmethylsilane	**f43**	1966: O&A
N-α-Naphthyl-1,2,3,6-tetrahydrophthalamic acid	**f10**	1967: M
Nitridodichloro bis(triphenylphosphine)rhenium(V)	**e8**	1967: D&I
Nitridodichloro tris(diethylphenylphosphine)rhenium(V)	**c28**	1967: C,D&I
9-Nitroanthracene	**g8**	1959: T
9,10-di-Nitroanthracene	**g9**	1959: T
4,4′-di-Nitrobiphenyl	**bI,5**	1943: vN; 1963: B
di-Nitrobiphenyl-biphenyl (3 : 1)	**bI,10**	1948: vN&S
di-Nitrobiphenyl-hydroxybiphenyl (3 : 1)	**bI,10**	1946: S
1,5-di-Nitro-4,8-dihydroxyanthraquinone	**g26**	1967: B&B
1,5-di-Nitronaphthalene	**f14**	1947: S,Z&U; 1948: S,Z&U; 1954: MC&A; 1960: T
1,8-di-Nitronaphthalene	**f15**	1947: Z&U; 1951: MC; 1965: A,K&S
tris(p-Nitrophenylethynyl) arsine	**c19**	1968: M&L
tri-p-Nitrophenylmethyl	**c5**	1967: A&K
Nitrosyl dicarbonyl bis(triphenylphosphine) manganese	**e18**	1967: E&I
Ovalene	**h26**	1949: D&R; 1953: D&R
di-Oxodinitrato bis(triphenylarsenicoxide) uranium(VI)	**e23**	1968: P,G,C,Z&B
μ_4-oxo-hexa-μ-chlorotetrakis [(triphenylphosphine oxide) copper(II)]	**e33**	1967: B

(*continued*)

BIBLIOGRAPHY TABLE, CHAPTER XVB-H (*continued*)

Compound	Paragraph	Literature
Oxopentachloropropionato bis (triphenylphosphine) dirhenium(IV)	**e16**	1968: C&F
trans-Oxotrichloro bis(di-ethylphenylphosphine)rhenium (V)	**bVI,6**	1963: E&O
Palladium bis(triphenylphosphine) carbon disulfide	**e10**	1968: K,Y,U,K,K,T&H
Palladium *o*-phenylene bis(*o*-dimethylarsinophenylmethylarsine)chloride perchlorate-benzene	**d33**	1967: B&P
Pentacene	**h10**	1961: C,R&T; 1962: C,R&T
1,9-5,10-di-Perinaphthylene-anthracene	**g5**	1959: R
Perylene	**h15**	1936: H&B; 1953: D,R&W; 1964: C&T
Perylene-fluoranil (1 : 1)	**h17**	1963: H
Perylene-tetracyanoethylene	**h16**	1967: I&K
Phenanthrene	**h2**	1921: B&J; 1929: H&M; 1935: B&C; 1948: B; 1950: B; 1961: M; 1963: T
Phenanthrene-chromium tricarbonyl	**h3,4**	1964: D&H; 1968: M,F&S
Phenoquinone	**c7**	1953: H&W; 1968: S
1,2-di-Phenoxyethane	**bV,2**	1967: Y,A&K
bi-Phenyl	**bI,1**	1929: H&M; 1930: C&P; 1931: C&P; 1932: D; 1933: P; 1961: H,R&T; R; T; 1962: H&R
N,*N*-di-Phenylacetamide	**bII,21**	1968: K,R&W
Phenylacetylene-ironcarbonyl	**c34**	1962: K
bis(di-Phenylacetylene)-ironoctacarbonyl	**d4,5**	1965: D&S
penta-Phenyl antimony	**e2**	1964: W
bis(di-Phenylarsenic)oxide	**d27**	1963: C&T
tris-(*o*-di-Phenylarsinophenyl) arsine ruthenium dibromide	**e31**	1965: M&P

(*continued*)

BIBLIOGRAPHY TABLE, CHAPTER XVB-H (*continued*)

Compound	Paragraph	Literature
tetra-Phenylarsonium bis(*N*-cyanodithiocarbimato) nickelate	e28	1968: C&H
tetra-Phenylarsonium *cis*-diaquotetrachlororuthenate monohydrate	d29	1966: H,Z,T&A
tetra-Phenylarsonium-3-fluoro-1,1,4,5,5-pentacyano-2-azapenta-dienide	d28	1966: P
tetra-Phenylarsonium iodide	d25	1940: M
tetra-Phenylarsonium oxotetrabromoaceto-nitrilerhenate(V)	d32	1966: C&L
tetra-Phenylarsonium oxotetrabromoaquomolybdate	d31	1967: S
tetra-Phenylarsonium tetrachloroferrate	d30	1957: Z&R
tetra-Phenylarsonium tetranitratocobaltate(II)	e26	1966: B&C
bis(tetra-Phenylarsonium) tri-μ-chlorooctachlorotrirhenate (III)	e27	1966: P&R
tetra-Phenylarsonium triiodide	d26	1959: S
p-di-Phenylbenzene	c1	1933: H&R; P
1,3,5-tri-Phenylbenzene	d1	1933: H&R; 1934: L; O&L; 1954: F
hexa-Phenylbenzene	e24	1968: B
N-Phenyl-*N*'-benzoylselenourea	bV,7	1965: H
tri-Phenyl bismuth	c20	1942: W; 1968: H&F
tri-Phenyl bismuth dichloride	c22	1968: H&F
N,*N*'-(2,2'-bi-Phenyl)bis (salicylaldiminato)copper(II)	d12	1966: C,H&W
tri-Phenyl bromomethane	c2	1953: S; 1961: S; 1965: S&P; 1966: S&P
cis,*cis*-1,2,3,4-tetra-Phenylbutadiene	d3	1965: K&D
tris(1-Phenyl-1,3-butanedionato) aquoyttrium(III)	c25	1968: C&L
di-Phenyldiacetylene	bV,1	1939: W; 1940: W
N,*N*'-di-Phenyl-1,5-diaminoanthraquinone	g27	1967: B&B

(*continued*)

BIBLIOGRAPHY TABLE, CHAPTER XVB-H (*continued*)

Compound	Paragraph	Literature
N,N'-di-Phenyl-1,8-diaminoanthraquinone	**g28**	1967: B&B
tris(Phenyldiethylphosphine) nonachlorotrirhenium(III)	**c27**	1964: C&M
di-Phenyl diselenide	**bIII,22**	1932: E,H&S; 1952: M
bi-Phenylene	**h1**	1944: W&L; 1962: M&T; 1966: F&T
tri-Phenylene	**h8**	1937: B&G; 1944: C; 1950: K; 1954: V&P; 1956: P,T&L; 1963: A&T
tris(*cis*-1,2-di-Phenylethylene-1,2-dithiolato)rhenium	**e7**	1965: E&I; 1966: E&I
tris(*cis*-1,2-di-Phenylethylene-1,2-dithiolato)vanadium	**e6**	1967: E&G
tris(Phenylethynyl)phosphine	**c11**	1967: M&S
tetra-Phenyl germanium	**d2**	1952: I
1,1,6,6-tetra-Phenylhexapentaene	**d14**	1953: W
di-Phenyliodonium chloride	**bVI,5**	1952: K&S; 1956: K
di-Phenyliodonium iodide	**bVI,5**	1952: K&S; 1956: K
tetra-Phenyl lead	**d2**	1927: G; 1938: G; 1949: Z&I; 1950: Z&I; 1967: B,M,S&DP
di-Phenyllead dichloride	**bII,14**	1967: M,B&DP
di-Phenyl mercury	**bII,1**	1948: K&G; 1950: H,P&R; 1962: Z; 1964: Z,M&K
tetra-Phenylmethane	**d2**	1950: S&ML
tri-Phenylmethylarsonium bis (toluene-3,4-dithiolato) cobaltate hemiethanolate	**e4**	1968: E,D,G&I
bis(tri-Phenylmethylarsonium) tetrachloro cobalt(II)	**e20**	1966: P
bis(tri-Phenylmethylarsonium) tetrachloro iron(II)	**e20**	1966: P
bis(tri-Phenylmethylarsonium) tetrachloro manganese(II)	**e20**	1966: P
bis(tri-Phenylmethylarsonium) tetrachloro nickel(II)	**e20**	1966: P
bis(tri-Phenylmethylarsonium) tetrachloro zinc(II)	**e20**	1966: P
tri-Phenylmethyl perchlorate	**c3**	1965: GdM,MG&E

(*continued*

BIBLIOGRAPHY TABLE, CHAPTER XVB-H (*continued*)

Compound	Paragraph	Literature
tri-Phenylmethylphosphonium bis (1,2-dicyanoethylene-1,2-dithiolato)nickelate(III)	**c26**	1966: F
tri-Phenylmethyl tetrafluoroborate	**c3**	1965: GdM,MG&E
1,8-di-Phenyl-1,3,5,7-octatetraene	**bV,16**	1953: D&W; 1955: D&W
tri-Phenyl phosphate	**c16**	1962: D&S; 1965: S&C
di-μ-di-Phenylphosphinatoacetylacetonato chromium(III)	**d24**	1965: W&J
bis(tri-Phenylphosphine) ethylene nickel(0)	**e9**	1968: D&D
tri-Phenylphosphine gold tetracarbonylmanganese trioxyphenylphosphine	**e17**	1967: M
tri-Phenylphosphine-iron carbonyl	**c33**	1968: D&J
tris(tri-Phenylphosphine) rhodium carbonyl hydride	**e30**	1963: LP&I; 1965: LP&I
di(tri-Phenylphosphinesulfide)-tri(iodine)	**c15**	1968: S&M
tetra-Phenylphosphonium bis (tetracyanoquinodimethanide)	**e5**	1968: G,S&T
tetra-Phenylphosphonium iodide	**d18**	1956: K&S
tri-Phenylphosphoranylideneketen	**c13**	1966: D&W
tri-Phenyl phosphoranylidenethioketen	**c14**	1967: D
tri-Phenyl phosphorus	**c10**	1950: H; H,P&R; 1952: I&R; 1954: H,L,R&W; 1963: D; 1964: D
penta-Phenyl phosphorus	**e1**	1964: W
2,2-di-Phenyl-1-picrylhydrazyl·benzene	**d11**	1966: W; 1967: W
bis(1,3-di-Phenyl-1,3-propanedionato)copper(II)	**d8**	1966: B,vT&R
bis(1,3-di-Phenyl-1,3-propanedionato)palladium	**d8**	1966: B,vT&R
bis(N-Phenylsalicylaldiminato) copper(II)	**d9**	1964: W,S&L

(continued)

BIBLIOGRAPHY TABLE, CHAPTER XVB-H (*continued*)

Compound	Paragraph	Literature
bis(*N*-Phenylsalicylaldiminato) nickel	**d9**	1964: W,S&L
di-Phenyl selenium dibromide	**bII,32**	1941: MC&H
di-Phenyl selenium	**bII,31**	1942: MC&H
di-Phenyl selenooxide	**bII,27**	1957: A
tetra-Phenyl silicon	**d2**	1938: G; 1949: Z&I; 1950: Z&I; 1952: I&Z
tri-Phenyl stibine dichloride	**c21**	1960: P&PK; 1966: P&PK
bis(Phenylsulfonyl) selenide	**bIV,14**	1954: F&O
bis(Phenylsulfonyl) sulfide	**bIV,14**	1948: D,M&R; 1949: M&R
di-Phenyl sulfoxide	**bII,27**	1957: A
di-Phenyltellurium dibromide	**bII,34**	1958: C&MC
1,2-di-Phenyltetrafluoroethane	**bIII,1**	1959: C,J&N
tetra-Phenyl tin	**d2**	1927: G; 1938: G; 1949: Z&I; 1950: Z&I; 1952: I&Z
tri-Phenyltin-penta-carbonylmanganese	**c32**	1967: W&B
tri-Phenyltin tetracarbonylmanganese triphenylphosphine	**e19**	1967: B
di-Phenyl trichlorostibine mono-hydrate	**bII,25**	1961: P&PK; 1967: P&PK
(1,3-di-Phenyl urea)·di(iron tricarbonyl)	**bIV,8**	1967: P,P&vM
Picric acid-1-bromo-2-aminonaphthalene	**f16**	1968: CO,G&H
N-Picryl-*p*-iodoaniline	**bII,18**	1948: G; 1949: G
Pyrene	**h6**	1935: D&G; 1947: R&W; 1953: D; 1965: C&T
Pyrene-tetracyanoethylene	**h7**	1966: K,I&A; 1968: I&K
Quaterphenyl	**h23**	1933: H&R
Quaterrylene	**h31**	1960: S&S
Rubidium tetraphenylboronate	**d17**	1961: O,V&I
N,N'-di-Salicylidene-ethylenediamine zinc monohydrate	**bVI,10**	1966: H&M
Sodium naphthionate tetrahydrate	**f39**	1966: B&C

(*continued*)

BIBLIOGRAPHY TABLE, CHAPTER XVB-H (*continued*)

Compound	Paragraph	Literature
Stilbene	**bIII,3**	1923: B&R; 1929: H&M; 1933: P; 1937: R&W
Tellurium catecholate	**bV,10**	1959: A&L; 1967: L
Tellurium dibenzenethiosulfonate	**bV,11**	1955: O&F; 1956: O&F
Tellurium-di-*p*-tolylthiosulfonate	**bV,12**	1955: F&O
Tetracene	**h10**	1961: R,S&T; 1962: C,R&T
Thorium tetrakis-dibenzoyl-methane	**d7**	1959: S&K; 1960: W&B
Tolane	**bIII,10**	1933: P; 1938: R&W; 1968: S,M&K
m-Tolidine	**bI,6**	1952: F
m-Tolidine dihydrochloride	**bI,7**	1950: F&H
di-Toluene chromium iodide	**bII,36**	1960: S&S; 1961: S&S
tri-*p*-Tolyarsine	**c17**	1963: T
bis(1-*m*-Tolylazo-2-naphtholato)nickel	**f42**	1968: A,S,P&K
di-*p*-Tolyl disulfide	**bIII,30**	1948: D,M&R; 1960: V,Z&Z; 1967: V,Z&Z
di-*p*-Tolyl selenide	**bII,26**	1955: B&A
di-*p*-Tolyl selenium dibromide	**bII,33**	1950: MC&M; 1951: N&MC
di-*p*-Tolyl selenium dichloride	**bII,33**	1950: MC&M; 1951: N&MC
tetra-*m*-Tolylsilane	**d16**	1947: S
di-*p*-Tolyl sulfide	**bII,26**	1943: T; 1955: B&A
bis-(*p*-Tolylsulfonyl) sulfide	**bIV,14**	1948: D,M&R; 1949: M&R
di-*p*-Tolyl telluride	**bII,26**	1955: B&A
tetra-(*p*-Tolyl)tin	**d2**	1949: Z&I; 1950: Z&I; 1952: I&Z
p-Tolyl(triphenylphos-phoranylidene)methyl sulfone	**d21**	1965: W
Uranium tetrakis-diben-zoylmethane	**d7**	1959: S&K; 1960: W&B
tri-*p*-Xylylarsine	**c18**	1963: T

BIBLIOGRAPHY, CHAPTER XVB–H

1921

Becker, K., and Jancke, W., "X-Ray Spectroscopic Investigations with Organic Compounds," *Z. Physik. Chem.*, **99**, 242.
Bragg, W. H., "The Structure of Organic Crystals," *Proc. Phys. Soc. London*, **34**, 33.

1923

Becker, K., and Rose, H., "X-Ray Spectroscopy of Organic Compounds," *Z. Physik*, **14**, 369.
Bragg, W. H., "The Crystalline Structure of Anthracene," *Proc. Phys. Soc. London*, **35**, 167.

1927

Allen, N. C. B., "The Crystal Structure of Benzil," *Phil. Mag.*, **3**, 1037.
George, W. H., "An X-Ray Study of Isomorphism in Simple Organometallic Series I. The Tetraphenyls," *Proc. Roy. Soc. (London)*, **113A**, 585.

1928

Robertson, J. M., "An X-Ray Investigation of the Structure of Some Naphthalene Derivatives," *Proc. Roy. Soc. (London)*, **118A**, 709.

1929

Hengstenberg, J., and Mark, H., "The Structure of Some Aromatic Compounds," *Z. Krist.*, **70**, 283.
Robertson, J. M., "An X-Ray Investigation of the Structure of Naphthalene and Anthracene," *Proc. Roy. Soc. (London)*, **125A**, 542; *Nature*, **125**, 456 (1930).

1930

Banerjee, K., "Structure of Naphthalene and Anthracene," *Nature*, **125**, 456; *Indian J. Phys.*, **4**, 557.
Clark, G. L., and Pickett, L. W., "Crystal Structure of Some Derivatives of Biphenyl," *Proc. Natl. Acad. Sci. U.S.*, **16**, 20.
Hertel, E., and Römer, G. H., "The Structure of Quinoid Compounds and of a Molecular Compound of the Quinhydrone Type," *Z. Physik. Chem.*, **11B**, 90.
Prasad, M., "An X-Ray Investigation of the Crystals of Azobenzene," *Phil. Mag.*, **10**, 306.

1931

Clark, G. L., and Pickett, L. W., "X-Ray Investigations of Optically Active Compounds II. Diphenyl and Some of its Active and Inactive Derivatives," *J. Am. Chem. Soc.*, **53**, 167.

1932

Caspari, W. A., "Crystallography of the Simpler Quinones," *Proc. Roy. Soc. (London)*, **136A**, 82.

Dhar, J., "X-Ray Analysis of the Structure of Diphenyl," *Indian J. Phys.*, **7**, 43.

Egartner, L., Halla, F., and Schacherl, R., "Determination of Structure of the Aromatic Disulfides and Diselenides," *Z. Physik. Chem.*, **18B**, 189.

1933

Hertel, E., and Römer, G. H., "The Crystal Structure of Terphenyl," *Z. Physik. Chem.*, **21B**, 292.

Hertel, E., and Römer, G. H., "The Fine Structure of the Isomeric Hydrocarbons Quaterphenyl and Triphenylbenzene," *Z. Physik. Chem.*, **23B**, 226.

Iball, J., and Robertson, J. M., "Structure of Chrysene and 1, 2, 5, 6-Dibenzanthracene in the Crystalline State," *Nature*, **132**, 750.

Pickett, L. W., "Crystal Structure of the Diphenyl Series," *Nature*, **131**, 513.

Pickett, L. W., "An X-Ray Study of *p*-Diphenylbenzene," *Proc. Roy. Soc. (London)*, **142A**, 333.

Prasad, M., "An X-Ray Investigation of Crystals of Stilbene and Tolane," *Phil. Mag.*, **16**, 639.

Robertson, J. M., "The Crystal Structure of Anthracene," *Z. Krist.*, **84**, 321; *Proc. Roy. Soc. (London)*, **140A**, 79.

Robertson, J. M., "The Crystalline Structure of Naphthalene. A Quantitative X-Ray Investigation," *Proc. Roy. Soc. (London)*, **142A**, 674.

Robinson, B. W., "The Reflection of X-Rays from Anthracene Crystals," *Proc. Roy. Soc. (London)*, **142A**, 422.

1934

Dhar, J., "X-Ray Analysis of the Crystal Structure of Dibenzyl," *Current Sci.*, **2**, 480; *Indian J. Phys.*, **9**, 1.

Iball, J., "X-Ray Analysis of the Structure of Chrysene," *Proc. Roy. Soc. (London)*, **146A**, 140.

Lonsdale, K., "Crystal Structure of 1-3-5 Triphenylbenzene," *Nature*, **133**, 67.

Orelkin, B., and Lonsdale, K., "The Structure of Symm. (1-3-5) Triphenylbenzene," *Proc. Roy. Soc. (London)*, **144A**, 630.

Palacios, J., and Salvia, R., "The Crystalline Structure of Naphthazarin," *Anales Soc. Españ. Fís. Quím.*, **32**, 49.

Robertson, J. M., "X-Ray Analysis of the Crystal Structure of Dibenzyl I. Experimental and Structure by Trial," *Proc. Roy. Soc. (London)*, **146A**, 473.

1935

Bernal, J. D., and Crowfoot, D., "The Structure of Some Hydrocarbons Related to the Sterols," *J. Chem. Soc.*, **1935**, 93.

Dhar, J., and Guha, A. C., "Crystal Structure of Pyrene," *Z. Krist.*, **91A**, 123.

Krishnan, K. S., and Banerjee, S., "The Magnetic Anisotropy and Crystal Structure of 1,2,5,6-Dibenzanthracene," *Z. Krist.*, **91A**, 173.

Robertson, J. M., "X-Ray Analysis of the Structure of Dibenzyl II. Fourier Analysis," *Proc. Roy. Soc. (London)*, **150A**, 348.

1936

Hertel, E., and Bergk, H. W., "The Structures of Condensed Aromatic Hydrocarbons and Their Molecular Compounds with Trinitrobenzene," *Z. Physik. Chem.*, **33B**, 319.

Iball, J., "X-Ray Analysis of the Orthorhombic Crystalline Modification of 1,2,5,6-Dibenzanthracene," *Nature*, **137**, 361.

Robertson, J. M., Prasad, M., and Woodward, I., "X-Ray Analysis of the Dibenzyl Series I. The Structure of Stilbene, Tolane, and Azobenzene," *Proc. Roy. Soc. (London)*, **154A**, 187.

1937

Banerjee, S., and Guha, A. C., "Crystal Structure of Triphenylene $C_{18}H_{12}$," *Z. Krist.*, **96A**, 107.

Robertson, J. M., and Woodward, I., "X-Ray Analysis of the Dibenzyl Series IV. Detailed Structure of Stilbene," *Proc. Roy. Soc. (London)*, **162A**, 568.

1938

Banerjee, K., and Haque, A., "Structure of Aromatic Compounds III. Benzophenone," *Indian J. Phys.*, **12**, 87.

Banerjee, K., and Sinha, K. L., "Structure of Aromatic Compounds II. Benzil," *Indian J. Phys.*, **11**, 409.

Giacomello, G., "X-Ray Studies of Metallo-organic Compounds I. Tetraphenyls of Silicon, Tin and Lead," *Gazz. Chim. Ital.*, **68**, 422.

Guha, B. C., "X-Ray Analysis of the Structure of Anthraquinone," *Phil. Mag.*, **26**, 213.

Iball, J., "The Crystal Structure of Condensed Ring Compounds VI. 1,2-Benzanthracene $C_{18}H_{12}$, 5-Methyl-1,2-benzanthracene $C_{19}H_{14}$, and 3-Methyl-1,2-benzanthraquinone $C_{19}H_{12}O_2$," *Z. Krist.*, **99A**, 230.

Robertson, J. M., and Woodward, I., "X-Ray Analysis of the Dibenzyl Series V. Tolane and the Triple Bond," *Proc. Roy. Soc. (London)*, **164A**, 436.

1939

Lange, J. J. de, Robertson, J. M., and Woodward, I., "X-Ray Crystal Analysis of *trans*-Azobenzene," *Proc. Roy. Soc. (London)*, **171A**, 398.

Neuhaus, A., "On the Isomorphous Replacement of H, OH, NH_2, CH_3, and Cl in the Structure of Naphthalene," *Z. Krist.*, **101A**, 177.

Robertson, J. M., "Crystal Structure and Configuration of the Isomeric Azobenzenes," *J. Chem. Soc.*, **1939**, 232.

Wiebenga, E. H., "The Crystal Structure of Diphenyldiacetylene," *Nature*, **143**, 980.

1940

Mooney, R. C. L., "An X-Ray Determination of the Crystal Structure of Tetraphenylarsonium Iodide, $(C_6H_5)_4AsI$," *J. Am. Chem. Soc.*, **62**, 2955.

Wiebenga, E. H., "The Crystal Structure of Diphenyldiacetylene," *Z. Krist.*, **102A**, 193.

1941

Hampson, G. C., and Robertson, J. M., "Bond Length and Resonance in the *cis*-Azobenzene Molecule," *J. Chem. Soc.*, **1941**, 409.

McCullough, J. D., and Hamburger, G., "The Crystal Structure of Diphenylselenium Dibromide," *J. Am. Chem. Soc.*, **63**, 803.

1942

McCullough, J. D., and Hamburger, G., "The Crystal Structure of Diphenylselenium Dichloride," *J. Am. Chem. Soc.*, **64**, 508.

Wetzel, J., "Crystal Structure Investigation of the Triphenyls of Bi, As, and Sb," *Z. Krist.*, **104A**, 305.

1943

Neuhaus, A., "Oriented Deposits of Alizarin and Other Anthracene Derivatives in Inorganic Supporting Lattices," *Z. Physik. Chem.*, **192A**, 309.

Niekerk, J. N. van, "The Crystal Structure of 4,4'-Dinitrodiphenyl," *Proc. Roy. Soc.*, (*London*), **181A**, 314.

Toussaint, J., "Crystallographic Investigation of Aromatic Sulfides," *Bull. Soc. Roy. Sci. Liège*, **12**, 153, 452, 533.

1944

Chorghade, S. L., "Crystal Structure and Space Groups of Some Aromatic Crystals II. Triphenylene," *Proc. Natl. Acad. Sci. India*, **14A**, 19.

Robertson, J. M., and White, J. G., "Crystal Structure of Coronene," *Nature*, **154**, 605.

Waser, J., and Lu, C.-S., "The Crystal Structure of Biphenylene," *J. Am. Chem. Soc.*, **66**, 2035.

1945

Jeffrey, G. A., "Structure of Dibenzyl," *Nature*, **156**, 82.

Kitaigorodskii, A. I., "Crystal Structure of the Naphthols," *Dokl. Akad. Nauk SSSR*, **50**, 319.

Robertson, J. M., and White, J. G., "Crystal Structure of Coronene," *J. Chem. Soc.*, **1945**, 607.

Sen, S. N., "Electron Density Map of Anthraquinone Crystal," *Indian J. Phys.*, **19**, 243.

1946

Saunder, D. H., "The Crystal Structure of Some Molecular Complexes of 4,4'-Dinitrodiphenyl I. The Complex with 4-Hydroxydiphenyl," *Proc. Roy. Soc. (London)*, **188A**, 31.

1947

Kitaigorodskii, A. I., "Crystallochemistry of Aromatic Compounds V. Crystal Structures of 2- and 2,6-Derivatives of Naphthalene," *Bull. Acad. Sci. URSS, Classe Sci. Chim.*, **1947**, 561.

Robertson, J. M., and White, J. G., "The Crystal Structure of Pyrene. A Quantitative X-Ray Investigation," *J. Chem. Soc.*, **1947**, 358.

Robertson, J. M., and White, J. G., "The Crystal Structure of the Orthorhombic Modification of 1,2,5,6-Dibenzanthracene. A Quantitative X-Ray Investigation," *J. Chem. Soc.*, **1947**, 1001.

Rogers, M. T., "The Magnetic Anisotropy of Coronene, Naphthazarin, and Other Crystals," *J. Am. Chem. Soc.*, **69**, 1506.

Ruston, W. R., and Rudorff, W., "The Crystallographic Structure of Coronene," *Bull. Soc. Chim. Belg.*, **56**, 97.

Sevast'yanov, N. G., Zhdanov, G. S., and Umanskii, M. M., "Crystal Structure of Dinitronaphthalenes II. X-Ray Determination of the Unit Cell and Space Group of 1,5-Dinitronaphthalene Crystals," *J. Phys. Chem. USSR*, **21**, 525.

Stehlik, B., "The Crystal Structure of Tetra-*m*-tolylsilane," *Coll. Czech. Chem. Comm.*, **12**, 6.

Zhdanov, G. S., and Umanskii, M. M., "Crystal Structure of Dinitronaphthalenes I. X-Ray Determination of the Unit Cell and Space Group of 1,8-Dinitronaphthalene Crystals," *J. Phys. Chem. USSR*, **21**, 523.

1948

Basak, B. S., "The Space Group of Phenanthrene," *Acta Cryst.*, **1**, 224.

Dawson, I. M., Mathieson, A. McL., and Robertson, J. M., "The Structure of Certain Polysulfides and Sulfonyl Sulfides I. A Preliminary X-Ray Survey," *J. Chem. Soc.*, **1948**, 322.

Grison, E., "Crystal Structure of Three Polymorphic Forms of N-Picryl-*p*-iodoaniline," *Mém. Services Chim. État (Paris)*, **34**, 59.

Kitaigorodskii, A. I., and Grdenich, D. R., "Crystal Structure of Diphenylmercury," *Bull. Acad. Sci. URSS, Classe Sci. Chim.*, **1948**, 262.

Niekerk, J. N. van, and Saunder, D. H., "The Crystal Structure of the Molecular Complex of 4,4'-Dinitrodiphenyl with Diphenyl," *Acta Cryst.*, **1**, 44.

Sen, S. N., "The Structure of Anthraquinone (A Quantitative X-Ray Investigation)," *Indian J. Phys.*, **22**, 347.

Sevast'yanov, N. G., Zhdanov, G. S., and Umanskii, M. M., "Crystal Structure of Dinitronaphthalenes III. Structure of the 1,5-Dinitronaphthalene Crystal ($C_{10}H_6N_2O_4$)," *Zh. Fiz. Khim.*, **22**, 1153.

Smare, D. L., "The Crystal Structure of 2,2'-Dichlorobenzidine," *Acta Cryst.*, **1**, 150.

White, J. G., "Crystal Structure of 1,12-Benzperylene: Quantitative X-Ray Investigation," *J. Chem. Soc.*, **1948**, 1398.

1949

Abrahams, S. C., Robertson, J. M., and White, J. G., "The Crystal and Molecular Structure of Naphthalene," *Acta Cryst.*, **2**, 233, 238.

Donaldson, D. M., and Robertson, J. M., "Crystal Structure of Ovalene," *Nature*, **164**, 1002.

Grison, E., "Crystal Structure of the Three Polymorphous Varieties of N-Picryl-p-iodoaniline," *Acta Cryst.*, **2**, 410.

Kitaigorodskii, A. I., "Crystallochemistry of Aromatic Compounds VII. Acenaphthene, α-Naphthol, γ-Hydroquinone," *Izv. Akad. Nauk SSSR, Otdel. Khim. Nauk*, **1949**, 263.

Mathieson, A. McL., and Robertson, J. M., "The Structure of Certain Polysulfides and Sulfonyl Sulfides III. The Crystal Structure of Bisphenylsulfonylsulfide," *J. Chem. Soc.*, **1949**, 724.

Stora, C., and Eller, G. van, "Radiocrystallographic Nature of Some Derivatives of Triphenylmethane," *Compt. Rend.*, **229**, 766.

Zhdanov, G. S., and Ismailzade, I. G., "X-Ray Determination of the Structures of Some Tetraaryl Compounds of Silicon, Tin and Lead," *Dokl. Akad. Nauk SSSR*, **68**, 95.

1950

Basak, B. S., "Crystal Structure of Phenanthrene," *Indian J. Phys.*, **24**, 309.

Fenimore, C. P., "The Crystal Structure of Dimeric Tribromonitrosobenzene," *J. Am. Chem. Soc.*, **72**, 3226.

Fowweather, F., and Hargreaves, A., "The Crystal Structure of *m*-Tolidine Dihydrochloride," *Acta Cryst.*, **3**, 81.

Howells, E. R., "Preliminary X-Ray Investigation of Nitrogen Triphenyl and Phosphorus Triphenyl," *Acta Cryst.*, **3**, 317.

Howells, E. R., Phillips, D. C., and Rogers, D., "The Probability Distribution of X-Ray Intensities II. Experimental Investigation and X-Ray Detection of Centers of Symmetry," *Acta Cryst.*, **3**, 210.

Klug, A., "The Crystal and Molecular Structure of Triphenylene," *Acta Cryst.*, **3**, 165.

McCullough, J. D., and Marsh, R. E., "The Crystal Structure of Di-*p*-tolylselenium Dichloride and Di-*p*-tolylselenium Dibromide," *Acta Cryst.*, **3**, 41.

Mathieson, A. McL., Robertson, J. M., and Sinclair, V. C., "The Crystal and Molecular Structure of Anthracene I. X-Ray Measurements," *Acta Cryst.*, **3**, 245.

Robertson, J. M., "Structures of Naphthalene and Anthracene," *J. Chim. Phys.*, **47**, 47.

Sinclair, V. C., Robertson, J. M., and Mathieson, A. McL., "The Crystal and Molecular Structure of Anthracene II. Structure Investigation by the Triple Fourier Series Method," *Acta Cryst.*, **3**, 251.

Sumsion, H. T., and McLachlan, D., Jr., "The Structure of Tetraphenylmethane," *Acta Cryst.*, **3**, 217.

Zhdanov, G. S., and Ismailzade, I. G., "X-Ray Investigation of Some Crystalline Tetraaryl Compounds of Silicon, Tin, and Lead," *Zh. Fiz. Khim.*, **24**, 1495.

1951

Keil, C., and Plieth, K., "The Valence Angle of the Sulfur Atom in Diphenyl Sulfone," *Naturwissenschaften*, **38**, 546.

McCrone, W. C., "Crystallographic Data. 1,8-Dinitronaphthalene," *Anal. Chem.*, **23**, 1188.

Neuerberg, G. J., and McCrone, W. C., "Crystallographic Data. Di-*p*-tolylselenium Dibromide," *Anal. Chem.*, **23**, 1042.

1952

Ahmed, F. R., and Cruickshank, D. W. J., "A Refinement of the Crystal and Molecular Structures of Naphthalene and Anthracene," *Acta Cryst.*, **5**, 852.

Fowweather, F., "The Crystal Structure of *m*-Tolidine," *Acta Cryst.*, **5**, 820.

Gol'der, G. A., Zhdanov, G. S., Umanskii, M. M., and Glushkova, V. P., "X-Ray Studies of Crystals of Some Nitro and Halogen Derivatives of Benzene and Naphthalene," *Zh. Fiz. Khim.*, **26**, 1259, 1434.

Ismailzade, I. G., "The Crystal Structure of Tetraphenylgermanium and the Analysis of the Structures of the Tetraphenyl Compounds of Elements in the Fourth Group," *Zh. Fiz. Khim.*, **26**, 1139.

Ismailzade, I. G., and Zhdanov, G. S., "X-Ray Analysis of the Crystal Structures of the Tetraphenyl Compounds of Silicon, Tin, and Lead," *Zh. Fiz. Khim.*, **26**, 1619.

Ivernova, V. I., and Rojtburd, T. M., "The Morphotropy in the Crystal Structure of Triphenyl Compounds of the Elements of the Fifth Group," *Zh. Fiz. Khim.*, **26**, 810.

Khotsyanova, T. L., and Struchkov, Y. T., "X-Ray Study of the Crystal Structure of Diphenyliodonium Iodide," *Zh. Fiz. Khim.*, **26**, 644.

Khotsyanova, T. L., and Struchkov, Y. T., "X-Ray Study of the Crystal Structure of Diphenyliodonium Chloride," *Zh. Fiz. Khim.*, **26**, 669.

Marsh, R. E., "The Crystal Structure of Diphenyl Diselenide," *Acta Cryst.*, **5**, 458.

McIntosh, A. O., Robertson, J. M., and Vand, V., "Crystal Structure of Dibenzo [3,4,5,6] Phenanthrene," *Nature*, **169**, 322.

Nanthey, W., Plieth, K., and Singewald, A., "The Bond Angle of the Keto-group Carbon Atom in Benzophenone," *Z. Elektrochem.*, **56**, 690.

1953

Cruickshank, D. W. J., and Robertson, A. P., "The Comparison of Theoretical and Experimental Determinations of Molecular Structures, with Applications to Naphthalene and Anthracene," *Acta Cryst.*, **6**, 698.

Donaldson, D. M., and Robertson, J. M., "The Crystal and Molecular Structure of Octamethyl Naphthalene: A Non-planar Naphthalene Derivative," *J. Chem. Soc.*, **1953**, 17.

Donaldson, D. M., and Robertson, J. M., "The Crystal and Molecular Structure of Ovalene: A Quantitative X-Ray Investigation," *Proc. Roy. Soc. (London)*, **220A**, 157.

Donaldson, D. M., Robertson, J. M., and White, J. G., "The Crystal and Molecular Structure of Perylene," *Proc. Roy. Soc. (London)*, **220A**, 311.

Drenth, W., and Wiebenga, E. H., "Crystal Data of 1,4-Diphenyl-1,3-butadiene, 1,6-Diphenyl-1,3,5-hexatriene and 1,8-Diphenyl-1,3,5,7-octatetraene," *Rec. Trav. Chim.*, **72**, 39.

Dutt, M. N., "Diffuse Spots of Pyrene at Higher Temperatures," *J. Chem. Phys.*, **21**, 380.

Hanson, A. W., "The Structure of the Free Radical, Bis(*p*-methoxyphenyl) Nitric Oxide," *Acta Cryst.*, **6**, 32.

Harding, T. T., and Wallwork, S. C., "The Structures of Molecular Compounds Exhibiting Polarization Bonding I. General Introduction and the Crystal Structure of Phenoquinone," *Acta Cryst.*, **6**, 791.

Kozhin, V. M., and Kitaigorodskii, A. I., "Anisotropic Thermal Expansion of Naphthalene," *Zh. Fiz. Khim.*, **27**, 534.

Kozhin, V. M., and Kitaigorodskii, A. I., "Anisotropic Thermal Expansion of Anthracene," *Zh. Fiz. Khim.*, **27**, 1676.

McCrone, W. C., "1-Naphthoic Acid. Crystallographic Data," *Anal. Chem.*, **25**, 1126.

Stora, C., "The Anomalous Structure of Triphenyl Bromomethane," *Bull. Soc. Chim. France*, **1953**, 1059.

Whittaker, E. J. W., "The Structure of 3,3-Dichloro-4,4'-dihydroxydiphenyl Methane," *Acta Cryst.*, **6**, 714.

Woolfson, M. M., "The Structure of 1:1:6:6 Tetraphenylhexapentaene," *Acta Cryst.*, **6**, 838.

1954

Brown, C. J., "The Crystal Structure of 4,4'-Dimethyldibenzyl," *Acta Cryst.*, **7**, 97.

Farag, M. S., "The Crystal Structure of 1,3,5-Triphenyl Benzene," *Acta Cryst.*, **7**, 117.

Furberg, S., and Øyum, P., "Crystal Structure of Selenium Dibenzene-sulfinate," *Acta Chem. Scand.*, **8**, 42.

Howells, E. R., Lovell, F. M., Rogers, D., and Wilson, A. J. C., "The Space Groups of Nitrogen Triphenyl and Phosphorus Triphenyl," *Acta Cryst.*, **7**, 298.

Krc, J., Jr., and Hinch, R. J., Jr., "Crystallographic Data. 2-Naphthoic Acid," *Anal. Chem.*, **26**, 780.

McCrone, W. C., and Andreen, J. H., "Crystallographic Data. 1,5-Dinitrona-phthalene," *Anal. Chem.*, **26**, 1390.

McIntosh, A. O., Robertson, J. M., and Vand, V., "Crystal Structure and Molecular Shape of 3,4,5,6-Dibenzophenanthrene," *J. Chem. Soc.*, **1954**, 1661.

Scanlon, J., and Collin, R. L., "Crystallographic Data on Some Diester Phosphate Compounds," *Acta Cryst.*, **7**, 781.

Vand, V., and Pepinsky, R., "The Crystal Structure of Triphenylene," *Acta Cryst.*, **7**, 595.

1955

Billy, C., "Structure of Naphthazarin (5,8-Dihydroxy-1,4-Naphthoquinone)," *Compt. Rend.*, **240**, 887.

Blackmore, W. R., and Abrahams, S. C., "The Crystal Structure of Di-p-tolyl Telluride," *Acta Cryst.*, **8**, 317.

Blackmore, W. R., and Abrahams, S. C., "The Crystal Structure of Di-p-tolyl Selenide," *Acta Cryst.*, **8**, 323.

Blackmore, W. R., and Abrahams, S. C., "The Crystal Structure of Di-p-tolyl Sulfide," *Acta Cryst.*, **8**, 329.

Drenth, W., and Wiebenga, E. H., "Crystal and Molecular Structure of 1,8-Diphenyl-1,3,5,7-octatetraene," *Acta Cryst.*, **8**, 755.

Foss, O., and Øyum, P., "The Structure of Tellurium Di-p-toluenethiosulfonate," *Acta Chem. Scand.*, **9**, 1014.

Keil, C., and Plieth, K., "The Structure of Bis(p-Iodophenyl) Sulfone and the Valency Angles at the Sulfur Atom in the Sulfone Group," *Z. Krist.*, **106**, 388.

Kozhin, V. M., and Kitaigorodskii, A. I., "The Naphthalene Crystal and Molecular Structure," *Zh. Fiz. Khim.*, **29**, 1897.

Murty, B. V. R., "The Space Group of Anthraquinone," *Acta Cryst.*, **8**, 113.

Øyum, P., and Foss, O., "The Structure of Tellurium Dibenzenethiosulfonate," *Acta Chem. Scand.*, **9**, 1012.

1956

Borgen, O., "X-Ray Investigation of Naphthazarin," *Acta Chem. Scand.*, **10**, 867.

Burns, D. M., and Iball, J., "The Bond Lengths in Chrysene," *Acta Cryst.*, **9**, 314.

Chaudhuri, B., and Hargreaves, A., "The Crystal and Molecular Structure of Bishydroxydurylmethane," *Acta Cryst.*, **9**, 793.

Cruickshank, D. W. J., "A Detailed Refinement of the Crystal and Molecular Structure of Anthracene," *Acta Cryst.*, **9**, 915.

Dunitz, J. D., and Rollett, J. S., "The Crystal Structure of Dibenzylphosphoric Acid," *Acta Cryst.*, **9**, 327.

Friedlander, P. H., and Sayre, D., "Crystal Structure of 1,2-Benzanthracene," *Nature*, **178**, 999.

Khotsyanova, T. L., "The Structures of Diphenyliodonium Chloride and Iodide," *Kristallografiya*, **1**, 524; *Dokl. Akad. Nauk SSSR*, **110**, 71.

Khotsyanova, T. L., and Struchkov, Y. T., "The Crystal Structure of Tetraphenylphosphonium Iodide and the Configuration of Tetraaryl Ions and Molecules," *Kristallografiya*, **1**, 669.

Øyum, P., and Foss, O., "Crystal Structure of Tellurium Dibenzenethiosulfonate," *Acta Chem. Scand.*, **10**, 279.

Pinnock, P. R., Taylor, C. A., and Lipson, H., "A Redetermination of the Structure of Triphenylene," *Acta Cryst.*, **9**, 173.

Robertson, J. M., and White, J. G., "The Crystal Structure of the Monoclinic Modification of 1,2,5,6-Dibenzanthracene. A Quantitative X-Ray Investigation," *J. Chem. Soc.*, **1956**, 925.

1957

Abrahams, S. C., "The Crystal Structure of Diphenyl Sulfoxide," *Acta Cryst.*, **10**, 417.

Cruickshank, D. W. J., "A Detailed Refinement of the Crystal and Molecular Structure of Anthracene," *Acta Cryst.*, **10**, 470.

Cruickshank, D. W. J., "A Detailed Refinement of the Crystal and Molecular Structure of Naphthalene," *Acta Cryst.*, **10**, 504.

Hargreaves, A., and Watson, H. C., "An X-Ray and Physical Study of β-Naphthol," Acta Cryst., **10**, 368.

Karle, I. L., Hauptman, H., Karle, J., and Wing, A. B., "Crystal and Molecular Structure of *p,p'*-Dimethoxybenzophenone by the Direct Probability Method," *Acta Cryst.*, **10**, 481.

Kruse, F. H., Marsh, R. E., and McCullough, J. D., "The Crystal Structure of *p,p'*-Dichlorodiphenyl Diselenide and *p,p'*-Dichlorodiphenyl Ditelluride," *Acta Cryst.*, **10**, 201.

Watase, H., Osaki, K., and Nitta, I., "Crystal Structure of a Monoclinic Form of Naphthazarin," *Bull. Chem. Soc. Japan*, **30**, 532.

Zaslow, B., and Rundle, R. E., "The Crystal Structure of Tetraphenylarsonium Tetrachloroferrate(III), $(C_6H_5)_4AsFeCl_4$," *J. Phys. Chem.*, **61**, 490.

1958

Bailey, M., "The Crystal Structure of 1,5-Dichloroanthraquinone," *Acta Cryst.*, **11**, 103.

Billy, C., "Structure of Naphthazarin," *Compt. Rend.*, **247**, 1019.

Christofferson, G. D., and McCullough, J. D., "The Crystal Structure of Diphenyl-tellurium Dibromide," *Acta Cryst.*, **11**, 249.

Gol'der, G. A., and Zhdanov, G. S., "X-Ray Study of Naphthazarin," *Dokl. Akad. Nauk SSSR*, **118**, 1131.

Hanson, A. W., and Ahmed, F. R., "The Crystal Structure and Absolute Configuration of the Monoclinic Form of *d*-Methadone Hydrobromide," *Acta Cryst.*, **11**, 724.

Karle, I. L., Hauptman, H., Karle, J., and Wing, A. B., "Crystal and Molecular Structure of *p,p'*-Dimethoxybenzophenone by the Direct Probability Method," *Acta Cryst.*, **11**, 257.

Kondrashev, Y. D., "Crystal Structure of Halogen Derivatives of Diazoamino-benzene," *Kristallografiya*, **3**, 229.

Padmanabhan, V. M., Shankar, J., and Khubchandani, P. G., "The Crystal Structure of *p*-Azotoluene," *Proc. Indian Acad. Sci.*, **47A**, 323.

Ramaseshan, S., and Venkatesan, K., "Crystal and Molecular Structure of 3,3'-Dibromobenzophenone," *Experientia*, **14**, 237.

Trotter, J., "The Crystal Structures of Some Anthracene Derivatives II. 9,10-Dibromoanthracene," *Acta Cryst.*, **11**, 803.

Watson, H. C., and Hargreaves, A., "The Crystal Structure of β-Naphthol," *Acta Cryst.*, **11**, 556.

1959

Antikainan, P. J., and Lundgren, G., "On the Crystal Structure of Tellurium(IV) Catecholate," *Suomen Kemistilehti*, **32B**, 175.

Cruickshank, D. W. J., Jeffrey, G. A., and Nyburg, S. C., "The Crystal Structure and Atomic Vibrations of 1,2-Diphenyltetrafluoroethane," *Z. Krist.*, **112**, 385.

Lipscomb, W. N., Robertson, J. M., and Rossmann, M. G., "Crystal Structure Studies of Polynuclear Hydrocarbons I. 2,3:8,9-Dibenzoperylene," *J. Chem. Soc.*, **1959**, 2601.

Rabaud, H., and Clastre, J., "Crystal Structure of 9-Cyanoanthracene," *Acta Cryst.*, **12**, 911.

Robertson, J. M., and Trotter, J., "Crystal Structure Studies of Polynuclear Hydrocarbons III. The α-Modification of Dinaphtho [7',1':1,13][1",7":6,8] Peropyrene," *J. Chem. Soc.*, **1959**, 2614.

Rossmann, M. G., "Crystal Structure Studies of Polynuclear Hydrocarbons II. 1,9:5,10-Diperinaphthylene Anthracene," *J. Chem. Soc.*, **1959**, 2607.

Shankar, J., and Kunchur, N. R., "The Crystal Structure of Cerium Tetra-kisdibenzoylmethane," *Acta Cryst.*, **12**, 940.

Slater, R. C. L. M., "The Triiodide Ion in Tetraphenyl Arsonium Triiodide," *Acta Cryst.*, **12**, 187.

Srivastava, S. N., "Preliminary Report on the Crystal Structure of Anthrone," *Indian J. Phys.*, **33**, 456.

Trotter, J., "The Crystal Structures of Some Anthracene Derivatives III. 9,10-Dichloroanthracene," *Acta Cryst.*, **12**, 54.

Trotter, J., "The Crystal Structures of Some Anthracene Derivatives IV. 9,10-Dinitroanthracene," *Acta Cryst.*, **12**, 232.

Trotter, J., "The Crystal Structures of Some Anthracene Derivatives V. 9-Nitroanthracene," *Acta Cryst.*, **12**, 237.

Trotter, J., "A Refinement of the 1:12-Benzperylene Structure," *Acta Cryst.*, **12**, 889.

Trotter, J., "The Crystal Structures of Some Anthracene Derivatives VI. 9-Anthraldehyde," *Acta Cryst.*, **12**, 922.

1960

Bacon, G. E., and Curry, N. A., "A Study of 4,4'-Dichloro Diphenyl Sulphone by Neutron Diffraction," *Acta Cryst.*, **13**, 10.

Burns, D. M., and Iball, J., "Refinement of the Structure of Chrysene," *Proc. Roy. Soc. (London)*, **257A**, 491.

Corradini, P., and Allegra, G., "X-Ray Structure of Bis-Tricarbonyl-Chromium-Biphenyl," *J. Am. Chem. Soc.*, **82**, 2075.

Hauw, C., "Crystal Structure of 9-Ethyl-10-Bromoanthracene," *Acta Cryst.*, **13**, 100.

Meerssche, M. van, and Leroy, G., "Structure of Cyanostilbenes III. X-Ray Crystallographic Determination of the Structure of 2-Bromo-4'-dimethyl-amino-α-cyanostilbene," *Bull. Soc. Chim. Belges*, **69**, 204.

Murty, B. V. R., "Refinement of the Structure of Anthraquinone," *Z. Krist.*, **113**, 445.

Polynova, T. N., and Porai-Koshits, M. A., "The Crystal Structure of Triphenyldi-chlorostibine," *Zh. Strukt. Khim.*, **1**, 159.

Rømming, C., "The Crystal Structure of the 1:1 Addition Compound Formed by Benzyl Sulfide and Iodine," *Acta Chem. Scand.*, **14**, 2145.

Sayre, D., and Friedlander, P. H., "Crystal Structure of 9,10-Dimethyl-1,2-Benzanthracene," *Nature*, **187**, 139.

Sime, J. G., and Abrahams, S. C., "The Crystal and Molecular Structure of 4,4'-Dichlorodiphenyl Sulfone," *Acta Cryst.*, **13**, 1.

Srivastava, H. N., and Speakman, J. C., "The Crystal and Molecular Structure of Quaterrylene," *Proc. Roy. Soc. (London)*, **257A**, 477.

Starovskii, O. V., and Struchkov, Y. T., "The Structure of Ditoluenechromium Iodide," *Dokl. Akad. Nauk SSSR*, **135**, 620.

Struchkov, Y. T., "Crystal and Molecular Structure of Phenylarsonic Acid," *Izv. Akad. Nauk SSSR, Otdel. Khim. Nauk*, 1960, 1962.

Trotter, J., "The Crystal Structure of 1,5-Dinitronaphthalene," *Acta Cryst.*, **13**, 95.

Trotter, J., "The Crystal Structure of 1-Naphthoic Acid," *Acta Cryst.*, **13**, 732.

Vorontsova, L. B., Zvonkova, Z. V., and Zhdanov, G. S., "X-Ray Diffraction Investigation of Di-*p*-Tolyl Disulfide," *Kristallografiya*, **5**, 698.

Wheatley, P. J., "An X-Ray Structural Investigation of Two Isomeric Forms of 1,2-Dimethyl-1,2-diphenyldiphosphine Disulfide," *J. Chem. Soc.*, **1960**, 523.

Wolf, L., and Bärnighausen, H., "The Spatial Configuration of β-Diketone Complexes of Quadrivalent Cations with the Co-ordination Number 8. The Crystal Structure of Cerium Tetrakis(dibenzoylmethane)," *Acta Cryst.*, **13**, 778.

1961

Allegra, G., "The Crystal Structure of Bis-Tricarbonyl Chromium Biphenyl," *Atti Accad. Nazl. Lincei, Rend., Classe Sci. Fis. Mat. Nat.*, **31**, 399.

Brown, I. D., and Dunitz, J. D., "The Crystal Structure of Diazoaminobenzene Copper(I)," *Acta Cryst.*, **14**, 480.

Campbell, R. B., Robertson, J. M., and Trotter, J., "The Crystal and Molecular Structure of Pentacene," *Acta Cryst.*, **14**, 705.

Davydova, M. A., and Struchkov, Y. T., "Crystal Structure of 1,4,5,8-Tetra-chloronaphthalene," *Zh. Strukt. Khim.*, **2**, 69.

Guilhem, J., "The Structure of Alizarin," *Acta Cryst.*, **14**, 88.

Hargreaves, A., Rizvi, S. H., and Trotter, J., "The Structure of Crystalline Biphenyl," *Proc. Chem. Soc.*, **1961**, 122.

Hospital, M., "Crystal Structure of 9-Chloro-10-bromoanthracene," *Acta Cryst.*, **14**, 76.

Jarvis, J. A. J., "The Crystal Structure of Copper(II)bis(benzeneazo-β-naphthol)," *Acta Cryst.*, **14**, 961.

Kitaigorodskii, A. I., Struchkov, Y. T., Avoyan, G. L., and Davydova, M. A., "The Crystal Structure of 1,4,5,8-Tetrachloronaphthalene," *Dokl. Akad. Nauk SSSR*, **130**, 607.

Kondrashev, Y. D., "Crystal and Molecular Structure of *p*-Dihalo Compounds of Diazoaminobenzene," *Kristallografiya*, **6**, 515.

Mason, R., "The Crystal Structure of Phenanthrene," *Mol. Phys.*, **4**, 413.

Métras, J. C., "The Crystal Structure of 2,3-Dichloro-1,4-Naphthoquinone," *Acta Cryst.*, **14**, 153.

Ozols, J., Vimba, S., and Ievins, A., "The Structure of Rubidium Tetraphenylbor-onate," *Latvijas PSR Zinatnu Akad. Vestis*, **1961**, 93.

Plieth, K., and Ruban, G., "X-Ray Diffraction Crystal Structure Analysis of *p,p'*-Dichlorodiphenylamine," *Z. Krist.*, **116**, 161.

Polynova, T. N., and Porai-Koshits, M. A., "Crystal Structure of Diphenyl-antimony Trichloride," *Zh. Strukt. Khim.*, **2**, 477.

Prat, M.-T., "Statistical Disorder in the Structure of 9-Bromo-10-methyl-anthracene," *Acta Cryst.*, **14**, 110.

Robertson, G. B., "Crystal and Molecular Structure of Biphenyl," *Nature*, **191**, 593.

Robertson, J. M., Sinclair, V. C., and Trotter, J., "The Crystal and Molecular Structure of Tetracene," *Acta Cryst.*, **14**, 697.

Robertson, J. M., and Trotter, J., "Crystal Structure Studies of Polynuclear Hydrocarbons V. 1,2:7,8-Dibenzocoronene," *J. Chem. Soc.*, **1961**, 1115.

Robertson, J. M., and Trotter, J., "Crystal Structure Studies of Polynuclear Hydrocarbons VI. 1,12:2,3:4,5:6,7:8,9:10,11-Hexabenzocoronene," *J. Chem. Soc.*, **1961**, 1280.

Sinha, U. G., "The Crystal Structure of Benzalazine," *Indian J. Phys.*, **35**, 374.

Srivastava, S. N., "Diffuse Layers in Anthrone Photographs," *Acta Cryst.*, **14**, 796.

Starovskii, O. V., and Struchkov, Y. T., "The Crystal Structure of Ditoluene-chromium Iodide," *Zh. Strukt. Khim.*, **2**, 162.

Stora, C., "The Structure of Triphenylmethyl Bromide," *Bull. Soc. Chim. France*, **1961**, 1512.

Treuting, R. G., and Abrahams, S. C., "Evaluation of the Electron Density in a General Plane: Application to 4,4'-Dichlorodiphenyl Sulfone," *Acta Cryst.*, **14**, 190.

Trotter, J., "The Crystal Structure of 2-Naphthoic Acid," *Acta Cryst.*, **14**, 101.

Trotter, J., "The Crystal and Molecular Structure of Biphenyl," *Acta Cryst.*, **14**, 1135.

Wallwork, S. C., "The Structure of *trans*-α,β-Dicyanostilbene," *Acta Cryst.*, **14**, 375.

1962

Bois d'Enghien, H., and Meerssche, M. van, "Structure of Cyanostilbenes VI. X-Ray Crystallographic Determination of the Structure of 2,6-Dibromo-α-cyanostilbene," *Bull. Soc. Chim. Belges*, **71**, 503.

Campbell, R. B., Robertson, J. M., and Trotter, J., "The Crystal Structure of Hexacene, and a Revision of the Crystallographic Data for Tetracene and Pentacene," *Acta Cryst.*, **15**, 289.

Chao, G. Y., and McCullough, J. D., "The Crystal Structure of Di-*p*-chloro-phenyltellurium Diiodide," *Acta Cryst.*, **15**, 887.

Davies, W. O., and Stanley, E., "The Crystal Structure of Triphenyl Phosphate," *Acta Cryst.*, **15**, 1092.

Davydova, M. A., and Struchkov, Y. T., "Steric Hindrance and Molecular Conformation VII. The Crystal and Molecular Structure of 1,4,5,8-Tetra-chloronaphthalene," *Zh. Strukt. Khim.*, **3**, 184.

Gafner, G., and Herbstein, F. H., "The Crystal and Molecular Structures of Overcrowded Halogenated Compounds III. 1,4,5,8-Tetrachloronaphthalene," *Acta Cryst.*, **15**, 1081.

Hargreaves, A., and Rizvi, S. H., "The Crystal and Molecular Structure of Biphenyl," *Acta Cryst.*, **15**, 365.

Iball, J., and Mackay, K. J. H., "The Crystal and Molecular Structure of 9:10-Anthrahydroquinone Dibenzoate," *Acta Cryst.*, **15**, 148.

King, G. S. D., "The Structure of the Iron Carbonyl Phenylacetylene Complex, $Fe_2(CO)_6(C_6H_5C_2H)_3$," *Acta Cryst.*, **15**, 243.

Mak, T. C. W., and Trotter, J., "Crystal and Molecular Structure of Biphenylene," *J. Chem. Soc.*, **1962**, 1.

Norment, H. G., and Karle, I. L., "The Crystal Structures of Deoxyanisoin and p,p'-Dimethoxybenzophenone," *Acta Cryst.*, **15**, 873.

Pascard-Billy, C., "Structure of the Three Crystalline Forms of Naphthazarin I. Structure of the B Form," *Bull. Soc. Chim. France*, **1962**, 2282.

Pascard-Billy, C., "Structure of the Three Crystalline Forms of Naphthazarin II. Structure of the A Form," *Bull. Soc. Chim. France*, **1962**, 2293.

Pascard-Billy, C., "Structure of the Three Crystalline Forms of Naphthazarin III. Structure of the C Form," *Bull. Soc. Chim. France*, **1962**, 2299; *Acta Cryst.*, **15**, 519.

Prakash, A., "The Crystal Structure of 1,8-Dihydroxy Anthraquinone," *Indian J. Phys.*, **36**, 654.

Srivastava, S. N., "Crystal Structure of Anthrone," *Z. Krist.*, **117**, 386.

Trotter, J., "The Crystal Structure of Bromodiphenylarsine," *J. Chem. Soc.*, **1962**, 2567.

Trotter, J., "Stereochemistry of Arsenic IV. Chlorodiphenylarsine," *Can. J. Chem.*, **40**, 1590.

Williams, D. E., Dunke, W. L., and Ruddle, R. E., "The Crystal Structure of Bis(m-Bromobenzoyl)methane," *Acta Cryst.*, **15**, 627.

Ziolkowska, B., "X-Ray Studies of Diphenyl Mercury," *Roczniki Chem.*, **36**, 1341.

1963

Ahmed, F. R., and Trotter, J., "The Crystal Structure of Triphenylene," *Acta Cryst.*, **16**, 503.

Avoyan, R. L., Kitaigorodskii, A. I., and Struchkov, Y. T., "The Crystal Structure of 5,6-Dichloro-11,12-diphenylnaphthacene," *Zh. Strukt. Khim.*, **4**, 633.

Boonstra, E. G., "The Crystal and Molecular Structure of 4,4'-Dinitrodiphenyl," *Acta Cryst.*, **16**, 816.

Brändén, C. I., "Crystal Structure of $HgCl_2 \cdot 2(C_6H_5)_3AsO$," *Acta Chem. Scand.*, **17**, 1363.

Chetkina, L. A., and Gol'der, G. A., "The Crystal Structure of 1,5-Diiodoanthraquinone," *Kristallografiya*, **8**, 582.

Chetkina, L. A., Gol'der, G. A., and Zhdanov, G. S., "Crystal Structure of 1,5-Dibromoanthraquinone," *Kristallografiya*, **8**, 194.

Cullen, W. R., and Trotter, J., "Stereochemistry of Arsenic XII. Bis(Diphenylarsenic) Oxide," *Can. J. Chem.*, **41**, 2983.

Daly, J. J., "The Structure of Phosphorus Triphenyl," *Z. Krist.*, **118**, 332.

Ehrlich, H. W. W., and Owston, P. G., "The Crystal Structure of *trans*-Oxotrichlorobisdiethylphenylphosphinerhenium(V)," *J. Chem. Soc.*, **1963**, 4368.

Ferrier, R. P., and Iball, J., "The Structure of Methyl-1-2-benzanthraquinone II. The Crystal and Molecular Structure of 5-Methyl-1,2-benzanthraquinone," *Acta Cryst.*, **16**, 269.

Ferrier, R. P., and Iball, J., "The Structure of Methyl 1:2-Benzanthraquinones III. The Crystal and Molecular Structure of 2′-Methyl-1:2-benzanthraquinone," *Acta Cryst.*, **16**, 513.

Gafner, G., and Herbstein, F. H., "Conformation of Octachloronaphthalene," *Nature*, **200**, 130.

Hanson, A. W., "The Crystal Structure of the 1 : 1 Perylene-fluoranil Complex," *Acta Cryst.*, **16**, 1147.

LaPlaca, S. J., and Ibers, J. A., "Metal-Hydrogen Bond: Structure of HRh-(CO)[P(C$_6$H$_5$)$_3$]$_3$," *J. Am. Chem. Soc.*, **85**, 3501.

Trotter, J., "Stereochemistry of Arsenic VI. Tri-*p*-tolylarsine," *Can. J. Chem.*, **41**, 14.

Trotter, J., "The Crystal and Molecular Structure of Phenanthrene," *Acta Cryst.*, **16**, 605.

Trotter, J., "Stereochemistry of Arsenic X. Tri-*p*-xylylarsine," *Acta Cryst.*, **16**, 1187.

1964

Akopyan, Z. A., and Struchkov, Y. T., "Steric Hindrances and Molecular Conformation X. Crystal Structure of 1,8-Dinitronaphthalene," *Zh. Strukt. Khim.*, **5**, 496.

Amano, Y., Osaki, K., and Watanabé, T., "Copper Chelates of Ephedrine and Related Compounds II. The Crystal Structure of the Bis-*l*-ephedrine Copper-(II) Chelate Benzene Clathrate," *Bull. Chem. Soc. Japan*, **37**, 1363.

Arthur, P., Jr., "Structural Data on Electrically Conductive Caesium Tetracyanoquinodimethanide," *Acta Cryst.*, **17**, 1176.

Avoyan, R. L., Kitaigorodskii, A. I., and Struchkov, Y. T., "Steric Hindrance and Molecular Conformation IX. The Crystal and Molecular Structure of 5,6-Dichloro-11,12-diphenyl Naphthacene," *Zh. Strukt. Khim.*, **5**, 420.

Bolton, W., "The Crystal Structure of Isoviolanthrone (Isodibenzanthrone)," *Acta Cryst.*, **17**, 1020.

Bolton, W., and Stadler, H. P., "The Crystal Structure of Violanthrone (Dibenzanthrone)," *Acta Cryst.*, **17**, 1015.

Brändén, C. I., "The Crystal Structure of [HgCl$_2$·(C$_6$H$_5$)$_3$AsO]$_2$," *Arkiv Kemi*, **22**, 485.

Bregman, J., Leiserowitz, L., and Osaki, K., "Topochemistry X. The Crystal and Molecular Structures of 2-Chloro-*N*-Salicylideneaniline," *J. Chem. Soc.*, **1964**, 2086.

Bregman, J., Leiserowitz, L., and Schmidt, G. M. J., "Topochemistry IX. The Crystal and Molecular Structures of *N*-5-Chlorosalicylideneaniline near 90 and 300°K.," *J. Chem. Soc.*, **1964**, 2068.

Breton, M., "Structure of 2-Chloro-3-methyl-1,4-naphthoquinone," *Compt. Rend.*, **258**, 3489.

Brown, D. S., Wallwork, S. C., and Wilson, A., "Molecular Complexes Exhibiting Polarization Bonding IV. The Crystal Structure of the Anthracene-s-Trinitrobenzene Complex," *Acta Cryst.*, **17**, 168.

Calleri, M., and Speakman, J. C., "The Crystal Structure of, and the Hydrogen Bonding in, Di-(*p*-Chlorophenyl) Hydrogen Phosphate," *Acta Cryst.*, **17**, 1097.

Camerman, A., and Trotter, J., "Crystal and Molecular Structure of Perylene," *Proc. Roy. Soc. (London)*, **279A**, 129.

Cotton, F. A., and Mague, J. T., "The Crystal and Molecular Structure of tris(Phenyldiethylphosphine)nonachlorotrirhenium(III)," *Inorg. Chem.*, **3**, 1094.

Daly, J. J., "The Crystal and Molecular Structure of Triphenylphosphorus," *J. Chem. Soc.*, **1964**, 3799.

Deuschl, H., and Hoppe, W., "Crystal and Molecular Structure of Phenanthrenechromiumtricarbonyl," *Acta Cryst.*, **17**, 800.

Engebretson, G. R., and Rundle, R. E., "The Crystal Structure of Bis(*m*-Chlorobenzoyl)methane," *J. Am. Chem. Soc.*, **86**, 574.

Gaultier, J., and Hauw, C., "Structure of α-Naphthoquinone," *Compt. Rend.*, **258**, 619.

Gaultier, J., and Hauw, C., "Structure of 2-Chloro-3-hydroxy-1,4-naphthoquinone," *Compt. Rend.*, **259**, 2845.

Guilhem, J., "Structure of Anthrarufine," *Compt. Rend.*, **258**, 617.

Iball, J., "A Refinement of the Structure of 9:10-Dimethyl-1:2-benzanthracene," *Nature*, **201**, 916.

Khotsyanova, T. L., and Struchkov, Y. T., "The Crystal and Molecular Structure of 2,6-Dichloronaphthalene," *Zh. Strukt. Khim.*, **5**, 404.

Kondrashev, Y. D., "Crystal Structure of *p*-Dimethyldiazoaminobenzene," *Kristallografiya*, **9**, 403.

Mason, R., "Crystallography of Anthracene at 95 and 290°K," *Acta Cryst.*, **17**, 547.

McCammon, C. S., and Trotter, J.," The Structure of *p*-Bromobenzoic Anhydride," *Acta Cryst.*, **17**, 1333.

Rérat, B., Dauphin, G., Gervais, H.-P., Kergomard, A., and Rérat, C., "Structure of N-(*p*-Bromophenyl)-*p*-chlorobenzenesulfonamide," *Compt. Rend.*, **259**, 4251.

Robinson, B., and Hargreaves, A., "The Crystal Structure of α-Naphthol," *Acta Cryst.*, **17**, 944.

Srivastava, S. N., "Three-dimensional Refinement of the Structure of Anthrone," *Acta Cryst.*, **17**, 851.

Wei, L., Stogsdill, R. M., and Lingafelter, E. C., "The Crystal Structure of Bis-(N-Phenylsalicylaldiminato)copper(II)," *Acta Cryst.*, **17**, 1058.

Wheatley, P. J., "The Crystal and Molecular Structure of Pentaphenyl Phosphorus," *J. Chem. Soc.*, **1964**, 2206.

Wheatley, P. J., "An X-Ray Diffraction Determination of the Crystal and Molecular Structure of Pentaphenyl Antimony," *J. Chem. Soc.*, **1964**, 3718.

Ziolkowska, P., Myasnikova, R. M., and Kitaigorodskii, A. I., "Crystal Structure of Diphenylmercury," *Zh. Strukt. Khim.*, **5**, 737.

1965

Akopyan, Z. A., Kitaigorodskii, A. I., and Struchkov, Y. T., "Steric Hindrance and Molecular Configuration XII. The Crystal and Molecular Structure of 1,8-Dinitronaphthalene," *Zh. Strukt. Khim.*, **6**, 729.

Alléaume, M., and Decap, J., "Structure of Diaminodiphenyl Sulfone," *Compt. Rend.*, **261**, 1693.

Beintema, J., "The Crystal Structure of 1,5-Dimethylnaphthalene," *Acta Cryst.*, **18**, 647.

Boetticher, H., Plieth, K., and Repmann, H., "Crystal Structure of 4,4'-Dimethoxy-α,β-diethylstilbene and 4,4'-Dichloro-α,β-Diethylstilbene," *Naturwissenschaften*, **52**, 390.

Breton, M., "Crystal Structure of 2-Bromo-3-Methyl-1,4-Naphthoquinone," *Compt. Rend.*, **260**, 5275.

Brown, C. J., and Sadanaga, R., "The Crystal Structure of Benzil," *Acta Cryst.*, **18**, 158.

Camerman, A., and Trotter, J., "The Crystal and Molecular Structure of Pyrene," *Acta Cryst.*, **18**, 636.

Davydova, M. A., and Struchkov, Y. T., "Steric Hindrance and Molecular Conformation XI. Crystal and Molecular Structure of 1,5-Dibromo-4,8-Dichloronaphthalene," *Zh. Strukt. Khim.*, **6**, 113.

Davydova, M. A., and Struchkov, Y. T., "Preliminary Data on the Crystal Structure of 1,4,5,8-Tetrabromonaphthalene," *Zh. Strukt. Khim.*, **6**, 922.

Dodge, R. P., and Schomaker, V., "The Molecular Structures of Two Isomers of $Fe_3(CO)_8(C_6H_5C_2C_6H_5)_2$," *J. Organometal. Chem.*, **3**, 274.

Eisenberg, R., and Ibers, J. A., "Structure of Hydridochlorobis(diphenylethylphosphine)platinum," *Inorg. Chem.*, **4**, 773.

Eisenberg, R., and Ibers, J. A., "Trigonal Prismatic Coordination. The Molecular Structure of Tris(*cis*-1,2,diphenylethene-1,2-dithiolato)rhenium," *J. Am. Chem. Soc.*, **87**, 3776.

Gaultier, J., and Hauw, C., "Structure of α-Naphthoquinone," *Acta Cryst.*, **18**, 179.

Gaultier, J., and Hauw, C., "Structure of 2-Bromo-1,4-naphthoquinone," *Acta Cryst.*, **18**, 604.

Gaultier, J., and Hauw, C., "Crystal Structure of the 2 and 2,3 Derivatives of 1,4-Naphthoquinone II. 2-Chloro-3-hydroxy-1,4-Naphthoquinone," *Acta Cryst.*, **19**, 580.

Gaultier, J., and Hauw, C., "Crystal Structure of the 2 and 2,3 Derivatives of 1,4-Naphthoquinone III. 2-Chloro-3-amino-1,4-naphthoquinone," *Acta Cryst.*, **19**, 585.

Gaultier, J., and Hauw, C., "Crystal Structures of the 2 and 2,3 Derivatives of 1,4-Naphthoquinone IV. Phthiocol Antagonism by Structural Analogy," *Acta Cryst.*, **19**, 919.

Gaultier, J., and Hauw, C., "Atomic Structure of 2-Bromo-3-amino-1,4-naphthoquinone," *Compt. Rend.*, **260**, 2831.

Gaultier, J., and Hauw, C., "Structure of Phthiocol, the Synthetic Vitamin K," *Compt. Rend.*, **260**, 3404.

Gaultier, J., and Hauw, C., "Crystal Structure of α-Naphthohydroquinone," *Compt. Rend.*, **261**, 3818.

Gaultier, J., and Hauw, C., "Crystal Structure of 2-Methyl-α-naphthohydroquinone," *Compt. Rend.*, **261**, 4109.

Gomes de Mesquita, A. H., MacGillavry, C. H., and Eriks, K., "The Structure of Triphenylmethyl Perchlorate at 85°C.," *Acta Cryst.*, **18**, 437.

Hope, H., "The Crystal Structure of N-Phenyl-N'-benzoylselenourea," *Acta Cryst.*, **18**, 259.

Karle, I. L., and Dragonette, K. S., "The Crystal and Molecular Structure of *cis,cis*-1,2,3,4-Tetraphenylbutadiene," *Acta Cryst.*, **19**, 500.

LaPlaca, S. J., and Ibers, J. A., "Structure of $IrO_2Cl(CO)[P(C_6H_5)_3]_2$, the Oxygen Adduct of a Synthetic Reversible Molecular Oxygen Carrier," *J. Am. Chem. Soc.*, **87**, 2581.

LaPlaca, S. J., and Ibers, J. A., "Crystal and Molecular Structure of Tristriphenylphosphine Rhodium Carbonyl Hydride," *Acta Cryst.*, **18**, 511.

LaPlaca, S. J., and Ibers, J. A., "A Five-coordinated d^6 Complex: Structure of Dichlorotris(triphenylphosphine)ruthenium(II)," *Inorg. Chem.*, **4**, 778.

Li, C.-T., and Caughlan, C. N., "The Crystal and Molecular Structure of Calcium 1-Naphthyl Phosphate Trihydrate," *Acta Cryst.*, **19**, 637.

Mais, R. H. B., and Powell, H. M., "The Crystal Structure of Tris-(*o*-diphenylarsinophenyl)arsineruthenium Dibromide," *J. Chem. Soc.*, **1965**, 7471.

Mak, T. C. W., and Trotter, J., "The Structure of the Adduct of $Ph_3P{=}N \cdot Ph$ and $MeO_2C \cdot {\equiv}C \cdot CO_2Me$," *Acta Cryst.*, **18**, 81.

Manojlović, L. M., and Edmunds, I. G., "The Crystal and Molecular Structure of 4,4'-Dihydroxythiobenzophenone Monohydrate," *Acta Cryst.*, **18**, 543.

Mariani, C., Mugnoli, A., and Casalone, G. L., "Crystal Structure of 1,1-Di-*p*-tolyl-2-bromoethylene," *Atti Accad. Nazl. Lincei, Rend., Classe Sci. Fis. Mat. Nat.*, **38**, 880.

Omel'chenko, Y. A., and Kondrashev, Y. D., "Crystal and Molecular Structure of 2,4-Dibromodiazoaminobenzene," *Kristallografiya*, **10**, 822.

Peerdeman, A. F., Holst, J. P. C., Horner, L., and Winkler, H., "Absolute Configuration of (+)-Methyl-*n*-propyl-phenyl-benzyl-phosphonium Bromide," *Tetrahedron Letters*, **1965**, 811.

Prakash, A., "The Crystal and Molecular Structure of Chrysazin," *Z. Krist.*, **122**, 272.

Rérat, B., Dauphin, G., Kergomard, A., and Rérat, C., "Structure of N-(1-Bromo-3,5-dimethylphenyl)benzenesulfonamide," *Compt. Rend.*, **261**, 139.

Stephens, F. S., "The Crystal and Molecular Structure of Benzoyl(triphenylphosphoranylidene)methyl Iodide," *J. Chem. Soc.*, **1965**, 5640.

Stephens, F. S., "The Crystal and Molecular Structure of Benzoyl(triphenylphosphoranylidene)methyl Chloride," *J. Chem. Soc.*, **1965**, 5658.

Stora, C., and Poyer, N., "Determination of the Fine Structure of Triphenylmethyl Bromide by the I.B.M. 704 Computer," *Compt. Rend.*, **260**, 1660.

Svetich, G. W., and Caughlan, C. N., "Refinement of the Crystal Structure of Triphenyl Phosphate," *Acta Cryst.*, **19**, 645.

Wheatley, P. J., "An X-Ray Determination of the Molecular Structure of a Wittig Reagent: *p*-Tolyl Triphenylphosphoranylidenemethyl Sulfone," *J. Chem. Soc.*, **1965**, 5785.

Wilkes, C. E., and Jacobson, R. A., "The Crystal Structure of Di-μ-diphenylphosphinatoacetylacetonatochromium(III)," *Inorg. Chem.*, **4**, 99.

1966

Albano, V., Bellon, P. L., and Scatturin, V., "Hydride Complexes of the Third Transition Series. Crystal and Molecular Structure of Trihydridobis-(triphenylphosphine)ethylenebis(diphenylphosphine)rhenium(III)," *Rend. Ist. Lombardo Sci. Lett.*, **100A**, 989.

Amit, A. G., and Hope, H., "Crystal and Molecular Structure of *trans-p,p′*-Dibromoazobenzene," *Acta Chem. Scand.*, **20**, 835.

Bergman, J. G., Jr., and Cotton, F. A., "The Crystal and Molecular Structure of Tetraphenylarsonium Tetranitratocobaltate(II); an Eight-coordinate Cobalt(II) Complex," *Inorg. Chem.*, **5**, 1208.

Bertrand, J. A., and Plymale, D. L., "Five-coordinate Complexes I. The Crystal and Molecular Structure of Dibromotris(diphenylphosphine)cobalt(II) and Related Compounds," *Inorg. Chem.*, **5**, 879.

Blackstone, M., Thuijl, J. van, and Romers, C., "The Crystal Structure of Bis(1,3-diphenyl-1,3-propanedionato)copper(II)," *Rec. Trav. Chim.*, **85**, 557.

Breton, M., "Crystalline Structure of 2,3-Dimethyl-1,4-naphthoquinone," *Compt. Rend.*, **263C**, 1129.

Breton, M., "Structure of 2,3-Dibromo-1,4-naphthoquinone," *Compt. Rend.*, **263C**, 1211.

Brown, C. J., "A Refinement of the Crystal Structure of Azobenzene," *Acta Cryst.*, **21**, 146.

Brown, C. J., "The Crystal Structure of *p*-Azotoluene," *Acta Cryst.*, **21**, 153.

Brown, C. J., and Corbridge, D. E. C., "The Crystal Structure of Sodium Naphthionate Tetrahydrate," *Acta Cryst.*, **21**, 485.

Casalone, G. L., Mariani, C., Mugnoli, A., and Simonetta, M., "Crystal, Molecular and Electronic Structure of 1,1-Diaryl 2-Haloethylenes II. Crystal and Molecular Structure of 2-Bromo-1,1-diphenyl-1-propene," *Atti Accad. Nazl. Lincei, Rend., Classe Sci. Fis. Mat. Nat.*, **41**, 245.

Cheeseman, T. P., Hall, D., and Waters, T. N., "The Color Isomerism and Structure of Some Copper Coordination Compounds XII. The Crystal Structure of N,N′-(2,2′-Biphenyl)bis(salicylaldiminato)copper(II)," *J. Chem. Soc.*, **1966A**, 1396.

Cotton, F. A., and Lippard, S. J., "Chemical and Structural Studies of Rhenium(V) Oxyhalide Complexes III. The Crystal and Molecular Structure of Tetraphenylarsonium Oxotetrabromoacetonitrilerhenate(V)," *Inorg. Chem.*, **5**, 416.

Daly, J. J., and Wheatley, P. J., "Structure of Triphenylphosphoranylideneketen," *J. Chem. Soc.*, **1966A**, 1703.

Eisenberg, R., and Ibers, J. A., "Trigonal Prismatic Coordination. The Crystal and Molecular Structure of Tris(*cis*-1,2-diphenylethene-1,2-dithiolato)-rhenium," *Inorg. Chem.*, **5**, 411.

Fawcett, J. K., and Trotter, J., "The Crystal and Molecular Structure of Coronene," *Proc. Roy. Soc. (London)*, **289A**, 366.

Fawcett, J. K., and Trotter, J., "A Refinement of the Structure of Biphenylene," *Acta Cryst.*, **20**, 87.

Fritchie, C. J., Jr., "The Crystal Structure of Triphenylmethylphosphonium Bis(1,2-dicyanoethylene-1,2-dithiolato)nickelate(III)," *Acta Cryst.*, **20**, 107.

Fritchie, C. J., Jr., and Arthur, P., Jr., "A Refinement of the Crystal Structure of Cesium Tetracyanoquinodimethanide," *Acta Cryst.*, **21**, 139.

Gaultier, J., and Hauw, C., "Structure of 3-Amino-1, 4-naphthoquinone," *Compt. Rend.*, **263C**, 925.

Gaultier, J., and Hauw, C., "Crystal Structure of the 2 and 2,3-Derivatives of 1,4-Naphthoquinone V. 2-Bromo-3-amino-1,4-naphthoquinone, Hydrogen Bonds, and the Nature of Halogen Substituents," *Acta Cryst.*, **20**, 620.

Gaultier, J., and Hauw, C., "Crystal Structure of the 2 and 2,3-Derivatives of 1,4-Naphthoquinone VI. 2-Methyl-3-amino-1,4-naphthoquinone Hydrate, $2C_{11}O_2NH_9 \cdot {}^1/_2H_2O$," *Acta Cryst.*, **21**, 694.

Hall, D., and Moore, F. H., "The Crystal Structure of NN'-Disalicylidene-ethylenediaminezinc(II) Monohydrate," *J. Chem. Soc.*, **1966A**, 1822.

Hall, D., and Nobbs, C. L., "The Crystal and Molecular Structure of 1,5-Dihydroxyanthraquinone," *Acta Cryst.*, **21**, 927.

Hopkins, T. E., Zalkin, A., Templeton, D. H., and Adamson, M. G., "The Crystal and Molecular Structure of Tetraphenylarsonium *cis*-Diaquo-tetrachlororuthenate Monohydrate," *Inorg. Chem.*, **5**, 1427.

Kuhn, H.-J., and Plieth, K., "Crystal Structure of *p*-Bromophenyldiphenyl-phosphine," *Naturwissenschaften*, **53**, 359.

Kuroda, H., Ikemoto, I., and Akamatu, H., "The Crystal Structure of the Pyrene-Tetracyanoethylene Complex," *Bull. Chem. Soc. Japan*, **39**, 547.

LaPlaca, S. J., and Ibers, J. A., "Structure of Chlorocarbonyl(sulfur dioxide) bis(triphenylphosphine)iridium," *Inorg. Chem.*, **5**, 405.

Lonsdale, K., Milledge, H. J., and El Sayed, K., "The Crystal Structure (at Five Temperatures) and Anisotropic Thermal Expansion of Anthra-quinone," *Acta Cryst.*, **20**, 1.

Mornon, J.-P., "An Application of the Symbolic Addition Method: Crystalline Structure of α-Naphthyl-4-chlorophthalamic Acid," *Compt. Rend.*, **263C**, 286.

Nesterova, Y. M., Porai-Koshits, M. A., Upadysheva, A. V., and Kazitsyna, L. A., "The Crystal Structure of *p-N,N*-Dimethylaminophenyl Diazonium Chlorozincate," *Zh. Strukt. Khim.*, **7**, 129.

Okaya, Y., and Ashida, T., "The Structures of α-Naphthylphenylmethylsilanes, Optically Active Silicon Compounds, and Their Absolute Configurations," *Acta Cryst.*, **20**, 461.

Orioli, P., Di Vaira, M., and Sacconi, L., "Crystal and Molecular Structure of the Five-coordinated Nickel(II) Complex with *N*-β-Diethylaminoethyl-5-chlorosalicylaldimine," *J. Am. Chem. Soc.*, **88**, 4383.

Palenik, G. J., "The Crystal Structure of Tetraphenylarsonium 3-Fluoro-1,1,4,5,5-pentacyano-2-azapentadienide," *Acta Cryst.*, **20**, 471.

Pauling, P., "The Crystal Structure of Bis(triphenylmethylarsonium)tetra-chloronickel(II), $[(C_6H_5)_3CH_3As]_2[NiCl]_4$, of the Isomorphous Compounds of Mn, Fe, Co, and Zn, and of the Corresponding Bromides," *Inorg. Chem.*, **5**, 1498.

Penfold, B. R., and Robinson, W. T., "The Crystal Structure of Bis(tetra-phenylarsonium) Tri-μ-Chloro-octachlorotrirhenate(III), $[(C_6H_5)_4As]_2-Re_3Cl_{11}$," *Inorg. Chem.*, **5**, 1758.

Polynova, T. N., and Porai-Koshits, M. A., "X-Ray Diffraction Study of Triphenylstibine Dichloride Crystals," *Zh. Strukt. Khim.*, **7**, 742.

Puckett, R. T., Pfluger, C. E., and Curtin, D. Y., "The Crystal Structure and Solid State Rearrangement of *p*-Bromobenzeneazotribenzoylmethane," *J. Am. Chem. Soc.*, **88**, 4637.

Rérat, B., Dauphin, G., Kergomard, A., and Rérat, C., "Structure of *N*-(4-Bromo-2-methylphenyl)benzenesulfonamide," *Compt. Rend.*, **262C**, 1318.

Romers, C., and Hesper, B., "Isomerism of Benzoquinone Monoximes (Nitro-sophenols) XIII. The Crystal Structure of *p*-Methoxyindophenol *N*-Oxide," *Acta Cryst.*, **20**, 162.

Saint-Gimiez, D. de, Laurent, A., Thuong, N. T., Rérat, C., and Chabrier, P., "Structure of Methyl Diphenylthiophosphinite Ph$_2$P(S)OMe," *Compt. Rend.*, **263C**, 1213.

Stam, C. H., and Sanseverino, L. R. di, "The Crystal Structures of 1,6-Di-*p*- and 1,6-Di-*o*-chlorophenyl-3,4-dimethylhexatriene," *Acta Cryst.*, **21**, 132.

Stora, C., "Stereochemical X-Ray Study of the Red Crystalline Tautomeric Form of 3,5-Dibromo-*p*-hydroxytriphenylmethanecarbinol," *Compt. Rend.*, **263C**, 1355.

Stora, C., and Poyer, N., "Refinement of the Structure of Triphenylmethyl Bromide by IBM 704," *Bull. Soc. Chim. France*, **1966**, 841.

Toman, K., and Očenášková, D., "The Crystal Structure of Some Conducting Organic Substances I. The Iodide of 4,4'-Bis(dimethylamino)diphenylamine Radical," *Acta Cryst.*, **20**, 514.

Watenpaugh, K., and Caughlan, C. N., "The Crystal and Molecular Structure of Dichlorodiphenoxytitanium(IV)," *Inorg. Chem.*, **5**, 1782.

Williams, D. E., "Structure of 2,2-Diphenyl-1-picrylhydrazyl Free Radical," *J. Am. Chem. Soc.*, **88**, 5665.

Williams, D. E., "Crystal Structure of Dibenzoylmethane," *Acta Cryst.*, **21**, 340.

1967

Andersen, P., and Klewe, B., "The Crystal Structure of the Free Radical Tri-*p*-nitrophenylmethyl," *Acta Chem. Scand.*, **21**, 2599.

Baenziger, N. C., Buckles, R. E., and Simpson, T. D., "Complexes of *p*-Anisylethylenes III. The Crystal Structure of the Dichloriodate(I) Salt of the Tetra-*p*-anisylethylene Dication," *J. Am. Chem. Soc.*, **89**, 3405.

Baikie, P. E., and Mills, O. S., "Carbon Compounds of the Transition Metals VI. The Reaction Product of Azobenzene with Iron Carbonyls: the Structure of μ-N,N'-Dehydrosemidinatobis(tricarbonyliron)," *Inorg. Chim. Acta*, **1**, 55.

Bailey, M., and Brown, C. J., "The Crystal Structure of 1,5-Dinitro-4,8-dihydroxyanthraquinone," *Acta Cryst.*, **22**, 392.

Bailey, M., and Brown, C. J., "The Crystal Structure of N,N'-Diphenyl-1,5-diaminoanthraquinone," *Acta Cryst.*, **22**, 488.

Bailey, M., and Brown, C. J., "The Crystal Structure of N,N'-Diphenyl-1,8-diaminoanthraquinone," *Acta Cryst.*, **22**, 493.

Berthou, J., Elguero, J., Jacquier, R., Marzin, C., and Rérat, C., "Structure of 4,4'-Dibromocinnamaldazine," *Compt. Rend.*, **265C**, 513.

Bertrand, J. A., "Five-Coordinate Complexes III. Structure and Properties of μ_4-Oxo-hexa-μ-chloro-tetrakis[(triphenylphosphine Oxide)copper(II)]," *Inorg. Chem.*, **6**, 495.

Blundell, T. L., and Powell, H. M., "The Crystal and Molecular Structure of [Pd(TPAS)Cl]ClO$_4$·C$_6$H$_6$ [TPAS = *o*-Phenylenebis(*o*-dimethylarsinophenylmethylarsine)]: The Stereochemistry of a Pentaco-ordinate Palladium Complex Ion," *J. Chem. Soc.*, **1967A**, 1650.

Breton-Lacombe, M., "Crystal Structure of the 2 and 2,3 Derivatives of 1,4-Naphthoquinone VII. Structure of 2,3-Dimethyl-1,4-Naphthoquinone," *Acta Cryst.*, **23**, 1024.

Breton-Lacombe, M., "Crystal Structure of the 2 and 2,3 Derivatives of 1,4-Naphthoquinone VIII. Structure of 2,3-Dibromo-1,4-Naphthoquinone," *Acta Cryst.*, **23**, 1031.

Brown, C. J., "The Crystal Structure of α,4-Dibromo-α-(4-methyl-2-nitrophenylazo)acetanilide," *J. Chem. Soc.*, **1967A**, 405.

Bryan, R. F., "Metal-Metal Bonding in Coordination Complexes I. The Crystal and Molecular Structure of Triphenyltin Tetracarbonyltriphenylphosphinemanganese," *J. Chem. Soc.*, **1967A**, 172.

Busetti, V., Mammi, M., Signor, A., and Del Pra, A., "A Refinement of the Crystal Structure of Tetraphenyl-lead," *Inorg. Chim. Acta*, **1**, 424.

Casalone, G. L., Mariani, C., Mugnoli, A., and Simonetta, M., "Crystal, Molecular and Electronic Structure of 1,1-Diaryl-2-halogenoethylenes I. 2-Bromo-1,1-di-p-tolylethylene," *Acta Cryst.*, **22**, 228.

Chetkina, L. A., "Crystalline and Molecular Structure of 1,5-Difluoroanthraquinone," *Kristallografiya*, **12**, 42.

Chetkina, L. A., and Gol'der, G. A., "Determination of Crystal Structure of p,p'-Dibromo-α,α'-Difluorostilbene," *Zh. Strukt. Khim.*, **8**, 106.

Chetkina, L. A., and Gol'der, G. A., "Least-Squares Refinement of the Crystal Structures of 1,5-Dibromo- and 1,5-Diiodoanthraquinone," *Kristallografiya*, **12**, 404.

Corfield, P. W. R., Doedens, R. J., and Ibers, J. A., "Studies of Metal-Nitrogen Multiple Bonds I. The Crystal and Molecular Structure of Nitridodichlorotris-(diethylphenylphosphine)rhenium(V), $ReNCl_2[P(C_2H_5)_2C_6H_5]_3$," *Inorg. Chem.*, **6**, 197.

Dale, D., "The X-Ray Crystallographic Determination of the Structure of Di-[3-Methyl-1(or 5)-phenyl-5(or 1)-p-tolylformazyl]nickel(II)," *J. Chem. Soc.*, **1967A**, 278.

Daly, J. J., "The Crystal and Molecular Structure of Triphenylphosphoranylidenethioketen," *J. Chem. Soc.*, **1967A**, 1913.

Degrève, Y., Meunier-Piret, J., Meerssche, M. van, and Piret, P., "Iron Carbonyl Complexes with Acetylene and its Derivatives III. Structure of $Fe_2(CO)_6 \cdot C_6H_5C_2C_6H_5$," *Acta Cryst.*, **23**, 119.

Doedens, R. J., and Ibers, J. A., "Studies of Metal-Nitrogen Multiple Bonds II. The Crystal and Molecular Structure of Nitridodichlorobis(triphenylphosphine)rhenium(V), $ReNCl_2[P(C_6H_5)_3]_2$," *Inorg. Chem.*, **6**, 204.

Doedens, R. J., Robinson, W. T., and Ibers, J. A., "The Crystal and Molecular Structure of μ-Hydrido-μ-diphenylphosphido-bis(tetracarbonylmanganese), $(CO)_4Mn(H)[P(C_6H_5)_2]Mn(CO)_4$," *J. Am. Chem. Soc.*, **89**, 4323.

Eisenberg, R., and Gray, H. B., "Trigonal-Prismatic Coordination: the Crystal and Molecular Structure of tris(cis-1,2-Diphenylethylene-1,2-dithiolato) vanadium," *Inorg. Chem.*, **6**, 1844.

Enemark, J. H., and Ibers, J. A., "The Molecular Structure of Nitrosyldicarbonylbis(triphenylphosphine)manganese, $Mn(NO)(CO)_2[P(C_6H_5)_3]_2$," *Inorg. Chem.*, **6**, 1575.

Fratini, A. V., Karle, I. L., and Karle, J., "The Crystal Structure of Auramine Perchlorate: Correlation of the Results with Specular Reflection Spectroscopy," *Appl. Opt.*, **6**, 2091.

Gaultier, J., Geoffre, S., and Hauw, C., "Crystalline Structure of 2,3-Dibromo-1,4-anthraquinone," *Compt. Rend.*, **264C**, 697.

Gaultier, J., and Hauw, C., "Crystal and Molecular Structure of 1,4-Naphtho-hydroquinone," *Acta Cryst.*, **23**, 1016.

Gaultier, J., Hauw, C., and Souron, C., "Crystalline and Molecular Structure of Vitamin K_5," *Compt. Rend.*, **264C**, 766.

Gerloch, M., and Mabbs, F. E., "The Crystal and Molecular Structure of Chloro-(N,N'-bis-(salicylidene)ethylenediamine)iron(III) as a Hexacoordinate Dimer," *J. Chem. Soc.*, **1967A**, 1900.

Gomes de Mesquita, A. H., "The Crystal Structure of 2,2'-Diaminodiphenyl Disulphide," *Acta Cryst.*, **23**, 671.

Guilhem, J., "Determination by X-Ray Diffraction of the Crystal Structure of Dihydroxyanthraquinones I. Crystal Structure of Anthrarufine," *Bull. Soc. Chim. France*, **1967**, 1656.

Guilhem, J., "Determination by X-Ray Diffraction of the Crystal Structure of Dihydroxyanthraquinones II. Crystal Structure of Alizarin," *Bull. Soc. Chim. France*, **1967**, 1666.

Guilhem, J., "Two Independent Determinations of the Crystal Structure of 1,5-Dihydroxyanthraquinone," *Acta Cryst.*, **23**, 330.

Ikemoto, I., and Kuroda, H., "The Crystal Structure of the Perylene-tetra-cyanoethylene Complex," *Bull. Chem. Soc. Japan*, **40**, 2009.

Kerr, K. A., Robertson, J. M., and Sim, G. A., "The Structure of the Benzil Monoximes: X-Ray Analysis of Benzil α-monoxime *p*-Bromobenzoate," *J. Chem. Soc.*, **1967B**, 1305.

Kumakura, S., Iwasaki, F., and Saito, Y., "The Crystal Structure of the 1:1 Complex of Naphthalene with 1,2,4,5-Tetracyanobenzene," *Bull. Chem. Soc. Japan*, **40**, 1826.

Kunz, V., and Nowacki, W., "Crystal and Molecular Structure of Naphthalene Chromium Tricarbonyl, $C_{10}H_8 \cdot Cr(CO)_3$," *Helv. Chim. Acta*, **50**, 1052.

Lindqvist, O., "The Crystal Structure of Tellurium(IV) Catecholate, $Te(C_6H_4O_2)_2$," *Acta Chem. Scand.*, **21**, 1473.

Mammi, M., Busetti, V., and Del Pra, A., "Crystal Structure of Diphenyllead-dichloride," *Inorg. Chim. Acta*, **1**, 419.

Mannan, K. A. I. F. M., "The Crystal Structure Determination of the Complex $(C_6H_5)_3P{\rightarrow}Au{\rightarrow}Mn(CO)_4P(OC_6H_5)_3$," *Acta Cryst.*, **23**, 649.

McGinnety, J. A., Doedens, R. J., and Ibers, J. A., "Structure of $IrIO_2(CO)$-$(P(C_6H_5)_3)_2 \cdot CH_2Cl_2$, the Oxygen Adduct of a Synthetic Irreversible Molecular Oxygen Carrier," *Inorg. Chem.*, **6**, 2243.

Mootz, D., and Sassmannshausen, G., "Crystal Structure of Tris(phenylethynyl)-phosphine," *Z. Anorg. Allgem. Chem.*, **355**, 200.

Mornon, J. P., "Application of the Symbolic Addition Method: Structure of α-Naphthyl-*cis*-tetrahydrophthalamic Acid," *Compt. Rend.*, **264C**, 192.

Mornon, J. P., "Application of the Symbolic Addition Method: Crystal Structure of N-α-Naphthyltetrahydro-1,2,3,6-phthalamic Acid," *Acta Cryst.*, **23**, 367.

Ohashi, Y., Iwasaki, H., and Saito, Y., "The Crystal Structure of the 1:1 Complex of N,N,N'N'-Tetramethyl-*p*-phenylenediamine and 1,2,4,5-Tetracyanobenzene," *Bull. Chem. Soc. Japan*, **40**, 1789.

Omel'chenko, Y. A., and Kondrashev, Y. D., "Crystal Structure of the Modifications of *p*-Bromodiazoaniline," *Kristallografiya*, **12**, 416.

Piron, J., Piret, P., and Meerssche, M. van, "Structure of Hexacarbonyl

(1,3-Diphenylurea)diiron, $Fe_2(CO)_6(C_6H_5NCONC_6H_5)$," *Bull. Soc. Chim. Belges*, **76**, 505.

Plieth, K., and Repmann, H., "The Crystal and Molecular Structure of 4,4'-Dichloro-α,β-diethylstilbene," *Z. Krist.*, **124**, 77.

Polynova, T. N., and Porai-Koshits, M. A., "X-Ray Structure Study of Diphenyltrichlorostibine," *Zh. Strukt. Khim.*, **8**, 112.

Prakash, A., "Refinement of the Crystal Structure of Anthraquinone," *Acta Cryst.*, **22**, 439.

Rérat, C., "Intramolecular van der Waals Interactions and Structure of *N-p-*Bromophenyl Benzenesulfonamide," *Acta Cryst.*, **23**, 856.

Rérat, B., Dauphin, G., Kergomard, A., and Rérat, C., "Structure of *N-p-*Bromophenyl Benzenesulfonamide," *Compt. Rend.*, **264C**, 500.

Sax, M., and McMullan, R. K., "The Crystal Structure of Dibenzoyl Peroxide and the Dihedral Angle in Covalent Peroxides," *Acta Cryst.*, **22**, 281.

Scane, J. G., "The Crystal Structure of Tetraphenylarsonium Oxotetrabromoaquomolybdate," *Acta Cryst.*, **23**, 85.

Silverman, J., and Yannoni, N. F., "The Crystal Structure of 10-Dicyanomethyleneanthrone, an Overcrowded Aromatic Compound," *J. Chem. Soc.*, **1967B**, 194.

Swaminathan, S., and Nigam, G. D., "Crystal Structure of 1:4-Dihydroxyanthraquinone," *Curr. Sci. (India)*, **36**, 541.

Toman, K., Očenášková, D., and Huml, K., "The Crystal Structure of Some Conducting Organic Substances III. Perchlorate of 4,4'-Bis(dimethylamino)-diphenylamine Radical," *Acta Cryst.*, **22**, 32.

Truter, M. R., and Watling, R. C., "The Crystal Structure of Trimethyl(salicylaldehydato)platinum(IV)," *J. Chem. Soc.*, **1967A**, 1955.

Vorontsova, L. G., Zvonkova, Z. V., and Zhdanov, G. S., "Refinement of the Crystal Structure of Di-*p*-tolyldisulfide," *Dokl. Akad. Nauk SSSR*, **172**, 584.

Vul, E. B., and Lobanova, G. M., "Preliminary Model of the Benzophenone Structure," *Kristallografiya*, **12**, 411.

Weber, H. P., and Bryan, R. F., "Metal-Metal Bonding in Coordination Complexes IV. The Structure of Triphenyltin Pentacarbonyl Manganese," *Acta Cryst.*, **22**, 822.

Wilford, J. B., and Powell, H. M., "The Crystal and Molecular Structure of the Pentacoordinate Complex Tricarbonyl(triphenylphosphine)-σ-tetrafluoroethylcobalt(I)," *J. Chem. Soc.*, **1967A**, 2092.

Williams, D. E., "Crystal Structure of 2,2-Diphenyl-1-picrylhydrazyl Free Radical," *J. Am. Chem. Soc.*, **89**, 4280.

Williams, R. M., and Wallwork, S. C., "Molecular Complexes Exhibiting Polarization Bonding IX. The Crystal and Molecular Structure of the Tetracyanoethylene-Naphthalene Complex," *Acta Cryst.*, **22**, 899.

Yasuoka, N., Ando, T., and Kuribayashi, S., "The Crystal Structure of *p,p'*-Dichlorodiphenoxy-1,2-ethane," *Bull. Chem. Soc. Japan*, **40**, 265.

Yasuoka, N., Ando, T., and Kuribayashi, S., "The Crystal Structure of 1,2-Diphenoxyethane," *Bull. Chem. Soc. Japan*, **40**, 270.

1968

Aimé, S., Gaultier, J., and Hauw, C., "Crystal and Molecular Structure of 4-Amino-1,2-naphthoquinone Hemihydrate," *Compt. Rend.*, **266C**, 1354.

Alcock, N. W., and Raspin, K. A., "Structures of Binuclear Ruthenium Complexes I. The Crystal and Molecular Structure of Tri-μ-chloro-chloropentakis-(diethylphenylphosphine)diruthenium(II)," *J. Chem. Soc.*, **1968A**, 2108.

Alcock, N. W., Spencer, R. C., Prince, R. H., and Kennard, O., "Crystal and Molecular Structure of Bis-(1-*m*-tolylazo-2-naphtholato)nickel(II)," *J. Chem. Soc.*, **1968A**, 2383.

Alcock, N. W., and Timms, R. E., "Organo-tin Acetates I. The Crystal and Molecular Structure of Tribenzyltin Acetate," *J. Chem. Soc.*, **1968A**, 1873.

Allmann, R., and Hellner, E., "Crystal Structure of Dimeric 3-Bromo-2,4,6-triphenylphenoxyls," *Chem. Ber.*, **101**, 2522.

Baker-Hawkes, M. J., Dori, Z., Eisenberg, R., and Gray, H. B., "The Crystal and Molecular Structure of the Tetra-*n*-butyl-ammonium Salt of the Dianionic Dimer of Bis(1,2,3,4-tetrachlorobenzene-5,6-dithiolato)cobaltate," *J. Am. Chem. Soc.*, **90**, 4253.

Bart, J. C. J., "The Crystal Structure of a Modification of Hexaphenylbenzene," *Acta Cryst.*, **24B**, 1277.

Bennett, M. J., Bratton, W. K., Cotton, F. A., and Robinson, W. R., "Reactions of the Octahalodirhenate(III) Ions VII. Structural Characterization of Dichlorotetrabenzoatodirhenium(III)," *Inorg. Chem.*, **7**, 1570.

Biryukov, B. P., Solodova, O. P., and Struchkov, Y. T., "Crystal Structures of Polynuclear Polymetallic Carbonyls III. Pentacarbonyl[diphenyl-(tetracarbonylcobalt)stannyl]manganese," *Zh. Strukt. Khim.*, **9**, 228.

Boer, J. L. de, and Vos, A., "The Crystal and Molecular Structure of N,N,N',N'-Tetramethyl-*p*-diaminobenzene-chloranil, TMPD-chloranil," *Acta Cryst.*, **24B**, 720.

Bright, D., and Ibers, J. A., "Studies of Metal-nitrogen Multiple Bonds III. The Crystal and Molecular Structures of Trichloro(*p*-methoxyphenylimino)-bis(diethylphenylphosphine)rhenium(V) and Trichloro(*p*-acetylphenylimino)bis(diethylphenylphosphine)rhenium(V)," *Inorg. Chem.*, **7**, 1099.

Carstensen-Oeser, E., Goettlicher, S., and Habermehl, G., "Crystal Structure of the π-Complex of Picric Acid and 1-Bromo-2-aminonaphthalene," *Chem. Ber.*, **101**, 1648.

Carter, O. L., McPhail, A. T., and Sim., G. A., "Metal-carbonyl and Metal-nitrosyl Complexes VII. X-Ray Analysis of 1-Aminonaphthalenetricarbonyl-chromium," *J. Chem. Soc.*, **1968A**, 1866.

Caticha-Ellis, S., and Abrahams, S. C., "The Crystal and Molecular Structure of 4,4'-Dibromo- and 4,4'-Dichloro-dibenzoyl Peroxide," *Acta Cryst.*, **24B**, 277.

Chawdhury, S. A., Hargreaves, A., and Sullivan, R. A. L., "The Crystal and Molecular Structure of 4,4'-Diamino-3,3'-dimethylbiphenyl (*o*-Tolidine)," *Acta Cryst.*, **24B**, 1222.

Chetkina, L. A., and Gol'der, G. A., "Crystal and Molecular Structures of p,p'-Dimethyl-α,α'-difluorostilbene," *Zh. Strukt. Khim.*, **9**, 250.

Cotton, F. A., and Foxman, B. M., "The Structure of Oxopentachloropropiona-tobis(triphenylphosphine)dirhenium(IV)," *Inorg. Chem.*, **7**, 1784.

Cotton, F. A., and Harris, C. B., "Structure of Tetraphenylarsonium Bis-(*N*-cyanodithiocarbimato)nickelate(II)," *Inorg. Chem.*, **7**, 2140.

Cotton, F. A., and Legzdins, P., "An Example of the Monocapped Octahedral Form of Heptacoordination. Crystal and Molecular Structure of Tris-(1-phenyl-1,3-butanedionato)aquoyttrium(III)," *Inorg. Chem.*, **7**, 1777.

Dahm, D. J., and Jacobson, R. A., "Crystal Structure of $Fe_3(CO)_{11}P(C_6H_5)_3$," *J. Am. Chem. Soc.*, **90**, 5106.

Davydova, M. A., and Struchkov, Y. T., "Crystal Structure of 1,4,5,8-Tetrabromonaphthalene," *Zh. Strukt. Khim.*, **9**, 258.

Davydova, M. A., and Struchkov, Y. T., "Conformation of the 1,5-Dibromo-4,8-dichloronaphthalene Molecule," *Zh. Strukt. Khim.*, **9**, 547.

Dreissig, W., and Dietrich, H., "The Crystal Structure of Bis(triphenylphosphine)-ethylene-nickel(0)," *Acta Cryst.*, **24B**, 108.

Eisenberg, R., Dori, Z., Gray, H. B., and Ibers, J. A., "The Crystal and Molecular Structure of the High-spin Square-planar Complex Triphenylmethylarsonium Bis(toluene-3,4-dithiolato)cobaltate-0.5-ethanol," *Inorg. Chem.*, **7**, 741.

Goldstein, P., Seff, K., and Trueblood, K. N., "The Crystal Structure of Tetraphenylphosphonium Bis(tetracyanoquinodimethanide)," *Acta Cryst.*, **24B**, 778.

Hanic, F., and Mills, O. S., "The Crystal Structure of Tricarbonylanthracene Chromium," *J. Organometal. Chem.*, **11**, 151.

Harrison, R. W., and Trotter, J., "The Structure of Acetyltriphenylgermane," *J. Chem. Soc.*, **1968A**, 258.

Hawley, D. M., and Ferguson, G., "The Stereochemistry of Some Organic Derivatives of Group V_B Elements. The Crystal and Molecular Structure of Triphenylbismuth," *J. Chem. Soc.*, **1968A**, 2059.

Hawley, D. M., and Ferguson, G., "The Stereochemistry of Some Organic Derivatives of Group V_B Elements II. The Crystal and Molecular Structure of Triphenylbismuth Dichloride," *J. Chem. Soc.*, **1968A**, 2539.

Ikemoto, I., and Kuroda, H., "The Refinement of the Crystal Structure of the Pyrene-Tetracyanoethylene Complex," *Acta Cryst.*, **24B**, 383.

Jarvis, J. A. J., Mais, R. H. B., and Owston, P. G., "The Stereochemistry of Complexes of Nickel(II) II. The Crystal and Molecular Structure of Dibromobis(triphenylphosphine)nickel(II)," *J. Chem. Soc.*, **1968A**, 1473.

Kashiwagi, T., Yasuoka, N., Ueki, T., Kasai, N., Kakudo, M., Takahashi, S., and Hagihara, N., "Addition of Some Simple Molecules to the Transition Metal-phosphine Complex and the Crystal and Molecular Structure of $Pd[(C_5H_5)_3P]_2(CS_2)$," *Bull. Chem. Soc. Japan*, **41**, 296.

Kondrashev, Y. D., "Crystal Structure and Tautomerism of *p*-Methyl- and *p*-Chlorodiazoaminobenzene," *Kristallografiya*, **13**, 622.

Krigbaum, W. R., Roe, R.-J., and Woods, J. D., "Crystal Structure of *N,N*-Diphenylacetamide," *Acta Cryst.*, **24B**, 1304.

Mills, O. S., and Redhouse, A. D., "Carbon Compounds of the Transition Metals XIII. Evidence for, and Crystal Structure of, Octacarbonyldiphenylvinylidenedi-iron," *J. Chem. Soc.*, **1968A**, 1282.

Mootz, D., and Look, W., "Crystal Structure of Tris(*p*-nitrophenylethynyl)-arsine," *Z. Anorg. Allgem. Chem.*, **356**, 244.

Motohashi, M., Amano, Y., and Uno, T., "The Crystal Structure of the Copper(II) Chelate of *N,N'*-Ethylene Bis(*l*-ephedrine)," *Bull. Chem. Soc. Japan*, **41**, 2007.

Muir, K. W., Ferguson, G., and Sim, G. A., "Crystal and Molecular Structure of Phenanthrenechromium Tricarbonyl," *J. Chem. Soc.*, **1968B**, 467.

Niimura, N., Ohashi, Y., and Saito, Y., "The Crystal Structure of the 1:1

Complex of 1,2,4,5-Tetracyanobenzene and Hexamethylbenzene," *Bull. Chem. Soc. Japan*, **41**, 1815.

Panattoni, C., Graziani, R., Croatto, U., Zarli, B., and Bombieri, G., "Chemistry of UO_2^{2+} Groups I. Molecular Structure of Dioxo-dinitratobis(triphenylarsine Oxide) Uranium(VI)," *Inorg. Chim. Acta*, **2**, 43.

Sakurai, T., "On the Refinement of the Crystal Structures of Phenoquinone and Monoclinic Quinhydrone," *Acta Cryst.*, **24B**, 403.

Samarskaya, V. D., Myasnikova, R. M., and Kitaigorodskii, A. I., "X-Ray Diffraction Study of Tolan-phase Crystals in a Tolan-diphenylmercury System," *Kristallografiya*, **13**, 616.

Schweikert, W. W., and Meyers, E. A., "The Crystal Structure of the Triphenylphosphine Sulfide-iodine Addition Complex," *J. Phys. Chem.*, **72**, 1561.

Shen, K. W., McEwen, W. E., LaPlaca, S. J., Hamilton, W. C., and Wolf, A. P., "Crystal and Molecular Structures of Methoxytetraphenylantimony and Dimethoxytriphenylantimony," *J. Am. Chem. Soc.*, **90**, 1718.

Shugam, E. A., Shkol'nikova, L. M., and Knyazeva, A. N., "Crystallochemical Data on Chelate Compounds of β-Diketones III. Crystal and Molecular Structures of Bis(dibenzoylmethyl)palladium," *Zh. Strukt. Khim.*, **9**, 222.

Stora, C., "Preliminary Refinement of the X-Ray Structure of 3,5-Dibromo-(hydroxy)triphenylmethanecarbinol Using an I.B.M. 740 Computer," *Compt. Rend.*, **266C**, 88.

Sutherland, H. H., and Hoy, T. G., "The Crystal Structure of 4-Acetyl-2'-chlorobiphenyl," *Acta Cryst.*, **24B**, 1207.

Williams, R. M., and Wallwork, S. C., "Molecular Complexes Exhibiting Polarization Bonding XI. The Crystal and Molecular Structure of 7,7,8,8-Tetracyanoquinodimethane-anthracene Complex," *Acta Cryst.*, **24B**, 168.

Young, D. W., Tollin, P., and Sutherland, H. H., "The Crystal Structure of 4-Acetyl-2'-fluorobiphenyl," *Acta Cryst.*, **24B**, 161.

NAME INDEX

FORMULA INDEX *

*In this bulk formula index carbon atoms come first and the formulas are arranged in order of the increasing number of these atoms. In each formula carbon is followed by hydrogen, oxygen, nitrogen, etc., each of which is ordered according to increasing number. Names are added only to distinguish between compounds having the same total formula.